CALCIUM CHANNELS:
STRUCTURE AND FUNCTION

ANNALS OF THE NEW YORK ACADEMY OF SCIENCES
Volume 560

CALCIUM CHANNELS: STRUCTURE AND FUNCTION

Edited by Dennis W.-Wray, Robert I. Norman, and Peter Hess

The New York Academy of Sciences
New York, New York
1989

Library of Congress Cataloging-in-Publication Data

Calcium channels : structure and function / edited by
  Dennis W.-Wray, Robert I. Norman, and Peter Hess.

  p.   cm.—(Annals of the New York Academy of
Sciences, ISSN 0077-8923 : v. 560)
  Based on a conference held July 18-20. 1988, in London,
U.K. sponsored by the New York Academy of Sciences.
  Includes bibliographies and index.
  ISBN 0-89766-519-8 (alk. paper).—ISBN 0-89766-520-1
(pbk. : alk. paper)
  1. Calcium channels—Congresses. I. W.-Wray, Dennis.
II. Norman, Robert I.   III. Hess, Peter. 1951-   .
IV. New York Academy of Sciences.   V. Series.
  [DNLM: 1. Calcium Channels—congresses. W1
AN626YL v. 560 / QH 601 C1445 1988]
Q11.N5 vol. 560
[QP535.L2]
500 s—dc20
[574.19'214]
DNLM/DLC
for Library of Congress   89-12284
                                                                   CIP

PCP
Printed in the United States of America
ISBN 0-89766-519-8 (cloth)
ISBN 0-89766-520-1 (paper)
ISSN 0077-8923

ANNALS OF THE NEW YORK ACADEMY OF SCIENCES

Volume 560
June 26, 1989

CALCIUM CHANNELS: STRUCTURE AND FUNCTION

Editors
DENNIS W.-WRAY, ROBERT I. NORMAN, AND PETER HESS

Advisory Board
M. LAZDUNSKI, D. BROWN, A. ENGEL, W. A. CATTERALL, AND D. J. TRIGGLE

CONTENTS

[a]This volume is the result of a conference entitled Calcium Channels: Structure and Function, which was held July 18-20, 1988 at The Royal Society, London, United Kingdom. This conference was sponsored by the New York Academy of Sciences.

Financial assistance was received from:

- BAYER UK LTD.
- BEECHAM PHARMACEUTICALS
- BRITISH HEART FOUNDATION
- CAMBRIDGE ELECTRONIC DESIGN
- FISONS PHARMACEUTICALS
- GLAXO S.p.A.
- ICI PHARMACEUTICALS GROUP
- IMPERIAL CHEMICAL INDUSTRIES LTD.
- MARION LABORATORIES, INC.
- MERCK SHARP & DOHME RESEARCH LABORATORIES
- NATIONAL HEART, LUNG AND BLOOD INSTITUTE/NIH
- NATIONAL INSTITUTE OF NEUROLOGICAL AND
  COMMUNICATIVE DISORDERS AND STROKE/NIH
- NATIONAL SCIENCE FOUNDATION
- PFIZER CENTRAL RESEARCH
- ROCHE PRODUCTS LIMITED
- RORER CENTRAL RESEARCH
- SANDOZ PHARMACEUTICALS
- SMITH KLINE AND FRENCH RESEARCH LTD.
- THE BIOCHEMICAL SOCIETY
- THE COUNCIL FOR TOBACCO RESEARCH USA
- THE ROYAL SOCIETY
- THE SQUIBB INSTITUTE FOR MEDICAL RESEARCH
- 3M HEALTH CARE
- UNITED STATES AIR FORCE OFFICE OF SCIENTIFIC RESEARCH
- UNITED STATES ARMY MEDICAL RESEARCH
- WYETH RESEARCH (UK) LTD.

Preface

D. W.-WRAY,[a] R. I. NORMAN,[b] AND P. HESS [c]

[a] Department of Pharmacology
Royal Free Hospital Medical School
London NW3 2PF, England

[b] Department of Medicine
University of Leicester
Leicester LE1 7RH, England

[c] Department of Cellular and Molecular Physiology
Harvard Medical School
Boston, Massachusetts 02115

In the last few years, much attention has been directed towards voltage-sensitive ion channels selective for calcium ions because of their important physiological roles in cell signaling. Advances in the physiological sciences, in particular the development of patch-clamping techniques, has facilitated detailed study of the functional properties of single calcium channels. Parallel biochemical advances, achieved by using radiolabeled calcium channel ligands and more recently cloned DNA techniques, have led to insights into the structural and molecular properties of calcium channels.

This volume contains the proceedings of an international conference on calcium channels that was held at the Royal Society, London, in July 1988. This meeting, convening some 300 participants from 22 countries, brought together many of the leading workers in calcium channel research from a variety of scientific disciplines with the aim of discussing recent developments in the understanding of structure and function of these channels. The following paragraphs should give the reader an idea of the exciting research covered in this volume.

Classification of calcium channels into subtypes has been on the basis of their electrophysiological and pharmacological properties. More specifically, subtypes have been differentiated according to their voltage threshold for activation and by their inactivation characteristics: low threshold inactivating (T), dihydropyridine-sensitive high-threshold noninactivating (L), and high-threshold inactivating (N). As further tissues are investigated, it is becoming clear that considerable variation exists within each of these subtype groups, and more subtypes have been suggested.

Further electrophysiological experiments have studied gating and ion selectivity of calcium channels. Differing subtypes of the channel share certain features. Thus gating kinetics can be understood in terms of several closed and open states, while selectivity of the channel for calcium appears to be due to reversible binding of calcium to sites within the pore together with ion-ion interactions.

Because of its relative abundance and sensitivity to dihydropyridines, the L-type calcium channel from skeletal muscle was the first to be purified, and its molecular properties and oligomeric structure (consisting of five distinct subunits α_1, α_2, β, γ and δ) are now well characterized. Important progress has been made in the study of molecular properties of cardiac and brain calcium channels. Besides dihydropyridines, calcium channel ligands such as ω-conotoxin are now being used, with the possibility of characterizing channel subtypes other than L-type.

Cloning and sequencing of cDNA encoding for the α_1 subunit have shown that the primary amino acid sequence is similar to that of the sodium channel. This subunit probably forms the major functional part of the L-type channel since cDNA for the α_1 subunit causes the expression of functioning calcium channels in skeletal muscle, while the purified α_1 subunit operates as a channel when reconstituted into artificial membranes.

L-type channels also appear to be essential for excitation-contraction coupling of skeletal muscle, acting as a voltage sensor. They are located in the cell membrane near to ryanodine receptors, which are in turn situated in the sarcoplasmic reticulum. The latter receptors form the calcium-release channels of the sarcoplasmic reticulum and have also been purified and reconstituted into artificial bilayers to form functioning channels.

Newer radiolabeled calcium channel drugs are becoming available, and it is now clear that benzothiazepines, phenylalkylamines, and dihydropyridines bind to three allosterically coupled binding sites on the α_1 subunit, located near the external surface. Structure-activity studies of a range of dihydropyridines are also providing useful information.

Other calcium channel ligands such as monoclonal as well as polyclonal antibodies are proving to be valuable probes against the channel, and have shown the structural as well as functional importance of the other subunits as well as the α_1. Antibodies against calcium channels (active zone particles) at motor nerve terminals are present in the plasma of patients with the Lambert-Eaton myasthenic syndrome; these antibodies appear to cross-react preferentially with specific subtypes of channels in neuronal tissues, and may be useful in the further characterization of calcium channel subtypes.

Voltage-dependent calcium channels can be modulated by neurotransmitters. For instance, β-adrenoreceptor agonists increase the probability of opening of L-type channels via their phosphorylation by cAMP-dependent protein kinase. On the other hand, α-adrenoreceptor agonists inhibit calcium currents, in some cases by acting on N-type channels, in others by a relatively indiscriminate action on several channel subtypes. Many receptors for drugs are linked to calcium channels by GTP binding proteins (G proteins), either directly or via second messengers or both. G proteins are also intimately concerned with the effects of dihydropyridine agonists and antagonists. Interestingly, other agents such as glucose can also modulate L-type activity in insulin-secreting cells.

We would like to acknowledge the advice and efficient organizational work of Ellen Marks and the staff of the New York Academy of Sciences, not forgetting Linda Mehta of the Editorial Department for helping to ensure rapid publication of these proceedings. We would also like to thank the sponsors for their financial support and, last but not least, we are grateful to the speakers and poster presenters for their cooperation.

Molecular Properties of Dihydropyridine-Sensitive Calcium Channels

WILLIAM A. CATTERALL, MICHAEL J. SEAGAR,[a]

MASAMI TAKAHASHI,[b] AND KAZUO NUNOKI

Department of Pharmacology
University of Washington
Seattle, Washington 98195

INTRODUCTION

Voltage-sensitive calcium channels constitute an essential link between transient changes in membrane potential and a variety of cellular responses including secretion of neurotransmitters and hormones, initiation of contraction in cardiac and smooth muscle, and activation of second-messenger responses in many cell types. Electrophysiological measurements have established the existence of multiple classes of calcium channels.[1-4] The most prominent channel type in muscle cells, termed Type II[1] or L-type,[4] mediate long-lasting calcium currents after strong depolarizations. In mammalian cells, the gating behavior of these channels is modulated by the 1,4-dihydropyridines. In single-channel recording experiments, calcium-channel "agonists" such as BAY K 8644 favor long openings whereas "antagonists" such as nimodipine stabilize the channel in an inactive state.[6] Radiolabeled dihydropyridine derivatives have been used as specific high-affinity probes of dihydropyridine-sensitive calcium channels.[7] Their binding is modulated by the phenylalkylamine calcium antagonists and the benzothiazepine calcium antagonists, which inhibit calcium channels by binding to different sites. This chapter reviews experiments from this laboratory that begin a biochemical analysis of the protein components of this calcium channel type.

[a] Permanent address: Michael J. Seagar, CNRS-INSERM, Laboratoire de Biochimie, Fac. Med. Nord, Marseille, France.

[b] Permanent address: Masami Takahashi, Mitsubishi-Kasei Institute of Life Sciences, Machida-Shi, Tokyo, Japan.

SKELETAL MUSCLE TRANSVERSE TUBULES AS A MODEL SYSTEM FOR STUDIES OF THE BIOCHEMICAL PROPERTIES OF DIHYDROPYRIDINE-SENSITIVE CALCIUM CHANNELS

Skeletal muscle fibers have substantial voltage-activated calcium currents that have been measured under voltage clamp.[8] These currents originate almost entirely in the transverse tubular system and are blocked by dihydropyridine calcium antagonists.[9] They have been presumed to play a role in excitation-contraction coupling although direct evidence for such a role has not been obtained. Transverse tubule membranes can be extensively purified from skeletal muscle by a combination of differential and density gradient centrifugation.[10] Antibodies prepared against the most highly purified fractions of T-tubule membranes stain only T-tubules and not sarcolemma or sarcoplasmic reticulum of intact muscle, indicating a high degree of purity of this preparation.[10] Analysis of [^{3}H]nitrendipine binding to membrane fractions from skeletal muscle reveals a specific localization of the calcium antagonist receptor in the T-tubule fraction.[11,12] These membranes contain a 10-fold greater concentration of calcium antagonist receptor than any other membrane preparation described to date. Because T-tubule membranes are the most enriched source of calcium antagonist receptors and they display a substantial voltage-activated calcium current that is blocked by dihydropyridines, they provide a favorable experimental preparation for examination of the molecular properties of the calcium antagonist receptor and its relationship to voltage-sensitive calcium channels. It is anticipated that the T-tubule calcium channel will resemble those of other tissues that are blocked by dihydropyridines and therefore that information on the molecular properties of this calcium channel will give insight into others.

CHARACTERIZATION OF THE DETERGENT-SOLUBILIZED CALCIUM ANTAGONIST RECEPTOR

Because it is likely to be an intrinsic membrane protein, the first essential step in purification and biochemical characterization of the calcium antagonist receptor is solubilization from an appropriate membrane source and characterization of the solubilized protein. After a survey of several detergents, we concluded that digitonin is the most effective detergent for solubilization of a specific [^{3}H]nitrendipine-receptor complex from brain and skeletal muscle transverse tubule membranes.[13,14] Up to 40% of the receptor-ligand complex is solubilized. The dissociation of bound [^{3}H]nitrendipine from the complex is accelerated by verapamil and slowed by diltiazem through allosteric interactions between the nitrendipine binding site and the binding sites for those ligands as previously observed in intact membranes. These results show that the three different binding sites for calcium antagonist drugs remain associated as a complex after detergent solubilization of the calcium antagonist receptor and provide further support for the conclusion that a specific receptor complex has been solubilized under conditions that allow retention of the functional allosteric regulation of dihydropyridine binding.

Sedimentation of the solubilized [^{3}H]nitrendipine-receptor-digitonin complex through sucrose gradients gives a single peak of specifically bound nitrendipine with a sedimentation coefficient of 19 to 20S.[13-15] Comparison of the sedimentation behavior of the solubilized complex of brain, heart, and skeletal and smooth muscle indicates that they have identical size. These results provide support for the view that the calcium antagonist receptor in different tissues is quite similar.

Many plasma membrane proteins are glycosylated during their synthesis and transport to the cell surface. The solubilized calcium antagonist receptor from brain or skeletal muscle is specifically adsorbed to and eluted from affinity columns with immobilized wheat germ agglutinin or other lectins.[13,14,16,17] Evidently, one or more of the subunits of the calcium antagonist receptor are glycoproteins.

PURIFICATION AND SUBUNIT COMPOSITION OF THE DIHYDROPYRIDINE-SENSITIVE CALCIUM CHANNEL FROM SKELETAL MUSCLE TRANSVERSE TUBULES

We have purified the calcium antagonist receptor, solubilized from T-tubule membranes by digitonin, 330-fold by affinity chromatography on wheat germ agglutinin Sepharose, ion-exchange chromatography on DEAE-Sephadex, and velocity sedimentation through sucrose gradients.[14] Analysis of the purified protein by NaDodSO$_4$/PAGE under alkylating conditions and silver staining (FIG. 1, lane 1) revealed three

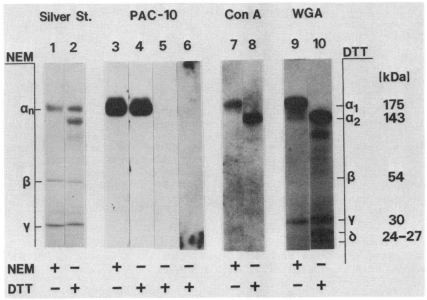

FIGURE 1. Polypeptide composition of the dihydropyridine-sensitive calcium channel. **Lanes 1 and 2:** Purified calcium channels were analyzed by NaDodSO$_4$/PAGE and silver staining with or without reduction of disulfide bonds as indicated beneath each lane. **Lanes 3 through 6:** Polypeptides separated by NaDodSO$_4$/PAGE, with or without reduction of disulfide bonds as indicated below each lane, were electrophoretically transferred to nitrocellulose strips and immunolabeled by incubation with PAC-10 (lanes 3 and 4), preimmune serum (lane 5), or PAC-10 that had been preadsorbed with purified calcium channel (lane 6), followed by incubation with [^{125}I]protein A, washing and autoradiography. **Lanes 7 and 8:** Calcium-channel subunits were transferred to a nitrocellulose sheet and labeled with [^{125}I]ConA. **Lanes 9 and 10:** Calcium-channel subunits were labeled directly in the gel with [^{125}I]WGA.

classes of polypeptide that we designated α (167 kDa), β (54 kDa), and γ (30 kDa). When disulfide bonds were cleaved with dithiothreitol, the α band split into two clearly resolved protein populations with molecular masses of 175 and 143 kDa (FIG. 1, lane 2). In the initial studies from this laboratory, the anomalous behavior of the α polypeptide was ascribed to partial cleavage and/or reformation of intrachain disulfide bonds resulting in a variable fraction of the protein with smaller apparent size.[14] The more recent use of a battery of specific labeling methods has now shown that that 175- and 143-kDa polypeptides are two distinct calcium channel subunits, α_1 and α_2, which have similar size but clearly different properties.[18]

A polyclonal antibody (PAC-10), obtained from the ascites fluids of a SJL/J mouse immunized with purified calcium channel selectively labeled the 167-kDa polypeptide before reduction of disulfide bonds (FIG. 1, lane 3), but only the 175-kDa polypeptide after reduction (FIG. 1, lane 4). No immunolabeling was observed with preimmune serum (FIG. 1, lane 5) or with PAC-10 that had been preadsorbed with purified calcium channel (FIG. 1, lane 6). These observations indicate that the 175-kDa and 143-kDa components are distinct polypeptides.

SUBUNIT GLYCOSYLATION

Solubilized [^{3}H]dihydropyridine receptors specifically bind to various immobilized lectins and affinity chromatography on wheat germ agglutinin-Sepharose is the most efficient purification step.[14,16,17] These results imply that at least one subunit is glycosylated. The oligosaccharide chains of the purified calcium channel were detected by separating subunits by NaDodSO$_4$/PAGE and probing the resolved polypeptides with [^{125}I]ConA or [^{125}I]WGA. Under alkylating conditions [^{125}I]ConA labeled the α polypeptide (FIG. 1, lane 7), after disulfide reduction [^{125}I]ConA labeled only the α_2 subunit (FIG. 1, lane 8). [^{125}I]WGA bound to both the α and γ protein bands in gels run under alkylating conditions (FIG. 1, lane 9), and to the α_2 and γ subunits after dithiothreitol treatment. Disulfide reduction led to the appearance of two new [^{125}I]WGA-labeled components at 24-27 kDa that are clearly distinct from the γ subunit. These polypeptides were also detected, but much less distinctly, by silver staining (FIG. 1, lane 2). They appear to be disulfide linked to the α_2 subunit under nonreducing conditions. The smaller polypeptide may be proteolytically derived from the larger so we refer to them collectively as the δ subunit. No labeling of α_1 or β was detected with either lectin. [^{125}I]ConA and [^{125}I]WGA binding to calcium channel subunits was blocked in the presence of 100 mM α-methylmannoside or N-acetylglucosamine, respectively (not shown).

To determine the extent of glycosylation and the core polypeptide size of the calcium-channel subunits, purified channel preparations were labeled with ^{125}I, incubated with glycosidases to remove oligosaccharide chains, and analyzed by NaDodSO$_4$/PAGE and autoradiography. Sequential deglycosylation with neuraminadase and endoglycosidase F caused a reduction in the apparent sizes of the α_2 and γ subunits reaching core polypeptide sizes of 105 kDa and 20 kDa, respectively. Poor iodination of the δ subunit prevented estimation of its carbohydrate content by this method. No shift in the mobility of the α_1 and β subunits was noted confirming the absence of N-linked carbohydrate in these two subunits.

COVALENT LABELING OF CALCIUM-CHANNEL SUBUNITS

[³H]PN200-110 and [³H]azidopine have been shown to covalently label a 145-170-kDa polypeptide in T-tubule membranes[19,20] and purified calcium channels[21] that presumably corresponds to one of the two α subunits. In our purified calcium-channel preparations, [³H]azidopine was incorporated by UV photolysis into a polypeptide

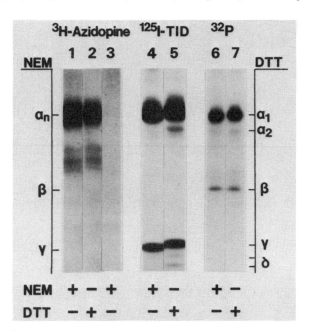

FIGURE 2. Differential labeling of calcium-channel subunits. **Photoaffinity labeling:** T-tubule membranes (0.4 mg/ml) in 25 mM Hepes and 1 mM CaCl$_2$ adjusted to pH 7.5 with Tris base were incubated with 6 nM [³H]azidopine in the absence (lanes 1 and 2) or presence (lane 3) of 2 μM PN200-110 and were irradiated for 15 minutes at 4°C with a 30-watt UV source (λmax, 356 nm). The membranes were solubilized in 1% digitonin, 10 mM Hepes, 185 mM NaCl, 0.5 mM CaCl$_2$, 0.1 mM PMSF, 1 μM pepstatin A adjusted to pH 7.5 with Tris base, and calcium channels were partially purified by chromatography on WGA-Sepharose and analyzed by NaDodSO$_4$/PAGE and fluorography. **Hydrophobic labeling:** [¹²⁵I]TID (15 Ci/mmol) was prepared and purified calcium channel was labeled with 100 μCi/ml [¹²⁵I]TID in a buffer containing 0.1% digitonin (lanes 4 and 5). **Phosphorylation:** Purified calcium channel was incubated with 0.3 μM cAMP-dependent kinase catalytic subunit and 0.12 μM carrier-free [γ-³²P]ATP for 15 minutes at 37°C (lanes 6 and 7).

that migrated as a band of 167 kDa before reduction of disulfide bonds (FIG. 2, lane 1) and 175 kDa after dithiothreitol treatment (FIG. 2, lane 2). The electrophoretic behavior of this polypeptide identifies it as the α_1 subunit. No labeling was observed in the presence of 2 μM PN200-110 (FIG. 2, lane 3).

Ion channel-forming polypeptides should contain transmembrane segments that may be detected using the hydrophobic probe [¹²⁵I]3-(trifluoromethyl)-3-(*m*-iodophenyl)diazirine (TID). This photoreactive compound partitions into free detergent micelles and detergent associated with the major hydrophobic domains of integral membrane proteins and is specifically incorporated into these regions by photolysis. The α_1 and γ subunits were prominently labeled by TID, with a much lower level of incorporation into α_2 and δ (FIG. 2, lanes 6 and 7). The β subunit was not detectably labeled. Quantitation of [^{125}I]TID in excised protein bands showed that the α_1 and γ subunits incorporated 10-fold more TID per unit mass than the α_2 or δ subunits, even though, as shown below, nearly all α_1 and γ subunits are associated with an α_2 subunit. These results indicate that the α_1 and γ subunits are the principal transmembrane components of the purified calcium channel complex.

The regulation of calcium channels via a pathway involving cAMP and protein phosphorylation is now well established,[22-25] although it is not known whether the site of phosphorylation is the calcium channel itself or another intracellular protein that in turn regulates channel function. As a step toward resolving this question, we have investigated whether subunits of the purified calcium antagonist receptor can serve as substrates for the catalytic subunit of cAMP-dependent protein kinase. Incubation with 33 μM [γ-^{32}P]ATP and 3 μM catalytic subunit led to stoichiometric ^{32}P incorporation into the α protein band and the β subunit at rates that are consistent with a physiologically significant phosphorylation reaction.[27] Comparison of the electrophoretic mobility of the phosphorylated bands before and after reduction of disulfide bonds (FIG. 2, lane 6 and 7) showed that the α_1 subunit is a good substrate for this enzyme while the α_2 and δ polypeptides are not labeled. The β subunit was more weakly labeled at the low ATP concentration used in this experiment (see legend). These results identify the α_1 and β subunits of the calcium channel as potential sites of regulation of the voltage-sensitive calcium channel by cAMP-dependent phosphorylation.

ANALYSIS OF NONCOVALENT SUBUNIT INTERACTIONS

By the use of several labeling techniques, we have established that α_1 has the properties expected of the calcium channel including a binding site for dihydropyridine calcium antagonists, at least one cAMP-dependent phosphorylation site, and extensive hydrophobic domains. It is important to determine whether other polypeptides present in the purified preparation are persistent impurities or specifically associated components of the oligomeric calcium-channel complex. The data presented in FIG. 1, lane 4 demonstrate that PAC-10 antibodies recognize only the α_1 subunit of Na-DodSO$_4$-denatured calcium channel. However, this polyclonal serum produced by immunization with native calcium channel may contain antibodies that bind only to native conformations of the $\alpha_2\delta$, β, or γ subunits. To eliminate any antibodies with this specificity, a nitrocellulose strip containing α_1 subunit was used as an affinity matrix to purify anti-α_1 antibodies. Purified anti-α_1 antibodies specifically precipitated [^{3}H]PN200-110-labeled calcium channel, while proteins from PAC10 that were non-specifically adsorbed to a bare nitrocellulose strip did not.

Immunoprecipitation of ^{125}I-labeled calcium channel (FIG. 3) showed that only α_1 was precipitated after NaDodSO$_4$ denaturation (lane 7). In contrast, α_1, α_2, β, and γ were precipitated as a complex in 0.5% digitonin (lane 1) or 0.1% CHAPS (lane 3), detergent conditions that are known to stabilize dihydropyridine binding, allosteric coupling of the three calcium antagonist receptor sites and ion conductance activity. A higher concentration of CHAPS (1%) caused dissociation of α_2 from the complex (data not shown). In addition, experiments in 1% Triton X-100 (lane 5) showed complete dissociation of the α_2 subunit. The β subunit and a small fraction of the γ subunit (not easily seen in FIG. 3) were coimmunoprecipitated with α_1 in Triton X-100. The results of these immunoprecipitation experiments have been confirmed by selective elution of α_2 subunits from immune complexes. Purified calcium channels were immunoprecipitated in 0.5% digitonin as in FIG. 3, lane 1. Resuspension in a buffer containing 1% Triton X-100 caused complete release of the $\alpha_2\delta$ dimer and partial release of γ without loss of β from the precipitate (data not shown). Complementary data supporting these observations have also been obtained using lentil lectin agarose that has the same specificity as ConA (see FIG. 1, lane 8) and is a selective probe for the α_2 subunit. Lentil lectin agarose binds all five of the calcium-channel subunits under conditions where activity is retained but binds only α_2 after denaturation in NaDodSO$_4$.[18]

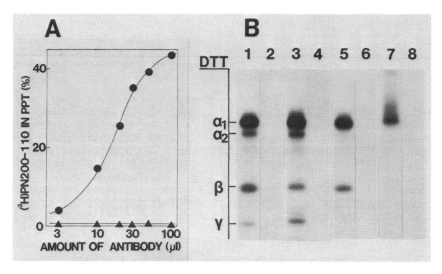

FIGURE 3. Immunoprecipitation of calcium-channel subunits by anti-α_1 antibodies. (**A**) Purified skeletal muscle calcium channels containing specifically bound [^{3}H]PN200-110 were immunoprecipitated with the indicated concentrations of purified anti-α_1 antibodies. (**B**) ^{125}I-labeled calcium channel was immunoprecipitated using affinity-purified anti-α_1 antibodies (lanes 1, 3, 5 and 7) or a control preparation (lanes 2, 4, 6, and 8) in an immunoassay buffer containing the indicated detergents: 0.5% digitonin, 0.1% CHAPS, and 1% Triton X-100, or 1% SDS for two minutes at 100°C followed by immunoassay in 0.5% digitonin detergent after exchange by gel filtration on a 2-ml Sephadex G-50 column.

AN OLIGOMERIC MODEL FOR THE DIHYDROPYRIDINE-SENSITIVE CALCIUM CHANNEL

On the basis of present knowledge of the structure of the dihydropyridine-sensitive calcium channel and in analogy with current models of the structure of voltage-sensitive sodium channels,[28] we have proposed a model (FIG. 4) based on a central ion channel-forming element interacting with three other noncovalently associated subunits. The α_1 subunit, which contains the calcium antagonist binding sites, cAMP-dependent phosphorylation sites, and the largest hydrophobic domains, is proposed to be the central ion channel-forming component of the complex. Its apparent molecular weight of 175 kDa from NaDodSO$_4$/PAGE is likely to be a reasonable approximation of the true polypeptide molecular weight since no N-glycosylation was detected. This calcium-channel subunit contains four homologous transmembrane domains analogous to those of the rat brain sodium-channel α subunit,[29] whose mRNA alone encodes a functional ion channel.[30,31]

The β subunit is also a substrate for cAMP-dependent kinase,[27] but hydrophobic labeling indicates that it does not interact with the membrane phase and it is not a glycoprotein. It is probably therefore tightly associated with an intracellular domain of α_1. The γ subunit of 30 kDa interacts independently with α_1, contains at least one transmembrane segment, and consists of approximately 30% carbohydrate. All these properties are similar to those of the β_1 subunit of the rat brain and skeletal muscle Na$^+$ channels.[28]

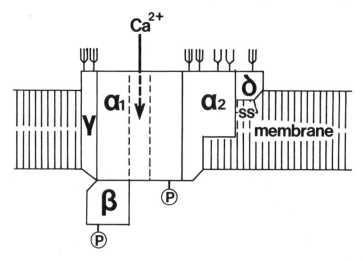

FIGURE 4. Proposed model for calcium-channel structure. Sites of cAMP-dependent phosphorylation (P), glycosylation, and interaction with the membrane are illustrated.

The $\alpha_2\delta$ dimer appears to interact with α_1, although the conditions necessary to achieve dissociation result in a loss of DHP binding activity. The 105-kDa core polypeptide of α_2 contains a heavily glycosylated extracellular domain but displays weak hydrophobic labeling, indicating a limited intramembrane domain. For this reason it seems unlikely that the ion channel is formed jointly by α_1 and α_2 at their zone of interaction.

The proposed model assumes a complex containing one of each subunit type. Our results suggest that each complex contains at least one α_1 and one α_2 subunit but do not specify the stoichiometry of any subunits. α_1 and α_2 appear to be present in approximately equal amounts on silver-stained gels, and the α_1 and β subunits incorporate approximately one mole of ^{32}P per mole of complex. The oligomeric molecular weight of the skeletal muscle calcium channel is 390 kDa, within reasonable range of the predicted size of the complex represented in FIGURE 4 (416 kDa). Thus, an assumption of one mole of each subunit in the complex is plausible, but requires direct experimental verification.

FUNCTIONAL PROPERTIES OF THE PURIFIED CALCIUM ANTAGONIST RECEPTOR IN PHOSPHOLIPID VESICLES

We purify the calcium antagonist receptor as a preformed complex with the calcium antagonists [³H]nitrendipine or [³H]PN200-110. After solubilization, it retains allosteric interactions among the separate binding sites for verapamil, diltiazem, and dihydropyridines. However, other aspects of calcium-channel function cannot be assessed in detergent solution. It is important, therefore, to return the purified calcium antagonist receptor to a membrane environment and to determine whether the purified protein is capable of mediating voltage-dependent calcium flux.

In order to maximize the probability of purification of the calcium channel in an active state, the calcium-antagonist receptors in T-tubule membranes were incubated with sufficient [³H]PN200-110 to label approximately 1% of the binding sites. This label was used to identify the calcium-antagonist receptor during purification as in previous studies.[14] The T-tubule membranes were then incubated in an excess of the specific calcium-channel activator BAY K 8644 so that the remaining 99% of dihydropyridine sites would be occupied by this agent. The calcium-antagonist receptors were then solubilized in digitonin and purified in the continued presence of Bay K 8644 using previously described procedures. Phosphatidylcholine for reconstitution was solubilized in the zwitterionic detergent CHAPS because it is poorly soluble in digitonin. Purified calcium-antagonist receptor dispersed in digitonin was then mixed with phosphatidylcholine dispersed in CHAPS and single-walled phospholipid vesicles were formed by removal of the detergents by molecular sieve chromatography. Analysis of the resulting vesicle preparations by sucrose density gradient sedimentation show that the purified calcium-antagonist receptors are quantitatively incorporated into phosphatidylcholine vesicles and 15 to 25% of vesicles contain at least one calcium antagonist receptor.[32] These preparations therefore provide a suitable system for analysis of the ion transport properties of this purified protein.

Initial rates of influx of $^{45}Ca^{2+}$ or $^{133}Ba^{2+}$ into reconstituted phosphatidylcholine vesicles were measured under counter transport conditions in which the intravesicular compartment contains a high concentration of unlabeled Ca^{2+} or Ba^{2+}.[32] These conditions greatly increase the amount of $^{45}Ca^{2+}$ or $^{133}Ba^{2+}$ uptake required to achieve isotopic equilibrium and greatly slow the approach to isotopic equilibrium as described previously for reconstituted sodium channels. They therefore maximize any ion flux mediated by reconstituted channels. Under these conditions, calcium influx into phosphatidylcholine vesicles containing reconstituted calcium-antagonist receptors was two- to three-fold greater than influx into protein-free phosphatidylcholine vesicles. This increase is completely blocked by verapamil at a concentration of 100 μM. If the calcium-channel activator BAY K 8644 is removed from the vesicle preparation by molecular sieve chromatography, $^{45}Ca^{2+}$ influx is markedly reduced. These results show that at least a fraction of the purified calcium antagonist receptors can function as calcium channels when incorporated into phosphatidylcholine vesicles.[32]

The inhibition of the reconstituted calcium channels by different concentrations of organic calcium-channel blockers from $10^{-9}M$ to $10^{-4}M$ was examined. Half-maximal inhibition was observed with approximately 1.5 μM verapamil, 1.0 μM D600, or 0.2 μM PN200-110. These concentrations are similar to those that give half-maximal inhibition of voltage-activated calcium currents in intact skeletal muscle fibers consistent with the conclusion that the calcium influx in reconstituted vesicles is mediated by functional purified calcium channels.

To determine whether the calcium influx stimulated by BAY K 8644 and blocked by PN200-110 and verapamil required the presence of the subunits of the calcium-antagonist receptor and not other detectable proteins, purified preparations were sedimented through sucrose gradients and each fraction was examined for bound [^{3}H]PN200-110, polypeptide composition, and ability to mediate ^{133}Ba^{2+} influx when incorporated into phosphatidylcholine vesicles. A close quantitative correlation was observed between the presence of the α, β and γ subunits of the calcium antagonist receptor and the ability to mediate barium influx.[32] Thus, these results are also consistent with the conclusion that the purified calcium antagonist receptor is capable of mediating ion flux with the pharmacologic characteristics expected of voltage-sensitive calcium channels.

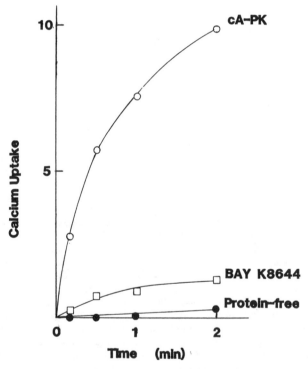

FIGURE 5. Activation of purified calcium channels by cAMP-dependent protein kinase. Calcium channels were radioactively labeled with PN200-110 at approximately 1% occupancy and saturated with BAY K8644. The channels were solubilized, purified, and reconstituted into phosphatidylcholine vesicles. The reconstituted vesicles were incubated under control conditions or with 1 mM ATP and 2 μM cAMP-dependent protein kinase for one hour at 4°C and the initial rate of ^{45}Ca^{2+} influx was measured as described previously.[32] Control experiments with protein-free vesicles were carried out in parallel.

Under these conditions, we found that less than 5% of the purified calcium channels were active in mediating $^{45}Ca^{2+}$ flux.[27] Since calcium channels in intact skeletal muscle[26] and in purified calcium channels reconstituted in planar phospholipid bilayers[33] are activated by cAMP-dependent protein kinase, it was of interest to examine the effects of phosphorylation of the purified calcium channels on their functional activity in reconstituted vesicles. As illustrated in FIGURE 5, incubation with cAMP-dependent protein kinase and ATP under conditions where the α and β_1 subunits are phosphorylated increases the initial rate and the final extent of $^{45}Ca^{2+}$ influx 8- to 10-fold. Because the extent of influx at equilibrium is increased, phosphorylation must allow activation of 8 to 10 times as many purified calcium channels and thereby increase the number of reconstituted calcium channels whose internal volume equilibrates with $^{45}Ca^{2+}$. These results provide further strong support for the concept that the isolated dihydropyridine receptors can be activated to yield functional calcium channels and direct phosphorylation of the α_1 and β subunits is sufficient to modulate the number of active calcium channels.[27,33]

IMMUNOSPECIFIC IDENTIFICATION OF CALCIUM-CHANNEL COMPONENTS IN OTHER TISSUES

In order to identify the molecular components of calcium channels in tissues other than skeletal muscle, the purified calcium channel from rabbit transverse tubules was injected into mice and polyclonal antisera were obtained after multiple injections. To determine the immunological cross-reactivity of this antiserum with calcium channels from heart and brain, radioimmune assays were performed. In each tissue, calcium channels were labeled with [^{3}H]PN200-110, solubilized with digitonin, and purified by chromatography on wheat germ agglutinin-Sepharose. PAC-10 antiserum that we have used to characterize the α_1 subunit of the calcium channel (see FIGS. 1 and 3) showed poor cross-reactivity with other tissues. However, another antiserum, designated PAC-2, recognized [^{3}H]PN200-110-labeled calcium channel in skeletal muscle, heart, and brain. The concentration dependence of immunoprecipitation was determined for each tissue. Each channel was precipitated to a similar extent by maximum concentrations of antiserum and the ratio of antiserum for half-maximal immunoprecipitation was 1.0 : 1.8 : 7.9 for skeletal muscle, heart, and brain. These results indicate that antigenic determinants present on the skeletal muscle calcium channel are also present on calcium channels in heart and brain.

Because these antibodies recognized calcium channels in heart and brain, it is possible to use them to identify and characterize the related proteins in these tissues. For this purpose, membrane preparations from heart and brain were solubilized with digitonin, and glycoproteins were purified by chromatography on wheat germ agglutinin-Sepharose. These purified glycoproteins were labeled with ^{125}I, and the components recognized by antibodies against the skeletal muscle calcium channel were isolated by immunoprecipitation with specific antiserum and a protein A-Sepharose immunoadsorbent. In each tissue, a polypeptide with the characteristics of the calcium channel α_2 subunit was identified.[15,34] The immunoprecipitated component had a molecular mass of 170 kDa in heart and 169 kDa in brain in nonreducing conditions and 141 kDa and 140 kDa, respectively, after reduction. In each case, immunoprecipitation was blocked by prior incubation of the antiserum with purified calcium channel. Thus, we conclude that dihydropyridine-sensitive calcium channels in skeletal muscle, heart, and brain are all associated with an α_2 subunit that is homologous, but not identical, in the three tissues. Polypeptides homologous to other subunits of the skeletal muscle calcium channel have not yet been detected using these techniques.

CONCLUSION

The molecular properties of dihydropyridine-sensitive calcium channels are now being elucidated by following a general strategy that was used previously in studies of voltage-sensitive sodium channels in this laboratory: identification by specific ligand binding and covalent labeling, solubilization and isolation by conventional protein purification methods, and reconstitution of channel function *in vitro*. The work reviewed here illustrates the substantial progress achieved with this approach to date. Future directions include further definition of the functional properties of the purified calcium-channel complex in reconstituted phospholipid vesicles, analysis of the mechanism of regulation of the channel by protein phosphorylation, determination of the primary structure of the channel subunits, extension of these molecular studies to calcium channels in heart, brain, and smooth muscle, and comparison of the structural features of calcium channels with those of voltage-sensitive sodium channels in order to define common structural themes underlying the function of voltage-sensitive ion channels in general.

REFERENCES

1. HAGIWARA, S. & L. BYERLY. 1981. Calcium channel. Ann. Rev. Neurosci. **4:** 69-125.
2. ARMSTRONG, C. M. & D. R. MATTESON. 1985. Two distinct populations of calcium channels in a clonal line of pituitary cells. Science **227:** 65-67.
3. CARBONE, E. & H. D. LUX. 1984. A low voltage-activated fully inactivating calcium channel in vertebrate sensory neurones. Nature **310:** 501-502.
4. NOWYCKY, M. C., A. P. FOX & R. W. TSIEN. 1985. 3 Types of neuronal calcium channel with different calcium agonist sensitivity. Nature **316:** 440-443.
5. NILIUS, M. C., P. HESS, J. B. LANSMAN & R. W. TSIEN. 1985. A novel type of cardiac calcium channel in ventricular cells. Nature **316:** 443-446.
6. HESS, P., J. B. LANSMAN & R. W. TSIEN. 1984. Different modes of calcium channel gating behavior favored by dihydropyridine calcium agonists and antagonists. Nature **311:** 538-544.
7. JANIS, R. & A. SCRIABINE. 1983. Sites of action of Ca^{2+} channel inhibitors. Biochem. Pharmacol. **32:** 3499-3507.
8. SANCHEZ, J. A. & E. STEFANI. 1978. Inward calcium current in twitch muscle fibers of the frog. J. Physiol. (London) **283:** 197-209.
9. ALMERS, W., R. FINK & P. T. PALADE. Calcium depletion in frog muscle tubules: The decline of calcium current under maintained depolarization. J. Physiol. (London) **312:** 177-207.
10. ROSEMBLAT, M., C. HIDALGO, C. VERGARA & N. IKEMOTO. 1981. Immunological and biochemical properties of transverse tubule membranes isolated from rabbit skeletal muscle. J. Biol. Chem. **256:** 8140-8148.
11. FOSSET, M., E. JAIMOVICH, E. DELPONT & M. LAZDUNSKI. 1983. [^{3}H]Nitrendipine receptors in skeletal muscle. J. Biol. Chem. **258:** 6086-6092.
12. GLOSSMANN, H., D. R. FERRY & C. B. BOSCHEK. 1983. Purification of the putative calcium

channel from skeletal muscle with the aid of [^{3}H]-nimodipine binding. Naunyn-Schmiedebergs Arch. Pharmacol. **323:** 1-11.

13. CURTIS, B. M. & W. A. CATTERALL. 1983. Solubilization of the calcium antagonist receptor from rat brain. J. Biol. Chem. **258:** 7280-7283.

14. CURTIS, B. M. & W. A. CATTERALL. 1984. Purification of the calcium antagonist receptor of the voltage-sensitive calcium channel from skeletal muscle transverse tubules. Biochemistry **23:** 2113-2118.

15. TAKAHASHI, M. & W. A. CATTERALL. 1987. Dihydropyridine-sensitive calcium channels in cardiac and skeletal muscle membranes: Studies with antibodies against the α subunits. Biochemistry **26:** 5518-5526.

16. GLOSSMANN, H. & D. R. FERRY. 1983. Solubilization and partial purification of putative calcium channels labeled with [^{3}H]-nimodipine. Naunyn-Schmiedebergs Arch. Pharmacol. **323:** 279-291.

17. BORSOTTO, M., J. BARHANIN, R. I. NORMAN & M. LAZDUNSKI. 1984. Purification of the dihydropyridine receptor of the voltage-dependent Ca^{2+} channel from skeletal muscle transverse tubules using (+) [^{3}H]PN200-110. Biochem. Biophys. Res. Commun. **122:** 1357-1366.

18. TAKAHASHI, M., M. J. SEAGAR, J. F. JONES, B. F. X. REBER & W. A. CATTERALL. 1987. Subunit structure of dihydropyridine-sensitive calcium channels from skeletal muscle. Proc. Natl. Acad. Sci. USA **84:** 5478-5482.

19. FERRY, D. R., A. GOLL & H. GLOSSMANN. 1984. Photoaffinity labeling of Ca^{2+} channels with [^{3}H]azidopine. FEBS Lett. **169:** 112-167.

20. GALIZZI, J. P., M. BORSOTTO, J. BARHANIN, M. FOSSET & M. LAZDUNSKI. 1986. Characterization and photoaffinity labeling of receptor sites for the Ca^{2+} channel inhibitors *d-cis*-diltiazem, (+)-bepridil, dismethoxyverapamil and (+) PN200-110 in skeletal muscle transverse tubule membranes. J. Biol. Chem. **261:** 1393-1397.

21. STRIESSNIG, J., K. MOOSBURGER, A. GOLL, D. R. FERRY & H. GLOSSMANN. 1986. Stereoselective photoaffinity labeling of the purified 1,4-dihydropyridine receptor of the voltage-dependent calcium channel. Eur. J. Biochem. **161:** 603-609.

22. REUTER, H. 1974. Localization of beta-adrenergic receptors, and effects of nonadrenaline and cyclic nucleotides on action potentials, ionic currents and tension in mammalian cardiac muscle. J. Physiol. (London) **242:** 429-451.

23. REUTER, H. 1983. Calcium channel modulation by neurotransmitters, enzymes and drugs. Nature **301:** 569-574.

24. TSIEN, R. W., W. GILES & P. GREENGARD. 1972. Cyclic AMP mediates the effects of adrenaline on cardiac Purkinjie fibres. Nature New Biol. **240:** 181-183.

25. BRUM, G., V. FLOCKERZI, W. OSTERRIEDER & W. TRAUTWEIN. 1983. Injection of catalytic subunit of cAMP-dependent protein kinase into isolated cardiac myocytes. Pfluger's Arch. **398:** 147-154.

26. ARREOLA, J., J. CALVO, M. C. GARCIA & J. A. SANCHEZ. 1987. Modulation of calcium channels of twitch skeletal muscle of the frog by adrenaline cyclic adenosine monophosphate fibers J. Physiol. (London) **393:** 307-330.

27. CURTIS, B. M. & W. A. CATTERALL. 1985. Phosphorylation of the calcium antagonist receptor of the voltage-sensitive calcium channel by cAMP-dependent protein kinase. Proc. Natl. Acad. Sci. USA **82:** 2528-2532.

28. CATTERALL, W. A. 1986. Molecular properties of voltage-sensitive sodium channels. Ann. Rev. Biochem. **55:** 953-985.

29. NUMA *et al.* Ann. N.Y. Acad. Sci. This volume.

30. NODA, M., T. IKEDA, H. SUZUKI, H. TAKESHIMA, T. TAKAHASHI, M. KUNO & S. NUMA. 1986. Expression of functional sodium channels from cloned cDNA. Nature **322:** 826-828.

31. GOLDIN, A. L., T. SNUTCH, H. LUBBERT, A. DOWSETT, J. MARSHALL, V. AULD, W. DOWNEY, L. C. FRITZ, H. A. LESTER, R. DUNN, W. A. CATTERALL & N. DAVIDSON. 1986. Messenger RNA coding for only the α subunit of the rat brain Na channel is sufficient for expression of functional channels in *Xenopus* oocytes. Proc. Natl. Acad. Sci. USA **83:** 7503-7509.

32. CURTIS, B. M. & W. A. CATTERALL. 1986. Reconstitution of the voltage-sensitive calcium channel purified from skeletal muscle transverse tubules. Biochemistry **25:** 3077-3083.

33. FLOCKERZI, V., H.-J. OEKEN, F. HOFMANN, D. PELZER, A. CAVALIE & W. TRAUTWEIN.
     1986. Purified dihydropyridine binding site from skeletal muscle T-tubules in a functional
     calcium channel. Nature **323:** 66-68.
34. TAKAHASHI, M. & W. A. CATTERALL. 1987. Identification of an α subunit of dihydro-
     pyridine-sensitive brain calcium channels. Science **236:** 88-92.

Biochemistry, Molecular Pharmacology, and Functional Control of Ca²⁺ Channels[a]

J. BARHANIN,[b] M. BORSOTTO,[b] T. COPPOLA,[b]
M. FOSSET,[b] M. M. HOSEY,[c] C. MOURRE,[b]
D. PAURON,[d] J. QAR,[e] G. ROMEY,[b] A. SCHMID,[b]
S. VANDAELE,[b] C. VAN RENTERGHEM,[b]
AND M. LAZDUNSKI [b,f]

[b] Centre de Biochimie du
Centre National de la Recherche Scientifique
Parc Valrose
06034 Nice Cedex, France

[c] Department of Biological Chemistry & Structure
University of Health Sciences / The Chicago Medical School
North Chicago, Illinois 60064

[d] Station de Nématologie et de Génétique Moléculaire des
Invertébrés, I.N.R.A., B. P. 78
06602 Antibes, France

[e] Marine Science Station of Aqaba
Aqaba, Jordan

Most of what we now know of the structure of the Ca²⁺ channel comes from work that has been carried out with skeletal muscle membranes. This is because it was shown a number of years ago that the richest source of receptors for Ca²⁺ channel blockers and particularly for 1,4-dihydropyridines (DHP) is the skeletal muscle T-tubule membrane system.[1] Two main types of Ca²⁺ channels are found in skeletal muscle. One of them is the low threshold, or T-type, Ca²⁺ channel, the other one is the high threshold or L-type Ca²⁺ channel.[2,3] Patch-clamp analysis has actually identified two different subtypes of L-type Ca²⁺ channels in mammalian skeletal muscle[2]; both are blocked by 1,4-dihydropyridines, but they have different voltage sensitivities of the activation process and different inactivation kinetics.

In muscular dysgenesis, a mouse muscle disease corresponding to an absence of contraction, receptor sites for Ca²⁺ channel blockers are nearly absent in muscle membranes and the functional expression of the Ca²⁺ channel is lacking.[4-6] *In vitro*

[a]This work was supported by the Centre de National de la Recherche Scientifique, the Fondation pour la Recherche Médicale, and the Association des Myopathes de France.

[f]Corresponding author.

innervation of diseased myotubes by spinal cord cells of normal animals has been shown to restore both Ca^{2+} channel activity and contraction.[7] The lack of Ca^{2+} channel and contractile activity in the mutant is due to an absence of intact triadic structures.[4,7] The diseased muscle only contains pseudo-diads.[4] Reinnervation restores triadic organization together with Ca^{2+} channel activity and contraction. Therefore an intact triadic organization is essential for the expression of both Ca^{2+} channel activity and contraction in skeletal muscle. Innervation seems to be essential to reach a correct stage of triad differentiation.

Ca^{2+} channel regulation has attracted a lot of interest in the last few years.[8] A particularly interesting regulation (because of the use of Ca^{2+} channel blockers in hypertension) concerns the effect of vasopressin on aortic cells. This effect is presented in FIG. 1A and B. Vasopressin strongly inhibits Ca^{2+} channel function. Since vasopressin activates a phospholipase C in this cell line, the effect of the peptide could be due either to diacylglycerol action *via* a C kinase phosphorylation or to inhibition of Ca^{2+} channel activity *via* an elevation of $[Ca^{2+}]_{in}$ due to IP_3 production. The inhibitory effect of IP_3 injection on Ca^{2+} currents in A7r5 aortic cells is presented in FIGURE 1C. It clearly suggests that vasopressin production of IP_3 leads to inhibition of the activity of Ca^{2+} channels.

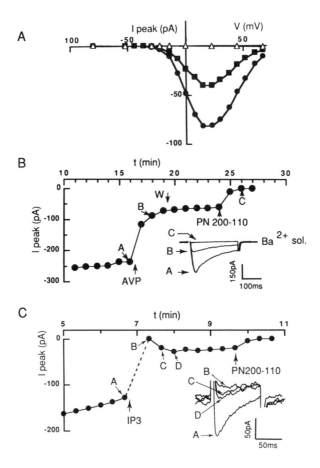

The Ca^{2+} channel blocker receptor of skeletal muscle has been photoaffinity-labeled and purified. These two approaches have shown that the two main components in its structure have a M_r of about 170 kDa.[8] One of these components (M_r 165 kDa) has been photoaffinity-labeled by diltiazem, bepridil, PN 200-110, azidopine and azidopamil.[9–15] Its mobility is unaltered upon reduction. The other component also has a M_r of about 170 kDa. It is cleaved by reduction into two components of 140 kDa and about 30 kDa, respectively.[16–18] The first component is often known as the α_1 component, while the other one is known as the α_2 component. The α_2 component is heavily glycosylated,[18] while the α_1 component is only very partially glycosylated.[10] The α_1 component that bears all binding sites for Ca^{2+} channel blockers is also the target of the cAMP-dependent phosphorylation (FIG. 2),[10,19] which is known to activate the skeletal muscle Ca^{2+} channel[20] in the same way as the cardiac Ca^{2+} channel.[21] The α_1 component is the subunit that has recently been cloned and sequenced and that has an extensive homology with the Na^+ channel protein.[22] Antibodies against the different polypeptide elements of the α_2 subunit reveal that these are also present in heart, brain, and smooth muscle.[17,23,24] The 140-kDa or 30-kDa subunits are not detected in tissues that lack Ca^{2+} blockers receptors. Moreover a monoclonal antibody against the α_2 subunit immunoprecipitates both DHP and phenylalkylamine binding activity.[25]

1,4-Dihydropyridines and phenylalkylamines have been important tools in gaining molecular access to the structure of skeletal muscle Ca^{2+} channels. Receptors for these different categories of drugs are also present in the central nervous system (CNS)[26–28] in which they have been localized by quantitative autoradiography at different stages of brain development,[28] but the function of the potential Ca^{2+} channels

---

FIGURE 1. Inhibition by Arg_8-vasopressin (AVP) and by IP_3 of an L-type calcium-channel activity in aortic smooth muscle cells (A7r5 cell line). (A) Characterization of a single type of depolarization-activated current in A7r5 cells. Plot of the peak value of the inward currents evoked by depolarizing pulses against the test potential in control conditions (●), then after application of AVP (100 nM) (■), then in the presence (△) of (+)PN 200-110 (100 nM) on the same cell. Holding potential: −80 mV. A leak current obtained by linear extrapolation from the hyperpolarized region has been subtracted (leak conductance: 50 pS). Note the absence of any depolarization-activated outward current. (B) Inhibition by AVP of the calcium current recorded with Ca^{2+} as charge carrier. The value of the peak PN 200-110-sensitive current evoked every minute by a pulse to 8 mV from a holding potential of −60 mV is plotted against the duration of the experiment since passage to the whole-cell configuration. AVP (100 nM) addition and removal, and addition of (+)PN 200-110 (200 nM) are indicated by vertical arrows. **Inset:** recordings of the currents evoked immediately before (A) and 1.8 min after (B) AVP addition, and after blockade by (+)PN 200-110 (C), at the times indicated in the plots. The recordings in B and C are presented without leak subtraction. (C) Inhibition of the calcium current by intracellular application of IP_3. Peak PN 200-110-sensitive calcium-current values are plotted against time since passage of the recording electrode to the whole-cell configuration. Holding potential: −60 mV. Test potential: +8 mV. IP_3 (200 μM) was injected through a second patch electrode, after the transition from the cell attached to the whole-cell configuration (vertical arrow, "IP_3"). Ba^{2+} was added to Ca^{2+} in the extracellular solution in order to block most of the potassium current evoked by IP_3 in these cells, so that a small outward current (60 pA) was present only at t_7 (not represented). **Inset:** recordings obtained 10 sec before (A), and 30 sec (B), 50 sec (C), and 70 sec (D) after the beginning of the IP_3 injection, at the times indicated on the plot. The currents were recorded from isolated cells using conventional patch voltage-clamp techniques. The pipette solution contained 150 mM KCl, 4 mM $MgCl_2$, 10 mM HEPES/NaOH, 0.5 mM EGTA, 3 mM ATP, 0.2 mM GTP, pH 7.2. The extracellular solution contained 140 mM NaCl, 10 mM $BaBl_2$ (A, B) or 5 mM $BaCl_2$ and 5 mM $CaCl_2$ (C), 1 mM $MgCl_2$, 5 mM KCl, 5 mM HEPES/NaOH, 5 mM glucose, pH 7.4.

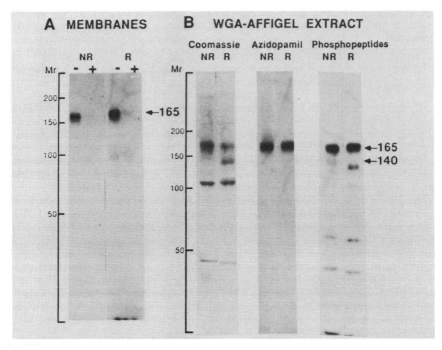

FIGURE 2. Photoaffinity labeling and phosphorylation of peptides present in membranes and partially purified preparations of DHP and phenylalkylamine-sensitive Ca^{2+} channels from rabbit skeletal muscle. (**A**) Transverse tubule membranes were photoaffinity labeled as described in the text in the absence (−) or presence (+) of 1 μM (±)verapamil, and the labeled peptides analyzed by gel electrophoresis and fluorography. The fluorogram shown was exposed for 14 days. (**B**) Partially purified preparations obtained after WGA-Affigel chromatography were analyzed by gel electrophoresis under reducing (R) or nonreducing (NR) conditions. Shown are the Coomassie blue stained gels depicting the peptide content, fluorograms showing the [^{3}H]azidopamil-labeled peptides, and autoradiograms depicting the peptides phosphorylated by cAMP-dependent protein kinase. The fluorogram was obtained after a five-day exposure using Kodak XAR film, while the autoradiogram was obtained after a six-hour exposure using Kodak X-Omat S film. Note that the small, reduction-induced increase in the apparent molecular weight of the 165-kDa peptide was not evident in the Coomassie-stained gel because of the overlap in mobility of the 170- and 165-kDa peptides under nonreducing conditions.

associated with these receptors is not yet clear.[29,30] Conversely, a polypeptide toxin, ω-conotoxin, has recently been shown to be a very active blocker of both L- and N-types of Ca^{2+} channels[31–33] in the CNS. This toxin forms very high affinity complexes with its receptor[34] (K_d in the pM range) and the stoichiometry of binding is particularly large in chick brain membranes (FIG. 3). The toxin dissociates slowly from its receptor ($k_{-1} = 9.4 \times 10^{-5}$ sec^{-1}, i.e. t $\frac{1}{2}$ = 2 hours) and the binding is antagonized by 10 mM Ca^{2+}. Cross-linking experiments[34] have shown that a large peptide of 210-220 kDa was labeled using the azidonitrobenzoyloxy derivative of ω-conotoxin while disuccinimidyl suberate specifically cross-linked the toxin to a 170-kDa component comprising a 140-kDa peptide disulfide linked to a 30-kDa peptide (FIGS. 4 and 5). The 210-220-kDa peptide might be the equivalent in the brain of the α_1 subunit identified for the skeletal muscle Ca^{2+} channel while the 170-kDa peptide (140 + 30 kDa) might be the equivalent of the α_2 component of the receptor for Ca^{2+} channel blockers that is present in T-tubules.

A particularly interesting case is found in the nervous system of *Drosophila*.[35] Binding studies using $(-)[^3H]D888$ and $(\pm)[^3H]$verapamil have identified a single class of very high-affinity binding sites for these ligands ($K_d = 0.1 - 0.4$ nM, $B_{max} = 1600\text{-}1800$ fmol/mg of protein) (TABLE 1). The most potent molecule in the phenylalkylamine series was $(-)$verapamil with K_d value as exceptionally low as 4.7 pM. Molecules in the benzothiazepine and diphenylbutylpiperidine series of Ca^{2+} channels blockers as well as bepridil inhibited $(-)[^3H]D888$ binding, suggesting a close correlation, as in the mammalian system, between these receptor sites and those recognized by phenylalkylamines. It is of particular interest that unlike what has been observed in the mammalian system, 1,4-dihydropyridines are without effect on phenylalkylamine binding. This observation strongly suggests that we have here a case of a protein that has the receptors for Ca^{2+} channel blockers, including very high-affinity receptors for phenylalkylamines but that lacks receptors for DHPs. The protein has been affinity labeled with a tritiated (arylazido)phenylalkylamine ($K_d = 0.24$ nM), and a protein of M_r 135 $\pm$ 5 kDa was found to be specifically labeled (FIG. 6). This protein would correspond to the α_1 subunit of the neuronal Ca^{2+} channel in *Drosophila*.

Another interesting case is found in plant membranes.[36] Ca^{2+} channel inhibitors of the phenylalkylamine and of the diphenylbutylpiperidine series, as well as other blockers such as bepridil, inhibit $^{45}Ca^{2+}$ influx into carrot protoplasts. The corresponding plasma membrane has a high density (120 pmol/mg of protein) of sites for the phenylalkylamine $(-)[^3H]D888$ ($K_d = 85$ nM). For 10 different Ca^{2+} channel blockers, there was a good correlation between efficacy of blockade of $^{45}Ca^{2+}$ influx into protoplasts and efficacy of binding of the $[^3H]$ligand to membranes. Binding sites

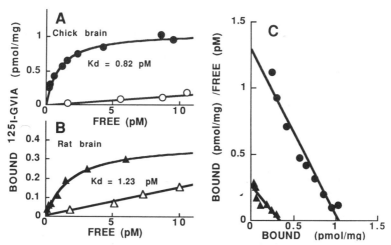

FIGURE 3. Equilibrium binding of $[^{125}I]$GVIA to chick and rat brain synaptosomes. Chick (**A**) or rat (**B**) brain synaptosomes (1 and 3 μg of protein per ml, respectively) were incubated with various concentrations of $[^{125}I]$GVIA for four hours at 25°C (final volume 4.5 ml). Each value is the mean of duplicate filtrations of 2 ml. Nonspecific binding ($\bigcirc$, $\triangle$) was measured in the presence of 10^{-9} M unlabeled toxin. Specific binding ($\bullet$, $\blacktriangle$) is the difference between total and nonspecific binding. (**C**) Scatchard plot of the data. ($\bullet$) chick; ($\blacktriangle$) rat. The specific activity of the $[^{125}I]$GVIA used was 2000 cpm per fmol.

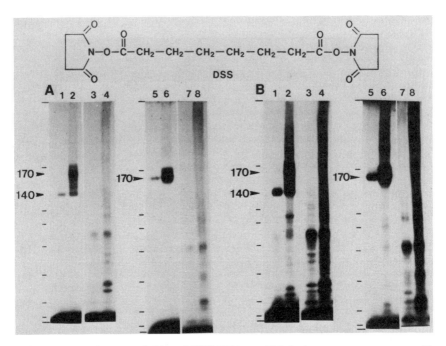

FIGURE 4. Covalent cross-linking of $[^{125}I]$GVIA to chick brain receptor components with DSS. Shown are the autoradiograms depicting the peptides labeled by $[^{125}I]$GVIA using 30 μM DSS (lanes 1, 3 and 5, 7) or 300 μM DSS (lanes 2, 4 and 6, 8). Lanes 3, 4 and 7, 8 were obtained for experiments performed in the presence of 10^{-9} M unlabeled GVIA. Gels were run under reducing conditions (lanes 1-4) and nonreducing conditions (lanes 5-8). **Part A:** autoradiograms exposed for 26 hours at $-70°C$ on Kodak X-OMAT films with a Cronex Hi-plus intensifying screen (Dupont). **Part B:** autoradiograms from the same gels as in A exposed for 140 hours. Positions of molecular weight markers (BioRad) are indicated by horizontal bars: myosin (200,000), β-galactosidase (116,500), phosphorylase B (96,500), bovine serum albumin (66,200), ovalbumin (45,000), carbonic anhydrase (31,000), and soybean trypsin inhibitor (21,500).

for DHPs are absent and no blockade of Ca^{2+} influx was observed with several molecules in this series such as $(+)PN$ 200-110, nifedipine, or nitrendipine.

DHPs have been very important compounds to open the way for a potent pharmacology of L-type Ca^{2+} channels. New molecules are emerging now that may have other properties. Neuroleptics of the diphenylbutylpiperidine series include molecules like fluspirilene, penfluridol, pimozide, and clopimozide. This class of drugs has the ability to relieve schizophrenic symptoms, but it is also known for its anti-anxiety properties in psychovegetative disorders. Fluspirilene at very low doses influences mental state in the direction of anxiolysis, relaxation, and increased self-confidence. Fluspirilene blocks L-type Ca^{2+} channel activity.[37,38] $[^3H]$Fluspirilene binds to skeletal muscle T-tubule membranes with a high affinity ($K_d = 0.1$ nM), and the fluspirilene receptor appears to be distinct from other well-identified receptors.[35] Diphenylbutyl-piperidines bind to heart, smooth muscle, and brain membranes with K_d values in the range of 10-100 nM. Binding activity and blockade of Ca^{2+} channels are parallel in smooth muscle cells.[38] It may be that anti-anxiety properties of fluspirilene that are observed at very low dosages of the drug are due to blockade of Ca^{2+} channels, with a very high affinity, of specific populations of neurons involved in the control of anxiety. The neuroleptic properties of these drugs are of course due to their action on D_2-receptors.

Benzolactams (FIG. 7) are another series of new compounds that also bind with high affinity to receptors associated with Ca^{2+}-channel function because they block Ca^{2+}-channel activity in a nM range of concentration.[39] This series of molecules, although it is competitive in binding with DHPs and although it has an action that, as that of DHPs, is voltage-dependent, binds to a receptor site that is new and allosterically related to receptors for all other Ca^{2+}-channel blockers.[39]

The voltage-sensitive Na^+ channel has multiple receptors for natural toxins and other compounds of pharmacological importance. At least six different receptor sites have been identified for natural toxins.[40] The situation may turn out to be the same for the L-type Ca^{2+} channel. If there are numerous receptors for drugs chemically synthesized by pharmaceutical companies, it may be that there are endogeneous ligands corresponding to each category of these receptors.

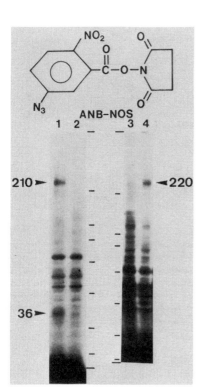

FIGURE 5. Analysis of the photoaffinity-labeled peptides using ANB-[^{125}I]GVIA on rat and chick brain membranes. Preparations labeled with ANB-[^{125}I]GVIA in the absence (lanes 1 and 4) or the presence (lanes 2 and 3) of 10^{-9} M unlabeled GVIA were analyzed under reducing conditions. Lanes 1 and 2: rat brain membranes; lanes 3 and 4: chick brain membranes. Molecular weight markers are the same as in FIGURE 4.

TABLE 1. Effects of Calcium-Channel Antagonists on $(-)[^3H]D888$ Binding to Insect Nervous System and Mammalian Skeletal Muscle

Antagonist	Insect Nervous System $K_I(nM)$	Mammalian Skeletal Muscle $K_I(nM)$
$(-)D888$	0.052	2
$(+)D888$	0.190	
$(-)$verapamil	0.005	20^a
$(+)$verapamil	2.3	
$(-)D600$	0.052	40^a
$(+)D600$	4.7	
Fluspirilene	12-14	0.4
$(-)$bepridil	17-26	20^a
$(+)$bepridil	17	
d-cis-diltiazem	190-270	60
l-cis-diltiazem	190	900

$^a K_{0.5}$ and K_I values obtained from racemic compound instead of pure enantiomers.

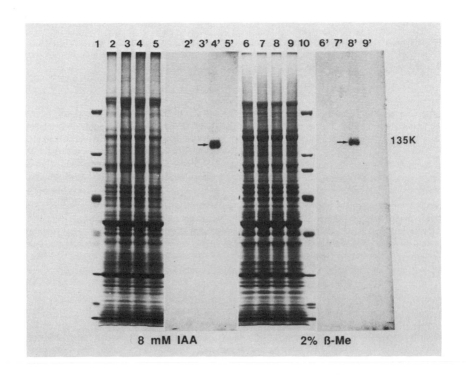

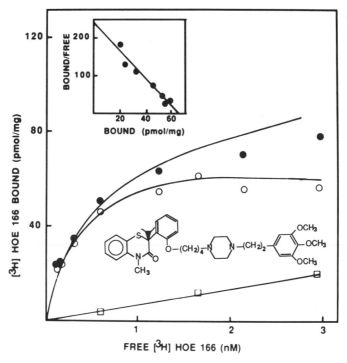

FIGURE 7. Direct binding of [³H]HOE 166 to T-tubule membranes at 20°C (0.008 mg of protein/ml). ●, total binding; □, nonspecific binding measured in the presence of 1 μ*M* unlabeled HOE 166; ○, specific binding. **Inset,** Scatchard plot of the specific binding.

ACKNOWLEDGMENTS

We thank C. Roulinat-Bettelheim for expert secretarial assistance.

FIGURE 6. Photoaffinity labeling of the phenylalkylamine receptor with [³H]LU49888. Labeling experiments were performed as described in Pauron *et al.*[35] Lanes 1 to 5 and 6 to 10: Coomassie blue staining of the gel. Lanes 2' to 5' and 6' to 9': autoradiography of the gel. Lanes 2, 4, 6, 8 and 2', 4', 6', 8': incubation in the absence of unlabeled (±)verapamil. Lanes 3, 5, 7, 9 and 3', 5', 7', 9': incubation in the absence of unlabeled (±)verapamil. Lanes 2, 3, 2', 3' and lanes 6, 7, 6', 7' are from incubations that where not irradiated. Lanes 1 and 10 are molecular weight markers from BioRad: (from the top to the front), myosin (200,000), β-galactosidase (116,500), phosphorylase B (96,500), bovine serum albumin (66,200), ovalbumin (45,000), carbonic anhydrase (31,000), soybean trypsin inhibition (24,500), and lysozyne (14,400). β-Me: β-mercaptoethanol (disulfide reducing conditions). IAA: iodoacetamide (nonreducing conditions). Despite the relatively large nonspecific binding of [³H]LU49888 after UV illumination, no nonspecifically labeled bands appeared in the autoradiography because all the nonspecific incorporation occurred in phospholipids and migrated ahead of the dye front of the gel.

REFERENCES

1. FOSSET, M., E. JAIMOVICH, E. DELPONT & M. LAZDUNSKI. 1983. [^{3}H]Nitrendipine receptors in skeletal muscle. J. Biol. Chem. **258:** 6083-6092.
2. COGNARD, C., M. LAZDUNSKI & G. ROMEY. 1986. Different types of Ca^{2+} channels in mammalian skeletal muscle cells in culture. Proc. Natl. Acad. Sci. USA **83:** 517-521.
3. COGNARD, C., G. ROMEY, J.-P. GALIZZI, M. FOSSET & M. LAZDUNSKI. 1986. Dihydropyridine-sensitive Ca^{2+} channels in mammalian skeletal muscle cells in culture: Electrophysiological properties and interactions with Ca^{2+} channel activation (Bay K 8644) and inhibitor (PN 200-100). Proc. Natl. Acad. Sci. USA **83:** 1518-1522.
4. PINCON-RAYMOND, M., F. RIEGER, M. FOSSET & M. LAZDUNSKI. 1985. Abnormal transverse tubule system and abnormal amount of receptors for Ca^{2+} channel inhibitors of the dihydropyridine family in skeletal muscle from mice with embryonic muscular dysgenesis. Dev. Biol. **112:** 458-446.
5. BEAM, K. G., C. M. KNUDSON & J. A. POWELL. 1986. A lethal mutation in mice eliminates the slow calcium current in skeletal muscle cells. Nature **320:** 168-170.
6. ROMEY, G., F. RIEGER, J.-F. RENAUD, M. PINCON-RAYMOND & M. LAZDUNSKI. 1986. The electrophysiological expression of Ca^{2+} channels and of apamin-sensitive Ca^{2+}-activated K^+ channels is abolished in skeletal muscle cells from mice with muscular dysgenesis. Biochem. Biophys. Res. Commun. **136:** 935-940.
7. RIEGER, F., R. BOURNAUD, T. SHIMAHARA, L. GARCIA, M. PINCON-RAYMOND, G. ROMEY & M. LAZDUNSKI. 1987. Restoration of dysgenic muscle contraction and calcium channel function by co-culture with normal spinal cord neurons. Nature **330:** 563-566.
8. HOSEY, M. M. & M. LAZDUNSKI. 1988. Calcium channels: Molecular pharmacology, structure and regulation. J. Membr. Biol. **104:** 81-105.
9. GALIZZI, J.-P., M. BORSOTTO, J. BARHANIN, M. FOSSET & M. LAZDUNSKI. 1986. Characterization and photoaffinity labeling of receptor sites for the Ca^{2+} channel inhibitors d-cis-diltiazem, (±)bepridil, desmethoxyverapamil, and (+)PN 200-110 in skeletal muscle transverse tubule membranes. J. Biol. Chem. **261:** 1393-1397.
10. HOSEY, M. M., J. BARHANIN, A. SCHMID, S. VANDAELE, J. PTASIENSKI, C. O'CALLAHAN, C. COOPER & M. LAZDUNSKI. 1987. Photoaffinity labelling and phosphorylation of a 165-kilodalton peptide associated with dihydropyridine and phenylalkylamine-sensitive calcium channels. Biochem. Biophys. Res. Commun. **147:** 1137-1145.
11. SHARP, A. H., T. IMAGAWA, A. T. LEUNG & K. P. CAMPBELL. 1987. Identification and characterization of the dihydropyridine binding subunit of the skeletal muscle dihydropyridine receptor. J. Biol. Chem. **262:** 12309-12315.
12. SIEBER, M., W. NASTAINCZYK, V. ZUBOR, W. WERNET & F. HOFMANN. 1987. The 165-KDa peptide of the purified skeletal muscle dihydropyridine receptor contains the known regulatory sites of the calcium channel. Eur. J. Biochem. **167:** 117-122.
13. STRIESSNIG, J., H. G. KNAUS, M. GRABNER, K. MOOSBURGER, W. SEITZ, H. LIETZ & H. GLOSSMANN. 1987. Photoaffinity labelling of the phenylalkylamine receptor of the skeletal muscle transverse-tubule calcium channel. FEBS Lett. **212:** 247-253.
14. TAKAHASHI, M., M. J. SEAGAR, J. F. JONES, B. F. X. REBER & W. A. CATTERALL. 1987. Subunit structure of dihydropyrydine-sensitive calcium channels from skeletal muscle. Proc. Natl. Acad. Sci. USA **84:** 5478-5482.
15. VAGHY, P. L., J. STRIESSNIG, K. MIWA, H. G. KNAUS, K. ITAGAKI, E. McKENNA, H. GLOSSMANN & A. SCHWARTZ. 1987. Identification of a novel 1,4-dihydropyridine- and phenylalkylamine-binding polypeptide in calcium channel preparations. J. Biol. Chem. **262:** 14337-14342.
16. BORSOTTO, M., J. BARHANIN, M. FOSSET & M. LAZDUNSKI. 1985. The 1,4-dihydropyridine receptor associated with the skeletal muscle voltage-dependent Ca^{2+} channel. Purification and subunit composition. J. Biol. Chem. **260:** 14255-14263.
17. SCHMID, A., J. BARHANIN, T. COPPOLA, M. BORSOTTO & M. LAZDUNSKI. 1986. Immunochemical analysis of subunit structures of 1,4-dihydropyridine receptors associated with voltage-dependent Ca^{2+} channels in skeletal, cardiac and smooth muscles. Biochemistry **25:** 3492-3495.
18. BARHANIN, J., T. COPPOLA, A. SCHMID, M. BORSOTTO & M. LAZDUNSKI. 1987. The

calcium channel antagonists receptor from rabbit skeletal muscle. Reconstitution after purification and subunit characterization. Eur. J. Biochem. **164:** 525-531.

19. Hosey, M. M., M. Borsotto & M. Lazdunski. 1986. Phosphorylation and dephosphorylation of dihydropyridine-sensitive voltage-dependent Ca^{2+} channel in skeletal muscle membranes by cAMP- and Ca^{2+}-dependent processes. Proc. Natl. Acad. Sci. USA **83:** 3733-3737.

20. Schmid, A., J.-F. Renaud & M. Lazdunski. 1985. Short-term and long-term effects of β-adrenergic effectors and cyclic AMP on nitrendipine-sensitive voltage-dependent Ca^{2+} channels of skeletal muscle. J. Biol. Chem. **260:** 13041-13046.

21. Reuter, H. 1983. Calcium channel modulation by neurotransmitters, enzymes and drugs. Nature **301:** 569-574.

22. Tanabe, T., H. Takeshima, A. Mikami, V. Flockerzi, H. Takahashi, K. Kangawa, M. Kojima, H. Matsuo, T. Hirose & S. Numa. 1987. Primary structure of the receptor for calcium channel blockers from skeletal muscle. Nature **328:** 313-318.

23. Schmid, A., J. Barhanin, C. Mourre, T. Coppola, M. Borsotto & M. Lazdunski. 1986. Antibodies reveal the cytolocalization and subunit structure of the 1,4-dihydropyridine component of the neuronal Ca^{2+} channel. Biochem. Biophys. Res. Commun. **139:** 996-1002.

24. Cooper, C. L., S. Vandaele, J. Barhanin, M. Fosset, M. Lazdunski & M. M. Hosey. 1987. Purification and characterization of the dihydropyridine-sensitive voltage-dependent calcium channel from cardiac tissue. J. Biol. Chem. **262:** 509-512.

25. Vandaele, S., M. Fosset, J.-P. Galizzi & M. Lazdunski. 1987. Monoclonal antibodies that coimmunoprecipitate the 1,4-dihydropyridine and phenylalkylamine receptors and reveal the Ca^{2+} channel structure. Biochemistry **26:** 5-9.

26. Cortes, R., P. Supavilai, M. Korobath & J. M. Palacios. 1984. Calcium antagonist binding sites in the rat brain: Quantitative autoradiographic mapping using the 1,4-dihydropyridines [^{3}H]PN 200-110 and [^{3}H]PY 108-068. J. Neural Trans. **60:** 169-197.

27. Quirion, R. 1985. Characterization of binding sites for two classes of calcium channel antagonists in human forebrain. Eur. J. Pharmacol. **117** 139-142.

28. Mourre, C., P. Cervera & M. Lazdunski. 1987. Autoradiographic analysis in rat brain of the postnatal ontogeny of voltage-dependent Na^+ channels, Ca^{2+}-dependent K^+ channels and slow Ca^{2+} channels identified as receptors for tetrodotoxin, apamin and $(-)$-desmethoxyverapamil. Brain Res. **417:** 21-32.

29. Miller, R. J. 1987. Multiple calcium channels and neuronal function. Science **235:** 46-52.

30. Hirning, L. D., A. P. Fox, E. W. McCleskey, B. M. Olivera, S. A. Thayer, R. J. Miller & R. W. Tsien. 1988. Dominant role of N-type Ca^{2+} channels in evoked release of norepinephrine from sympathetic neurons. Science **239:** 57-61.

31. Olivera, B. M., J. M. McIntosh, L. J. Cruz, F. A. Luque & W. R. Gray. 1984. Purification and sequence of a presynaptic peptide toxin from *Conus geographus* venom. Biochemistry **23:** 5087-5090.

32. Reynolds, I. J., J. A. Wagner, S. H. Snyder, S. A. Thayer, B. M. Olivera & R. J. Miller. 1986. Brain voltage-sensitive calcium channel subtypes differentiated by ω-conotoxin fraction GVIA. Proc. Natl. Acad. Sci. USA **83:** 8804-8807.

33. McCleskey, E. W., A. P. Fox, D. H. Feldman, L. J. Cruz, B. M. Olivera, R. W. Tsien & D. Yoshikami. 1987. ω-Conotoxin: Direct and persistent blockade of specific types of calcium channels in neurons but not muscle. Proc. Natl. Acad. Sci. USA **84:** 4327-4331.

34. Barhanin, J., A. Schmid & M. Lazdunski. 1988. Properties of structure and interaction of the receptor for ω-conotoxin, a polypeptide active on Ca^{2+} channels. Biochem. Biophys. Res. Commun. **150:** 1051-1062.

35. Pauron, D., J. Qar, J. Barhanin, D. Fournier, A. Cuany, M. Pralavorio, J.-B. Berge & M. Lazdunski. 1987. Identification and affinity labeling of very high affinity binding sites for the phenylalkylamine series of Ca^{2+} channel blockers in the *Drosophila* nervous system. Biochemistry **26:** 6311-6315.

36. Graziana, A., M. Fosset, R. Ranjeva, A. Hetherington & M. Lazdunski. 1988. Ca^{2+} channel inhibitors that bind to plant cell membranes block Ca^{2+} entry into protoplasts. Biochemistry **27:** 764-768.

37. Galizzi, J.-P., M. Fosset, G. Romey, P. Laduron & M. Lazdunski. 1986. Neuroleptics

of the diphenylbutylpiperidine series are potent calcium channel inhibitors. Proc. Natl. Acad. Sci. USA **83:** 7513-7517.

38. QAR, J., J.-P. GALIZZI, M. FOSSET & M. LAZDUNSKI. 1987. Receptors for diphyenylbutylpiperidine neuroleptics in brain, cardiac, and smooth muscle membranes. Relationship with receptors for 1,4-dihydropyridines and phenylalkylamines with Ca^{2+} channel blockade. Eur. J. Pharmacol. **141:** 261-268.

39. QAR, J., J. BARHANIN, G. ROMEY, R. HENNING, U. LERCH, R. OEKONOMOPULOS, H. URBACH & M. LAZDUNSKI. 1988. A novel high affinity class of Ca^{2+} channel blockers. Mol. Pharmacol. **33:** 363-369.

40. LAZDUNSKI, M., C. FRELIN, J. BARHANIN, A. LOMBET, H. MEIRI, D. PAURON, G. ROMEY, A. SCHMID, H. SCHWEITZ, P. VIGNE & H. P. M. VIJVERBERG. 1987. Polypeptide toxins as tools to study voltage-sensitive Na^{+} channels. Ann. N.Y. Acad. Sci. **479:** 204-220.

L-Type Calcium Channels in Cardiac and Skeletal Muscle

Purification and Phosphorylation[a]

M. MARLENE HOSEY, FONG C. CHANG, CLIFF M.
O'CALLAHAN, AND JUDY PTASIENSKI

Department of Molecular and Cellular Biochemistry
University of Health Sciences / The Chicago Medical School
North Chicago, Illinois 60064

Voltage-dependent calcium channels in cardiac cells are important in supplying the Ca that is directly or indirectly used in the contractile process. The predominant type of voltage-dependent Ca channel in the heart is an L-type,[1] slow or dihydropyridine (DHP)-sensitive Ca channel. The critical role of L-type Ca channels in cardiac cells is attested to by the observation that blockade of the channels with Ca-channel inhibitors such as the dihydropyridines or phenylalkylamines (PAA) results in a decreased force of contraction of the heart and can result in total suppression of activity (see Morad & Cleemann[2] and Fleckenstein[3]). Cardiac L-type Ca channels are of particular interest biochemically because they are *regulated* by the stimulatory neurotransmitter norepinephrine and the inhibitory transmitter acetylcholine via mechanisms believed to involve phosphorylation and dephosphorylation, respectively, of the channel protein.[4–6] However, this important regulation of Ca channels is very poorly understood from a biochemical perspective. This is largely due to the fact that the proteins that comprise cardiac Ca channels have been difficult to purify because they are present in very low amounts in cardiac membranes; the number of biochemically identifiable L channels (high-affinity DHP receptors) in cardiac membranes is 1-3 pmols/mg protein.

L-type Ca channels are also found in skeletal muscle and appear to be present in high density in skeletal muscle transverse tubule membranes where the number of DHP receptors is 50-100-fold higher than in cardiac membranes.[7,8] Consequently, skeletal muscle membranes have been extensively used as a model system for the biochemical characterization and purification of L-type Ca channels.[9–14] However, while the biochemical studies of L channels from skeletal muscle have provided information concerning the characteristics of the channel protein(s), it is known that the properties of DHP-sensitive L channels are different in various tissues. For example, the kinetics of opening and closing of L channels in cardiac cells are quite different from those in skeletal muscle.[15,16] In addition, there are differences in the pharmacology

[a] This work was supported by National Institutes of Health Grant HL 23306. MMH was an Established Investigator of the American Heart Association.

27

of L channels.[1,17] Thus there appear to be subtypes of L channels, and structural differences may account for the observed differences in properties. In order to understand the molecular basis for the differences in properties of L-type Ca channels in cardiac and skeletal muscle, it is necessary to isolate and characterize the corresponding proteins. We have undertaken the purification of cardiac L channels from chick heart. Chick heart is the most suitable tissue for these purification studies because it contains the highest density of high-affinity DHP receptors of different cardiac membranes tested.[18] We have purified putative L-type Ca channels from chick cardiac muscle by virtue of their associated dihydropyridine and phenylalkylamine receptors and have compared the properties of the isolated peptides with those previously purified from skeletal muscle.

PURIFICATION AND CHARACTERIZATION OF THE CARDIAC DIHYDROPYRIDINE-PHENYLALKYLAMINE RECEPTORS

The approach we have used for the purification of cardiac L channels has been to prelabel the DHP or PAA receptors with corresponding radiolabeled derivatives, solubilize the receptors with digitonin, and purify the channels as high-affinity DHP or PAA receptors using standard chromatographic techniques. Our first attempt at purifying L channels from chick heart[18] resulted in the purification of small peptides of 60,000, 54,000, and 32,000, and, despite the fact that many protease inhibitors were used, we believe that proteolysis was a problem. We therefore modified our purification protocol and have obtained a preparation that was highly enriched in DHP-binding activity and larger peptides. After one cycle of lectin affinity chromatography and sucrose density gradient centrifugation, the bulk of the protein did not migrate with the peak of DHP-binding activity (FIG. 1A). However, analysis of the peptide content of the peak of DHP-binding activity indicated that a peptide of ~ 185 kDa appeared to comigrate with the peak of binding activity (FIG. 1A). After further purification, the most highly purified fractions corresponding to the peak of DHP-binding activity were highly enriched in a peptide that exhibited a M_r of $\sim 185,000$ under reducing (FIG. 1B) and nonreducing (data not shown) conditions. In addition, the purified fractions contained an approximately equal amount of a peptide of 140,000 daltons (FIG. 1B), whose M_r was increased to $\sim 180,000$ under nonreducing conditions. This peptide appeared to be similar[20] to the α_2 peptide observed in preparations of L channels purified from skeletal muscle.[9-14] In order to directly compare the purified cardiac preparations with those purified from skeletal muscle, we purified skeletal muscle channels using a similar purification protocol and analyzed for peptide content under the same SDS gel electrophoresis conditions. As can be seen in FIGURE 1C, the skeletal muscle preparations were enriched, as expected,[9-14] in peptides of 165 (α_1) and 140 kDa (α_2).

In order to more fully characterize the purified cardiac preparation, and to identify the component bearing the DHP and PAA receptors, we performed photoaffinity labeling with either the DHP [³H]azidopine or the PAA [³H]azidopamil. These ligands were found to specifically bind to the chick cardiac dihydropyridine and phenylalkylamine receptors with K_d values of 0.3 nM and 3 nM, respectively, similar to values previously reported.[9,19] When cardiac membranes were photolabeled with these ligands before purification, a single, specifically labeled component of 185,000-190,000 daltons was present in the purified fractions (FIG. 2A and B). The mobility of this component was unchanged by reducing or nonreducing conditions

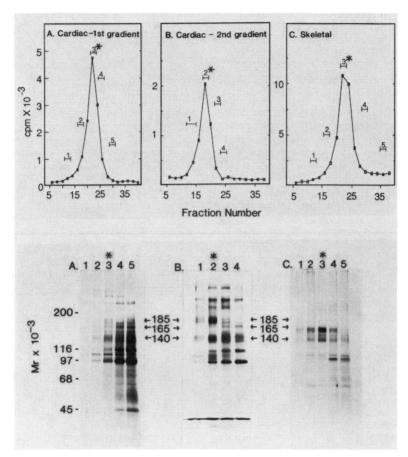

FIGURE 1. Purification of dihydropyridine-sensitive Ca channels from cardiac and skeletal muscle membranes. Purified chick cardiac membranes and transverse tubule membranes from rabbit skeletal muscle were prelabeled with 5-6 nM (+)[^{3}H]PN 200-110 and the dihydropyridine receptors were purified using DEAE and lectin affinity chromatography and sucrose density gradient centrifugation. **Top panel:** Profiles of (+)[^{3}H]PN 200-110 binding activity on 5-20% sucrose gradients. **A:** binding activity in a cardiac preparation after purification through a sucrose gradient step. **B:** binding activity in a cardiac preparation after purification through a second sucrose gradient step. **C:** binding activity in a rabbit skeletal muscle preparation purified through WGA-Sepharose and one sucrose gradient. In A-C, the numbers 1-5 above the small bars represent the fractions that were pooled for analysis by NaDosSO$_4$-gel electrophoresis. The * represents the pool containing the peak of dihydropyridine binding. **Bottom panel:** A-C are silver-stained NaDodSO$_4$-gels depicting the peptide composition of the pooled fractions indicated in the top panel. The gels contained a gradient of 5-15% polyacrylamide, and the samples were electrophoresed under reducing conditions. (In B, the peptide migrating at ~40 kDa is the catalytic subunit of cAMP-dependent protein kinase, which was added for the experiments described in FIG. 4).

(not shown). For each ligand, the incorporation of radioactivity into the 185-190-kDa peptide was prevented by coincubation with an excess of the appropriate unlabeled ligand during the labeling reactions (FIG. 2A and B, right). In contrast, photolabeling of the skeletal muscle T-tubule membranes with these ligands led to the expected

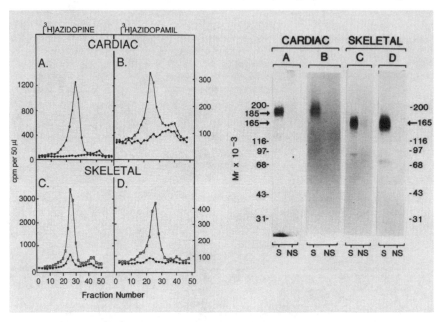

FIGURE 2. Photoaffinity labeling of dihydropyridine and phenylalkylamine receptors associated with cardiac and skeletal muscle Ca channels. Purified chick cardiac membranes or rabbit skeletal muscle transverse tubule membranes were incubated with 10-12 nM of either [³H]azidopine or [³H]azidopamil in the presence or absence of either 1 μM nitrendipine or LU 47781, respectively. The prelabeled membranes were photolyzed and the photolabeled proteins were purified as described in FIGURE 1. **Left panel:** A-D depict profiles of either [³H]azidopine or [³H]azidopamil-labeled material after sucrose density gradient centrifugation. **A & B:** cardiac preparations labeled with [³H]azidopine or [³H]azidopamil, respectively, and purified as in FIGURE 1A. **C & D:** skeletal muscle preparations labeled with [³H]azidopine and [³H]azidopamil, respectively, and purified as in FIGURE 1C. The open squares depict total binding, while the filled circles show nonspecific binding. **Right panel:** A-D are fluorograms obtained after NaDodSO₄-gel electrophoresis of the samples from the left panel containing the peak of binding activity. S and NS represent specific (from fractions depicted on the left in open squares) and nonspecific (from fractions depicted on the left with closed circles), respectively. (In other experiments we ascertained that there were no cardiac components specifically labeled with [³H]azidopamil in the region of ~100 kDa where the gel shown cracked during drying). Other conditions were as described in the legend to FIGURE 1. Exposure times were 30 days for A and B, and five days for C and D.

labeling of a smaller component of 165 kDa (FIG. 2, C and D). Thus the 185-kDa peptide observed in the purification of cardiac L channels appears to possess the cardiac DHP and PAA receptors. This peptide would be analogous to the α_1 peptide purified from skeletal muscle;[9-14] however, it is significantly larger than its skeletal muscle counterpart.

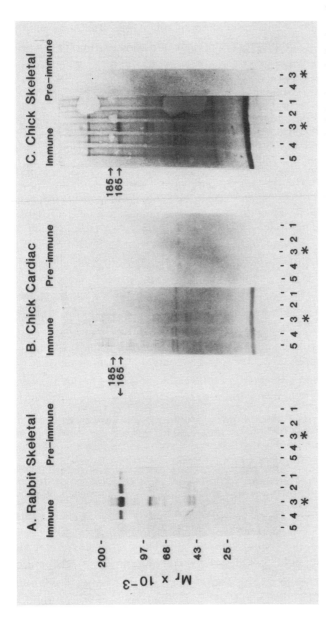

FIGURE 3. Immunoblotting of peptides associated with cardiac and skeletal muscle Ca-channels using anti-rabbit skeletal muscle Ca-channel antibodies. Dihydropyridine receptors were purified from chick cardiac membranes, rabbit skeletal muscle transverse tubule membranes, and chick skeletal muscle microsomes as described. The peptides that separated on sucrose density gradients were subjected to NaDodSO₄-gel electrophoresis under reducing conditions and were subsequently transblotted onto nitrocellulose filters. The peptides were immunostained with either preimmune serum or with serum containing antibodies against the 165-kDa dihydropyridine receptor of rabbit skeletal muscle. **A, B,** and **C** represent immunoblots obtained with rabbit skeletal muscle, chick cardiac and chick skeletal muscle preparations, respectively. Fractions 1–5 correspond to the pooled fractions obtained from the sucrose density gradients (as described for FIG. 1), and in each case the * depicts the pooled fraction containing the peak of dihydropyridine binding activity.

To further compare the properties of this 185-kDa cardiac DHP/PAA receptor to the smaller 165-kDa DHP/PAA receptor previously purified from skeletal muscle, we performed immunoblotting analyses using antibodies raised against the 165-kDa skeletal muscle DHP/PAA receptor. These antibodies produced a strong reaction toward the skeletal muscle 165-kDa DHP/PAA receptor (FIG. 3A), but only poorly recognized the cardiac 185-190-kDa component, if at all (FIG. 3B). The inability of the antibody to recognize the cardiac 185-kDa peptide was not due to lack of cross reactivity across species, as the antibody readily identified the 165-kDa peptide present in chick skeletal muscle microsomes (not shown) or partially purified preparations of chick skeletal dihydropyridine receptors (FIG. 3C).

A third difference in the properties of cardiac and skeletal muscle dihydropyridine/phenylalkylamine receptors was revealed in studies using cAMP-dependent protein kinase to phosphorylate the purified preparations. It has been demonstrated that the skeletal muscle α_1 DHP/PAA receptor can be phosphorylated by cAMP-dependent protein kinase.[12,13,21,27] However, when phosphorylation experiments were performed with the purified cardiac L channel preparations, a surprising finding was that, in contrast to the α_1 peptide from skeletal muscle which was readily phosphorylated by the kinase under the same conditions (FIG. 4B), the 185-kDa peptide present in the purified fractions obtained from cardiac muscle did not undergo phosphorylation by cAMP-dependent protein kinase (FIG. 4A). This was not due to a difference in the source of Ca channels used (chick vs. rabbit), as the 165-kDa DHP receptor in preparations purified from chick skeletal muscle membranes was readily phosphorylated by the kinase (FIG. 4C).

The results show that the DHP and PAA receptors in cardiac muscle are contained in a 185-190-kDa peptide that is significantly larger than, and structurally and immunologically different from, its skeletal muscle counterpart. The fact that the purified peptide was structurally different from the skeletal muscle α_1 peptide may be related to the known differences in properties between L channels in cardiac and skeletal muscle. The copurification of the cardiac α_1 peptide with a peptide of 140 kDa suggests that these may be subunits of cardiac Ca channels, but further studies are required to determine if this is so.

PHOSPHORYLATION OF L-TYPE Ca CHANNELS

As introduced earlier, L-type Ca channels are regulated by receptor-dependent processes.[3-5] Our studies to date (with the exception of those indicated above) to elucidate the biochemical events responsible for regulation of Ca channels by phosphorylation have concentrated on characterizing the phosphorylation of skeletal muscle Ca-channel components. A direct demonstration that peptides associated with L-type Ca channels are phosphorylated by cAMP-dependent protein kinase *in intact cells* has not yet been achieved. However, we have found that the skeletal muscle α_1 peptide is an excellent substrate *in vitro* for cAMP-dependent and other protein kinases. The 165-kDa component can be phosphorylated in its native state in rabbit skeletal muscle T-tubule membranes by cAMP-dependent protein kinase and a multifunctional Ca/CaM-dependent protein kinase *in vitro* (FIG. 5). The initial rates of phosphorylation of the skeletal muscle α_1 peptide by both protein kinases compared favorably to the rates of phosphorylation of established substrates of the enzymes (FIG. 5). Phosphorylation of the skeletal muscle α_1 peptide by both protein kinases was additive.[21]

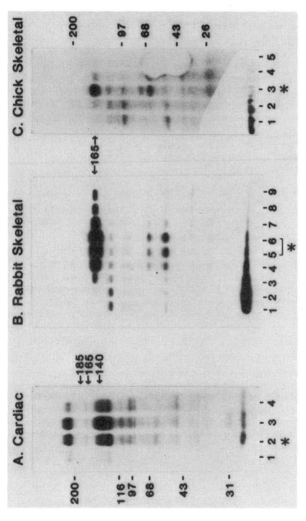

FIGURE 4. Phosphorylation of the purified dihydropyridine receptors from cardiac and skeletal muscle with cAMP-dependent protein kinase. Dihydropyridine receptors were purified from chick cardiac membranes, rabbit skeletal muscle transverse tubule membranes, and chick skeletal muscle microsomes, and the fractions obtained after sucrose density gradient centrifugation were phosphorylated with 0.12 μM cAMP-dependent protein kinase. The samples were subjected to NaDodSO$_4$-gel electrophoresis under reducing conditions, transblotted onto nitrocellulose filters and exposed to Kodak X-Omat film for 12–14 hours with intensifying screens. A, B, and C are autoradiograms prepared from the chick cardiac, rabbit skeletal, and chick skeletal preparations, respectively. The numbers under the lanes correspond to the pooled fractions obtained from the sucrose density gradients (as described in FIG. 1) and the * depicts the fractions containing the peak of dihydropyridine binding activity. The fractions shown in A correspond to those shown in FIGURE 1B, while those shown in panel C correspond to those shown in FIGURE 3C.

cAMP-dependent protein kinase phosphorylated the α_1 peptide preferentially at serine residues, while the Ca/CaM-dependent protein kinase induced formation of slightly more phosphothreonine than phosphoserine,[21] The phosphorylation could be reversed by the Ca-dependent phosphatase calcineurin, which dephosphorylated the skeletal muscle membrane-bound α_1 peptide that had been previously phosphorylated by either protein kinase.[22]

The Ca and phospholipid-dependent protein kinase, protein kinase C, has been shown to modulate the activity of Ca channels in a variety of cells.[23–26] We have found

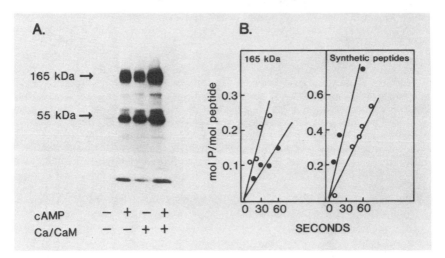

FIGURE 5. Phosphorylation of the 165-kDa DHP receptor in skeletal muscle transverse-tubule membranes by cAMP-dependent and Ca/calmodulin-dependent protein kinases. **A:** Autoradiogram depicting phosphorylation of the 165-kDa DHP receptor by each protein kinase alone and together. Transverse tubule membranes were phosphorylated with the purified catalytic subunit of cAMP-dependent protein kinase or with partially purified preparations of a Ca/calmodulin-dependent protein kinase. DHP receptors were purified from the phosphorylated membranes and electrophoresed on 5–15% gels. Shown is an autoradiogram of the resulting gel. **B:** Initial rates of phosphorylation of the 165-kDa peptide and two synthetic peptides by cAMP- and Ca/CaM-dependent protein kinases. The left-hand part of the figure shows the rate of incorporation of ^{32}P into the membrane-bound 165-kDa peptide in the presence of either 0.1 μM catalytic subunit of cAMP-dependent protein kinase (●), or 6 μg/ml (approximately 1-2 nM) of Ca/CaM-dependent protein kinase (○). The right-hand part of the figure shows the initial rates of phosphorylation of the synthetic peptide "kemptide" (25 nM) by cAMP-dependent protein kinase (0.1 μM) (●), and the Ca/CaM-dependent protein kinase substrate analogue (400 μM) by the Ca/CaM-dependent protein kinase (6 μg/ml) (○).

that the skeletal muscle α_1 peptide is an efficient substrate for protein kinase C when the peptide is phosphorylated in its membrane-bound state (FIG. 6). Up to 1.5-2.0 mol phosphate/mol peptide was incorporated within two minutes in incubations carried out at 37°C. In contrast, the purified protein was a poorer substrate for the kinase; this may reflect that the protein is not in the proper confirmation to serve as a substrate in detergent solution.[27] Phosphoamino acid analysis indicated that protein kinase C phosphorylated the α_1 peptide at both serine and threonine residues.[27] Phosphopeptide mapping experiments showed that cAMP- and Ca/CaM-dependent

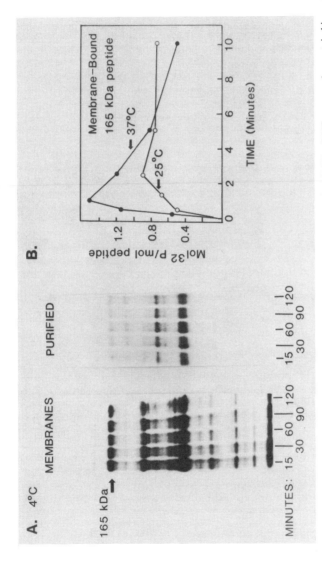

FIGURE 6. Time courses of phosphorylation of the membrane-bound and purified skeletal muscle 165-kDa DHP receptor by protein kinase C. Transverse tubule membranes (**A**, left panel, and **B**) or partially purified DHP receptors (**A**, right panel) were phosphorylated by protein kinase C (10 n*M*) at 4°C (**A**), 25°C (○, **B**), or 37°C (●, **B**). After electrophoresis of the samples on SDS gels, autoradiograms were produced and the 165-kDa phosphopeptide localized. Panel **A** show autoradiograms comparing the phosphorylation of the membrane-bound and purified 165-kDa peptide by protein kinase C at 4°C. For panel **B**, the 165-kDa peptide was excised from the dried gels and the ^{32}P incorporation was quantitated by scintillation counting.

protein kinases and protein kinase C phosphorylated unique, as well as shared sites, in the α_1 peptide (FIG. 7). The unique site for protein kinase C was peptide 1, while those for cAMP- and Ca/CaM-dependent protein kinases were peptides 2 and 3 and 4 and 5, respectively (FIG. 7). As the α_1 peptide may be the major functional unit of DHP-sensitive Ca channels, the results suggest that the phosphorylation-dependent modulation of Ca-channel activity by neurotransmitters may involve phosphorylation of the α_1 peptide at multiple sites. The results are in agreement with the suggestion from the predicted amino acid sequence for the α_1 peptide from skeletal muscle that indicates there are multiple potential phosphorylation sites in the peptide.[28] All sites are predicted to be located cytoplasmically.

Phosphorylation of the 55-kDa, putative β subunit, has also been observed in *in vitro* studies with cAMP- and Ca/CaM-dependent protein kinases, as well as by protein kinase C (FIGS. 5,6). Future studies are needed to elucidate which of the putative channel components may be phosphorylated in the intact cell and contribute to activation of the channels.

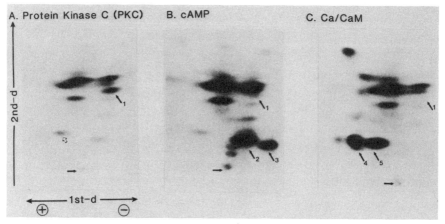

FIGURE 7. Two-dimensional phosphopeptide maps obtained after phosphorylation of the 165-kDa skeletal muscle DHP receptor by protein kinase C, cAMP-dependent protein kinase and/or a multifunctional Ca/calmodulin-dependent protein kinase. The 165-kDa peptide was phosphorylated in transverse tubule membranes for five minutes at 25°C by the protein kinases indicated. After electrophoresis on SDS gels, the bands corresponding to the 165-kDa phosphopeptide were excised and processed for phosphopeptide mapping.

DISCUSSION

This chapter documents new information concerning the peptides associated with L-type Ca channels in cardiac muscle, and the regulation of Ca channels by phosphorylation. First, purification of the putative cardiac L-type Ca channels as high-affinity dihydropyridine receptors demonstrated for the first time that the purified fractions were highly enriched in a 185-kDa peptide. The 185-kDa peptide copurified with dihydropyridine binding activity and with a previously characterized peptide[20] that migrates as a peptide of 140 kDa under reducing conditions and ~ 180 kDa under nonreducing conditions. Second, the results of photoaffinity labeling studies have demonstrated that the peptide of 185 kDa contains the receptors for both the dihydropyridine and phenylalkylamine Ca-channel inhibitors. Third, structural and immunological differences were identified in the 185-kDa dihydropyridine/phenylalkylamine receptor isolated from cardiac muscle and its 165-kDa counterpart previously purified from skeletal muscle.

A point raised by these results concerns the question of whether there are subunits of L-type Ca channels. The copurification of the cardiac (FIG. 1) and skeletal muscle DHP/PAA receptors[9-14] with 140-kDa peptides may suggest that both components might be subunits of L-type Ca channels. If the peptides were unrelated, it would seem unlikely that they would copurify from avian cardiac and mammalian skeletal muscle using the several different types of purification protocols that have been described. Further studies using reconstitution of the separated peptides or expression of mRNAs for each peptide are needed in order to further understand the relationship of these peptides and their possible role in L-type Ca channel function.

Another comment concerning the results is related to the finding that the purified 185-kDa cardiac dihydropyridine/phenylalkylamine receptor did not serve as a substrate for cAMP-dependent protein kinase. This result was surprising in that it is well established that dihydropyridine-sensitive, L-type Ca channels in cardiac muscle are regulated by a cAMP-dependent phosphorylation event (see Refs. 3-6), and, as we have demonstrated, the 165-kDa skeletal muscle dihydropyridine receptor is an excellent substrate for cAMP-dependent protein kinase and other protein kinases (FIG. 5-7). The inability of the purified cardiac DHP peptide to serve as a substrate for cAMP-dependent protein kinase may indicate that either: (i) the peptide is not a substrate for this enzyme, (ii) that it is partially proteolyzed and has lost its phosphorylatable site(s), or (iii) that it is not in the proper conformation in detergent solution to serve as a substrate. The latter situation has been observed with the beta-adrenergic receptor, in that the receptor can only be phosphorylated by the beta-adrenergic receptor kinase when it is reconstituted out of detergent solution and into phospholipid vesicles.[29] In that situation, it appears that the receptor in digitonin solution cannot assume the proper conformation to serve as a substrate for its kinase. While further studies will be necessary to determine why the 185-kDa cardiac dihydropyridine/phenylalkylamine receptor in digitonin solution does not serve as a substrate for cAMP-dependent protein kinase, caution must be exercised in interpreting the present results. The results should be considered too preliminary in nature to be taken as evidence that the well-known regulation of cardiac L-type Ca channels by a cAMP-dependent phosphorylation event does not occur via the phosphorylation of this putative channel component.

ACKNOWLEDGMENTS

We are indebted to Dr. Martin Traut (Knoll AG) for the generous supply of $(-)[^3H]$azidopamil and LU 47781 and are also grateful to Dr. A. Scriabine (Miles) for a generous supply of nitrendipine.

REFERENCES

1. McCLESKEY, E. W., A. P. FOX, D. FELDMAN & R. W. TSIEN. 1986. J. Exp. Biol. **124:** 177-190.

2. MORAD, M. & L. CLEEMANN. 1987. J. Mol. Cell. Cardiol. **19:** 527-553.
3. FLECKENSTEIN, A. 1983. Circ. Res. **52**(suppl I): 3-16.
4. HOSEY, M. M. & M. LAZDUNSKI. 1988. J. Membr. Biol. **104:** 81-105.
5. REUTER, H. 1983. Nature **301:** 569-574.
6. TSIEN, R. W., B. P. BEAN, P. HESS, J. B. LANSMAN, B. NILIUS & M. NOWYCKY. 1986. J. Mol. Cell. Cardiol. **18:** 691-710.
7. FOSSET, M., E. JAIMOVICH, E. DELPONT & M. LAZDUNSKI. 1983. J. Biol Chem. **258:** 6086-6092.
8. GLOSSMANN, H., D. R. FERRY & C. B. BOSHCEK. 1983. Naunyn-Schmiedeberg's Arch. Pharmacol. **323:** 1-11.
9. STRIESSNIG, J., H.-G. KNAUS, M. GRABNER, K. MOOSBURGER, W. SEITZ, H. LEITZ & H. GLOSSMANN. 1987. FEBS Lett. **212:** 247-253.
10. SIEBER, M., W. NASTAINCZYK, V. ZUBOR, W. WERNET & F. HOFMANN. 1987. Eur. J. Biochem. **167:** 117-122.
11. LEUNG. A. T., T. IMAGAWA & K. P. CAMPBELL. 1987. J. Biol. Chem. **262:** 7943-7946.
12. TAKAHASHI, M., M. J. SEAGAR, J. F. JONES, B. F. X. REBER & W. A. CATTERALL. 1987. Proc. Natl. Acad. Sci. USA **84:** 5478-5482.
13. HOSEY, M. M., J. BARHANIN, A. SCHMID, S. VANDAELE, J. PTASIENSKI, C. O'CALLAHAN, C. COOPER & M. LAZDUNSKI. 1987. Biochem. Biophys. Res. Commun. **147:** 1137-1145.
14. VAGHY, P. L., J. STRIESSNIG, K. MIWA, H.-G. KNAUS, K. ITAGAKI, E. MCKENNA, H. GLOSSMANN & A. SCHWARTZ. 1987. J. Biol. Chem. **262:** 14337-14342.
15. COGNARD, C., M. LAZDUNSKI & G. ROMEY. 1986. Proc. Natl. Acad. Sci. USA **83:** 517-521.
16. ROSENBERG, R. L., P. HESS, J. P. REEVES, H. SMILOWITZ & R. W. TSIEN. 1986. Science **231:** 1564;-1566.
17. GLOSSMANN, H., D. R. FERRY, A. GOLL & M. ROMBUSCH. 1984. J. Cardiovasc. Pharmacol. **6:** S608-S621.
18. RENGASAMY, A., J. PTASIENSKI & M. M. HOSEY. 1985. Biochem. Biophys. Res. Commun. **126:** 1-7.
19. STRIESSNIG, J., K. MOOSBURGER, A. GOLL, D. R. FERRY & H. GLOSSMANN. 1986. Eur. J. Biochem. **161:** 603-609.
20. COOPER, C. L., S. VANDAELE, J. BARHANIN, M. FOSSET, M. LAZDUNSKI & M. M. HOSEY. 1987. J. Biol. Chem. **262:** 509-512.
21. O'CALLAHAN, C. M. & M. M. HOSEY. 1988. Biochemistry **27:** 6071-6077.
22. HOSEY, M. M., M. BORSOTTO & M. LAZDUNSKI. 1986. Proc. Natl. Acad. Sci. USA **83:** 3733-3737.
23. MARCHETTI, C. & A. M. BROWN. 1988. Am. J. Physiol. **254:** C206-C210.
24. DOSEMECI, A., R. S. DHALLAN, N. M. COHEN, W. J. LEDERER & T. B. ROGERS. 1988. Circ. Res. **62:** 347-357.
25. RANE, S. G. & K. DUNLAP. 1986. Proc. Natl. Acad. Sci. USA **83:** 184-188.
26. DIVIRGILIO, F., T. POZZAN, C. B. WOLHEIM, L. M. VINCENTINI & J. MELDOLESI. 1986. J. Biol. Chem. **261:** 32-35.
27. O'CALLAHAN, C. M., J. PTASIENSKI & M. M. HOSEY. 1988 J. Biol. Chem. **263:** 17342-17349.
28. TANABE, T., H. TAKESHIMA, A. MIKAMI, V. FLOCKERZI, H. TAKAHASHI, K. KANGAWA, M. KOJIMA, H. MATSUO, T. HIROSE & S. NUMA. 1987. Nature **328:** 313-318.
29. BENOVIC, J. L., R. H. STRASSER M. G. CARON & R. J. LEFKOWITZ. 1986. Proc. Natl. Acad. Sci. USA **83:** 2797-2801.

Solubilization, Partial Purification, and Properties of ω-Conotoxin Receptors Associated with Voltage-Dependent Calcium Channels from Rat Brain Synaptosomes[a]

ROBERT L. ROSENBERG,[b] JEFFREY S. ISAACSON,[c]
AND RICHARD W. TSIEN[d]

Department of Cellular and Molecular Physiology
Yale University School of Medicine
New Haven, Connecticut 06510

INTRODUCTION

The biochemical characterization of voltage-dependent calcium channels has progressed rapidly in recent years. Skeletal muscle transverse tubules provide the richest source of 1,2-dihydropyridine (DHP) receptors,[1] and therefore most experiments directed toward the biochemical isolation of Ca channels have focused on those from this tissue. The purified DHP receptors from skeletal muscle transverse tubules are comprised of at least four subunits, with molecular masses 175 kDa (135-150 kDa after reduction of disulfide bonds), 170 kDa (both before and after reduction), 50-52 kDa, and 32-33 kDa.[2-8] The 170-kDa peptide contains the binding sites for DHPs[3-6] and verapamil analogues.[6,8] Its primary sequence has features similar to those of voltage-dependent sodium channels.[9] The 175-kDa peptide is heavily glycosylated.[2-4,10] Purified DHP receptors have been incorporated into planar lipid bilayers and the Ca channels recorded had properties very similar to those from skeletal muscle t-tubules.[11-13] For complete reviews, see articles by Catterall, Lazdunski, Glossmann, Hosey, and Campbell in this volume.

Much less is known about the structures of voltage-dependent Ca channels from

[a] This work was supported by National Institutes of Health grants to RWT and a fellowship from the Muscular Dystrophy Association to RLR.

[b] Present address: Department of Pharmacology, The University of North Carolina at Chapel Hill, CB# 7365, Chapel Hill, NC 27599.

[c] Present address: Department of Physiology, The University of California-San Francisco, San Francisco, CA 94143.

[d] Present address: Department of Molecular and Cellular Physiology, Stanford University, Stanford, CA 94305.

neuronal cells. DHP receptors have been solubilized from rat brain membranes,[14] and antibodies to purified DHP receptors from skeletal muscle identify a 175-kDa glycopeptide from rat brain.[15] However, the low density of DHP receptors in brain synaptosomes, together with technical problems in resolving specific binding of the hydrophobic DHPs after membrane solubilization in detergents, have made it difficult to extend the characterization of DHP receptors from neuronal sources.

Recently, a peptide toxin from *Conus geographus,* ω-CgTX VIA,[16] was found to block N-[17,18] and L-type[17] Ca channels from nerve cells, and its affinity and specificity suggested that the toxin could be used as a biochemical probe for neuronal Ca channels. Radio-iodinated[19-22] and ^{3}H-propionylated[23] derivatives of the peptide were found to bind with high affinity to sites in brain synaptosomes. The binding of the peptide toxin was not affected by DHPs,[19] and the density of toxin binding sites was four to five times that of DHPs.[21]

In addition to identifying binding sites in intact neuronal membranes, [^{125}I]ω-CgTX was a good candidate as an effective marker for neuronal Ca channels after solubilization and purification steps. Because the toxin is hydrophilic, it is easier to measure specific toxin binding in the presence of detergents and lipid. Also, the relatively large number of [^{125}I]ω-CgTX receptors in neuronal membranes helps identify receptor sites in dilute solutions after solubilization and purification. Because ω-CgTX blocks both N- and L-type Ca channels in neurons,[17] [^{125}I]ω-CgTX probably labels both types of channels in membranes and in solubilized preparations. This article describes experiments directed toward the solubilization and physiochemical characterization of ω-CgTX receptors from rat brain synaptosomal membranes (see also Martin-Moutot *et al.,* this volume). A preliminary report of some of these results has appeared in abstract form.[25]

EXPERIMENTAL METHODS

Synaptosomes from whole rat brain were prepared essentially as described previously.[26] The homogenization buffer contained 0.32 M sucrose, 5 mM NaH$_2$PO$_4$, pH 7.4, and a mixture of protease inhibitors (1 mM PMSF, 2 μM leupeptin, 1 μg/ml pepstatin A). The brains of six Sprague-Dawley rats (100-200 g body weight) were homogenized in ice-cold buffer (10 ml per g wet weight) with 10 strokes of a motor-driven teflon-glass homogenizer. The homogenate was centrifuged at 1000 $\times$ g for 10 minutes. The supernatant was retained, and the pellet was homogenized and centrifuged as before. The supernatants were combined and centrifuged at 17,000 $\times$ g for 60 minutes. The pellets were resuspended in the sucrose buffer (approximately 2.5 ml per g wet weight of brain) with a tight-fitting glass-glass homogenizer. This material was layered on sucrose gradients of 10 ml each of 1.2, 1.0, and 0.8 M sucrose (containing 5 mM NaH$_2$PO$_4$ and the protease inhibitors). These were centrifuged at 100,000 $\times$ g in an SW28 rotor (Beckman Instruments) for 90 minutes. The membranes within the 1 M sucrose and at the 1.0-1.2 M sucrose interface were collected, diluted slowly with 5 mM NaH$_2$PO$_4$ to a final concentration of 0.32 M sucrose, and centrifuged at 100,000 $\times$ g for 45 minutes. The pellets were resuspended in the sucrose buffer to a final concentration of ~15 mg protein per ml, frozen in liquid nitrogen, and stored at −70°C.

The binding of [^{125}I]ω-CgTX to synaptosomal membranes was done in a method similar to that described.[19] The assay buffer contained 145 mM NaCl, 5 mM KCl, 10 mM HEPES-NaOH, pH 7.4, and 0.2 mg/ml bovine serum albumin (BSA). The free

Ca^{2+} concentration in this buffer was buffered to ~ 1 μM with 1 mM HEDTA and 0.3 mM $CaCl_2$, except where noted. Synaptosomal membranes were diluted in the assay buffer to a concentration of 0.5 mg protein/ml in a volume of 0.2 ml. $[^{125}I]\omega$-CgTX was added, and the mixture was incubated at room temperature for 60 minutes. Determination of nonspecific binding was done in the presence of 200 nM unlabeled ω-CgTX, added before the $[^{125}I]\omega$-CgTX. The mixtures were then diluted with 4 ml of wash buffer (same as the assay buffer, but containing 1 mg/ml BSA), and immediately collected on glass fiber filters (Whatman GF/C). The filters were washed four times with 4 ml of ice-cold wash buffer and counted for bound radioactivity. The $[^{125}I]\omega$-CgTX was radiolabeled to a specific activity of 2000 Ci/mmol, but for most of the binding experiments the specific activity was reduced to 100 Ci/mmol with unlabeled toxin to conserve material. Membrane protein concentrations were determined as described[27] using bovine serum albumin as a standard.

Detergent-solubilized extracts of synaptosomal membranes were prepared as follows. A 1 : 1 mixture of Triton X-100 and CHAPS (10% wt/vol total detergent concentration) was added to the membranes (~ 15 mg protein/ml) to a final concentration of 1% or 1.5% (wt/vol) detergent. This was incubated on ice for 60 minutes with occasional gentle mixing. The mixture was centrifuged at $100,000 \times g$ for 45-60 minutes and the supernatants were collected, with care taken to avoid a loose cloudy material on top of the pellets. Protein concentrations in detergent extracts and partially purified samples were determined as described.[27,28] Small systematic differences in the assay standard curves in the presence of the detergents were accounted for.

The assay of $[^{125}I]\omega$-CgTX binding to detergent-solubilized extracts and partially purified ω-CgTX receptors used rapid gel filtration to separate bound from unbound ligand.[29] Sephadex G-50 was prepared in 50 mM NaH_2PO_4, 0.1% (wt/vol) Lubrol PX, pH 7.4. Small columns of 2.1 ml Sephadex G-50 were prepared in 3-ml disposable syringes. $[^{125}I]\omega$-CgTX was added to 125 μl of the solubilized material, and the mixtures were incubated on ice for 60 minutes. Nonspecific binding was obtained in the presence of 200 nM unlabeled ω-CgTX, as before. The mixture (100 μl) was applied to the syringe columns, and these were immediately centrifuged at $\sim 250 \times g$ (1500 rpm in a desktop centrifuge) for 45 second. The eluate was collected and counted.

RESULTS AND DISCUSSION

Binding of $[^{125}I]\omega$-CgTX to Synaptosomal Membranes

The binding of $[^{125}I]\omega$-CgTX to synaptic membranes has been studied in detail.[19-24] Because we used a synthetic ω-CgTX peptide[30] in some experiments, we compared the binding properties of the mono-iodinated derivative of the purified natural toxin[19] (kindly provided by L. J. Cruz and B. M. Olivera), and the mono-iodinated derivative of synthetic ω-CgTX[30] (New England Nuclear) under the same experimental conditions. FIGURE 1 shows binding isotherms of natural (FIG. 1A) and synthetic (FIG. 1B) $[^{125}I]\omega$-CgTX to rat brain synaptosomes. The amount of toxin bound was clearly a saturable function of the toxin concentration. Because the dissociation rate of the toxin is slow,[20] binding equilibrium was probably not reached during the one-hour incubation used. However, determination of the apparent K_ds is useful for comparative

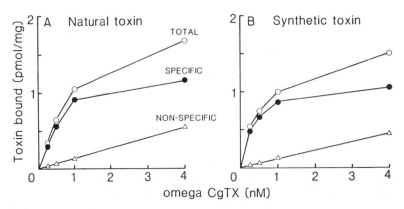

FIGURE 1. Concentration dependence of the binding of natural [^{125}I]ω-CgTX (**A**) and synthetic [^{125}I]ω-CgTX (**B**) to rat brain synaptosomes. Synaptosomes were diluted into assay buffer (see EXPERIMENTAL METHODS) to a final concentration of 0.5 mg protein per ml in a volume of 0.2 ml, and incubated with the indicated concentration of [^{125}I]ω-CgTX for one hour at room temperature. Nonspecific binding was obtained in the presence of 200 nM unlabeled natural (**A**) or synthetic (**B**) ω-CgTX. Specific binding is the difference between total and nonspecific binding.

purposes; for both natural and synthetic [^{125}I]ω-CgTX the apparent K_d was approximately 0.4 nM, in agreement with other experiments.[19–21,24] For both natural and the synthetic toxin, there were ~1 pmol of binding sites per mg membrane protein. The amount of nonspecific binding of the synthetic ω-CgTX was approximately the same as for the natural toxin, representing 10-15% of the total amount of toxin bound at 1 nM.

The close similarity of the binding properties of natural and synthetic toxin indicates that the synthetic product is a good marker for the ω-CgTX receptor. This was true even though the mono-iodinated natural toxin is labeled only at Tyr-22[21] but mono-iodinated synthetic toxin is labeled at either Tyr-22 or Tyr-12 (R. Garlick, NEN, personal communication).

We have found no evidence of the very high affinity ω-CgTX binding site described by Barhanin et al.[22] ($K_d \sim 1$ pM), even when the concentration of membrane protein in the assay mixture is reduced. At the concentrations of [^{125}I]ω-CgTX routinely used (0.25-1.0 nM), there was some depletion of the free toxin due to binding to the receptors, but not enough to account for a two to three order of magnitude difference in binding affinity. Also, as illustrated in FIGURE 1, the binding data at the lower concentrations of [^{125}I]ω-CgTX extrapolated to a very low y-intercept, not to a large positive value expected if many binding sites with a K_d of 1 pM were present. The data do not rule out the possible contribution of a small proportion of very high affinity sites, as reported by Yamaguchi et al.[23] It is important to note that a binding site with a K_d in the nanomolar range under these conditions (low Ca^{2+}) is more consistent with the concentrations of toxin required to block Ca^{2+} currents (0.1-1.0 μM).[31]

Ca^{2+} Inhibits [^{125}I]ω-CgTX Binding

FIGURE 2 shows, as described previously, that Ca^{2+} is a strong inhibitor of [^{125}I]ω-CgTX binding.[20,22,23] In our experiments, 1 mM Ca^{2+} produced a ~50% reduction in specific binding of the toxin measured at concentrations up to 1 n*M*. This agrees closely with values for Ca^{2+} inhibition obtained by Abe *et al.*[20] but is a lower concentration than that found by Cruz and Olivera[19] or Barhanin *et al.*[22] The inhibition of toxin binding by Ca^{2+} may account for the micromolar levels of ω-CgTX needed for a 50% block of Ca currents in electrophysiological recordings where the divalent ion concentration was 2-10 m*M*.[17,18,31]

Solubilization of Receptors for [^{125}I]ω-CgTX

The first step in the isolation and characterization of ω-CgTX receptors is to solubilize them in a functional form from intact synaptosomal membranes. FIGURE 3 shows binding isotherms of [^{125}I]ω-CgTX to intact synaptosomes (FIG. 3A) and to a detergent extract of those membranes (FIG. 3B). As with the membrane-bound receptor, the solubilized receptors bound [^{125}I]ω-CgTX with an apparent affinity of about 0.4 n*M*. The signal-to-noise ratio of the binding of [^{125}I]ω-CgTX to the solubilized receptors was good; at 1 n*M* total [^{125}I]ω-CgTX concentration, nonspecific binding accounted for only ~10% of the total counts bound. The specific binding activity in the detergent extracts was only about 50-70% of that in the initial membranes, indicating that solubilization of functional ω-CgTX receptors was somewhat less successful than that of total protein. The solubilized receptors were relatively stable at 4°C; detergent extracts could be maintained on ice for several days with only a small loss of binding activity (not shown).

Recently, Yamaguchi *et al.*[23] showed that ω-CgTX receptors could be solubilized from bovine brain membranes with digitonin. They state that deoxycholate, *n*-octyl glucoside and, to a lesser extent, CHAPS were also effective, but that Triton X-100 and Lubrol PX were not. We have found that functional [^{125}I]ω-CgTX binding sites

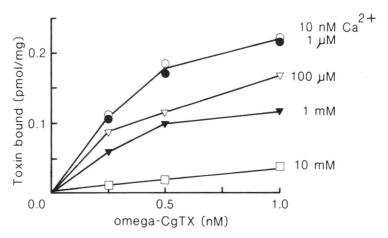

FIGURE 2. Inhibition by $CaCl_2$ of natural [^{125}I]ω-CgTX binding to rat brain synaptosomes. Specific binding is plotted against [^{125}I]ω-CgTX concentration. The assay was as described in EXPERIMENTAL METHODS, except that the $CaCl_2$ concentration in the assay buffer was changed to that indicated. (○) 1 m*M* EDTA, 0.33 m*M* $CaCl_2$, ~10 n*M* free Ca^{2+}, (●) 1 m*M* HEDTA, 0.26 m*M* $CaCl_2$, ~1 μ*M* free Ca^{2+}.

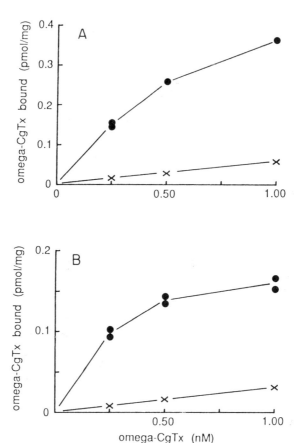

FIGURE 3. Binding of natural $[^{125}I]\omega$-CgTX to rat brain synaptosomes (**A**) and detergent extract (**B**). (●), Specific binding; (×) nonspecific binding. The detergent extract was from a 1% detergent solubilization obtained as described in EXPERIMENTAL METHODS.

from rat brain synaptosomes can be solubilized in either Triton X-100 or CHAPS, and that a mixture of Triton X-100 and CHAPS is highly effective in the solubilization of ω-CgTX receptors, releasing about 25% of the receptors from the membranes. The efficiency of solubilization was maximal at a detergent concentration of 1-1.5% (wt/vol) when the membrane protein concentration was ~15 mg/ml.

Ca^{2+} reduced binding to detergent extracts of brain membranes, as it did to the intact synaptosomes (not shown, see Yamaguchi et al.[23]). Additionally, NaCl at relatively low concentrations (100-500 mM) inhibited binding to detergent extracts (not shown), although much higher concentrations were required to inhibit toxin binding to intact synaptosomes.[19,20]

Sucrose Gradient Sedimentation

In order to characterize the hydrodynamic properties of the solubilized ω-CgTX receptor, we performed sucrose gradient sedimentation analysis (FIG. 4). Detergent extracts of synaptosomal membranes were fractionated on a linear 12-30% sucrose gradients. ω-CgTX receptors sedimented faster than the bulk of the protein and were found in a single peak located at a position corresponding closely to that of thyroglobulin, indicating that the sedimentation coefficient of the solubilized ω-CgTX receptor was about 19 S. This is very close to similar determinations of the sedimentation coefficient of the DHP receptors from brain,[14] heart,[32] and skeletal muscle[33] solubilized in digitonin but is significantly larger than the value of ~15.5 S reported for DHP receptors solubilized in CHAPS.[32] Because there was relatively good separation of the peaks of ω-CgTX receptors and total protein, sucrose gradient fractionation could be used to purify the ω-CgTX receptors two- to fivefold over unfractionated detergent extracts.

Sepharose 6B Chromatography

In an effort to purify further the ω-CgTX receptor and to determine the molecular size of the protein, we used gel exclusion chromatography on a column of Sepharose

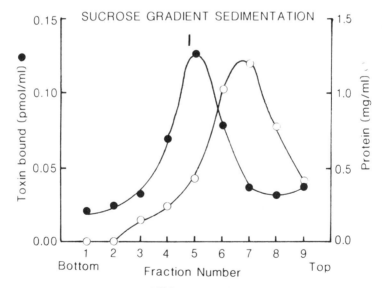

FIGURE 4. Specific binding of natural [^{125}I]ω-CgTX (●) and total protein concentration (○) in fractions from a linear 12-30% (wt/vol) sucrose gradient. Gradient solutions also contained 5 mM NaH$_2$PO$_4$, 1 mM HEDTA, 0.3 mM CaCl$_2$, 0.1% (wt/vol) Triton X-100/CHAPS (1 : 1), 0.2 mg/ml egg phosphatidylcholine. Detergent extract (2 ml) was applied to the top of a 36-ml gradient. This was centrifuged at 195,000 × g in a VTi50 rotor (Beckman Instruments) for three hours. Fraction volumes were 6 ml (#1), 4 ml (#2-#8), and 2 ml (#9). The vertical bar indicates the peak of thyroglobulin (19.2 S) sedimented in an identical gradient.

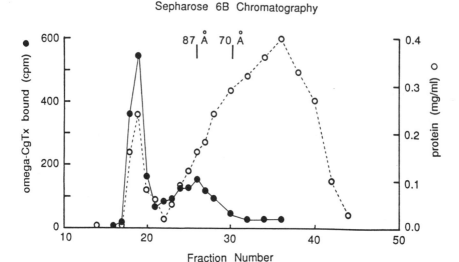

Sepharose 6B Chromatography

FIGURE 5. Sepharose 6B fractionation of [^{125}I]ω-CgTX receptors (●) and total protein (○) from a 1.5% (wt/vol) Triton X-100/CHAPS (1 : 1) extract of rat brain synaptosomes. Extract (1.5 ml) was applied to the top of a 50-ml column. The column was equilibrated in and eluted with 145 mM NaCl, 5 mM KCl, 20 mM HEPES, pH 7.4, 1 mM HEDTA, 0.3 mM CaCl$_2$, 0.1% (wt/vol) Triton X-100/CHAPS (1 : 1), 0.2 mg/ml egg phosphatidylcholine. Fractions were 1.7 ml, and the flow rate was ~0.25 ml/min. The lines indicate the position of the peaks of thyroglobulin (87 Å) and β-galactosidase (70 Å) in the same column buffer.

6B (Fig. 5). The ω-CgTX receptor appears in two well-defined peaks. The first peak is very sharp and appears at the excluded volume of the column. The second peak is relatively broad, and appears in fractions that correspond to the elution of thyroglobulin, with a Stokes radius of 87 Å. This is similar to the value determined by Horne *et al.*[32] for cardiac DHP receptors. The bulk of the total protein appears later in the elution, well-separated from both peaks of ω-CgTX receptors. This technique could be used to gain an approximately three-fold enrichment of ω-CgTX receptors, but because the elution was relatively slow, the overall recovery of functional ω-CgTX receptors was only ~30%.

The sharp peak of ω-CgTX receptors at the column exclusion volume represents material with a very large molecular size (>4000 kDa). The nature of the receptors in these large particles is difficult to determine with certainty. The large particle size probably indicates that the receptors are in multimolecular aggregates of proteins, lipids, and detergent. These particles could have aggregated after solubilization of the synaptosomal membranes, perhaps due to the low ionic strength during solubilization. Alternatively, these particles might be clusters of ω-CgTX receptors that never became completely dispersed in the detergent. In keeping with this possibility, higher detergent concentrations during solubilization tended to shift the ω-CgTX receptors from the "breakthrough" peak and to the 87-Å peak. The fact that no very heavy population of ω-CgTX receptors was found in sucrose gradient fractions probably indicates that these large particles contained large amounts of lipid, because the low density of the lipid would tend to balance the high density of the protein and detergent, reducing the sedimentation velocity in the sucrose gradients.

Wheat Germ Agglutinin-Sepharose 4B Chromatography

Lectin affinity chromatography is a powerful purification technique that has been used successfully as a step toward the purification of DHP receptors from skeletal muscle,[33,34] brain,[14] and heart.[35] We used a column of wheat germ agglutinin-Sepharose 4B to determine if the ω-CgTX receptors from rat brain were also glycosylated in a form that binds to WGA, and to purify the receptors.

FIGURE 6 shows the elution of ω-CgTX receptors and total protein from the lectin-affinity resin. A rinse of the resin eluted a large amount of the total protein and some of the ω-CgTX receptors (fractions 1-5). After the resin was fully rinsed, 100 m*M* *N*-acetylglucosamine was added to the column buffer to elute any bound glycoproteins. This caused the immediate elution of a large amount of ω-CgTX receptors and protein (fractions 15-20). Because the fractions following addition of *N*-acetyl glucosamine were relatively rich in total protein, indicating a large amount of glycoproteins in this preparation, the overall purification of ω-CgTX receptors was only 1.5-2-fold.

In this experiment, the ω-CgTX receptors that eluted in the early rinse fractions were not bound to the resin, probably because it was overloaded with a large concentration of ω-CgTX receptors. However, in another experiment where the protein concentration of the extract was reduced fivefold before incubation with the WGA-Sepharose, essentially the same elution profile resulted. This suggests that not all of the ω-CgTX receptors from rat brain are glycosylated in a form recognized by WGA.

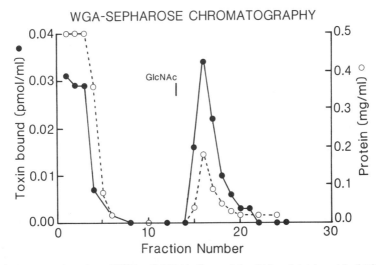

FIGURE 6. Fractionation of [^{125}I]ω-CgTX binding activity (●) and total protein (○) on a column of wheat germ agglutinin-Sepharose 4B. The column resin was prepared by coupling 80 mg of WGA (United States Biochemicals) with 10 ml swollen CNBr-activated Sepharose 4B (*Pharmacia*) as described in the *Pharmacia* literature. Detergent extract (2 ml) was mixed with 5 ml buffer (145 m*M* NaCl, 5 m*M* KCl, 20 m*M* HEPES, pH 7.4, 1 m*M* HEDTA, 0.3 m*M* CaCl$_2$) and ~6 ml of the wet WGA-Sepharose 4B. This was incubated on ice for 60 minutes, and then the mixture was loaded into a small column (0.5 cm diameter). Fractions were collected as the column settled, and it was then eluted with the column buffer (see legend to FIG. 5). At the position indicated, *N*-acetylglucosamine (100 m*M*) was added to the column buffer. Fractions were 4.2 ml (#1-#12) and 2.0 ml (#13-#34).

Unsuccessful Ion-Exchange Chromatography

Anion exchange chromatography was used to great advantage to purify DHP receptors from skeletal[33,34] and cardiac[35] muscle. However, we were unable to use this technique to purify ω-CgTX receptors solubilized from brain; the receptors bound to DEAE Sephadex in low ionic strength buffers, but functional ω-CgTX receptors could not be recovered from the columns after stepwise increases in the ionic strength up to 500 mM, despite the recovery of close to 100% of the total protein. ω-CgTX receptors solubilized in Triton X-100 and CHAPS may be fragile, and interactions with the highly charged anion exchanger may cause dissociation of subunits that must be together for the peptide toxin to bind. Initial efforts to recover activity by selective combination of fractions were unsuccessful.

Chemical Cross-Linking of [^{125}I]ω-CgTX

Cross-linking of ω-CgTX derivatives using chemical[21,22] and photoaffinity[22-24] methods has been used to characterize peptide composition of the ω-CgTX receptor in intact synaptosomal membranes from chick,[21,22] rat[22,24] and bovine[23] brain. Chemical cross-linking of [^{125}I]ω-CgTX to the receptor proteins in chick synaptosomes labeled a peptide of ~140 kDa under reducing gel conditions,[21,22] and the apparent molecular weight increased to 170 kDa under nonreducing conditions.[22] This behavior is similar to the 175/140 kDa peptide of skeletal muscle DHP receptors. However, photoaffinity derivatives of ω-CgTX labeled peptides of ~300 kDa, ~230 kDa, and 34 kDa,[23,24] or ~210 kDa[22] (under disulfide-reducing conditions), with no labeling of a 170/140 kDa subunit. Thus, the peptide composition of the ω-CgTX receptor in neuronal membranes is not yet certain.

In the hope that solubilized, partially purified preparations of ω-CgTX receptors might provide a clearer answer to the question of peptide composition, we used disuccinimidyl suberate (DSS, Pierce Chemical) to covalently incorporate [^{125}I]ω-CgTX into receptors after solubilization, Sepharose 6B gel exclusion chromatography, and WGA-Sepharose affinity chromatography (FIG. 7). Partially purified receptors were incubated with 1 nM [^{125}I]ω-CgTX in the absence or presence of 250 nM unlabeled ω-CgTX, and the samples were treated with DSS as described.[36] A peptide of ~300 kDa (under reducing conditions) was heavily, specifically labeled, as indicated by autoradiography of the polyacrylamide gels. A 66-kDa protein was also heavily labeled, but in a nonspecific manner. Some specific labeling was found at the top of the stacking gel (not shown), suggesting that some of the [^{125}I]ω-CgTX receptors had become covalently linked into large aggregates. There was slight specific labeling of peptide of ~170, ~140, and ~120 kDa.

The ~300-kDa peptide clearly identified in these experiments appears similar to those identified by Yamaguchi et al.[23] and Abe and Saisu,[24] using photoaffinity derivatives of ω-CgTX. The 170-, 140-, and 120-kDa peptides that are slightly labeled may be similar to those reported by Cruz et al.[21] and Barhanin et al.,[22] but we could not observe any differences in peptide mobility or autoradiographic density when samples were electrophoresed under nonreducing conditions.[22] We have found no strong indications of peptides of 200-230 kDa that were identified by photoaffinity labeling.[22-24]

SUMMARY AND CONCLUSIONS

These experiments provide a starting point for biochemical characterization of Ca channels from neuronal membranes, using ω-CgTX as a specific marker. The purification of the ω-CgTX receptors is far from complete. Each of the purification steps described results in only a two- to fivefold enrichment of the receptor proteins, and is accompanied by a loss of receptor concentration and stability, so the maximal specific activity achieved by a combination of these steps falls several orders of magnitude short of that of a large, homogeneous, active protein.

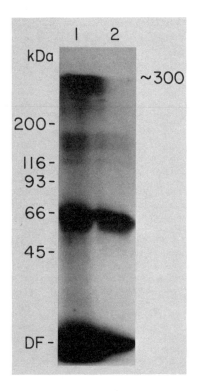

FIGURE 7. Autoradiogram of an SDS polyacrylamide gel showing the peptides covalently labeled by synthetic [^{125}I]ω-CgTX. Partially purified ω-CgTX receptors were incubated with 1 nM [^{125}I]ω-CgTX (2000 Ci/mmol) in the absence (Lane 1) and presence (Lane 2) of 250 nM unlabeled ω-CgTX for 60 minutes on ice. DSS was added (to 1 mM from a stock solution of 100 mM in DMSO), and the samples were rapidly mixed and incubated on ice. After 15 minutes, the samples were eluted through columns of Sephadex G-50 (equilibrated in 5 mM NaH$_2$PO$_4$) to remove DSS and desalt the samples. Protein-containing fractions were lyophilized, dissolved in SDS-PAGE sample buffer containing 1% (vol/vol) β-mercaptoethanol, and electrophoresed on 5-12% acrylamide gradient gels. Gels were fixed, stained for protein, and dried. Autoradiography was for seven days with Kodak X-AR film and an intensifying screen. The molecular weights and positions of standard proteins are shown at the left.

Nevertheless, these studies have yielded important information about the ω-CgTX receptor. The Stokes' radius, determined from gel exclusion chromatography, is ~87 Å, and the sedimentation coefficient, determined from sucrose gradient sedimentation, is ~19 S. These values are similar to those found for the DHP receptors[14,32,33] solubilized in digitonin. We have also found that at least some of the ω-CgTX receptors have complex carbohydrate moieties that are recognized by WGA, together with evidence of heterogeneity of receptor glycosylation.

Additionally, we have been able to use the solubilized, partially purified receptors in cross-linking experiments to tentatively identify the molecular weights of the ω-CgTX targets from rat brain. A large peptide of ~300 kDa, similar to that identified in photoaffinity studies,[23,24] is very clearly labeled by the chemical incorporation of

[^{125}I]ω-CgTX into partially purified receptor preparations, but some ambiguity remains because of the faint labeling of peptides in the 120-170-kDa range. The ~300-kDa peptide is much larger than any single peptide component of DHP receptors from skeletal muscle, and it may be related to a molecular combination of the 170-kDa and 135-kDa subunits of the DHP receptor.

Because [^{125}I]ω-CgTX presumably labels both N- and L-type neuronal Ca channels, both channel types will probably be found in the purified preparations. Thus, at some time, it will be necessary to separate DHP-sensitive L-type channels from preparations of L- and N-type channels identified by ω-CgTX binding.

ACKNOWLEDGMENTS

The authors are indebted to B. M. Olivera and L. J. Cruz for helpful discussions and for providing the natural ω-CgTX and [^{125}I]ω-CgTX used in these experiments.

REFERENCES

1. FOSSET, M., E. JAIMOVICH, E. DELPONT & M. LAZDUNSKI. 1983. [^{3}H]Nitrendipine receptors in skeletal muscle. Properties and preferential localization in transverse tubules. J. Biol. Chem. **258**: 6086-6092.
2. LEUNG, A. T., T. IMAGAWA & K. P. CAMPBELL. 1987. Structural characterization of the 1,4-dihydropyridine receptor of the voltage-dependent Ca^{2+} channel from rabbit skeletal muscle. Evidence for two distinct high molecular weight subunits. J. Biol. Chem. **262**: 7943-7946.
3. TAKAHASHI, M., M. J. SEAGAR, J. F. JONES, B. F. X. REBER & W. A. CATTERALL. 1987. Subunit structure of dihydropyridine-sensitive calcium channels from skeletal muscle. Proc. Natl. Acad. Sci. USA **84**: 5478-5482.
4. SHARP, A. H., T. IMAGAWA, A. T. LEUNG & K. P. CAMPBELL. 1987. Identification and characterization of the dihydropyridine-binding subunit of the skeletal muscle dihydropyridine receptor. J. Biol. Chem. **262**: 12309-12315.
5. VAGHY, P. L., J. STRIESSNIG, K. MIWA, H.-G. KNAUS, K. ITAGAKI, E. MCKENNA, H. GLOSSMANN & A. SCHWARTZ. 1987. Identification of a novel 1,4 dihydropyridine- and phenylalhylamine-binding polypeptide in calcium channel preparatons. J. Biol. Chem. **262**: 14337-14342.
6. SIEBER, M., W. NASTAINCZYK, V. ZUBOR, W. WERNET & F. HOFMANN. 1987. The 165-kDa peptide of the purified skeletal muscle dihydropyridine receptor contains the known regulatory sites of the calcium channel. Eur. J. Biochem. **167**: 117-122.
7. MORTON, M. E. & S. C. FROEHNER. 1987. Monoclonal antibody identifies a 200-kDa subunit of the dihydropyridine-sensitive calcium channel. J. Biol. Chem. **262**: 11904-11907.
8. HOSEY, M. M., J. BARHANIN, A. SCHMID, S. VANDAELE, J. PTASIENSKI, C. O'CALLAHAN, C. COOPER & M. LAZDUNSKI. 1987. Photoaffinity labelling and phosphorylation of a 165-kilodalton peptide associated with dihydropyridine and phenylalkylamine-sensitive calcium channels. Biochem. Biophys. Res. Commun. **147**: 1137-1145.
9. TANABE, T., H. TAKESHIMA, A. MIKAMI, V. FLOCKERZI, H. TAKAHASHI, M. KANGAWA, H. MATSUO, T. HIROSE & S. NUMA. 1987. Primary structure of the receptor for calcium channel blockers from skeletal muscle. Nature **328**: 313-318.
10. BARHANIN, J., T. COPPOLA, A. SCHMID, M. BORSOTTO & M. LAZDUNSKI. 1987. The calcium channel antagonists receptor from rabbit skeletal muscle. Reconstitution after purification and subunit characterization. Eur. J. Biochem. **164**: 525-531.

11. FLOCKERZI, V., H.-J. OEKEN, F. HOFMANN, D. PELZER & W. TRAUTWEIN. 1986. Purified dihydropyridine-binding site from skeletal muscle t-tubules is a functional calcium channel. Nature **323:** 66-68.

12. SMITH, J. S., E. J. MCKENNA, J. MA, J. VILVEN, P. L. VAGHY, A. SCHWARTZ & R. CORONADO. 1987. Calcium channel activity in a purified dihydropyridine-receptor preparation of skeletal muscle. Biochemistry **26:** 7182-7188.

13. TALVENHEIMO, J. A., J. F. WORLEY III & M. T. NELSON. 1987. Heterogeneity of calcium channels from a purified dihydropyridine receptor preparation. Biophys. J. **52:** 891-899.

14. CURTIS, B. M. & W. A. CATTERALL. 1983. Solubilization of the calcium antagonist receptor from rat brain. J. Biol. Chem. **258:** 7280-7283.

15. TAKAHASHI, M. & W. A. CATTERALL. 1987. Identification of an α-subunit of dihydropyridine-sensitive brain calcium channels. Science **236:** 88-91.

16. OLIVERA, B. M., W. R. GRAY, R. ZEIKUS, J. M. MCINTOSH, J. VARGA, J. RIVIER, V. DE SANTOS & L. J. CRUZ. 1985. Peptide neurotoxins from fish-hunting cone snails. Science. **230:** 1338-1343.

17. MCCLESKEY, E. W., A. P. FOX, D. H. FELDMAN, L. J. CRUZ, B. M. OLIVERA, R. W. TSIEN & D. YOSHIKAMI. 1987. ω-Conotoxin: Direct and persistent blockade of specific types of calcium channels in neurons but not muscle. Proc. Natl. Acad. Sci. USA **84:** 4327-4331.

18. KASAI, H., T. AOSAKI & J. FUKUDA. 1987. Presynaptic Ca-antagonist ω-conotoxin irreversibly blocks N-type Ca channels in chick sensory neurons. Neurosci. Res. **4:** 228-235.

19. CRUZ, L. J. & B. M. OLIVERA. 1986. Calcium channel antagonists. ω-Conotoxin defines a new high affinity site. J. Biol. Chem. **261:** 6230-6233.

20. ABE, T., K. KOYANO, H. SAISU, Y. NISHIUCHI & S. SAKAKIBARA. 1986. Binding of ω-conotoxin to receptor sites associated with the voltage-sensitive calcium channel. Neurosci. Lett. **71:** 203-208.

21. CRUZ, L. J., D. S. JOHNSON & B. M. OLIVERA. 1987. Characterization of the ω-conotoxin target. Evidence for tissue-specific heterogeneity in calcium channel types. Biochemistry **26:** 820-824.

22. BARHANIN, J., A. SCHMID & M. LAZDUNSKI. 1988. Properties of structure and interaction of the receptor for ω-conotoxin, a polypeptide active on Ca^{2+} channels. Biochem. Biophys. Res. Commun. **150:** 1051-1062.

23. YAMAGUCHI, T., H. SAISU, H. MITSUI & T. ABE. 1988. Solubilization of the ω-conotoxin receptor associated with voltage-sensitive calcium channels from bovine brain. J. Biol. Chem. **263:** 9491-9498.

24. ABE, T. & H. SAISU. 1987. Identification of the receptor for ω-conotoxin in brain. Probable components of the calcium channel. J. Biol. Chem. **262:** 9877-9882.

25. ROSENBERG, R. L., J. S. ISAACSON, B. M. OLIVERA, W. S. AGNEW & R. W. TSIEN. 1987. Solubilization and partial purification of ω-conotoxin binding sites from rat brain synaptosomes. Soc. Neurosci. Abstr. **13:** 1011.

26. TAMKUN, M. M. & W. A. CATTERALL. 1981. Ion flux studies of voltage-sensitive sodium channels in synaptic nerve-ending particles. Molec. Pharmacol. **19:** 78-86.

27. PETERSON, G. L. 1977. A simplification of the protein assay of Lowry *et al.* which is more generally applicable. Anal. Biochem. **83:** 346-356.

28. BRADFORD, M. M. 1976. A rapid and sensitive method for the quantitation of microgram quantities utilizing the principle of protein dye-binding. Anal. Biochem. **72:** 248-254.

29. LEVINSON, S. R., C. J. CURATALO, J. REED & M. A. RAFTERY. 1979. A rapid and precise assay for tetrodotoxin binding to detergent extracts of excitable tissues. Anal. Biochem. **99:** 72-84.

30. NISHIUCHI, Y., K. KUMAGAYE, Y. NODA, T. X. WATANABE & S. SAKAKIBARA. 1986. Synthesis and secondary-structure determination of ω-conotoxin GVIA: A 27-peptide with three intramolecular disulfide bonds. Biopolymers **25:** S61-S68.

31. OYAMA, Y., Y. TSUDA, S. SAKAKIBARA & N. AKAIKE. 1987. Synthetic ω-conotoxin: A potent calcium channel blocking neurotoxin. Brain Res. **424:** 58-64.

32. HORNE, W. A., G. A. WEILAND & R. E. OSWALD. 1986. Solubilization and hydrodynamic characterization of the dihydropyridine receptor from rat ventricular muscle. J. Biol. Chem. **261:** 3588-3594.

33. CURTIS, B. M. & W. A. CATTERALL. 1984. Purification of the calcium antagonist receptor of the voltage-sensitive calcium channel from skeletal muscle transverse tubules. Biochemistry **23:** 2113-2118.

34. BORSOTTO, M., J. BARHANIN, M. FOSSET & M. LAZDUNSKI. 1985. The 1,4-dihydropyridine receptor associated with the skeletal muscle voltage-dependent Ca^{2+} channel. Purification and subunit composition. J. Biol. Chem. **260:** 14255-14263.

35. COOPER, C. L., S. VANDAELE, J. BARHANIN, M. FOSSET, M. LAZDUNSKI & M. M. HOSEY. 1987. Purification and characterization of the dihydropyridine-sensitive voltage-dependent calcium channel from cardiac tissue. J. Biol. Chem. **262:** 509-512.

36. PILCH, P. F. & M. P. CZECH. 1980. The subunit structure of the high affinity insulin receptor. J. Biol. Chem. **255:** 1722-1731.

Properties of the Calcium Channel Associated ω-Conotoxin Receptor in Rat Brain

N. MARTIN-MOUTOT, B. MARQUEZE, F. AZAIS,
M. SEAGAR, AND F. COURAUD

Laboratoire de Biochimie
CNRS UA 1179- INSERM U 172
Faculte de Medecine Secteur Nord
Bd. Pierre Dramard
13326 Marseille Cedex 15, France

Omega-conotoxin GVIA (ω-CgTx) is a 27 amino acid neurotoxic peptide purified from the venom of the marine snail *Conus geographus.*[1] ω-CgTx produces sustained blockade of transmitter release in certain synapses. This action has been attributed to persistent inhibition of the classes of high-threshold calcium channels that have been designated L and N in chick dorsal root ganglion neurons.[2] ω-CgTx does not appear to interact with muscle channels and promises to be a useful specific probe for the biochemical assay of neuronal calcium-channel components.

RESULTS AND DISCUSSION

Binding Parameters

Synthetic ω-CgTx (Peptide Inst., Osaka, Japan) was iodinated by the lactoperoxidase method and monoiodinated species were isolated by reverse-phase HPLC.[3] [^{125}I]ω-CgTx was used to identify binding sites on synaptic membranes from the adult rat brain. Addition of a large excess of unlabeled ω-CgTx to membranes containing receptor-bound [^{125}I]ω-CgTx at 4°C did not lead to any ligand dissociation. An apparent slow dissociation detected at 37°C was shown to be due to receptor denaturation. The ω-CgTx/receptor interaction is therefore irreversible and can only be characterized by the association rate constant and the binding site capacity which were: $k_a = 6 \times 10^6 \, M^{-1}sec^{-1}$ and $B_{max} = 0.65$ pmol/mg protein at 37°C and pH 7.2. [^{125}I]ω-CgTx was also shown to bind irreversibly to intact primary cultured neurons from rat brain ($k_a = 3 \times 10^6 \, M^{-1} \, sec^{-1}$ and $B_{max} = 60$ fmol/mg protein).

[^{125}I]ω-CgTx binding to synaptic membranes was unaffected by the addition of organic calcium antagonists such as nifedipine (1 μM), verapamil (100 μM), or diltiazem (10 μM). Binding was, however, inhibited by fairly low concentrations of

53

tri- and divalent cations with the following order of potency: $La^{3+} = Gd^{3+}(18$ $\mu M) > Cd^{2+}(450 \mu M) > Co^{2+}(700 \mu M) > Ca^{2+}(1.8 \text{ m}M) > Sr^{2+}(3.5 \text{ m}M) > Ba^{2+}(4.5$ mM). (The concentrations in parentheses gave 50% inhibition of $[^{125}I]\omega$-CgTx binding.) Control experiments with NaCl indicated that inhibition was not due to increasing ionic strength. The fact that ions that block calcium channels (La, Gd, Cd, Co) in physiological experiments are more potent inhibitors than permeant ions (Ca, Sr, Ba) may point to some functional significance of this ionic site.

Photoaffinity Labeling

Mono-$[^{125}I]\omega$-CgTx was modified with azidonitrophenylaminoacetyl (ANPAA)-succinimidyl ester to produce a photosensitive ligand that could be used to identify the ω-CgTx binding polypeptide(s).[3]

A 218K component that was clearly labeled in synaptic membranes (FIG. 1, lane 1) was protected from photoincorporation by unlabeled ω-CgTx (lane 2). A lower molecular weight, specifically labeled band of 25-35K was identified in some preparations but not in others. In cultured neurons a single high molecular weight band was specifically labeled (FIG. 1, lanes 3 & 4). The diffuse nature of the band made it

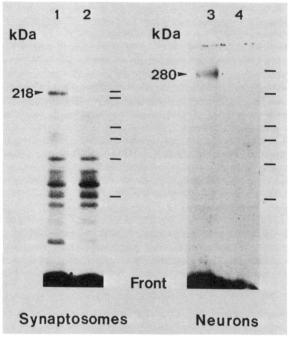

FIGURE 1. Photoaffinity labeling of the ω-conotoxin receptor. Synaptic membranes (lanes 1 and 2) or intact primary cultured neurons (lanes 3 and 4) were incubated with 0.1 nM $[^{125}I]$ANPAA ω-CgTx in the absence (lanes 1 and 3) or presence (lanes 2 and 4) of 0.1 μM native ω-CgTx. Samples were then submitted to UV photolysis. After denaturation in the presence of β-mercaptoethanol, membrane preparations were analyzed by SDS-PAGE and autoradiography.

difficult to determine the molecular weight with precision. It varied between 250 and 300K but was always of higher molecular weight than the polypeptide identified in synaptic membranes. The disparity in molecular weight observed between synaptic membranes and cultured neurons may be due either to developmental differences (i.e., a longer translation product or greater degree of glycosylation in the embryonic culture system) or to proteolysis occurring more readily in synaptosomal membrane preparations. Similarly the variably detectable 25-35K chain may also be a proteolytic product or a second calcium channel subunit similar to the γ subunit of the skeletal muscle dihydropyridine receptor.[4]

REFERENCES

1. OLIVERA, B. M., W. R. GRAY, R. ZEIKUS, J. M. MCINTOSH, J. VARGA, J. RIVIER, V. DE SANTOS & L. J. CRUZ. 1985. Science **230:** 1338-1343.
2. MCCLESKEY, E. W., A. P. FOX, D. H. FELDMAN, L. J. CRUZ, B. M. OLIVERA, R. W. TSIEN & D. YOSHIKAMI. 1987. Proc. Natl. Acad. Sci. USA **84:** 4327-4331.
3. MARQUEZE, B., N. MARTIN-MOUTOT, C. LEVEQUE & F. COURAUD. 1988. Mol. Pharmacol. **34:** 87-90.
4. TAKAHASHI, M., M. SEAGAR, J. JONES, B. F. X. REBER & W. A. CATTERALL. 1987. Proc. Natl. Acad. Sci USA **84:** 5478-5482.

The Detection and Purification of a Verapamil-Binding Protein from Plant Tissue

H. J. HARVEY,[a] M. A. VENIS,[b] AND A. J. TREWAVAS[a]

[a] Botany Department
University of Edinburgh
Edinburgh EH9 3JH, Scotland

[b] Institute of Horticultural Research
Kent ME19 6BJ, England

Experimental evidence suggests that calcium channels do exist in plant cells.[1-3] Calcium-channel antagonists have been demonstrated to disrupt a number of plant functions.[4] Nifedipine and verapamil have been shown to bind specifically to plant membranes *in vitro*.[5,6] Various phenylalkylamines have been shown to inhibit $^{45}Ca^{2+}$ influx into carrot protoplasts.[7]

In this research, [^{3}H]verapamil has been used to investigate calcium channels in maize coleoptile membranes. The K_D for specific binding of [^{3}H]verapamil to crude membrane fractions was 72 nM, B_{max} = 135 pmol/mg protein. For solubilized membrane fractions the K_D = 158 nM, B_{max} = 78 pmol/mg protein. In both cases the Scatchard plots were linear, indicating a single class of binding sites. Binding of verapamil took two hours to reach equilibrium at pH 7.5 at 20°C. Verapamil was bound in a saturable manner and the binding was reversible.

[^{3}H]Verapamil binding to crude membrane fractions could not be displaced by unlabeled D888 or by nifedipine, but could be displaced by unlabeled verapamil and

TABLE 1. Purification of the Verapamil-Binding Protein

Purification Step	Verapamil Bound		Protein		Specific Activity	Fold Purification
	c.p.m.	%	mg	%		
CHAPS Extract	32240	100	25	100	1305	1
DEAE	10304	32	4	16	2525	2
Sephacryl S400	23800	74	0.24	0.108	99166	76
Phenyl Sepharose	1264	4	0.008	0.032	15800	121

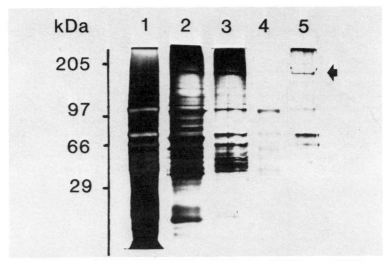

FIGURE 1. SDS polyacrylamide gel electrophoresis of verapamil-binding fractions from maize coleoptile membranes at different stages of purification.
     Lane 1. CHAPS-solubilized extract (5 μg protein). **Lane 2.** Pooled DEAE fraction (5 μg protein). **Lane 3.** Pooled sephacryl S400 fraction (5 μg protein). **Lane 4.** Sephacryl S400 fraction with no verapamil binding activity (5 μg protein). **Lane 5.** Phenyl sepharose fraction (1 μg protein).

D600, with IC_{50} values of 100 nM and 700 nM, respectively. This suggests that plant calcium channels have specific binding sites for the different calcium-channel antagonists.
     A protein that binds verapamil has been partially purifed from maize coleoptile membranes. The membrane-bound protein was solubilized in the detergent CHAPS and then purified using a combination of ion exchange, gel filtration, and hydrophobic interaction chromatography. This resulted in up to 120-fold purification of the verapamil-binding protein. (TABLE 1). SDS PAGE analysis of the purification steps suggests that there has been extensive purification of the verapamil-binding protein (FIG. 1). The final purification step contains four silver-stained bands of M_r 169,000, M_r 100,000, M_r 70,000, and M_r 66,000 (FIG. 1, Lane 5). A band of M_r 169,000 is enriched in the final verapamil-binding fraction (Lane 5, see arrow). This band is absent from the protein profiles of inactive fractions (Lane 4). A protein of M_r 100,000 is observed throughout the verapamil-binding and nonverapamil-binding fractions from the S400 column and the breakthrough fractions from the DEAE column. A 70-kDa protein 70 is observed in some of the nonverapamil-binding fractions of the S400 elution profile (Lane 4). The 66-kDa band is believed to be a staining artifact. On this basis, the protein of M_r 169,000 is most likely to contain the verapamil-binding site.
     The M_r 169,000 component of the verapamil-binding protein in maize coleoptile membranes reported here is comparable to the major subunit of M_r 142,000 found for the dihydropyridine receptor in rabbit skeletal muscle and to the polypeptide of M_r 165,000 found for the phenylalkylamine receptor, also in rabbit skeletal muscle.[8,9] Further work is in progress, using the photoaffinity label [³H]LU49888 to verify whether the protein of M_r 169,000 is the phenylalkylamine receptor in plant tissue.

REFERENCES

1. WILLIAMSON, R. E. & C. C. ASHLEY. 1982. Nature **296:** 647-651.
2. HAYAMA, T., T. SHIMMEN & M. TAZAWA. 1979. Protoplasma **99:** 305-321.
3. SAUNDERS, M. J. & P. K. HEPLER. 1982. Science **217:** 943-945.
4. LEHTONEN, J. 1984. Plant Sci. Lett. **33:** 53-60.
5. HETHERINGTON, A. M. & A. J. TREWAVAS. 1984. Plant Sci. Lett. **35:** 109-113.
6. ANDREJAUSKAS, E., R. HERTEL & D. MARME. 1985. J. Biol. Chem. **260:** 5411-5414.
7. GRAZIANA, A., M. FOSSET, R. RANJEVA, A. M. HETHERINGTON & M. LAZDUNSKI. 1987. Biochemistry **27:** 764-768.
8. HOSEY, M. M., J. BARHANIN, A. SCHMID, S. VANDAELE, J. PTASLIENSKI, C. O'CALLAHAN, C. COOPER & M. LAZDUNSKI. 1987. Biochem. Biophys. Res. Commun. **147:** 1137-1145.
9. GALIZZI, J. P., M. BORSOTTO, J. BARHANIN, M. FOSSET & M. LAZDUNSKI. 1986. J. Biol. Chem. **261:** 1393-1397.

Studies on the Purification of the 1,4-Dihydropyridine Receptor Complex from Skeletal and Cardiac Muscle Using Mab#78

RENJIE CHANG AND HENRY SMILOWITZ

Department of Pharmacology
University of Connecticut Health Center
Farmington, Connecticut 06032

Basic agreement now exists as to the molecular composition of a complex that copurifies with the 170K, 1,4 dihydropyridine (DHP) binding subunit of the skeletal muscle calcium channel.[1-6] A model has been recently presented in which the complex consists of at least 170K (α_1), 140K (α_2), 55K (β), and 30K (γ) polypeptides (reducing conditions) in a 1 : 1 : 1 : 1 ratio.[3] It is not yet clear whether all of these polypeptides are bona fide components of the calcium channel or whether some are proteins that copurify with the channel.

We have isolated a monoclonal antibody (Mab#78) that is directed to the 170-kDa subunit of the skeletal muscle calcium channel.[7] Our antibody immunoprecipitates DHPR (FIG. 1B) and immunoblots to a 170K polypeptide which (a) does not shift molecular weight upon reduction (FIG. 1A), (b) does not contain a detectable level of N-linked sugars, (c) is phosphorylated by cyclic-AMP-dependent protein kinase, and (d) is photoaffinity labeled by [^{3}H]azidopine (FIG. 1C, D) as shown by two-dimensional SDS PAGE and immunoblotting and is completely distinct from the 140K glycoprotein under reducing conditions. As indicated, some of these observations are documented in FIGURE 1 and are consistent with the criteria used to describe the DHPR of skeletal muscle.[1-6]

Mab#78 is an IgGl subtype with high affinity for the DHPR of rabbit skeletal muscle; the concentration of the antibody that gives a half-maximal response on an ELISA assay is approximately 1 nM (FIG. 2A). As expected Mab#78 cross-reacts with membranes with a potency proportional to the amount of DHPR present. Mab#78 cross-reacts well with rabbit and pig skeletal DHPR (and dog cardiac DHPR) but very weakly with chicken and frog skeletal DHPR, suggesting the antibody recognizes mammalian DHPR preferentially.

We have constructed an affinity column with Mab#78, and we have developed methods for the purification of milligram quantities of the rabbit skeletal muscle DHPR in a single step. Shown in FIGURE 2B is Mab#78 immunoaffinity-purified DHPR run on SDS PAGE under reducing conditions. Note that α_1 runs as a doublet; we do not know if this represents isoforms or proteolytic fragmentation of α_1. In addition to α_1, α_2, β, and γ, we see contaminating bands at about 100K, 70K, and 60K (calsequestrin, Cal).

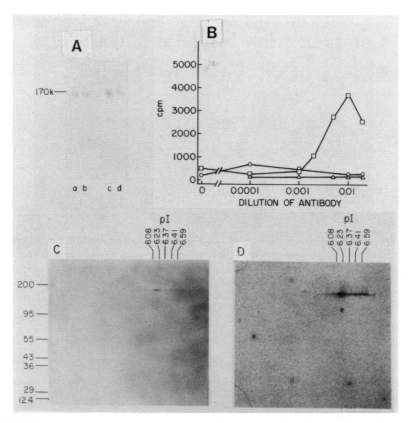

FIGURE 1. (A) Immunoblot with Mab#78. Lanes a and c and lanes b and d contain DHPR complex with 500-1000 and 1000-2000 pmol/mg protein PN200-110 binding sites, respectively. Lanes a, b, nonreducing conditions; lanes c, d reducing conditions. Antibody incubations: first antibody (1/1000 dilution of Mab#78 for one hour at 37°C); second antibody (1/1000 dilution of HRP-labeled GAM (Hyclone).

(**B**) Antibody #78 immunoprecipitated DHPR. T-Tubule membranes (0.2 mg/ml in 50 mM Tris containing 32 pmol/mg protein DHP binding sites) were labeled with 10 nM [^{3}H]PN 200-110 for 30 minutes at 4°C in the dark in the absence ($\square$) (total labeling) and in the presence ($\triangle$) of 4 μM unlabeled PN 200-110 (nonspecific labeling). Membrane extraction and immunoprecipitation using protein A Sepharose reagent was done according to Takahashi and Catterall.[10] Maximal immunoprecipitation was 30% of bound PN 200-110 over 15 hours. (O), nonspecific mouse IgG.

(**C,D**) Photoaffinity labeling of antibody #78 immunoreactive material with [^{3}H]azidopine. After irradiation, DHPR was partially purified by wheat germ affinity chromatography. The preparation was subjected to two-dimensional SDS PAGE immunoblotting with antibody #78. C: immunoblot, D: autoradiograph (exposure time was two weeks).

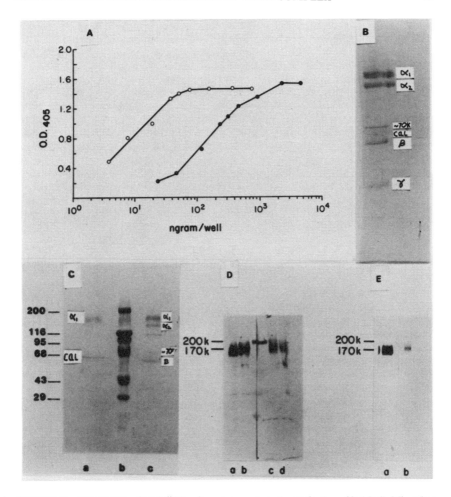

FIGURE 2. (A) ELISA of Mab#78 culture supernate and protein A purified IgG. Microtiter wells were coated overnight (37°C) with 0.5 µg DHPR complex. Mab#78 concentration was varied as indicated. (●), Hybridoma culture supernatant (1% fetal calf serum); (○), Protein A purified Mab#78.

(B) DHPR purified by Mab#78 affinity chromatography. Membranes were extracted with 1% digitonin or 2.5% CHAPS (final) in buffer A containing 25 mM HEPES, 185 mM KCl, 1 mM EDTA, and a protease-inhibitor cocktail. Detergent to protein ratio was 5 : 1 (digitonin) and 6-10 : 1 (CHAPS). Detergent extracts were clarified by ultracentrifugation and incubated with antibody-linked affigel-10 overnight at 4°C. The resin was then packed onto a column, washed with 20 ml buffer A (1% detergent) and 20 ml buffer A (0.3% detergent) and eluted with BioRad buffer (pH 3.0, 0.3% detergent). Fractions (0.5 ml) were immediately neutralized with 65 µl 1 M Tris, pH 9.0.

(C) Comparison of DHPR purified by monoclonal antibody affinity chromatography after CHAPS (Lane a) and digitonin (Lane c) extraction. Lanes a and c contain 5 µg protein; reducing conditions. Lane b, molecular weight markers.

(D,E) Immunoblot of cardiac and skeletal DHPR with Mab#78. Lanes a and b, skeletal. Lanes c and d, cardiac. Lanes a and c, nonreduced. Lanes b and d, reduced. E. Lane a, skeletal. Lane b, cardiac.

We have used the antibody affinity column to compare the subunit composition of the DHPR in different detergents. We have shown that the α_2 component as well as some of the β subunit dissociates from the complex in the detergent CHAPS (FIG. 2C). Both digitonin and CHAPS preparations contain the α_1 subunit as revealed by immunoblot with Mab#78. The β subunit is recovered in the digitonin preparation but to a lesser extent in the CHAPS preparation; the major 60K band in the CHAPS preparation is calsequestrin (Cal); much less calsequestrin contaminates the digitonin preparation. These results resolve a conflict in the literature (Takahashi, et al.[3] versus Leung et al.[8]) regarding digitonin and CHAPS preparations and have been written up with Dr. Susan Hamilton, who has independently obtained corroborating data.

We have also demonstrated that the canine cardiac DHPR contains an approximately 170-kDa component that migrates similarly under both reducing and nonreducing conditions. FIGURE 2D,E shows (using two different preparations) that the cardiac α_1 may be slightly larger than the skeletal α_1, but additional experiments are needed to verify this and to identify the other components of the cardiac DHPR complex.

REFERENCES

1. LEUNG, A. T., T. IMAGAWA & K. P. CAMPBELL. 1987. J. Biol. Chem. **262:** 7943-7946.
2. SHARP, A. H., T. IMAGAWA, A. T. LEUNG & K. P. CAMPBELL. 1987. J. Biol. Chem. **262:** 12309-12315.
3. TAKAHASHI, M., M. J. SEAGER, J. F. JONES, B. F. X. REBER & W. A. CATTERALL. 1987. Proc. Natl. Acad. Sci. USA **84:** 5478-5482.
4. VAGHY, P. L., J. STRIESSNIG, M. KUNIHISA, H. G. KRAVS, K. ITAGAKI, E. MCKENNA, H. GLOSSMANN & A. SCHWARTZ. 1987. J. Biol. Chem. **262:** 14337-14342.
5. NAISTANZYK, W., A. RÖHRKASTEN, M. SIEBER, C. RUDOLPH, C. SCHÄCHTELE, D. MARME & F. HOFMANN. 1987. J. Biol. Chem. **169:** 137-142.
6. MORTON, M. E. & S. C. FROEHNER. 1987. J. Biol. Chem. **262:** 11904-11907.
7. SMILOWITZ, H., R. J. CHANG & C. BOWIK. 1987. Soc. Neurosci. Abstr.
8. LEUNG, A. J., T. IMAGAWA, B. BLOCK, C. FRANZINI-ARMSTRONG & K. P. CAMPBELL. 1988. J. Biol. Chem. **263:** 994-1001.
9. CHANG, R. J. & H. SMILOWITZ. 1988. Life Sci. **43:** 1055-1061.
10. TAKAHASHI, M. & W. A. CATTERALL. 1987. Science **236:** 88-91.

Gating and Permeation of Different Types of Ca Channels

P. KOSTYUK,[a] N. AKAIKE,[b] YU. OSIPCHUK,[a]
A. SAVCHENKO,[a] AND YA. SHUBA [a]

[a] *Bogomoletz Institute of Physiology*
Kiev, USSR

[b] *Department of Physiology*
Kyushu University
Fukuoka, Japan

When calcium transmembrane currents were first separated and measured in various excitable tissues, it seemed that they were carried by similar molecular structures. However, introduction of techniques allowing a more detailed analysis of membrane phenomena, especially single-channel recordings, showed definitely that calcium conductance is maintained by a whole family of different electrically operated channels. At first, data about participation of two types of calcium channels in the production of calcium inward current was obtained on rat sensory neurons in our laboratory[1,2] and then independently in the laboratories of Lux[3–5] and Tsien.[6] Later on, the data were confirmed on many cellular structures: cultured sensory neurons,[7] cardiomyocytes,[8,9] smooth[10] and striated[11,12] muscle fibers, cloned pituitary cells,[13] and even ciliates.[14] The main difference between the described channel types is their functioning in quite separate membrane potential regions; therefore, they can be classified as "low-threshold" and "high-threshold" types (T and L channels). Subsequently calcium channels were also differentiated according to their behavior during long-lasting membrane potential displacements; some of them underwent more or less rapid inactivation, others did not, thus creating a steady influx of Ca^{2+} into the cell.[15–18] The pharmacological sensitivity also turned out to be extremely different for different types of calcium channels and even for the channels of similar type located in different cellular structures.

All these data raised a series of new questions about the molecular organization of calcium-conducting pathways as well as about their role in the functioning of excitable cells. This paper summarizes new data obtained along these lines in our laboratories on different types of nerve cells.

SEPARATION OF DIFFERENT TYPES OF CALCIUM CHANNELS

Measurements were made with the use of the patch-clamp technique on cell-attached and isolated membrane fragments.[19] Cultured neurons from 12–14-day-old mouse embryos were used as the main object. The technique of their isolation, culturing, and recording is described elsewhere.[18] In the membranes of these cells, all three types of electrically operated calcium channels are well expressed, and their comparative analysis can be easily made. However, a high density of different calcium channels creates problems for a detailed analysis of each channel type. Therefore, we searched for cells in which only one type of calcium channel was dominant. In fact, N1E-118 neuroblastoma cells demonstrated only those calcium channels whose characteristics are very close to low-threshold inactivating (LTI) channels of mouse sensory neurons, thus being very convenient for analysis. LTI-channels were found by Kaneda, Nakamura, and Akaike[20] to also be predominant in the somatic membrane of freshly isolated neurons from rat ventromedial hypothalamus.

High-threshold noninactivating (HTN) channels could be effectively studied on PC-12 pheochromocytoma cells. They demonstrated typical super-long bursts of openings,[15] although their inactivation was somewhat more prominent than in sensory neurons.

For a quantitative description of channel-activation kinetics, a model has been used that includes a two-stage transition from resting state R via intermediate closed state C into activated closed state A. These transitions can be described by potential-dependent Hodgkin-Huxley kinetics using a square power of the m-variable (see Kostyuk et al.[21]). Then a rapid potential-independent transition into open state occurs:

$$R \underset{\beta}{\overset{2\alpha}{\rightleftharpoons}} C \underset{2\beta}{\overset{\alpha}{\rightleftharpoons}} A \underset{b}{\overset{a}{\rightleftharpoons}} O$$

The presence of inactivation was not taken into account at this stage of the analysis because of its slow time-course. Theoretical analysis of the proposed model has shown that, based on experimental data about distributions of channel open and closed times, duration of "bursts" of openings, and waiting time until first opening, all transition rate constants (α, β, a and b) can be calculated in a simple way[22]:

$$\tau_{op} = 1/b; \quad \tau_{cl}^{(f)} = 1/a; \quad \tau_{bu}^{(s)} = a/2b\beta; \quad (2\,\tau_s\tau_{vs})^{\frac{1}{2}} = 1/\alpha$$

where τ_{op} is mean channel life time in open state, $\tau_{cl}^{(f)}$ is mean time of intraburst closings; $\tau_{bu}^{(s)}$ is mean duration of bursts of openings; τ_s and τ_{vs} are time constants of two exponents, the difference of which describes the waiting time distribution.

Our measurements have shown that activation of all three types of calcium channels can be satisfactorily described by this model; however, the transition rate constants and their potential dependence substantially differ for different channels. For HTN channels the presence of two open states with different mean lifetimes should be accepted.

FIGURE 1 presents examples of the activity of all three channel types recorded from mouse sensory neuron, as well as holding and test potential shifts necessary for their activation. FIGURE 2 demonstrates recordings of the LTI channel from neuroblastoma and HTN-channel from pheochromocytoma cells.

PERMEATION THROUGH DIFFERENT TYPES OF CHANNELS

Permeability of different types of calcium channels has been compared by plotting single-channel current-voltage characteristics using different permeant cations. Such an approach was important because when comparing integral currents, their maximum

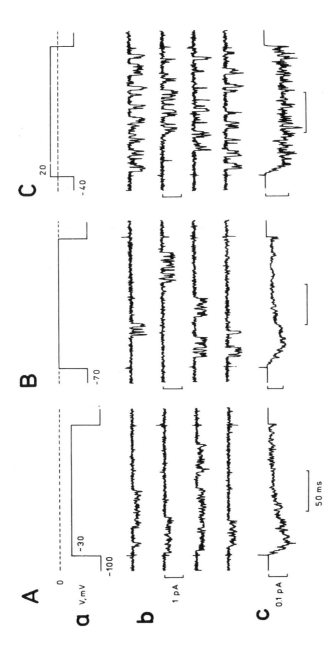

FIGURE 1. Three types of single calcium channel activity in the membrane of mouse DRG neurons. Voltage protocols (**a**), original single-channel currents (**b**), and corresponding averages of idealized records (**c**) for LTI (**A**), HTI (**B**), and HTN (**C**) channels. 60 m*M* Sr.$^{2+}$ as a charge carrier. Cell-attached configuration. Data from different cells.

amplitude may be determined not only by permeability of individual channels but also by mean duration of their open state (which may depend on the type of permeating ion, as will be shown later). To avoid uncertainty in the determination of resting potential level, cells were usually immersed in high-potassium solution shifting the membrane potential close to zero level.

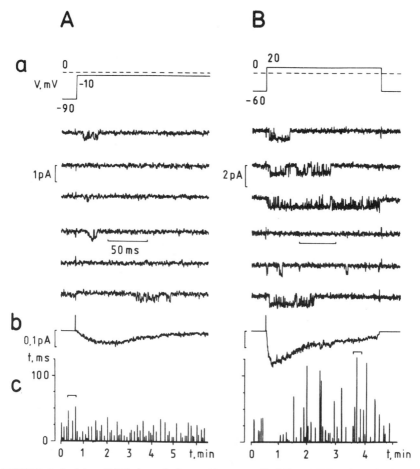

FIGURE 2. Activity of LTI channel of neuroblastoma cells (**A**) and HTN channel of pheochromocytoma cells (**B**). (**a**) Voltage protocols and original current traces, (**b**) corresponding averages of idealized records, and (**c**) dependence of total time that channel spent in open state at every consecutive depolarization of 400 msec (**A**) and 160 msec (**B**) duration (height of the columns) upon the time of recording. Horizontal bars in **c** indicate records plotted in **a**. 60 mM-Ba^{2+} as a charge carrier. Cell-attached configuration.

TABLE 1 presents conductance values for all three types of channels and three types of permeant ions: Ca^{2+}, Sr^{2+}, and Ba^{2+}. For LTI channels the conductance for all of them (taken at 60 mM concentration) was almost identical (7.2 pS). Nevertheless, the maximum integral barium current produced by these channels was considerably

TABLE 1. Conductances (pS) of Different Types of Calcium Channels

Charge Carrier (60 mM)	DRG Neurons			Neuroblastoma	Pheochromocytoma
	LTI (T)	HTI (N)	HTN (L)	LTI (T)	HTN (L)
Ca^{2+}	6.3	7.2	7.6	7.2	7.2
Sr^{2+}	5.7	7.2	9.0	7.2	10.3
Ba^{2+}	7.2	11.4	18.4	7.2	17.0
Na^+				18.0	17.2

less than the calcium or strontium one, because of a lower channel activity in barium solution. For HTI and HTN channels, the conductance was identical in Ca and Sr solutions, but 1.5-2 times higher in a barium solution. Thus, conductance differences between the channels are rather quantitative in nature. No principal differences were observed in blocking of different channels by inorganic ions (Co^{2+}, Mn^{2+}, Cd^{2+}). Fox et al.[16] noticed that in chick dorsal root ganglion (DRG) cells LTI (T) channels can be separated by sensitivity to different blocking ions; they are more resistant to Cd^{2+} and more sensitive to Ni^{2+}. However, we did not find such a difference in mouse sensory neurons. LTI channels in rat hypothalamic neurons (see FIG. 8A) could be most effectively blocked by La^{2+} ($K_d \sim 7 \times 10^{-7} M$).

The "selectivity filter" in all types of channels has a remarkable property: It loses selectivity for divalent cations and changes into a sodium-selective channel when the concentration of divalent cations in the external medium is decreased below $10^{-7}M$. This effect was first detected on neuronal membranes[23] and later confirmed on several other structures (smooth, striated, and cardiac muscle fibers). It was suggested recently that such a transformation can be induced by lowering the pH of the external medium[24]; if this is the case, it may explain the proton-induced sodium conductance described earlier in sensory neurons.[25] Two different explanations of the mechanism of this transformation have been proposed. We suggested that a special high-affinity binding site for Ca^{2+} ions is present in the mouth of the calcium channel. The presence of bound ion is essential for the maintenance of channel selectivity for divalent cations; its removal produces conformational changes altering channel selectivity.[26] A simpler explanation was proposed by Almers et al.[27] and shared by Hess et al.[28]: The presence of two closely located binding sites inside the channel is suggested. If divalent cations are present in the medium, both sites are occupied and ion transition through the channel is facilitated by their electrostatic repulsion. If both sites are empty, the channel loses its selectivity and starts to pass monovalent cations. However, with this model it is difficult to explain some peculiarities of the functioning of modified channels (see Kostyuk and Mironov[29]).

Channel transformation is followed by a substantial increase in its conductance (see TABLE 1). The selectivity sequence for modified LTI channels in neuroblastoma cells was $P_{Na} : P_{Li} : P_K = 1 : 0.55 : 0.37$. A qualitatively similar sequence was obtained for high-threshold channels in mollusc neurons.[30]

GATING IN CHANNELS OF DIFFERENT TYPES

FIGURE 3A presents the potential dependence of the mean lifetime in open state and mean duration of intraburst closing for the LTI channel, corresponding to 1/a and 1/b values in the accepted model. The type of potential dependence and absolute values are close for different permeant ions. FIGURE 3B shows similar dependences for the HTI channel: in this case both characteristics proved to be almost potential independent. An important feature was found for the HTN channel: The distribution of open-state lifetimes could be adequately described by the sum of two exponents, indicating the presence of two open states with different transition rate constants. In 60 mM barium solution, the time constant of the "slow" state practically did not change with increase of depolarization (mean value 1.9 ± 0.2 msec), but that of the "fast" state demonstrated a clear potential dependence (changing from 0.7 msec at V = 0 mV to 0.2 msec at V = 30 mV).

Using the proposed model, the values and potential dependence of the α and β constants were also determined for LTI and HTI channels. In all cases an exponential dependence on membrane potential was observed. If plotted in semilogarithmic scale, the intersection of both lines corresponded well to channel half-activation potential, and their shifts for different permeant ions fitted the shifts of the corresponding

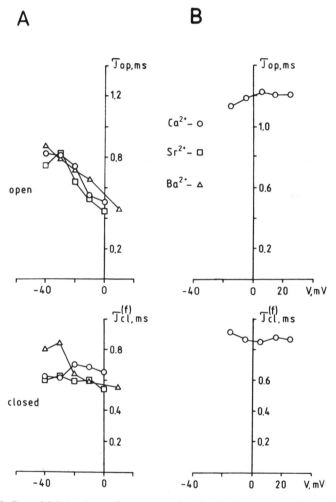

FIGURE 3. Potential dependence of mean open time (τ_{op}) and mean time of intraburst gaps ($\tau_{cl}^{(f)}$) for single LTI calcium channel in neuroblastoma cells (**A**) and HTI calcium channel in mouse DRG neurons (**B**). All charge carriers were used in 60 m*M* concentration.

activation curves. As the two initial kinetic stages in the model corresponded formally to intramembrane shifts of two charged gating particles, the logarithm of relation α/β should correspond to the effective valency of one m particle during its shift in the course of channel activation.

FIGURE 4A presents potential dependences of α and β constants for the LTI channel, showing a definite dependence of their slopes on the type of penetrating ion. This can be explained by the assumption that penetrating ions influence the effective charge of gating particles probably by changing the relief of transmembrane potential fall after ion binding in the channel. FIGURE 4B shows similar dependences for the HTI channel; the calculated effective valency of the m particle is 2.2 if Ca^{2+} is used as a charge carrier. This value corresponds well to that obtained from the measurements of asymmetric displacement currents.[31] Similar measurements for HTN channels are in progress.

After channel transformation into a sodium, two drastic changes in gating appear; they were studied in most detail on neuroblastoma LTI channels. (1) The dependences $\tau_{op}(V)$ and $\tau_{cl}^{(0)}(V)$ shift in the hyperpolarizing direction for about 70 mV corresponding

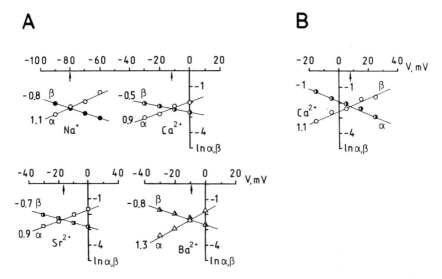

FIGURE 4. Potential dependence of rate constants α and β for LTI calcium channel of neuroblastoma cells (**A**) and HTI calcium channel of mouse DRG neurons (**B**). Corresponding charge carriers (all in 60 mM concentration) and semilogarithmic slopes of α and β are shown near each graph. Arrows indicate potentials of intersection for α and β lines that correspond to half-activation of the channel.

to a shift of activation curve. (2) The τ_{op} value diminishes considerably, depending on both the type and concentration of the permeant ion. With increased concentration a decrease in open lifetime is observed within the whole potential range studied. In other words, the mean open time of the channel becomes current dependent. Closed time distribution is much less affected upon channel transformation.

The described current dependence of channel gating is quite difficult to understand on the basis of the usual assumptions about independent selectivity and gating mechanisms. Obviously, some new approaches are required in order to describe ion transition through the channel. One approach is to assume that ion transition through the open channel produces local displacements of charged molecular groups lining the wall of its steric region. This would lead to formation of dipoles along the trajectories of ion movement that exist during dipole relaxation time (about 10^{-7} sec). For divalent cations the calculated mean dipole moments, which also depend on the rate of ion transitions through the channel, are insignificant and may have no considerable effect on the potential dependence of the channel opening. However, after channel modification, which induces strong fluxes of monovalent ions, a quite different situation emerges. The frequency of ion transitions increases drastically and becomes comparable with frequencies of dipole relaxation. The contribution of dipole moments formed by the ion flux may become significant, which would lead to a decrease in the part of the external electric field affecting the gating region of the channel (due to a large increase in the dielectric constant of this region). At high Na^+ concentrations, $\tau_{op} = 0.2$ msec and is completely potential independent.

DEPENDENCE OF CHANNEL ACTIVITY ON MEMBRANE—CYTOPLASMIC CONNECTIONS

The described types of calcium channels largely differ in their dependence on the maintenance of connections between the membrane and the cytoplasm. FIGURE 5A shows recordings of LTI-channel activity in barium solution before and after separation of the membrane patch from the cell. Practically no change in activity was observed except for some decrease in the amplitude of unitary current during "inside-out" recording.

On the contrary, the activity of high-threshold channels rapidly decreased after patch separation; this is true for both inactivating and noninactivating channels, although HTN channels seem to be more sensitive in this respect. Examples are shown in FIGURE 5B.

On the basis of experiments showing a preventive effect of the intracellular introduction of $cAMP + ATP + Mg^{2+}$ on the "wash-out" of high-threshold calcium currents[32-34] and on acceleration of inactivation of such currents after intracellular injection of phosphatases,[35,36] the suggestion seems to be justified that high-threshold calcium channels need continuous phosphorylation mediated by cAMP-dependent phosphatases for normal functioning. After dephosphorylation the channel enters into an inactive "sleeping" state. There is no such necessity for the low-threshold channels that resemble the TTX-sensitive sodium channels in their ability to function in isolated membrane fragments.

The described difference in the functioning of low- and high-threshold calcium channels coincides with a principal difference in the mechanism of their inactivation. In all investigated cases, the inactivation of LTI channels (contrary to that of high-threshold ones) was potential dependent; it developed mainly in the region of negative membrane potentials in which no calcium inward current could be activated. A corresponding example for HTI channels in isolated hypothalamic neurons is shown in FIGURE 6 (see also Refs. 2, 3, and 37).

PHARMACOLOGICAL PROPERTIES OF CALCIUM CHANNELS

Until recently the low sensitivity of neuronal calcium channels to organic blockers has been considered their dominant feature, making them substantially different from calcium channels in other excitable cells, particularly striated and cardiac muscle

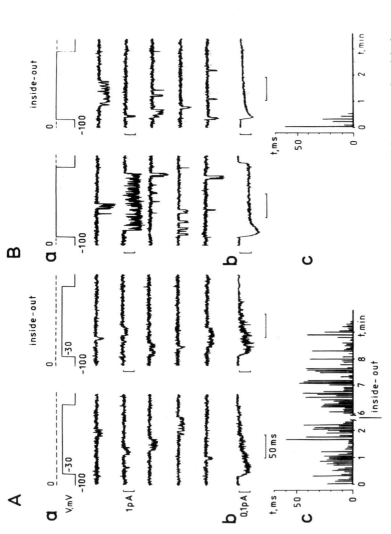

FIGURE 5. Effect of patch excision on LTI and HTI + HTN Ca channels of mouse DRG neurons. Voltage protocols, original current traces (**a**), and corresponding averages (**b**) for LTI (**A**) and HTI+HTN (**B**) channels in cell-attached and inside-out configurations. Ac shows the changes of total LTI channel open time at every consecutive depolarization of 160-msec duration before and after patch excision. Bc demonstrates the diminishing of total HTI channel open time after isolation of the patch. Activity of HTN channel in excised patch was completely absent. 60 mM Ba^{2+} as a charge carrier. Resting membrane potential was zeroed with high external K$^+$.

fibers. In fact, in experiments on cultured mouse and rat sensory neurons, as well as on cloned blastomatic cells, a 50% depression of calcium current under the action of phenylalkylamines, benzodiazepines, and dihydropyridines was observed only at concentrations several orders of magnitude higher than those for muscle fibers.[2,7,38–40]

A comparison of the sensitivity of different types of calcium channels in the membrane of the same cultured neuron has shown that high-threshold channels are more effectively blocked by organic blockers than the low-threshold ones. This difference is also true for a newly detected antagonist, ω-conotoxin. Our comparison of the action of organic blockers on all three types of calcium channels in the membrane of cultured mouse sensory neurons has shown that HTN channels can be almost completely blocked by $2.5 \times 10^{-5}M$ D-600. However, LTI and HTI channels do not substantially change their activity. Calcium-channel agonist BAY K8644 exerted some

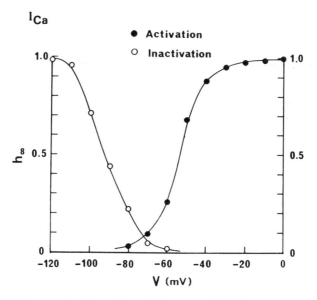

FIGURE 6. Stationary activation and inactivation curves for LTI Ca current in isolated rat hypothalamic neurons.

potentiating effect only on HTN channels, prolonging the time constants of both exponential components in the open time distribution (from 0.5 and 3.3 msec in control to 1.3 and 4.2 msec after drug application). Thus, the first ("short") open state of the channel changed more significantly. At the same time, the mean time of intraburst closings somewhat decreased (from 0.5 to 0.4 msec). From these data we did not get the impression that under BAY K8644 a special mode appears in calcium channel kinetics absent before its action, as was suggested by Nowycky, Fox, and Tsien.[15] Some examples of BAY K8644 action on the activity of single HTN channels are shown in FIGURE 7.

To our great surprise, our recent detailed investigation of pharmacological sensitivity of freshly isolated rat hypothalamic neurons, which reveal a high density of LTI channels, gave results completely different from those already reported. In this case substances belonging to all three main groups of calcium-channel antagonists effectively blocked the LTI channels. They include D-600, diltiazem and its derivative TA3090, nimodipine, nifedipine, nicardipine, and a new drug, flunarizine, which is a diphenylpiperazine and difluoro derivative of cynnarizine. The latter exerts a slowly developing inhibition of calcium currents in smooth muscle fibers.[41,42] The order of effectiveness was as follows: diltiazem < D-600 < TA3090 < nimodipine < nifedipine < nicardipine < flunarizine. K_d values obtained for corresponding Langmuir's isotherms were: $7 \times 10^{-5}M$ (diltiazem), $5 \times 10^{-5}M$ (D-600), $2 \times 10^{-5}M$ (TA3090), $7 \times 10^{-6}M$ (nimodipine), $5 \times 10^{-6}M$ (nifedipine), $3.5 \times 10^{-6}M$ (nicardipine) and $7 \times 10^{-7}M$ (flunarizine) (see FIG. 8B). It should be mentioned that BAY K8644 in this case also depressed the calcium current with $K_d = 1 \times 10^{-5}M$.

The time course and reversibility of action differed for different substances. The effect of dihydropyridines developed most rapidly and was completely reversible. The effect of flunarizine developed slowly (about 100 sec) and was only partly reversible. This also occurred when LTI channels were previously inactivated by lowering the holding potential level.

The reason for such extreme differences in pharmacological sensitivity of kinetically identical calcium channels is yet unclear; it may represent a genuine difference in channel structure. On the other hand, even short-term culturing of cells produces some changes in the outer part of the channel-forming protein, resulting in drastic changes in binding of antagonists to corresponding receptor groups. The fact that no changes in selectivity or kinetics occur in parallel may indicate that such groups are of little importance for the maintenance of channel functions.

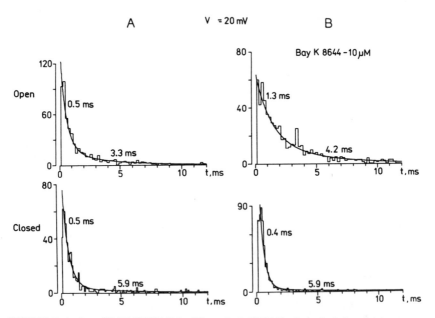

FIGURE 7. Action of BAY K8644 (10 μM) on single HTN Ca channel of pheochromocytoma cells. Histograms of open and closed times before (**A**) and after (**B**) application of BAY K8644. Smooth lines: biexponential fitting of experimental histograms. Corresponding time constants are shown near each experimental component.

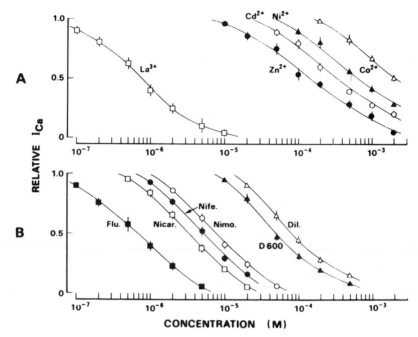

FIGURE 8. Blocking action of inorganic (**A**) and organic (**B**) Ca-channel antagonists on LTI Ca current of isolated rat hypothalamic neurons. Points—experimental data, smooth curves—Langmuir's isotherms. Corresponding K_ds are given in text.

The obtained data can be of practical importance: The action of calcium-channel antagonists turns out to be much more complex and effective than that predicted from experiments on cultured cells, especially considering the data indicating a prominent role for low-threshold calcium conductance in the membrane of certain central neurons.

CONCLUSIONS

The presented data indicate that different types of calcium channels share some important basic features, thus providing evidence of their genetic similarity. The gating mechanism of all channels includes several "preopen" states with transitions that can be satisfactorily described by the Hodgkin-Huxley model with two charged independent "gating" particles; channels of different types differ mainly in rate constants and degree of their potential dependence. Then a rapid transition into the open state occurs. The rate constants of this transition differ in different channels; two open states with different lifetimes may occur in a single channel.

The permeability of different types of calcium channels has important common features. As any calcium channel can be modified into a sodium channel by eliminating divalent cations from the external medium, it is obvious that the fundamental mechanism creating ionic selectivity is similar in all types of channels. It is based on binding of the penetrating ion with one or several specific binding sites inside the channel. The structure of these sites in different channels has some minor variations.

In the light of recent data it is obviously necessary to abandon the simplified idea about the selective filter of the calcium channel as a rigid structure not subjected to any dynamic changes during passage of ions through the channel. Passing ions interact with the channel wall and produce changes in it, most probably changing its dielectric characteristics, thus affecting the function of the gating mechanism. Such interaction is especially prominent during large fluxes of monovalent cations appearing after modification of channel selectivity.

Organic antagonists or agonists of calcium channels differ considerably in their effect on the channels of different types as well as on similar channels in different membranes. Obviously, the primary receptor sites for such substances are located on the parts of the channel that are not vital for determination of channel kinetics and selectivity.

Considering the functional characteristics of different calcium channels, one cannot avoid questioning the functional meaning of their diversity. Definitely, special kinetic features of LTI channels substantially influence the participation of the corresponding current in cellular activity. It can be effectively controlled by small membrane potential changes switched off even by slight depolarization or put into action again by membrane hyperpolarization. Therefore, in combination with calcium-dependent potassium currents, they may create an effective mechanism for generation of membrane potential oscillations and autorhythmicity. In our laboratory a statistical analysis of the expression of different types of calcium conductances in sensory neurons has been made during postnatal development of mice (from birth until three months). A significant decrease in the percentage of cells exhibiting LTI current was found; in three-month-old animals, it could be observed in only 1% of the cells studied.[43] In parallel, sodium currents of a special embryonic (slow TTX-insensitive) type also disappeared. Thus, we suggest that LTI channels reflect some embryonic features of cell function disappearing after final cell differentiation—maybe inclination to spontaneous activity. However, in the brain of adult rats, some neurons retain LTI channels as a dominant mechanism of calcium conductance. This conclusion has been made for olivary neurons by Llinas and Yarom[44] on the basis of action potential analysis; we have shown this directly for hypothalamic neurons.

The main difference in the functional significance of high-threshold calcium channels arises from the necessity of strong membrane potential shift for activation, which is possible during generation of full-sized action potentials or very effective synaptic or endogenous depolarizations, for instance in chromaffin cells[45] or pancreatic β-cells.[46] For this reason high-threshold channels may serve as generators of effective calcium signals coupling membrane and intracellular activities; later these signals can be potentiated or modulated by addition of systems mobilizing Ca from the intracellular stores, by activation of calcium-dependent enzymatic systems, and so forth.

The function of some types of calcium channels may change during cell development. For instance, synaptic terminals may be considered as final stages of the development of the growth cone; their membrane has a high density of calcium channels of a poorly defined type. Such channels inactivate, and they are insensitive to organic blockers and remain active in isolated membrane fragments.[47,48] Because of these properties, they may be considered derivatives of embryonic LTI channels. However, other views have also been expressed.[17]

ACKNOWLEDGMENTS

We want to thank Kyowa Hakko Kogyo Co., Ltd. for a gift of flunarizine.

REFERENCES

1. VESELOVSKY, N. S. & S. A. FEDULOVA. 1983. Two type of calcium channels in the somatic membrane of rat dorsal root ganglion neurons. Dokl. Akad. Nauk SSSR (Moscow) **268:** 747-750.
2. FEDULOVA, S. A., P. G. KOSTYUK & N. S. VESELOVSKY. 1985. Two types of calcium channels in the somatic membrane of newborn rat dorsal root ganglion neurones. J. Physiol. (London) **359:** 431-446.
3. CARBONE, E. & H. D. LUX. 1984. A low voltage-activated, fully inactivating Ca channel in vertebrate sensory neurones. Nature **310:** 501-503.
4. CARBONE, E. & H. D. LUX. 1987. Single low-voltage-activated calcium channels in chick and rat sensory neurones. J. Physiol. (London) **386:** 571-601.
5. CARBONE, E. & H. D. LUX. 1987. Kinetics and selectivity of a low-voltage-activated calcium current in chick and rat sensory neurones. J. Physiol. (London) **386:** 547-570.
6. NOWYCKY, M. C., A. P. FOX & R. W. TSIEN. 1984. Two components of calcium channel current in chick dorsal root ganglion cells. Biophys. J. **45:** 36a.
7. BOSSU, J.-L., A. FELTZ & J. M. THOMANN. 1985. Depolarization elicits distinct calcium currents in vertebrate sensory neurones. Pflügers Arch. Ges. Physiol. **403:** 360-368.
8. BEAN, B. P. 1985. Two kinds of calcium channels in canine atrial cells. Differences in kinetics, selectivity, and pharmacology. J. Gen. Physiol. **86:** 1-30.
9. BONVALLET, R. 1987. A low threshold calcium current recorded at physiological Ca concentrations in single frog atrial cells. Pflügers Arch. Ges. Physiol. **408:** 540-542.
10. LOIRAND, G., P. PACAUD, C. MIRONNEAU & J. MIRONNEAU. 1986. Evidence for two distinct calcium channels in rat vascular smooth muscle cells in short-term primary culture. Pflügers Arch. Ges. Physiol **407:** 566-568.
11. COGNARD, C., M. LAZDUNSKI & G. ROMEY. 1986. Different types of Ca^{2+} channels in mammalian skeletal muscle cells in culture. Proc. Natl. Acad. Sci. USA **83:** 517-521.
12. CAVALIE, A., V. FLOCKERZI, F. HOFMANN, D. PELZER & W. TRAUTWEIN. 1987. Two types of calcium channels from rabbit fast skeletal muscle transverse tubules in lipid bilayers: Differences in conductance, gating behaviour and chemical modulation. J. Physiol. (London) **390:** 82P.
13. MATTESON, D. R. & C. M. ARMSTRONG. 1986. Properties of two types of calcium channels in clonal pituitary cells. J. Gen. Physiol. **87:** 161-182.
14. DEITMER, J. W. 1986. Voltage dependence of two inward currents carried by calcium and barium in the ciliate *Stylonychia mytilus.* J. Physiol. (London) **380:** 551-574.
15. NOWYCKY, M. C., A. P. FOX & R. W. TSIEN. 1985. Three types of neuronal calcium channel with different calcium agonist sensitivity. Nature **316**(6027): 440-443.
16. FOX, A. P., M. C. NOWYCKY & R. W. TSIEN. 1987. Kinetic and pharmacological properties distinguishing three types of calcium currents in chick sensory neurones. J. Physiol. (London) **394:** 149-172.
17. FOX, A. P., M. C. NOWYCKY & R. W. TSIEN. 1987. Single-channel recordings of three types of calcium channels in chick sensory neurones. J. Physiol. (London) **394:** 173-200.
18. KOSTYUK, P. G., YA. M. SHUBA & A. N. SAVCHENKO. 1987. Three types of calcium channels in the membrane of mouse sensory neurones. Biol. Membr. (Moscow) **4:** 366-373.
19. HAMILL, G. P., A. MARTY, E. NEHER, B. SAKMANN & F. SIGWORTH. 1981. Improved patch clamp technique for high-resolution current recordings from cell and cell-free patches. Pflügers Arch. Ges. Physiol. **391:** 85-100.
20. KANEDA, M., H. NAKAMURA & N. AKAIKE. 1988. Mechanical and enzymatic isolation of mammalian CNS neurons. Neurosci. Res. **5:** 299-315.
21. KOSTYUK, P. G., O. A. KRISHTAL & YU. A. SHAKHOVALOV. 1977. Separation of sodium and calcium currents in the somatic membrane of mollusc neurones. J. Physiol. (London) **270:** 545-568.
22. SHUBA, YA. M. & V. I. TESLENKO. 1987. Kinetic model for activation of single calcium channels in mammalian sensory neurone membrane. Biol. Membr. (Moscow) **4:** 315-329.
23. KOSTYUK, P. G. & O. A. KRISHTAL. 1977. Effects of calcium and calcium-chelating agents on the inward and outward currents in the membrane of mollusc neurones. J. Physiol. (London) **270:** 569-580.

24. KONNERTH, A., H. D. LUX & M. MORAD. 1987. Proton-induced transformation of calcium channel in chick dorsal root ganglion cells. J. Physiol. (London) **386**: 603-633.

25. KRISHTAL, O. A. & V. I. PIDOPLICHKO. 1981. A receptor for protons in the membrane of sensory neurons may participate in nociception. Neuroscience **6**: 2599-2601.

26. KOSTYUK, P. G., S. L. MIRONIV & YA. M. SHUBA. 1983. Two ion-selecting filters in the calcium channel of the somatic membrane of mollusc neurones. J. Membr. Biol. **76**: 83-93.

27. ALMERS, W., E. W. MCCLESKEY & P. T. PALADE. 1986. The mechanism of ion selectivity in calcium channels of skeletal muscle membrane. Fortschr. Zool. **33**: 61-73.

28. HESS, P., J. B. LANSMANN & R. W. TSIEN. 1986. Calcium channels selectivity for divalent and monovalent cations. Voltage and concentration dependence of single channel current in ventricular heart cells. J. Gen. Physiol. **88**: 293-319.

29. KOSTYUK, P. G. & S. L. MIRONOV. 1986. Some predictions concerning the calcium channel model with different conformational states. Gen. Physiol. Biophys. **6**: 649-659.

30. KOSTYUK, P. G. & YA. M. SHUBA. 1982. A study of monovalent cation selectivity of calcium EDTA-modified channels. Neurophysiology (Kiev) **14**: 491-498.

31. KOSTYUK, P. G., O. A. KRISHTAL & V. I. PIDOPLICHKO. 1981. Calcium inward current and related charge movements in the membrane of snail neurones. J. Physiol. (London) **310**: 403-421.

32. FEDULOVA, S. A., P. G. KOSTYUK & N. S. VESELOVSKY. 1981. Calcium channels in the somatic membrane of the rat dorsal root ganglion neurons. Effect of cAMP. Brain Res. **214**: 210-214.

33. DOROSHENKO, P. A., P. G. KOSTYUK & A. E. MARTYNYUK. 1982. Intracellular metabolism of adenosine 3',5'-cyclic monophosphate and calcium inward current in perfused neurones of *Helix pomatia*. Neuroscience **7**: 2125-2134.

34. DOROSHENKO, P. A., P. G. KOSTYUK, A. E. MARTYNYUK, M. D. KURSKY & Z. D. VOROBETZ. 1984. Intracellular protein kinase and calcium inward currents in perfused neurones of the snail *Helix pomatia*. Neuroscience **11**: 263-267.

35. CHAD, J. E. & R. ECKERT. 1986. An enzymatic mechanism for calcium current inactivation in dialysed *Helix* neurones. J. Physiol. (London) **378**: 31-51.

36. ECKERT, R., J. E. CHAD & D. KALMAN. 1986. Enzymatic regulation of calcium current in dialysed and intact molluscan neurons. J. Physiol. (Paris) **81**: 318-324.

37. DUPONT, J.-L., J.-L. BOSSU & A. FELTZ. 1986. Effect of internal calcium concentration on calcium currents in rat sensory neurones. Pflügers Arch. Ges. Physiol. **406**: 433-435.

38. BEAN, B. P. 1984. Nitrendipine block of cardiac calcium channels: High-affinity binding to the inactivated state. Proc. Natl. Acad. Sci. USA **81**: 6388-6392.

39. KONGSAMUT, S., S. B. FREEDMAN & R. J. MILLER 1985. Dihydropyridine-sensitive calcium channels in a smooth muscle cell line. Biochem. Biophys. Res. Commun. **127**: 71-79.

40. COGNARD, C., G. ROMEY, J.-P. GALIZZI, M. FOSSET & M. LAZDUNSKI. 1986. Dihydro-pyridine-sensitive Ca^{2+} channels in mammalian skeletal muscle cells in culture: Electro-physiological properties and interactions with Ca^{2+} channel activator (Bay K8644) and inhibitor (PN 200-110). Proc. Natl. Acad. Sci. USA **83**: 1518-1522.

41. TAKEDA, K., YU. OHYA, K. KITAMURA & H. KURIYAMA. 1986. Actions of flunarizine, a Ca^{2+} antagonist, on ionic currents in fragmented smooth muscle cells of the rabbit small intestine. J. Pharmacol. Exp. Ther. **240**: 978-983.

42. ITOH, T., S. SATOH, T. ISHIMATSU, T. FUJIWARA & YU. KANMURA. 1987. Mechanism of flunarizine-induced vasodilatation in the rabbit mesenteric artery. Circ. Res. **61**: 446-454.

43. KOSTYUK, P. G., S. A. FEDULOVA & N. S. VESELOVSKY. 1986. Changes in ionic mech-anisms of electrical excitability of the somatic membrane of rat dorsal root ganglion neurons during ontogenesis. Distribution of ionic channels of inward current. Neuro-physiology (Kiev) **18**: 813-820.

44. LLINAS, R. & Y. YAROM. 1981. Properties and distribution of ionic conductances generating electro responsiveness of mammalian olivary neurones in vitro. J. Physiol. (London) **315**: 569-584.

45. SCHNEIDER, A. S., H. T. CLINE, K. ROSENHECK & M. SONENBERG. 1981 Stimulus-secretion coupling in isolated adrenal chromaffin cells: Calcium channel activation and possible role of cytoskeletal elements. J. Neurochem. **37**: 567-575.

46. FINDLAY I. & M. J. DUNNE. 1985. Voltage-activated Ca^{2+} currents in insulin-secreting cells. FEBS Lett. **189:** 281-285.
47. SUSZKIW, J. B., M. E. O'LEARY, M. M. MURAWSKY & T. WANG. 1986. Presynaptic calcium channels in rat cortical synaptosomes: Fast-kinetics of phasic calcium influx, channel inactivation, and relationship to nitrendipine receptors. J. Neurosci. **6:** 1349-1357.
48. VASSILEV. P. M., M. P. KANARISKO & H. TIEN. 1987. Ca^{2+} channels from brain microsomal membranes reconstituted in patch-clamped bilayers. Biochim. Biophys. Acta: Biomembr. **897(M146):** 324-330.

Mechanisms of Interaction of Permeant Ions and Protons with Dihydropyridine-Sensitive Calcium Channels[a]

PETER HESS, BLAISE PROD'HOM, AND
DANIELA PIETROBON

Department of Cellular and Molecular Physiology and
Program in Neuroscience
Harvard Medical School
Boston, Massachusetts 02115

INTRODUCTION

Single-channel recordings have made it possible over the last several years to record the interactions between single dihydropyridine-sensitive (L-type) Ca channel molecules and individual permeant and blocking ions.[1-4] These measurements have led to detailed kinetic descriptions of the mechanisms underlying entry and exit of ions into and out of the channel pore. The merging view is that of the open channel as a narrow pore, in which ions cannot cross each other. As ions move through the pore, they reversibly bind to at least two sites. Relative permeabilities and selectivity are obtained by the different affinities of the permeant ions for these binding sites. Ions that bind tightly can lead to simultaneous occupancy of the two binding sites. Under this condition the two ions experience mutual repulsion, which reduces their apparent affinity, speeds up the rate of exit from the pore, and assures that, for instance for Ca and Ba ions, the throughput rate (unitary conductance) remains high despite dissociation constants (K_ds) of ~ 1 μM to the individual binding site (for review see 5). This view of Ca-channel permeation,[4,6,7] governed primarily by interactions of ions with intrapore binding sites and by ion-ion interactions in the multiply occupied pore, contrasts with a previous model by Kostyuk and coworkers,[8] who include a modulatory, external site with a K_d for Ca of ~ 1 μM, which allosterically changes the selectivity of the channel pore. According to Kostyuk et al.,[8] the pore itself does not contain a high-affinity Ca-binding site.

We have recently obtained evidence for an allosteric modulation of the channel conductance by protons interacting with an external site on the L-type Ca channel.[9,10] Here we present some of this data and ask to what extent this mechanism can influence

[a] This work was supported by National Institutes of Health Grant HL37124, fellowships from EMBO and Fogarty (D.P.), and a fellowship from the Muscular Dystrophy Association (B.P.).

80

channel selectivity under physiological conditions of permeant ions. Although our observations are not directly related to the allosteric model proposed by Kostyuk *et al.,*[8] the demonstration of such allosteric modulation of channel conductance makes it necessary to reconsider their model. We test for, and confirm, the presence of a high-affinity Ca-binding site in the channel pore by using the criterion that such a site, unlike an external site, must be accessible both from the outside and from the inside of the channel, in an approach similar to that of Ma and Coronado[4] for t-tubular Ca channels incorporated into lipid bilayers. We conclude that even though allosteric modulation of Ca-channel conductance can be demonstrated under certain ionic conditions, high-affinity intra-pore binding sites and ion-ion interactions in multiply occupied pores form the basis of Ca-channel selectivity under physiological conditions.

METHODS

The methods for obtaining single-cell preparations, recording and analysis of single-channel currents were as previously described.[1,9,10] Cell-attached and inside-out patch recordings[11] were obtained from freshly dissociated guinea pig ventricular myocytes[12] or undifferentiated PC-12 cells. Single L-type Ca channels were identified by their voltage dependence, conductance, and sensitivity to the dihydropyridine Ca-channel agonist (+)-S-202-791 (gift from Dr. Hof, Sandoz Co., Basle, Switzerland), which was present at 0.5-1 μM in all single-channel experiments.

For cell-attached recordings with monovalent charge carriers, the pipette solution contained 150 mM of the Cl-salt of Li, Na, or K, 5 mM EDTA and 5 mM of the pH buffer MES (pK 6.15), HEPES (pK 7.55), or TAPS (pK 8.4). The solutions were titrated to the desired pH with Li$^-$, Na$^-$, or KOH. Thus the final pipette concentration of cation was approximately 160 mM.

In all cell-attached recordings, the bathing solution was the high-K solution used to zero the cell membrane potential outside the patch.[12] It contained (in mM): 140 K-Aspartate, 10 EGTA, 10 Hepes, 20 MgCl$_2$ titrated to pH 7.4 with KOH.

Inside-out patches were excised into a Ca- and Mg-free solution containing 140 Na-aspartate, 10 mM EDTA and 5 mM Hepes, titrated to pH 7.5. The pipette solutions for inside-out patches were similar to those used for cell-attached recordings, except that Cl ions were substituted with aspartate ions.

RESULTS

Protons Induce Unitary Current Fluctuations between Two Conductance Levels in Unitary Currents Carried by Monovalent Ions

In the absence of divalent ions, Ca channels become highly permeable to monovalent ions.[2,7,13–18] Unitary currents carried by Na, K, and Cs through L-type Ca channels are shown in FIGURE 1. The top traces show the typical elementary currents of L-type Ca channels in the presence of the dihydropyridine Ca channel agonist (+)-(S)-202-791[19,20] when Cs$^+$ is the charge carrier. The opening of the channel is followed

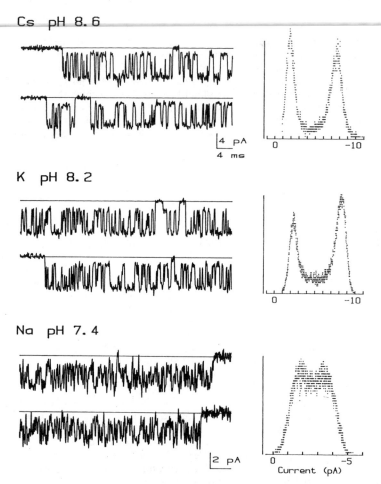

FIGURE 1. Cell-attached patch recordings (left panels) and amplitude histograms (right panels) of single L-type Ca^{2+}-channel currents carried by Cs^+ at pH 8.6 (top), K^+ at pH 8.2 (middle), or Na^+ at pH 7.4 (bottom). Level of closed channel is indicated by solid line in current traces and marked as 0 in amplitude histograms. Current segments of closed channels were not included in amplitude histograms, which therefore represent only current fluctuations between the two conducting levels. L-type Ca channels were activated by depolarizations from a holding potential of −90 to −110 mV and channel activity was recorded either during the activating pulse (at −40 mV with Na^+ as the charge carrier) or as unitary tail currents following repolarization to varying test potentials (−70 mV and −90 mV with K^+ or Cs^+, respectively). In these voltage protocols, which are dictated by the voltage dependence of L-type Ca channel gating in the absence of divalent ions[2,9,12] and the negative reversal potentials when K^+ or Cs^+ are the charge carriers (see FIG. 3), we took advantage of the absence of voltage dependence of the protonation kinetics.[9,21] The bath solution contained 1 μM (+)-(S)-202-791 (see methods for ionic compositions). The currents were digitized at 25 kHz and filtered at 5 kHz (−3dB, 8-pole Bessel filter). Temperature: 20-22°C. (From Pietrobon *et al.*[10] with permission of *Nature*.)

by transitions between two different conducting levels. We have previously concluded that these transitions reflect fluctuations of the channel conductance induced by protons at a single protonation site because the fraction of time the channel spends at the low conductance increases with pH and the mean dwell time at the high conductance level, obtained as the time constant of the exponentially distributed lifetimes, decreases linearly with increasing H^+ concentration.[9,21] In our first report[9] we have assumed that the two conductance states represent the nonprotonated and protonated states of the channel, respectively. From detailed analysis of the pH dependence of the lifetimes of the high and low conductance, we have concluded recently[21] that the channel can be protonated and deprotonated at either conductance, resulting in a set of a minimum of four states, in which the conductance change itself results from a conformational change of the channel protein. However, at the high conductance the protonation equilibrium favors the unprotonated state, whereas at the low conductance the apparent protonation equilibrium is shifted by about three pH units in the alkaline direction.[21] Thus, to a first approximation, the high and low conductances still represent the channel with the protonation site in its unprotonated and protonated state, respectively.

A similar pattern of single-channel currents is seen when K^+ is the charge carrier (FIG. 1, middle traces), but the lifetime of the low conductance state is shorter than with Cs^+. With Na^+ as the charge carrier (FIG. 1 bottom traces), transitions between the two levels are so fast that they can no longer be resolved at the bandwidth of our recording system (5 kHz) and become apparent as large, open-channel noise.[9] The pH_o values for the traces in FIGURE 1 were chosen such that with each permeant ion the open channel spends approximately equal time in the protonated and unprotonated state, as seen most directly from the similar peaks at the two levels in the amplitude histogram for each ion (FIG. 1). Therefore the equilibrium between the two conductances not only depends on pH but also on the ionic condition, and it is clear that the equilibrium is shifted by more than 1 pH unit as Na^+ replaces Cs^+ as the permeant ion.

The Permeant Ion Changes the Equilibrium of the Proton-Induced Fluctuations by Destabilizing the Low-Conductance State

We analyzed the mean dwell times at each conductance level under the three ionic conditions shown in FIGURE 1. For K and Cs, we could measure the durations directly. The distributions of lifetimes were well fit by a single decaying exponential, from which we extracted the time constant by a nonlinear χ^2 minimization, as previously published.[9] In the case of Na as permeant ion, the rate constants were obtained by reconstruction of the original signal, which at our filter setting gave rise to the observed noise.[9] This procedure involved simulation of stochastic first-order transitions between two current levels. The rate constants for the simulations were changed in 5% increments until the resulting traces, when filtered and sampled like real experimental traces and added to background noise, reproduced the observed currents. The accuracy of the simulations was judged by superposition of amplitude histograms from the experimental and the simulated traces. The assumption that the current noise observed with Na as permeant ion is caused by similar but faster fluctuations than those seen in the presence of K and Cs seems justified, because even with Na, discrete transitions between two conductance states could be resolved when the fluctuations where slowed down by substitution of H_2O by D_2O.[9]

A plot of the transition rate constants, defined as

$$k_{on} = (1/\tau_h)/a_H \quad \text{and} \quad k_{off} = 1/\tau_l$$

with each of the three ions (FIG. 2A) shows that the shift of the equilibrium results mainly from an acceleration of k_{off}. The average time the channels spends at the low-conductance level decreases from 480 μsec in the presence of Ca^+ to 60 μsec when Na^+ is the charge carrier.

When only two states (protonated and unprotonated) are considered,[9] k_{on} is the proton association rate constant and k_{off} the deprotonation rate constant. Quantitative analysis within the framework of a four-state model[21] shows that while the protonation rate constant is the rate-limiting forward step for the transition to the low conductance, not every protonation event must be followed by a conductance change, because the deprotonation rate constant at the high conductance is very rapid, such that the proton can leave again before the channel has a chance to switch conductance. Thus for a four-state model the measured values of k_{on} represent a lower limit estimate of the true proton association rate constant. In such a model the termination of a dwell time at the low conductance, and thus k_{off}, is mainly determined by the rate constant for the conformational change underlying the transition from the low to the high conductance and not by the deprotonation rate constant.

An Ion's Effect on k_{off} Is Correlated with the Ion's Relative Affinity for a Permeation Site

Because the L-type Ca channel distinguishes between permeant ions primarily on the basis of the ion's affinity to intrachannel binding sites (see INTRODUCTION), we explored the possibility that the value of k_{off} with each permeant ion was related to the ion's binding affinity for a permeation site. We judged the relative affinity of the channel for each ion by its conductance and zero-current potential. Representative unitary current voltage (IV) relations are plotted in FIGURE 3. Because all recordings shown were obtained from cell-attached patches, only the inward currents are carried by the ion of interest and are shown here. The upper and lower panels show the IVs for the high and low conductance, respectively. The channel conductance increases in the order $Na^+ < K^+ < Cs^+$, but as the conductance increases, the extrapolated zero-current potential shifts toward more negative values. This combination of conductance and zero-current potential is the expected finding for selectivity by affinity[12-22]: the ion that binds most tightly (in this case Na^+) has the lowest mobility (lowest conductance) but the most positive zero current potential, because it competes most effectively for channel occupancy with the internal K^+ present in these cell-attached recordings. The affinity thus increases in the order $Cs^+ < K^+ < Na^+$, the same order in which Cs^+, K^+, and Na^+ destabilize the low-conductance state. The low-conductance channel maintains its selectivity among monovalent cations as shown by the similar sequence of conductances and extrapolated zero current potentials.

The mean values of conductances and zero current potentials for each ion are shown in FIGURE 2B underneath the values of k_{on} and k_{off} to emphasize the correlation between increasing affinity to the permeation path and increasing value of k_{off}.

A higher affinity for a pore site means that at equal concentration of permeant

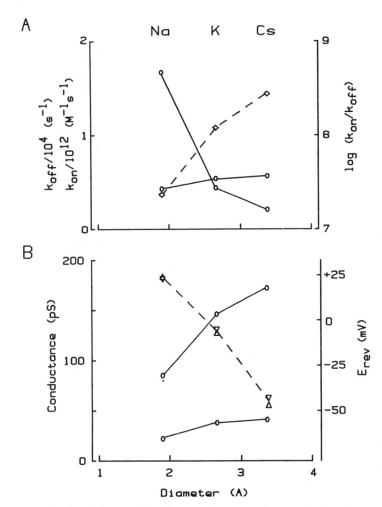

FIGURE 2. Kinetics of the transitions between the two conductance levels and permeation parameters with Na$^+$, K$^+$, or Cs$^+$ as the charge carrier. (**A**) Values of k_{on} (open circles), k_{off} (filled circles), and log k_{on}/k_{off} (diamonds) with 160 mM Na$^+$, K$^+$, or Cs$^+$ in the pipette are plotted as a function of the diameter of the unhydrated ion (see text for definition and meaning k_{on} and k_{off}). (**B**) Unitary conductances (circles) and zero-current potentials (triangles) of the unprotonated (filled symbols) and protonated channel (open symbols).

The values for k_{on}, k_{off}, and pK with Cs and K were obtained at pH 8.2 and with Na$^+$ at pH 7.0-7.4. The conductances for K$^+$ and Cs$^+$ were measured at pH values ranging from 8.4 to 8.6. The conductances for Na$^+$ were obtained at pH 9 for the unprotonated channel and at pH 6 for the protonated channel. The zero-current potentials were corrected for liquid junction potentials. All symbols are means of three to nine determinations. Standard errors of the mean are not shown because they are always smaller than the size of the symbols. (From Pietrobon *et al.*[10] with permission of *Nature.*)

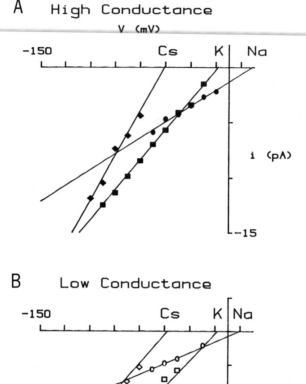

FIGURE 3. Current voltage (IV) relations obtained in cell-attached patches with Na, K, or Cs in the pipette. (**A**) IV of high-conductance level. (**B**) IV of low-conductance level. The conductances for K^+ and Cs^+ were measured at pH values ranging from 8.4 to 8.6. The conductances for Na^+ were obtained at pH 9 for the high conductance and at pH 6 for the low conductance. The IVs are not corrected for liquid junction potentials. The corrected mean values are shown in FIGURE 2B.

ion the channel should be occupied a greater fraction of time in the presence of the more tightly binding ion. Thus our results suggest that channel occupancy may influence k_{off}. Another way to change channel occupancy is to change the concentration of permeant ion. As expected, we found, that k_{off} increases by ~60% when the K^+ concentration is increased from 60 to 160 mM.[10]

A Ca Ion in the Pore Destabilizes the Low-Conductance State

Ca^{2+} and Ba^{2+} ions bind much more strongly to the intrapore sites than monovalent ions.[4,6,7] At micromolar external Ca^{2+} concentrations, the dwell time of an individual Ca^{2+} ion in the channel is long enough to block flow of monovalent ions for a measurable time.[1] The resulting elementary current traces at two Ca^{2+} concentrations are shown in FIGURE 4A and compared with a control trace with K^+ in the absence of Ca^{2+}. As expected, in the presence of Ca^{2+} the long channel openings are now interrupted by discrete transitions to the fully closed level, which we interpret as individual sojourns of a Ca^{2+} ion in the channel pore.[1] As long as the Ca^{2+} ion is bound in the pore, we cannot judge its effect on the protonation kinetics. However our hypothesis that occupancy of the binding site destabilizes the low-conductance state predicts that immediately following departure of the Ca^{2+} ion, the channel should be in its high-conductance state. Inspection of the traces with Ca^{2+} in FIGURE 4A shows that indeed almost all Ca^{2+}-unblocking transitions occur from the zero-current level to the high-conductance state. This is confirmed in a more rigorous way in FIGURE 4B, where we display the time course of the mean current just preceding entry of a Ca^{2+} ion and that just following Ca^{2+} exit, obtained by lining up and averaging several hundred individual blocking and unblocking events. The current following the departure of the Ca^{2+} ion reaches an initial peak from which it relaxes back to the steady-state value halfway between the two current levels, which reflects the fact that in the absence of Ca^{2+} at pH 8.2 the open channel spends approximately equal time at the two conductance levels (FIG. 1, FIG. 4A, top trace). This relaxation shows unequivocally that binding of a Ca^{2+} ion to a site within the conduction pathway has biased the equilibrium toward the high-conductance state. The site to which Ca^{2+} binds must be located within the pore because (1) the presence of a Ca^{2+} ion completely blocks current flow and (2) the departure rate constant is speeded up by hyperpolarization, indicating that the exit occurs toward the inside of the channel (Hess *et al.*[12] and unpublished data). Thus the experiment in FIGURE 4 directly confirms the hypothesis that it is the occupation of an intrapore binding site that increases k_{off} and makes it very likely that the effects of the monovalent cations on k_{off} are also mediated via an intrachannel site.

In contrast to the initial relaxation of the averaged current following Ca^{2+} exit from the pore, little deviation from the steady-state level is seen in the averaged current preceding a Ca^{2+}-blocking event (FIG. 4b). This means that Ca^{2+} entry occurs with roughly equal rate into the high-conductance (unprotonated) and low-conductance (protonated) channel. We conclude that the protonation site is far enough from the permeation pathway that an entering ion does not experience the different local potential associated with the protonated or unprotonated site. Thus, the different conductances of the protonated and unprotonated channel cannot be explained by a protonation-dependent surface potential at the channel entrance that changes the local concentration of permeant ions.[9] A direct competition between protons and permeant cations for the same site can be excluded as well, because in this case a Ca^{2+} ion would only be able to enter the unprotonated site. Further arguments against competition of the two ions for a common site include the finding that protons do not block the permeation pathway completely and that the effects of permeant ions on the channel lifetimes are mainly on k_{off}.

We conclude that protons and permeant cations interact allosterically at different sites. Occupancy of the cation-permeation site changes the protein conformation in a way that destabilizes the proton-induced low conductance and accelerates deprotonation.[10] In turn, protonation triggers a conformational change that reduces the channel conductance for monovalent ions with little change in selectivity.[10]

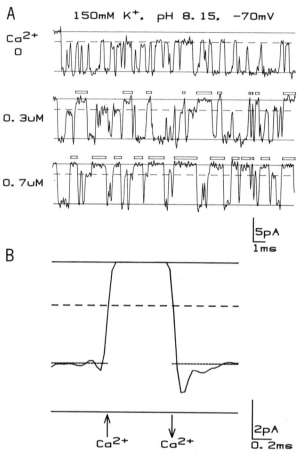

FIGURE 4. Binding of a Ca^{2+} ion to an intrachannel site perturbs the protonation equilibrium. (**A**) Current traces from three different patches obtained with 160 mM K$^+$ in the pipette at pH 8.2 and 0 (top trace), 0.3 μM (middle), and 0.7 μM Ca^{2+}. The pipette Ca^{2+} was buffered to the desired level with HEDTA ($K_{D,Ca}$ = 0.33 μM at pH 8.2). The horizontal lines indicate the zero-current level and the two conductance levels of the protonated and unprotonated channel. In the presence of Ca^{2+} the records are interrupted by frequent transitions to the zero-current level. These sojourns of individual Ca^{2+} ions in the pore are indicated above the current records by horizontal bars. Patch potential -70mV. Records filtered at 5 kHz and digitized at 25 kHz. (**B**) Mean current just preceding the entry (left part) and just following the exit of a Ca^{2+} ion. Idealization of individual blocking and unblocking events was used to determine the time of alignment of the original events for the construction of the mean currents, which represent averages of several hundred events. The continuous horizontal lines indicate the three current levels of the unitary recordings. The interrupted dotted line indicates the level of average (steady state) proton block, which as expected at pH 8.2 and 160 mM K$^+$ is about 50% (see Figs. 1,2). The transient of the average current following the Ca^{2+}-unblocking event is attenuated because of filtering and poor temporal synchronization of the alignment of events due to noise. When simulated events with obligatory first transitions to the fully open level were filtered, added to experimental background noise, and analyzed with the same programs used to obtain the averages from the real traces, the resulting transient following the simulated unblocking event was very similar to that shown in Figure 4b and also fell short of reaching the fully open level. (From Pietrobon *et at.*[10] with permission of *Nature.*)

The large differences in the affinity for protons in the high- and low-conductance conformation suggests short-range electrostatic interactions between the protonated group and its local environment (see, e.g., Warshel[23]). These electrostatic interactions appear to be very sensitive to changes in channel conformation, thus allowing us to probe the conformational changes associated with the coordination of permeant ions within the channel.

In an effort to assess the physiological relevance of the proton-induced conductance fluctuations, we measured whole-cell L-type Ca currents between pH 7.5 and 6.5 with physiological concentrations of Ca and Na (data not shown). We found that although the peak currents were decreased at pH 6.5 in the negative range of test potentials, this could be fully accounted for by the well-documented shift of activation toward more positive potentials, as recently shown by Krafte and Kass.[24] The macroscopic conductance on the descending limb of the IV curve remained unchanged. Thus, even millimolar concentrations of Ca seem high enough to fully prevent the proton-induced conductance changes between pH 6.5 and 7.5, in line with our conclusion that Ca, which at these concentrations occupies the channel continuously at at least one site,[6,7] very effectively biases the equilibrium between the two conductances toward the high-conductance conformation.

Monovalent Unitary Currents in Excised Patches

The demonstration of an allosteric effect of an external protonation site on channel conductance made us reinvestigate the possibility that the high-affinity binding site for Ca, which creates selectivity for divalent over monovalent ions (see INTRODUCTION), is not an intrapore site, but also exerts its effect allosterically from an external location, as originally proposed by Kostyuk *et al.*[8] The missing piece of evidence so far is a direct demonstration that the high-affinity site can also be accessed from the cytoplasmic side of the channel, and that the voltage dependence of the rates of binding and unbinding is as expected for a site located in the pore itself. Steady-state block of Na-carrying Ca channels by internal Ca has been demonstrated in t-tubular Ca channels from skeletal muscle,[4] although half-maximal inhibition required a significantly higher Ca concentration (35 mM at 0 mV, see Ma & Coronado[4]) than that necessary to block cardiac L-type Ca channels from the external side.[2,6,7]

To gain access to the cytoplasmic side of the patch, we measured single L-type Ca-channel activity in inside-out patches. In confirmation of most previous studies,[25–29] we found that L-type Ca channels remain active in excised patches for only very brief periods of time, ranging from seconds to a few minutes at best. In order to observe outward currents through L-type Ca channels, we had to eliminate all internal divalent ions with the use of EDTA. We used Cl-free solutions to exclude the possibility of overlapping outward Cl-currents.

FIGURE 5 shows samples of unitary currents and the corresponding IV with symmetrical Na-Aspartate. The pipette pH was 8.5 in order to eliminate as much as possible the pH-induced fluctuations. The IV is linear over the entire range and the conductance is 90 pS, very similar to that of inward Na currents recorded from cell-attached patches (see FIGURE 3.). While we observed no rectification under conditions of symmetrical [Na currents carried by Cs showed marked rectification. With symmetrical [Cs], the conductance of the outward currents was significantly lower than that of the inward currents (unpublished data). Thus unlike Ba- (see Rosenberg *et al.*[30]) and Na-permeation, Cs-permeation appears to indicate functional asymmetry of the permeation pathway.

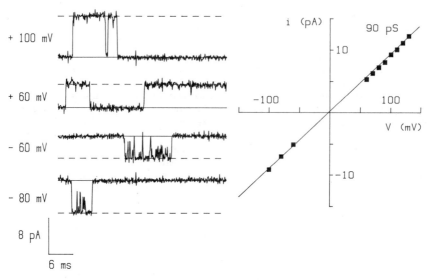

FIGURE 5. IV relation with symmetrical Na. Recordings from inside-out patch. Pipette and bath solutions contain 150 mM Na-Aspartate and 5 mM EDTA. Pipette pH 8.5, buffered with TAPS, bath pH 7.4, buffered with HEPES. Bath contains 1 μM (+)-(S)-202-791. Filtered at 5 kHz, sampled at 25 kHz.

A High-Affinity Ca-Binding Site in the Pore Can Be Reached from Both Sides

FIGURE 6 shows the result of the test for access to a high-affinity Ca-binding site from both sides. The three traces are outward currents at +50 mV in symmetrical [Na], in the absence of Ca on either side (top trace) or with 2 μM Ca either in the pipette (middle) or on the cytoplasmic side (bottom). In the absence of Ca a long opening, interrupted only by a few brief closings, is seen. With 2 μM Ca on either side, the opening is interrupted by many transitions to the zero current level. These records show the expected current pattern for unitary blocking events caused by individual Ca ions interacting with a high-affinity binding site (see Lansman *et al.*[1] and FIG. 4). Even though the limited duration of activity of L-type Ca channels in excised patches has so far prevented us from a full kinetic investigation of the blocking transitions, we feel that this result strongly suggests an intra-pore location for at least one high-affinity Ca-binding site, accessible from both sides of the channel. The block by internal Ca is voltage dependent in a manner consistent with an intrachannel site, in that it is relieved by further depolarization, apparently mainly by a shortening of the blocked times (unpublished data). However, in contrast to the block from the outside, which is characterized by an essentially voltage-independent blocking rate constant,[1] the blocking rate constant from the inside is also voltage dependent, since very few blocking events are seen on inward currents (unpublished data). Thus, while the high-affinity site(s) can be reached from both sides, the different voltage dependence of its access implies asymmetry in the channel structure.

DISCUSSION

The proton-induced current fluctuations and the sensitivity of their kinetics to the species and concentration of permeant ion provide new insight into a complex interaction between an externally located protonation site and intrapore binding sites in the L-type Ca channels. The results presented here, and quantitative analysis of the kinetics of the transitions between the conductance levels and their dependence on pH, ionic strength, and concentration of permeant ion have led us to conclude that the protonation site must be represented as a minimum of four states, with an unprotonated and protonated state at each conductance.[21] The conductance change represents an intrinsic conformational change of the channel protein, but its occurrence is greatly enhanced once the channel is protonated. Channel occupancy favors the conformation of high-conductance and tightly binding ions like Ca may completely prevent the transition to the low conductance. It is as if the physiological ion forms a "complex" with the channel that enhances its stability and prevents it from assuming conformations that might impair its normal function. Thus proton-induced current fluctuations are not observed with physiological concentrations of Ca and therefore do not contribute to the modulation of Ca currents by external protons in the physiological pH range.

The importance of the conformational changes induced by permeant ions, as demonstrated in our case by the destabilization of the low conductance (mainly

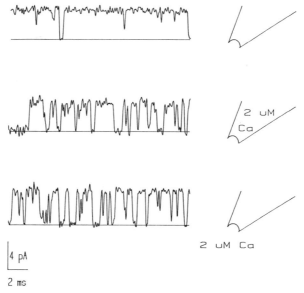

FIGURE 6. Outward currents recorded from inside-out patch in the presence and absence of Ca on either side of the membrane. Test potential +50 mV, symmetrical Na-aspartate (150 mM, pH 7.4). Top trace: [Ca] < 1 nM on both sides of the patch, conditions as in FIG. 5. Middle trace: [Ca] = 2 μM in pipette, buffered with HEDTA (see legend to FIG. 4). Bottom trace: [Ca] = 2 μM in bath, buffered with HEDTA. Bath contains 1 μM (+)-(S)-202-791. Filtered at 5 kHz, sampled at 25 kHz.

protonated) state, cannot be assessed conclusively. It remains possible that specific ion-channel interactions create unfavorable conditions for the passage of a competing ion, thus contributing to channel selectivity.

The best documented mechanism for channel selectivity, however, remains the one involving differential affinities of permeant ions to intra-channel binding sites (see INTRODUCTION). Our results strongly support this conclusion by showing that at least one high-affinity Ca-binding site can be reached from both sides of the channel, and its access from the cytoplasmic side is voltage dependent. This places it within the electric field of the channel pore and makes assumptions about other, externally located high-affinity Ca-binding sites with possible allosteric effects on L-type Ca channels unnecessary.

ACKNOWLEDGMENTS

Part of the data in this paper are reproduced from a recent publication[10] with the permission of Macmillan Journals. We thank Dr. Mark Plummer for help with tissue culturing.

REFERENCES

1. LANSMAN, J. B., P. HESS & R. W. TSIEN. 1986. J. Gen. Physiol. **88:** 321-347.
2. MATSUDA, H. 1986. Pfluegers Arch. **407:** 465-475.
3. NELSON, M. T. 1986. J. Gen. Physiol. **87:** 201-222.
4. MA, J. & R. CORONADO. 1988. Biophys. J. **53:** 556a.
5. TSIEN, R. W., P. HESS, E. W. MCCLESKEY & R. L. ROSENBERG. 1987. Ann. Rev. Biophys. Biophys, Chem. **16:** 265-290.
6. ALMERS, W. & E. W. MCCLESKEY. 1984. J. Physiol. (London) **353:** 585-608.
7. HESS, P. & R. W. TSIEN. 1984. Nature **309:** 453-456.
8. KOSTYUK, P. G., S. L. MIRONOV & Y. M. SHUBA. 1983. J. Membr. Biol. **76:** 83-93.
9. PROD'HOM, B., D. PIETROBON & P. HESS. 1987. Nature **329:** 243-246.
10. PIETROBON, D., B. PROD'HOM & P. HESS. 1988. Nature **333:** 373-376.
11. HAMILL, O., A. MARTY, E. NEHER, B. SAKMANN & F. J. SIGWORTH. 1981. Pflugers Arch. **391:** 85-100.
12. HESS, P., J. B. LANSMAN & R. W. TSIEN. 1986. J. Gen. Physiol. **88:** 293-319.
13. GARNIER, D., O. ROUGIER, Y. M. GARGOUIL, Y. M. & E. CORABOEUF. 1969. Plfuegers Arch. **313:** 321-342.
14. KOSTYUK, P. G., S. L. MIRONOV & P. A. DOROSHENKO. 1982. J. Membr. Biol. **70:** 181-189.
15. ALMERS, W., E. W. MCCLESKEY & P. T. PALADE. 1984. J. Physiol. (London) **353:** 565-583.
16. FUKUSHIMA, Y. & S. HAGIWARA. 1985. J. Physiol. (London) **358:** 585-608.
17. LEVI, R. & L. J. DEFELICE. 1986. Biophys. J. **50:** 5-9.
18. CORONADO, R. & H. AFFOLTER. 1986. J. Gen. Physiol. **87:** 933-953.
19. HOF, R. P., U. T. RUEGG, A. HOF & A. VOGEL. 1985. J. Cardiovasc. Pharmacol. **7:** 689-693.
20. KOKUBUN, S., B. PROD'HOM, C. BECKER, H. PORZIG & H. REUTER. 1986. Mol. Pharmacol. **30:** 571-584.
21. PROD'HOM, B., D. PIETROBON & P. HESS. Submitted for publication.
22. CORONADO, R. & J. S. SMITH. 1987. Biophys. J. **51:** 497-502.
23. WARSHEL, A. 1987 Nature **330:** 15-16.

24. KRAFTE, D. S. & R. S. KASS. J. Gen. Physiol. **91:** 641-657.
25. FENWICK, E. M., A. MARTY & E. NEHER. 1982. J. Physiol. (London) **331:** 599-635.
26. CAVALIE, A., R. OCHI, D. PELZER & W. TRAUTWEIN. 1981. Pfluegers Arch. **398:** 284-297.
27. NILIUS, B., P. HESS, J. B. LANSMAN & R. W. TSIEN. 1985. Nature **316:** 443-446.
28. ARMSTRONG, D. & R. ECKERT. 1987. Proc. Natl. Acad. Sci. USA **84:** 2518-2522.
29. YATANI, A., J. CODINA, Y. IMOTO, J. P. REEVES, L. BIRNBAUMER & A. M. BROWN. 1987. Science **238:** 1288-1292.
30. ROSENBERG, R. L., P. HESS, J. P. REEVES, H. SMILOVITZ & R. W. TSIEN. 1986. Science **231:** 1564-1566.

Block of Na⁺ Ion Permeation and Selectivity of Ca Channels*a*

H. D. LUX,[b] E. CARBONE,[c] AND H. ZUCKER [b]

[b] *Abteilung Neurophysiologie*
Max-Planck-Institut für Psychiatrie
D-8033 Planegg, Federal Republic of Germany

[c] *Dipartimento di Anatomia e Fisiologia Umana*
Corso Raffaello 30
I-10125 Torino, Italy

INTRODUCTION

In typical extracellular media, Ca channels select for their duty Ca ions against a majority of monovalent cations such as Na. Indeed, the channels become permeable to Na and other alkaline ions only if the external level of free Ca is below micromolar values (Refs. 1-7, for review see also Hagiwara & Byerly[8] and Tsien *et al.*[9]). Current views of the mechanisms[5,6,10] by which Ca ions interfere with the flux of alkaline ions through Ca channels employ specifically strong binding of Ca ions to channel sites. Such selection by affinity involves differences in energy minima represented by well depths for different ions. Strong attraction and thus pausing at one site makes the channel unavailable for other, even more mobile, ions in single-file passage. But this view needs to be reconciled with observed significant Ca fluxes (on the order of 10^6 ions per second) through the open channel. Alternatively, undesirable ions could be rejected at a molecular sieve which is represented by heights of energy barriers. Energetically unfavorable conditions can prevent one kind of ion (Na) to enter the channel while others could pass. However, Na currents through single Ca channels (in the absence of external Ca) are even larger in amplitude than Ca currents. Thus a sieve, if it applies, necessitates the presence of external Ca ions.

Ion selection by minima or maxima of energy conditions evidently present extreme points of view. Although combined schemes are possible, they would complicate a discussion of the implications for selective ion passage. Pure selection by affinity would demand a mechanism by which the binding is removed to enable the specific passage of blocking ion. This occurs at external divalent ion concentrations beyond those that are observed to block monovalent ion passage with apparent K_Ds in the millimolar range. An attractive assumption is that the channel in this range of $[Ca^{2+}]_o$ becomes more frequently simultaneously occupied at multiple sites.[5,6] Repelling forces between bound ions could then offset binding to increase permeation. Repulsion between two

a This work was partially supported by NATO (grant No. 0576/87) and by LIRCA Synthélabo S.p.A.

divilent ions should indeed be particularly strong as compared with situations of monovalent or mixed occupation of the sites.

Two independent binding sites, one representing the K_D of block by Ca, the other that of Ca flux, were previously proposed.[4,10] The postulated external site of high affinity for Ca is first conceived to screen ions. Thus, facilitated ion passage at higher Ca concentrations would imply some ion-channel interaction at this "regulatory" site.[10] Ca-binding triggers conformational changes that reduce the channel's Na permeability. Some predictions of the so far proposed models will be discussed.

PERMEABILITY SEQUENCES AND CONSEQUENCES OF ION BINDING

If ions move independently of each other through the channel, the relative magnitudes of the fluxes of different ions are directly measured by the reversal potentials (E_{rev}). Permeability sequences under assumption of independence differ greatly from those derived from unitary conductance measurements of the same channels (Hess *et al.*,[11] see also Tsien *et al.*[9]) as shown in TABLE 1. This is already true if ions of the

TABLE 1. Selectivity of Two Types of Ca Channels for Alkaline Metal and Alkaline Earth Cations[a]

Crystal Diameter (Å)	Ion	Cardiac L-Ca Channel[9,17]		Chick DRG lva-Ca Channel	
		P_{ion}/P_{Cs}	Unitary Conductance (pS)	Unitary Conductance (pS)	P_{ion}/P_{Cs}
1.30	Mg^{2+}	~0	–	–	~0
1.98	Ca^{2+}	4,200	9	5.5	400
2.26	Sr^{2+}	2,800	9	6	200
2.70	Ba^{2+}	1,700	25	5	300
1.20	Li^+	9.9	45	22	7.5
1.90	Na^+	3.6	85	24	8
2.66	K^+	1.4	–	–	1.5
3.38	Cs^+	1.0	–	4	1.0

[a] Apparent permeability ratios from reversal potential measurements using Cs ions as internal reference were determined as described in Tsien *et al.*[9]

same charge are compared, particularly in the case of the cardiac L-type Ca channel. With the exception of Mg^{2+} (and Sr^{2+} for the T-type channel), the permeability of ions of the same charge tends to decrease with increasing ion size. Contrary to expectations from E_{rev} measurements, the unitary conductances of Na and other monovalent ions are larger than those of divalent cations (see FIG. 1). Because flux rates behave inversely to measured permeabilities as derived from E_{rev}, the assumption of independent ion movement is not tenable. This property is, however, expected for specific ion-binding or, generally, for ion-channel interaction during ion passage. Ions that interact more strongly will generally have the lowest mobility, which results in a block for the other ions in single-file passage.

Because the conductance (mobility) data on monovalent cations were received in

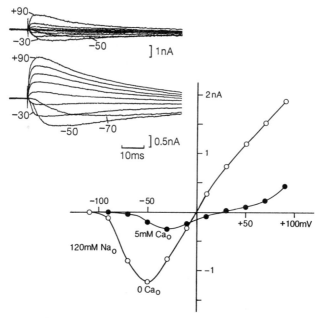

FIGURE 1. Ca and Na currents through the lva channel of chick dorsal root ganglionic (DRG) cells. Current-voltage relationship in the presence (●) and the absence (○) of external Ca^{2+}. Holding potential was -110 mV. Note the reduced Na outward currents and the shifted reversal potential with 5 mM [Ca^{2+}]$_o$. Inset, upper traces: Currents with 5 mM [Ca^{2+}]$_o$ successively recorded from -50 to $+90$ mV with voltage increments of 20 mV. Some membrane potentials indicated when superimposed currents overlap. Below: Na currents of same cell in the presence of 10 mM external EGTA with nominally zero [Ca^{2+}]$_o$. External [Na^+] as well as internal [Na^+] (perfusing patch pipette) was 120 mM in both situations. ω-Conotoxin (5 μM) was added to all external solutions to dissect lva Ca currents, see Carone & Lux, this volume.

the absence of blocking divalent cations (but see below), it is not possible to conclude that the actual occupation of a site by the more tightly binding earthalkaline ions excludes undesirable ions such as Na.

VOLTAGE-DEPENDENT BLOCK

To reach or leave its blocking site within the channel, the ion experiences the electric field applied across the pore. Because a permeant ion such as Ca can exit the pore in either direction, the block exerted by Ca on Na passage should be relieved by either hyperpolarization or depolarization of the membrane. For an impermanent ion such as Mg^{2+} (for review see Tsien et al.[9]), the block will be relieved if the inside of the cell is made more positive (FIG. 2). The results are consistent with the assumption that in both cases a single blocking site is located deep within the pore at an electric distance of about 0.5 from the external mouth of the channel. This applies to the neuronal lva[12] or T-type[13,14] Ca channel of FIGURE 2 in line with previous

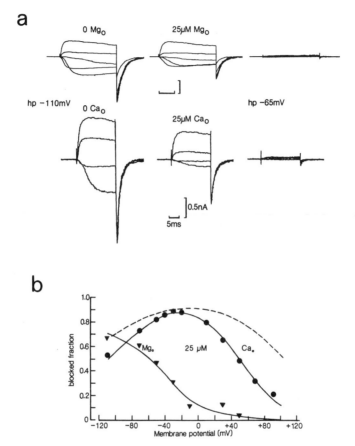

FIGURE 2. Blocking of Na currents through lva channels by divalent ions. (**a**) Na currents through lva channels in the presence (middle) of 25 μM free $[Mg^{2+}]_o$ (upper panel) and 25 μM $[Ca^{2+}]_o$ (lower panel, different cell) during successively applied voltage pulses, for Mg: to -70, -50, -30, $+10$, $+50$ mV, for Ca: to -50, -10, $+50$, $+90$ mV membrane potential. Controls in the absence of external divalent ions (left). Note the predominant effect of Mg_o on the current trace at -70 mV and on tail currents at -110 mV. Ca primarily affects Na currents during intermediate depolarizing steps. Currents (at 0 $[Ca^{2+}]_o$) are inactivated (right) at lowered holding potential (V_h -65 mV). (**b**) Voltage dependence of the block of Na currents through lva Ca channels by $[Ca^{2+}]_o$ and $[Mg^{2+}]_o$. Block by 25 μM $[Ca^{2+}]_o$ (●) and Mg^{2+}_o (▼). The block is displayed (full block = 1) by normalizing the blocked peak currents to full currents in the absence of Ca and Mg, see FIGURE 2a. Dashed line displays voltage dependence of block at 25 μM $[Ca^{2+}]_o$ of a two-binding site model with parameters as given in Almers and McCleskey[5] and an additionally assumed repulsion of an entering ion by an occupied site. This repulsion force was calculated for a distance between entering and bound ion twice as long as that between the two sites.

results on a similar Ca channel type of mouse neoplastic β-lymphocytes.[7] A single binding site model was chosen because a significant separation[5,6] of two (or more) binding sites produces a weaker voltage dependency of the block in either direction than that observed. This is because the closer the ion is to either mouth of the channel, the weaker is the force exerted on the ion by the electric field across the membrane. A symmetrical barrier model is strongly suggested by data on unitary Ba conductance (with external application of the blocking ion) in the L-type channel also predicts a maxium block nearer to zero voltage than observed on the lva Ca channel.

In the case of the lva channel, the two binding site model implies that the sites are close neighbors at probably not more than a few nm (see also Hess & Tsien[6]). This in turn would favor ion-ion repulsion and thus the transition from blocking behavior to relief of the block. If only exit rates would display repulsive forces, the voltage dependence of the block could become almost as strong as in the case of a single central binding site. However, introducing the reasonable assumption that an entering ion also experiences these forces makes the channel block by Ca entry less efficient, because there is a considerable probability that the channel is already occupied by at least one ion (see Hess & Tsien[6]). This considerably reduces the voltage sensitivity of the blocking process (see dashed curve in FIG. 2b). Only at positive membrane potential (beyond $+50$ mV) with external blocker application, the block would become more strongly voltage-dependent under this otherwise reasonable assumption. It should be noted that repulsion-dependent entry rates as proposed above for a channel with two binding sites strongly reduce the probability of Ca-permeable states of the channel. This effect could be compensated for by lowering the entry barriers, which leads to energy levels considerably below the diffusion-limited values.

At strongly positive membrane potentials ($+80$ mv), external Ca concentrations even beyond (up to about 10 times) the maximum half-blocking value for Na and Li currents (near 2 μM at -30 mV for the lva Ca channel) had only weak effects on the outward Na current measured with symmetric $[Na^+]$ (see FIG. 2b). If a regulatory site, responsible for the exclusion of Na ions, was present at a site near the external mouth[10] apart from a second internally located permeation site of lower affinity, the voltage dependence of the block would mostly disappear. At best, with reduced affinity of the regulatory site, the blocking/voltage curve would approach constant values at extreme positive potentials. Lva Ca channels give an indication of such behavior. It thus appears that the process of ion selection essentially takes place inside this Ca channel. The voltage-dependent block on Na currents by Mg is also in line with an internally located regulatory site and shows that actual permeation of the blocking ion through the channel is not a prerequisite.

It appears reasonable to assume that the maximum of the block is due to the longest average sojourn of the blocking ions inside the channel. This assumption of a potential-dependent balance between rates of entry (from the external side) and exits (inward or outward) was used to describe the potential dependence of the block in FIGURE 2b.

At any given membrane potential, the block of Na currents through this Ca channel by $[Ca^{2+}]_o$ appeared to be approximated by a 1 : 1 stoichiometry (see also Fukushima & Hagiwara[7]) in line with the assumption of binding of a single divalent ion to a specific blocking site inside the channel. Ca ions are, in general, more powerful than Mg ions in blocking Na (or Li) currents through the lva Ca channel, but a quantitative comparison can only be made for given membrane potentials. At potentials with maximum inward currents (-55 to -40 mV), the blocking efficacy of Ca/Mg is about 5. An important finding was the insensitivity of the Na currents to internally applied Ca and Mg concentrations.[15] To excert a Na current block in this way, about 100-fold stronger Ca concentrations than those effective at the outside were found necessary. Millimolar $[Mg^{2+}]_i$ was similarly ineffective. To account for this behavior, asymmetric barriers with higher energy levels on the inside have to be assumed.

OBSERVATIONS ON SINGLE-CHANNEL ACTIVITIES IN LVA CA CHANNELS

At $[Ca^{2+}]$ below 0.1 μM, the single-channel Na⁺ currents (FIG. 3) lasted on average 0.26 ± 0.06 msec, a far shorter open time than that of unitary Ca^{2+} currents.[12,15] Openings occurred frequently in bursts with intermediate closures of similar mean duration. The average amplitude of the single-channel events during depolarizations to −40 mV was 1.85 ± 0.35 pA, which was about five times larger than that

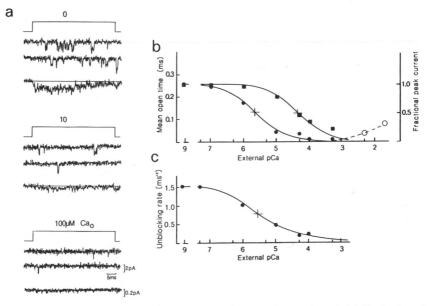

FIGURE 3. Ca^{2+} blockade of Na⁺ currents through a single lva Ca channel. (a) Single-channel Na⁺ currents in an outside-out patch of chick DRG recorded during successive depolarizations to −40 mV in the presence of nominally zero (upper panel), 1×10^{-5} (middle), and 1×10^{-4} M $[Ca^{2+}]_o$ (lower panel). Bottom traces are sample averages of the currents recorded during 25 (0 $[Ca^{2+}]_o$) to 73 (100 μM $[Ca^{2+}]_o$) trials. Note the increase in silent periods already at 10 μM $[Ca^{2+}]_o$ with little effect on open durations. Short-lasting openings are rarely observed with 100 μM $[Ca^{2+}]_o$, and traces without activity became particularly frequent with higher $[Ca^{2+}]_o$. Holding potential was −110 mV in all cases. Cut-off frequency was 4 kHz. (b) Mean values (4 to 7 experiments each) of normalized currents (●) and open times (■) as a function of −log $[Ca^{2+}]_o$. Data at 0.5 mM are means from two experiments with sufficient collections of openings. Half value of block (cross) is at 2 μM and half value of mean open time is at 45 μM $[Ca^{2+}]_o$ with assumed 1:1 stoichiometry (solid lines). Relief of block beyond 500 μM $[Ca^{2+}]_o$ is due to increasing Ca currents (dashed line), analogous to anomalous mole fraction behavior. (c) Mean values of unblocking rate calculated from measured closed times of experiments similar to those in (a).

recorded with 20 mM [Ca^{2+}]$_o$, at corresponding potentials. Generally, these depolarizations produced sufficiently large currents as seen in sample averages or whole-cell recordings (see FIG. 1).

Increasing [Ca^{2+}]$_o$ from nanomolar to 10 micromolar values strongly reduced the current amplitude of averaged samples. However, the amplitude of single-channel currents remained unchanged (see FIG. 3a). The main effect on unitary currents was a decrease of mean open times with their values halved at around 50 μM [Ca^{2+}]$_o$ (FIG. 3b). This concentration is about 20 times larger than that required (2 μM) to produce a half block of the averaged currents of samples and of whole cells (FIG. 3b) at this membrane potential. The primary reason for the reduction of sample averages of unitary currents with increased [Ca^{2+}]$_o$ is an increase of the duration of closures (FIG. 3c, see also Refs. 13,16) such that the reduction of averaged Na currents is reflected by a decreased number of openings.

Ba ions were found to act like Ca ions at similar concentrations, but Mg was less effective at these membrane potentials in accordance with the results on whole cell currents. Membrane patches subjected to high internal free Ca^{2+} (100 to 500 μM) behaved like patches dialyzed with low free [Ca^{2+}] ($< 10^{-9}$ M) showing the same changes in the distributions of closed and open states (see also Carbone & Lux[15]). In line with results on whole-cell currents (see above), internal Ca ions are thus rather ineffective in blocking Na currents through single lva Ca channels.

The half-reduced average open time at a [Ca^{2+}]$_o$ near 50 μM can be taken as a measure of the average time interval by which a blocking Ca ion would enter the channel. At sufficiently negative membrane potentials, the number of entering ions would roughly correspond to that of transported Na ions. The observation that the ratio of [Ca]$_o$ and [Na]$_o$ compares with that of blocking (Ca) and transported (Na) ions could indicate similar entry conditions for the ion species at the voltage used (-40 mV) and supports the view of a single ion block. This conclusion (which applies to unsaturated sites) can also be drawn from data on voltage-dependent block of Li currents through unitary L-type Ca channels[17] at micromolar [Ca]$_o$.

The asymptotic value of the mean open time in lva channels (0.26 msec at 10^{-9} M [Ca^{2+}]$_o$) and the Ca^{2+} concentration at the half open time leads to an estimate of 7×10^7 M^{-1} sec^{-1} for the rate of Ca^{2+} entry. A minimum for the rate by which Ca^{2+} leaves the binding site of about 3×10^3 sec^{-1} can be deduced from the fast time constant of the closed states in low and micromolar [Ca^{2+}]$_o$. Although this results in a somewhat higher apparent dissociation constant for Ca^{2+} binding than estimated from the K_D of the block of whole cell currents, the value is too small to account for the unitary Ca currents under this condition. They would necessitate an exit rate of about 6×10^6 sec^{-1} ions.

Increased blocking rates with [Ca]$_o$, that is, a decrease of the mean open time and rather [Ca]$_o$-insensitive unblocking rates (mean closed times) were observed in the L-type Ca channel.[17] This behavior contrasts with the results on lva Ca channels. Part of this divergence is due to the long-lasting open states of the dihydropyridine-treated L-type Ca channel, which put more weight on closing reactions in determining conductive states. Near constancy of mean closed times also agrees with a two-binding-site model as proposed. This follows from the fact that the channel once blocked by a Ca ion stays blocked in a Ca-independent manner. Ca-dependent exit becomes significant only at [Ca]$_o$ much beyond micromolar concentrations when Ca-Ca occupation relieves the block by repulsion.

A particular mechanism to account for an increase of mean closed times with increased [Ca]$_o$ such as in the lva channels is unavailable. Ca-dependent increases in mean closed times are achieved, however, by a model where an entering Ca ion not only blocks but makes the channel Ca-selective. Ca-induced conformational change, see also Läuger *et al.*[18] and Läuger,[19] would overlast the existence of the blocking ion, providing time for an eventual subsequent Ca entry. Thus an increase in [Ca]$_o$ would increase the probability of finding the channel in a Ca-selective conformation that appears as a block of the Na current.

CONCLUSIONS

Ionic selectivity of the Ca channel is better accounted for by models that incorporate specific binding properties of the channel for the ion to be selected. Ca selectivity through ion sieving by a rigid energy barrier appears unlikely for reasons stemming from the physical properties of earthalkaline ions as compared with those of Na ion and other monovalent ions (see Tsien *et al.*[9]). It is also inconsistent with a significant passage of monovalent ions in the absence of the preferred divalent ions.

Multiple binding site models and especially those that allow for transport facilitation by ion-ion interaction[5,6] appear capable of describing the basic features of ion selectivity and specific transport. Eventual insufficiencies of such models are mostly of a quantitative nature and could be removed by introducing different modes of ion permeation. This allows for the possibility that the apparently blocked channel is specifically conductive for the blocking ion. This could indeed be the case because blocking and conducting states cannot be distinguished as long as the concentration of the blocking ion is too low to produce measurable unitary currents.

Ion-channel interaction leading to a Ca-specific conformation of the channels appears to be suitable to explain apparent channel block due to prolonged closures that is the dominant feature in the case of the lva Ca channel. This does not invalidate the virtues of multiple binding site models but can provide a means of understanding the extraordinary capability of the Ca channel to select its ion.

REFERENCES

1. GARNIER, D., O. ROUGIER, Y. M. GARGOUIL & E. CORABOEUF. 1969. Analyse électrophysiologique du plateau des réponses myocardiques, mise en évidence d'un courant lent entrant en absence d'ions bivalents. Pflügers Arch. **313:** 321-342.
2. REUTER, H. & H. SCHOLZ. 1977. A study of the ion selectivity and kinetic properties of calcium-dependent slow inward current in mammalian cardiac muscle. J. Physiol. (London) **264:** 17-47.
3. KOSTYUK, P. G. & O. A. KRISHTAL. 1977. Effects of calcium and calcium-chelating agents on the inward and outward current in the membrane of mollusc neurones. J. Physiol. (London) **270:** 569-580.
4. KOSTYUK, P. G., S. L. MIRONOV & P. A. DOROSHENKO. 1982. Energy profile of the calcium channel in the membrane of mollusc neurons. J. Membr. Biol. **70:** 181-189.
5. ALMERS, W. & E. W. McCLESKEY. 1984. Non-selective conductance in calcium channels of frog muscle: Calcium selectivity in a single-file pore. J. Physiol. (London) **353:** 585-608.
6. HESS P. & R. W. TSIEN. 1984. Mechanism of ion permeation through calcium channels. Nature (London) **309:** 453-456.
7. FUKUSHIMA, Y. & S. HAGIWARA. 1985. Currents carried by monovalent cations through calcium channels in mouse neoplastic B-lymphocytes. J. Physiol. (London) **358:** 255-284.
8. HAGIWARA, S. & J. BYERLY. 1981. Calcium channel. Ann. Rev. Neurosci. **4:** 69-125.

9. TSIEN, R. W., P. HESS, E. W. MCCLESKEY & R. L. ROSENBERG. 1987. Calcium channels: Mechanisms of selectivity, permeation, and block. Ann. Rev. Biophys. Biophys. Chem. **16:** 265-290.

10. KOSTYUK, P. G., S. L. MIRONOV & Y. M. SHUBA. 1983. Two ion-selecting filters in the calcium of the somatic membrane of mollusc neurons. J. Membr. Biol. **76:** 83-93.

11. HESS, P., J. B. LANSMAN & R. W. TSIEN. 1986. Calcium channel selectivity for divalent and monovalent cations. Voltage and concentration dependence of single channel current in ventricular heart cells. J. Gen. Physiol. **88:** 293-319.

12. CARBONE, E. & H. D. LUX. 1984. A low voltage-activated, fully inactivating Ca channel in vertebrate sensory neurones. Nature (London) **310:** 501-502.

13. CARBONE, E. & H. D. LUX. 1986. External Ca^{2+} ions block unitary Na^+ currents through Ca^{2+} channels of cultured chick sensory neurones by favouring prolonged closures. J. Physiol. (London) **382:** 124P.

14. FOX, A. P., M. C. NOWYCKY & R. W. TSIEN. 1987. Single-channel recordings of three types of calcium channels in chick sensory neurones. J. Physiol. (London) **394:** 173-200.

15. CARBONE, E. & H. D. LUX. 1987. Single low voltage-activated calcium channels in chick and rat sensory neurones. J. Physiol. (London) **386:** 571-601.

16. LUX, H. D. & E. CARBONE. 1987. External Ca ions block Na conducting Ca channel by promoting open to closed transitions. *In* Receptors and ion channels. Y. A. Ovchinnikov & E. Hucho, Eds.: 149-155. Walter de Gruyter. Berlin/New York.

17. LANSMAN, J. B., P. HESS & R. W. TSIEN. 1986. Blockade of Ca current through single calcium channels by Cd^{2+}, Mg^{2+}, and Ca^{2+}: Voltage and concentration dependence of calcium entry into the pore. J. Gen. Physiol. **88:** 321-347.

18. LÄUGER, P., W. STEPHAN & E. FREHLAND. 1980. Fluctuations of barrier structure in ionic channels. Biochim. Biophys. Acta **602:** 167-180.

19. LÄUGER, P. 1985. Ionic channels with conformational substates. Biophys. J. **47:** 581-590.

Voltage-Dependent Calcium Conductances in Mammalian Neurons

The P Channel

RODOLFO R. LLINÁS, MUTSUYUKI SUGIMORI,
AND BRUCE CHERKSEY

Department of Physiology and Biophysics
New York University Medical Center
New York, New York 10016

Calcium plays several critical roles in the electrophysiology of mammalian central neurons. As a charge carrier, it is capable of generating either action potentials or graded voltage responses.[1] As a second messenger, calcium is involved in such events as transmitter release,[2] the activation of ionic channels (e.g., the calcium-dependent potassium conductance[3]), and the phosphorylation of molecules, which in turn modulate ionic conductances.[4] Here we plan to review briefly some aspects of the voltage-dependent calcium conductance in the neurons of the mammalian brain with particular emphasis on the conductance present in dendrites of cerebellar Purkinje cells.

Before describing in detail the voltage-dependent channel properties of Purkinje cells, we will review in general terms the main calcium-dependent electroresponsiveness encountered in mammalian neurons. The first description of calcium-dependent spikes in vertebrate central nervous system (CNS) neurons was obtained by direct recordings from avian cerebellar Purkinje cell dendrites.[5] This was later confirmed in mammilary neurons.[6] The existence of more than one voltage-dependent calcium conductance was originally encountered in the inferior olive (IO). These cells demonstrated two types of responses, the high- and the low-threshold spikes (HTS and LTS, respectively).[7,8] Because the most common electroresponsiveness observed in intracellular recordings from different types of central neurons fall into these categories,[1] we will continue to describe them as stated above.

By contrast the single channel responsible for the low-threshold calcium conductance was first described by Carbone and Lux.[9] Later the calcium channels were grouped into three categories[10]: (a) the T channels, which we now believe correspond to the low-threshold calcium conductance; (b) the N channels, which correspond to a certain extent to the high-threshold calcium conductance; and (c) the L calcium channel, which does not seem to be very commonly represented in the CNS, but which would also fall into this category of high-threshold calcium conductance. A simple incorporation of our nomenclature into the framework proposed above[10] may prove to be problematic because the criteria for these two characterizations are so different. The single-channel criteria are based on direct measurements and do not consider cable properties. Ours was developed from the standpoint of neuronal integration, which takes into account parameters such as the spatial location of the channels on the soma-dendritic plasmalemma. For instance, as we will see below, the intradendritic threshold for dendritic calcium spiking is quite different from that observed in a simultaneously implemented somatic recording.[11] Thus, high-threshold relates to the firing level for a calcium spike in a living cell and is measured from the soma as opposed to that for single channels in a patch-clamp pipette. Indeed, while both criteria are relevant they do represent different approaches to the problem.

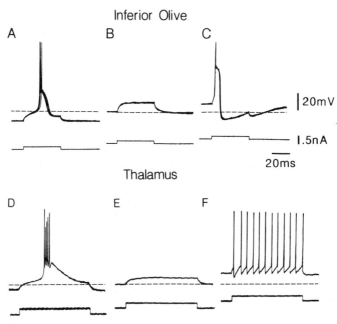

FIGURE 1. Intracellular recording from the inferior olive (IO) and the thalamus in *in vitro* guinea pig brain slices. In the IO the low-threshold calcium spike is shown in **A** and the high-threshold in **C**. In **B** the same current injections as in **A** and **C**, but from the rest membrane potential (dash line). Hyperpolarization of the cell, as in **A**, deinactivates the low-threshold calcium channels generating a spike. Depolarizing the cell activates the high-threshold calcium spike. The initial part of the action potential is sodium-dependent. Similar recordings from the thalamus (**D-F**), demonstrate that, as in the IO, a hyperpolarization of the cell produces the activation of a low-threshold calcium spike, which itself generates four distinct sodium-dependent action potentials (superimposed). If the cell is depolarized from rest (**F**), the sodium stimulus, as in **E**, produces repetitive firing. In this cell the high-threshold calcium spike is small and contributes, by activating a calcium-dependent potassium conductance, to the amplitude of the after-hyperpolarization. Modified from Llinás and Yarom[15] and from Jahnsen and Llinás.[29]

LOW-THRESHOLD CALCIUM SPIKES

The LTSs similar to those described in the inferior olive (IO) have been encountered in many different types of neurons.[1] As shown in FIGURE 1A and B for IO and thalamic neurons, these LTSs characteristically occur at membrane potentials negative to rest and thus are generated, for the most part, as a rebound potential following a hyperpolarization. Indeed, transient IPSPs or the afterhyperpolarization, that follows spikes are often sufficient to activate the rebound potential and to produce an oscillatory type of activity.[12]

The currents that generate the LTS can be examined using the single-electrode voltage-clamp approach. Examples of voltage-clamp results from inferior olivary and thalamic cells are shown in FIGURES 2A and 2E, respectively. The typical voltage current relation shown here indicates that this low-threshold calcium current is activated at -65 to -70 mV and reaches a peak at 30-40 mV for both types of cells (FIG. 1A, C, and G). This latter measurement is more negative than that reported for the "typical" T channel.[13]

The time course of these conductances is in many ways similar to that observed for other low-threshold conductances in other systems.[13] In the IO and in the thalamus for a 50-mV voltage step, the peak of this current occurs at approximately 15-20 msec and falls with a time constant near 20 msec. For the most part, a tail calcium current is not seen for voltage steps longer than 80 msec.[14] Physiologically this current generates a low-threshold calcium spike.

PHARMACOLOGICAL BLOCKAGE OF LOW-THRESHOLD SPIKES

Recent findings indicate that alcohol has marked effect on these conductances in the CNS. In particular, the addition of octanol at a dose of 50 μmol produces not only a block of the LTS, but also a reduction of the inward current as seen in voltage-clamp experiments.[14,15] This was encountered in both the IO and thalamus. The reduction of the inward current by octanol in the IO and thalamus are illustrated in FIGURES 2C and 2F, respectively. Note that there are differences in the degree of block observed in these two cell types. It is also worth mentioning that octanol has no effect on the high-threshold calcium spike (as illustrated in FIG. 2D). When octanol was injected into adult rats, a clear reduction of inferior olivary activity was observed as indicated by the cessation of harmaline-induced tremor.[16] Extracellular recording from IO neurons *in vitro* indicates that this low-threshold calcium conductance is present at the soma level, because LTSs are accompanied by a clear negative extracellular field.[8]

HIGH-THRESHOLD CALCIUM SPIKES

The HTSs were first encounted in Purkinje cells[5,6,17] and have been analyzed *in vitro* in many central neurons.[1,18] In mammalian Purkinje cells, at least two types of voltage responses were encountered, a plateau-like graded depolarization that could last for several seconds and all-or-none calcium-dependent spikes.[6,17] Direct confirmation of the dendritic location of these conductances was obtained using calcium-sensitive dyes, such as Arsenazo III[19] and more recently, Fura II.[20] In the latter study,

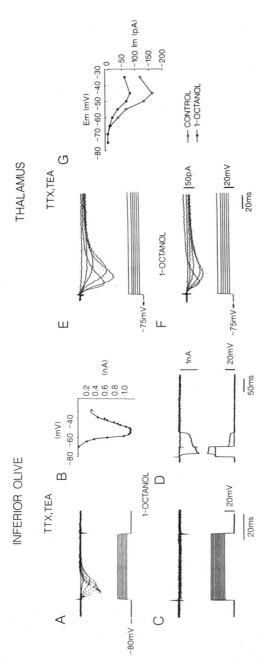

FIGURE 2. Low-threshold calcium current in the IO and thalamus and its blocking by alcohol. In **A**, inward current produced by depolarizing voltage steps from a holding potential of −80 mV. In **B**, current-voltage relation. Addition of 50 μM octanol (**C,D**) blocked inward calcium current (**C**), but an all-or-none calcium-dependent spike could be observed with high-amplitude voltage steps (**D**) indicating that octanol does not block high-threshold calcium spikes. In **E**, similar voltage clamp as in **A** but from a thalamic neuron from a holding potential of −75 mV. Current-voltage relation is shown in **G** (open circles). In **F**, after the application of octanol (50 μM), there is a clear reduction of the amplitude of the inward calcium current as plotted in **G** (closed circles).

Records **A-D**, from Llinás and Yarom, unpublished observations; **E-G**, Geijo-Barrientos and Llinás, unpublished observations.

imaging techniques demonstrated that the channels responsible for the graded calcium plateau potentials are located mainly in the fine dendritic branches (the spiny branchlets) of the Purkinje cell while the all-or-none spikes are generated in the thicker, more proximal dendritic tree.

CALCIUM-DEPENDENT ELECTRORESPONSIVENESS IN PURKINJE CELLS

Direct depolarization of Purkinje cells *in vitro* by square current pulses produced rapid sodium-dependent firing and prolonged and repetitive Ca^{2+}-dependent spikes

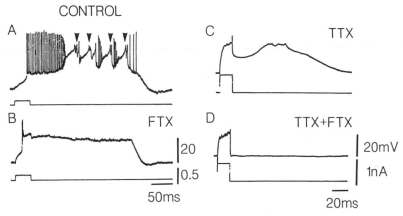

FIGURE 3. Intracellular recording from the guinea pig Purkinje cell. In **A,** activation of a Purkinje cell by a small current pulse (lower trace) showing repetitive activity and the generation of voltage-dependent calcium spikes (arrows). After the application of FTX, the calcium-dependent spikes are abolished but a plateau potential remains that outlasts the duration of the current step and that can be blocked by TTX. In **C,** the slow calcium-dependent action potential in a Purkinje cell is shown after the blockage of sodium conductances by TTX. In **D,** block of the calcium spike after the application of FTX. Modified from Llinás *et al.*[22]

(arrows, FIG. 3A). Following bath application of funnel-web spider toxin (FTX), the Ca^{2+}-dependent spikes disappeared, and the duration of the afterpotential hyperpolarization was reduced (FIG. 2B).[20-22] This blocking occurred 5-10 minutes after bath application of the toxin at a 0.5 μM (calculated) final concentration and was partly reversible after washing for 30 to 40 minutes. Both the sodium-dependent plateau potentials, which occur at the somata of Purkinje cells,[6] and the initial fast spikes were blocked by the addition of TTX (10^{-6} g/ml) to the bath (FIG. 3C). This indicated that following FTX administration only the voltage-dependent sodium conductances remained. That is, all voltage-dependent calcium conductance was blocked by this toxin fraction. Nearly identical results were obtained with FTX from the three different types of funnel-web spiders tested.

In a second set of experiments, the converse sequence of pharmacological blockage was implemented. Recordings were made from the Purkinje cell soma after TTX administration. Under these conditions, Purkinje cell stimulation is known to elicit a slow Ca^{2+}-dependent plateau potential. In this case, a short, direct depolarization of the soma initiated a prolonged dendritic potential that outlasted the pulse by about 20 msec. However, after application of FTX ($\approx 0.5 \ \mu M$), this Ca^{2+}-dependent conductance was blocked,[22] as was the oscillatory property of these neurons.[20]

ISOLATION OF VOLTAGE-DEPENDENT CALCIUM CHANNELS FROM THE MAMMALIAN CEREBELLUM

Functional calcium channels were isolated from guinea pig cerebellum[22-24] using an FTX affinity gel. The functional activity of the isolated protein was assessed using

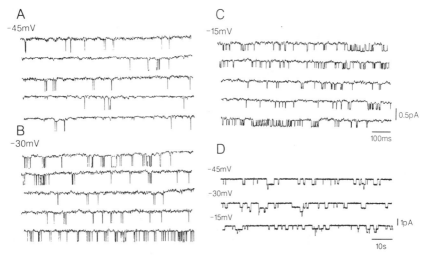

FIGURE 4. Functional properties of channels isolated from guinea pig cerebellum by an FTX affinity gel. In **A-C,** single-channel recording with 80 mM external Ba^{2+} obtained for protein reconstituted in lipid vesicles and fused with lipid bilayer. Holding potentials: -45 mV (**A**), -30 mV (**B**), and -15 mV (**C**). In **D,** single-channel recordings obtained with 80 mM Ba^{2+} on the cytoplasmic face of the channel. Holding potentials of -45, -30, and -15 mV are shown. In this particular experiment, three identical channels had fused with the bilayer. The long duration of the open time seems to be an effect of the high amount of internal Ba^{2+}, as it can be converted into the short-duration channels (**A-C**) when Ba^{2+} is removed from the cytoplasmic side. Modified from Llinás et al.[22]

the lipid-bilayer technique. Electrical activity was measured in asymmetric solutions containing 80 mM $BaCl_2$ in the bath or in the patch pipette. An increase in the conductance of the bilayer was usually detected within 10 minutes of adding vesicles containing the purified protein to the bilayer chamber. The fusion-promoting effect of divalent cations often made it difficult to obtain recordings from only a single channel when vesicles were added to the Ba^{2+}-containing solutions.

Two types of single-channel activity were recorded. The first type, seen when Ba^{2+} was present on the extracellular face of the channel protein, was closed at -60 mV and was characterized by short-duration openings to depolarizations (FIG. 4A). A set of results obtained at -45, -30, and -15 mV holding potentials are shown in FIGURES 4A to C. The maximum open probability (0.6-0.65) occurred at holding potentials greater than 0 mV and the mean open time was also voltage-dependent, ranging from under 0.6 to 5 msec. A conductance of 10-12 pS in Ba^{2+} and 6-8 pS in 100 mM Ca^{2+} was obtained for this type of channel. The extrapolated reversal potential, between -90 and -120 mV, is within the theoretical value for a Ba^{2+}-permeable channel.

When the solutions were reversed and the high-Ba^{2+} solution was on the cytoplasmic face of the channel, a second type of channel-like activity was found (FIG. 4D). The results illustrated were obtained at holding potentials of -45, -30, and -15 mV. In the experiment illustrated in FIGURE 4D three channels with identical conductances had fused with the bilayer. Note the predominance of openings longer than one second. The estimated unitary channel conductance for this channel type was 20 pS. Replacement of the Ba^{2+} with Cs^+, resulted in conversion of their activity to the short-duration type illustrated in FIGURE 4A to C. Both types of reconstituted channel activity were blocked by both Cd^{2+} and Co^{2+} in concentrations of less than 100 μM and by the microliter addition of a 1 : 10 dilution of FTX.

Measured Ca^{2+} currents in neurons represent contributions from both the individual currents and the opening probabilities of the channels. Thus, the macroscopic current may be approximated by multiplying the single-channel currents by the opening probabilities at each potential as has been done for the "fast" Ca^{2+} channel.[25] The voltage and time dependence of this calculated current is comparable to the results from current clamp experiments in Purkinje cell soma.[17] However, in the absence of voltage-clamp measurements for Purkinje cell dendrites, only a qualitative statement may be offered.

CONCLUSIONS

The results summarized above indicate that at least two types of calcium spikes may be observed in central neurons that are most likely the product of the activations of at least four types of calcium channels. These are the L, N, and T channels[10] and the P channel described in Purkinje cells.[22] L channels, although observed in hippocampal neurons,[26] do not appear to be present in Purkinje cells.[21] By contrast, thalamic or inferior olivary cells do not seem to express P channels.[14]

In comparing the voltage-current relations of the low- and high-threshold spikes, other differences between this particular nomenclature and that derived from single-channel studies must be considered. In the case of the low-threshold spike, we have at least two types of cells where low-threshold calcium conductances may be separable pharmacologically. Indeed, harmaline increases the transient conductance in the IO[15] but does not affect thalamic neurons.[14] This suggests then, that the low-threshold spikes in different neuron types may be generated by slightly different sets of T-like channels, due perhaps to the same type of molecular diversity encountered for potassium channels.[27]

THE P CHANNEL

The presence of a voltage-dependent calcium conductance in Purkinje cells that does not resemble pharmacologically or biophysically the characteristics described for the L, N, or T channels has led us to propose the existence of yet another channel type. We have named it the P channel[22] because it was first identified in Purkinje cells. It has been demonstrated pharmacologically to be present in the squid giant synapse and has also been isolated from the squid optic lobe.[22] The P channel isolated from guinea pig cerebellum shows some resemblance to the N channel. However, there are clear and definitive differences between the P and N channels. The P channel is activated at -50 mV as demonstrated most particularly in the presynaptic terminal in squid.[22] Pharmacologically, Ω-conotoxin, which blocks the N channel,[13] does not block Purkinje cell calcium spikes, or the single calcium channels isolated from cerebellar tissue.[22-24] Nor, does it block the presynaptic calcium current, or single-channel activity in lipid bilayers.[23] In short, on the basis of biophysical and pharmacological evidence, we propose the existence of a channel that we have named the P channel.[23]

From the point of view of nomenclature, it must be noted that both the calcium-dependent plateau potential and the dendritic spikes that are blocked by FTX[20,21] were initially described as parts of the high-threshold response in Purkinje cells.[6] However, as stated above, high threshold is sometimes taken to mean responses mediated by L channels.[13] This is not the case here. Pharmacologically, dendritic conductances in Purkinje cells are not blocked by dihydropyridines[22] or enhanced by BAY K 8644.[28] Ultimately the conclusion to be drawn from the results presented here seems evident—a wealth of calcium-channel types and voltage-dependent calcium conductances await further explorations in the CNS of invertebrates and vertebrates alike.

REFERENCES

1. LLINÁS, R. 1988. Science **242:** 1654-1664.
2. KATZ, B. 1969. The Release of Neural Transmitter Substances. C. C. Thomas. Springfield, IL.
3. CONNOR, J. A. & C. F. STEVENS. 1971. J. Physiol. (London) **213:** 1.
4. NESTLER, E. J. & P. GREENGARD. 1984. Protein Phosphorylation in the Nervous System. Neurosciences Institute. J. P. Wiley. New York.
5. LLINÁS, R. & R. HESS 1976. Soc. Neurosci. Abstr. 2.
6. LLINÁS, R. & M. SUGIMORI. 1980. J. Physiol. (London) **305:** 171-195.
7. LLINÁS, R. & Y. YAROM. 1981. J. Physiol. (London) **315:** 549-567.
8. LLINÁS, R. & Y. YAROM. 1981. J. Physiol. (London) **315:** 569-584.
9. CARBONE, E. & H. D. LUX. 1987. J. Physiol. **386:** 547-570.
10. NOWYCKY, M. C., A. P. FOX & R. W. TSIEN. 1985. Nature **316:** 440-443.
11. LLINÁS, R. & M. SUGIMORI. 1984. Soc. Neurosci. Abstr. **10:** 659.
12. STERIADE, M. & R. LLINÁS. 1988. Physiol. Rev. **68:** 649-742.
13. TSIEN, R. W., D. LIPSCOMBE, D. V. MADISON, K. R. BLEY & A. P. FOX. 1988. TINS. **11:** 431.
14. GEIJO-BARRIENTOS, E. & R. LLINÁS. Unpublished observation.
15. LLINÁS, R. & Y. YAROM. 1986. J. Physiol. (London). **376:** 163-182.
16. SINTON, C. M., B. KROSSER, K. D. WALTON & R. LLINÁS. 1989. Pflugers Arch. In press.
17. LLINÁS, R. & M. SUGIMORI. 1980. J. Physiol. (London) **305:** 197-213.
18. MILLER, R. J. 1985. TINS **8:** 45-47.

19. ROSS, W. N. & R. WERMAN. 1986. J. Physiol. **389:** 319-336.
20. TANK, D. W., M. SUGIMORI, J. A. CONNOR & R. R. LLINÁS. 1988. Science **242:** 633-828.
21. SUGIMORI, M. & R. LLINÁS. 1987. Soc. Neurosci. Abstr. **13.**
22. LLINÁS, R., M. SUGIMORI, J.-W. LIN & B. CHERKSEY. 1989. Proc. Natl. Acad. Sci. USA **86:** 1689-1693.
23. CHERKSEY, B., M. SUGIMORI, B. RUDY & R. LLINÁS. 1988. Soc. Neurosci. Abstr. **14:** 901.
24. CHERKSEY, B., M. SUGIMORI, J.-W. LIN & R. LLINÁS. 1989. J. Biophysiol. **55:** 438a.
25. HILLE, B. 1984. Ionic Channels of Excitable Membranes. Sinauer. Sunderland, MA.
26. BROWN, D. *et al.* 1989. Ann. N.Y. Acad. Sci. This volume.
27. RUDY, B. 1988. Neuroscience **25:** 729.
28. LLINÁS, R. & M. SUGIMORI. Unpublished observations.
29. JAHNSEN, H. & R. LLINÁS. 1984. J. Physiol. (London) **349:** 205-226.

Calcium Currents in Isolated Taste Receptor Cells of the Mudpuppy

SUE C. KINNAMON,[a] THOMAS A. CUMMINGS,[a]
STEPHEN D. ROPER,[a] AND KURT G. BEAM[b]

[a] Department of Anatomy and Neurobiology
Colorado State University
Fort Collins, Colorado 80523
and
Rocky Mountain Taste and Smell Center
University of Colorado Health Sciences Center
Denver, Colorado 80262

[b] Department of Physiology
Colorado State University
Fort Collins, Colorado 80523

Recent studies have shown that taste receptor cells are electrically excitable and possess voltage-dependent K^+, Na^+, and Ca^{2+} currents.[1] Although taste cells generate action potentials in response to a variety of chemical stimuli, the role of voltage-dependent conductances in taste transduction remains unclear. In this study we have characterized the voltage-dependent Ca^{2+} current (I_{Ca}) because it represents an important link between the receptor potentials and transmitter release in taste cells.

Taste cells were isolated from the surrounding lingual epithelium as previously described.[2] Whole-cell I_{Ca} was studied by bathing cells in amphibian saline containing 10 mM tetraethylammonium (TEA) and 1 μm TTX, and by using an intracellular pipette solution containing CsCl. Current was activated by step depolarizations to $\geq$ -10 mV from a holding potential of -100 mV. I_{Ca} reached a peak in 5-15 msec and slowly inactivated with a time constant of approximately 60 msec (FIG. 1A). Substitution of Ba^{2+} for Ca^{2+} caused the current to increase in magnitude and shifted the activation approximately 20 mV in the hyperpolarizing direction. Macroscopic kinetics were similar in Ca^{2+} and Ba^{2+}, implying that inactivation is voltage-dependent, rather than Ca^{2+}-dependent.

The effect of brief (2-sec) depolarizing prepulses on inactivation is illustrated in FIGURE 1B. I_{Ca} was approximately half-inactivated by a prepulse to -40 mV and completely inactivated by a prepulse to 0 mV. When the half-inactivated current was scaled to the same magnitude as the control current (trace 3, FIG. 1B), it decayed at approximately the same rate as the control current. This observation suggests that I_{Ca} in taste cells is mediated by a single type of Ca^{2+} channel.

I_{Ca} was reversibly blocked by 150 μM verapamil (80% block), 100 μM cadmium, and by substitution of Mg^{2+} for Ca^{2+}. I_{Ca} was little affected by bath addition of ω-conotoxin VIC (1 μM), 8-bromo-cAMP (1 mM) or 8-bromo-cGMP (1 mM). The effects of dihydropyridine drugs are illustrated in FIGURE 2. When taste cells were held at a potential that approximately half-activated I_{Ca}, both $(+)$ PN 200-110 and BAY K 8644 (10 μM) blocked I_{Ca} by approximately 80%. Little effect of these

A

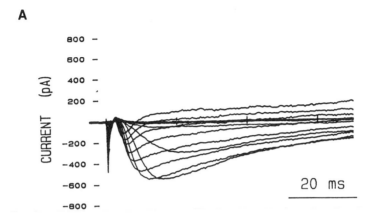

B

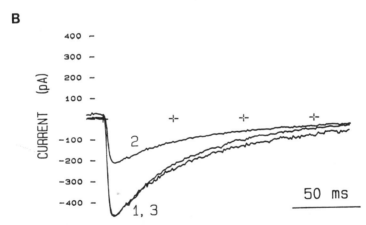

FIGURE 1. I_{Ca} in isolated taste cells. (**A**) The holding potential was -100 mV and the membrane was stepped from -40 to $+80$ mV in 10 mV steps. (**B**) Effect of a two-second depolarizing prepulse on I_{Ca}. The holding potential was -100 mV and the test potential was $+20$ mV. Bath contained 8 mM Ca^{2+}. Trace 1: control (no prepulse). Trace 2: I_{Ca} after a prepulse to -40 mV. Trace 3: same as trace 2, except the current is scaled to the same magnitude as the control current (trace 3 inactivates slightly slower than the control current). Pipette contained 80 mM CsCl, 10 mM NaCl, 10 mM HEPES (pH 7.2), 2 mM MgCl$_2$, 1 mM BAPTA, 0.09 mM CaCl$_2$, 5 mM ATP, 0.2 mM GTP, and 100 μM leupeptin. Leak and linear capacitative currents were scaled and subtracted in this and the next figure.

dihydropyridines was observed when cells were held at -100 mV. Taken together, the data suggest that taste cell I_{Ca} does not fit into the standard T-, N-, and L-type classification scheme. Even with Ba^{2+} as the charge carrier, the current inactivates completely, thus resembling N- but not L-type currents. However, I_{Ca} is modulated

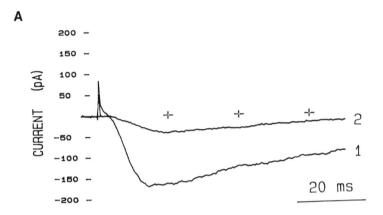

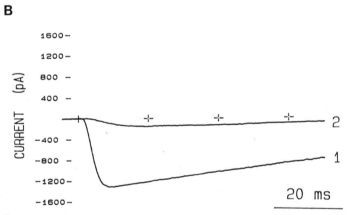

FIGURE 2. Effect of dihydropyridines on Ca^{2+} channel currents in taste cells. (**A**) Effect of 10 μM (+) PN 200-110. (**B**) Effect of 10 μM Bay K 8644. In each figure, trace 1 is the control trace and trace 2 is the effect of the drug. Bath contained 8 mM Ba^{2+}. The holding potential was -60 mV and the test potential was 0 mV. Same pipette solution as FIGURE 1.

by dihydropyridines, which in other cells affects L- but not N-type current. The pharmacology differs from typical L-type current, however, in that BAY K 8644 blocks, rather than enhances, current.

The distribution of Ca^{2+} channels between apical and basolateral membranes was determined by using a combination of whole-cell and loose-patch recording techniques. A whole-cell pipette was used to voltage-clamp the taste cell membrane and record whole-cell Ca^{2+} current. A loose-patch pipette was used to record currents flowing across small patches of membrane in different regions of the cells. The data revealed that Ca^{2+} channels are present in the apical membrane as well as in the basolateral membrane. This differs from the distribution of voltage-gated K^+ channels, which are restricted to the apical membrane, in accordance with their hypothesized role in taste transduction.[3]

In summary, taste cell Ca^{2+} channels appear to be of a single type, with properties intermediate between N- and L-type Ca^{2+} channels. Strong, transient depolarization is required for activation, suggesting that signaling between taste cells and primary sensory neurons may require action potentials. The presence of channels in nonsynaptic regions (i.e., apical membrane) suggests that Ca^{2+} channels in taste cells may have important functions in addition to mediation of neurotransmitter release.

REFERENCES

1. KINNAMON, S. C. & S. D. ROPER. 1988. Membrane properties of isolated taste cells. J. Gen. Physiol. **91:** 351-371.
2. KINNAMON, S. C., T. A. CUMMINGS & S. D. ROPER. 1988. Isolation of single taste cells from lingual epithelium. Chem. Senses **13:** 355-366.
3. KINNAMON, S. C., V. E. DIONNE & K. G. BEAM. 1988. Apical localization of K^+ channels in taste cells provides the basis for sour taste transduction. Proc. Natl. Acad. Sci. USA **85:** 7023-7027.

T-Type Calcium Channels in Swiss 3T3 Fibroblasts

ANTONIO PERES

Dipartimento di Fisiologia e Biochimica Generali
University of Milan
Milano, Italy

Mouse Swiss 3T3 fibroblasts were voltage-clamped using the whole-cell technique. Under physiological conditions of extracellular solution and temperature, depolarizations from holding potentials (V_h) < -60 mV caused the appearance of a transient inward current.[1,2] As the depolarizations increased the current eventually became outward with a reversal potential around $+40$ mV in 2 mM Ca^{2+} (FIG. 1). Both inward and outward current transients were not seen when the V_h was > -60 mV.

The current was unaffected by removal of external Na^+ and increased when Ca^{2+}_o was increased.[3] FIGURE 1 shows that the current inactivates completely even when it is outward and it has an inactivation range between -90 and -60 mV. In addition, the current is strongly but not completely blocked by 1 mM Cd^{2+}, and it is almost unaffected by dihydropyridines.

All these observations allow us to attribute this current to Ca^{2+} channels of the low-voltage-activated "T" type.[4,5]

The calcium channels of Swiss 3T3 fibroblasts also share the peculiar features already described in several other preparations.[6,7] In the complete absence of divalent cations (chelated with 1 mM EDTA) a 10-fold increase in the current occurs together with a negative shift of the activation curve and of the reversal potential (which moves to $+10$ mV). In 0 external Ca^{2+} but with 2.4 mM Mg^{2+}, the inward current is blocked and a large, inactivating outward current is observable at potential > 0 mV.

These observations may be explained by the presence in the channel structure of sites having high affinity for calcium. A consequence of this is that the channel is selective for Ca^{2+} only when Ca^{2+} is present (that is, $> 10^{-6}$ M), but it is otherwise easily permeable to monovalent cations.

The voltage-dependent calcium channel of 3T3 fibroblasts therefore appears to be identical to those described in other cells with respect to the selectivity mechanisms.

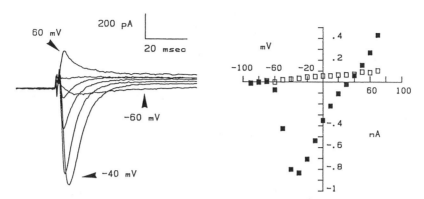

FIGURE 1. Shown on the left are current traces obtained by depolarizations from $V_h = -100$ mV to potentials between -60 and $+60$ mV in 20-mV steps (some values are indicated). The bath solution contained (in mM) 125 NaCl, 5 KCl, 1.2 MgSO$_4$, 2 CaCl$_2$, 25 mM HEPES-NaOH at pH 7.3. The pipette contained (in mM): 140 KCl, 1.2 MgCl$_2$, 1 EGTA, 2 ATP, 0.5 GTP, 10 HEPES-KOH at pH 7.2. Temperature: 35°C. On the right are the I-V curves for the peak current (filled squares) and for the steady-state current (open squares).

REFERENCES

1. CHEN, C., M. J. CORBLEY, T. M. ROBERTS & P. HESS. 1988. Science **239:** 1024-1026.
2. PERES, A., R. ZIPPEL, E. STURANI & G. MOSTACCIUOLO. 1988. Pfluegers Arch. **411:** 554-557.
3. PERES, A., E. STURANI & R. ZIPPEL. 1988. J. Physiol. **401:** 639-655.
4. CARBONE, E. & H. D. LUX. 1984. Nature **310:** 501-502.
5. MILLER, R. J. 1987. Science **235:** 46-52.
6. FUKUSHIMA, Y. & S. HAGIWARA. 1985. J. Physiol. **358:** 255-284.
7. PERES, A. 1987. J. Physiol. **391:** 573-588.

Transformation Changes the Functional Expression of Calcium Channels in Fibroblasts

CHINFEI CHEN,[a] MICHAEL J. CORBLEY,[b] THOMAS
M. ROBERTS,[b] AND PETER HESS[a]

[a] Department of Physiology and Biophysics
Neuroscience Program
and
[b] Dana Farber Cancer Institute
Harvard Medical School
Boston, Massachusetts 02115

Calcium channels in excitable cells like neurons, secretory cells, and cardiac and smooth muscle have been shown to be involved in the control of transmitter release, secretion, and contraction. In contrast, much less is known about the existence and possible role of Ca^{2+} channels in nonexcitable cells, despite recent recognition of an important role of intracellular Ca^{2+} in such fundamental cellular processes as proliferation and cell division,[1-4] cell motility and cytoskeletal rearrangment,[5] and phagocytosis.[6]

With the patch-clamp technique, we found two types of voltage-sensitive Ca^{2+} channels in 3T3 fibroblasts that had characteristic properties of L- and T-type Ca^{2+} channels.[7] While the two Ca^{2+}-channel types were consistently present in normal cells, 3T3 fibroblasts that were transformed by retrovirus-mediated transfer of an activated oncogene (c-H-*ras,* EJ-*ras,* v-*fms,* v-*src,* or polyoma middle T)[8,9] displayed an altered population of calcium channels. Although these transformed cells contained L-type Ca^{2+} current densities similar to normal fibroblasts, they all specifically lacked the T-type Ca^{2+} current (FIG. 1). The functional suppression of the T-type Ca^{2+} channel was seen with all transforming oncogenes that code for membrane proteins. In contrast, fibroblasts transformed by oncogenes whose gene products are primarily located in the nucleus (polyoma large T, SV40 large T)[10] still contained T-type Ca^{2+} channels, but at a significantly lower density than control, nontransformed 3T3 cells (FIG. 1).

In an effort to correlate the time course of the functional suppression of T-type Ca^{2+} currents with the establishment of the transformed phenotype, we used a cell line that contained the polyoma middle T oncogene under the control of a dexamethasone inducible promoter.[11,12] In this cell line, which has a nontransformed phenotype in the absence of glucocorticoids, maximal levels of middle T antigen and morphological transformation can be detected within 24-48 hours of addition of 1 μM dexamethasone to the culture medium. FIGURE 2 shows that T-current density was not significantly decreased one week following the induction of transformation indicating that the disappearance of T-type Ca^{2+} currents is most likely a consequence, and not a cause, of transformation. However, after a month, only 50% of the cells

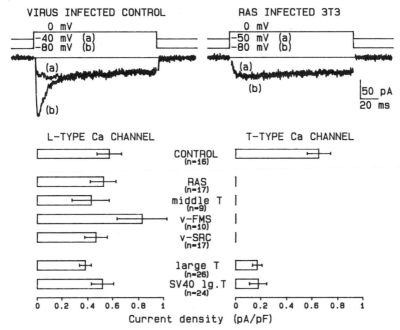

FIGURE 1. Current densities through Ca channels in normal and transformed 3T3 fibroblasts. **A,B:** Superposition of whole-cell Ba-currents elicited from two holding potentials. Voltage protocol shown above the current traces. **A:** Control virus-infected NIH 3T3 cell has both components of Ca-channel current. Cell capacitance 95 pF. **B:** The two current traces recorded from different holding potentials in a NIH 3T3 cell transformed by the activated c-H-*ras* oncogene superimpose perfectly, indicating a complete lack of T-type current. Capacitance, 76 pF. **C:** Mean values $+/-$ SEM of the current densities of the two Ca-channel types in control and transformed cells, measured in whole-cell recordings like those illustrated in panels A and B. The number of cells (n) studied is indicated for each group. The mean densities of the L-type current did not differ significantly between control and transformed cells. Standard procedures were used for whole-cell patch-clamp recordings.[13] External solution (in mM): 20 Ba-Acetate TEA-aspartate, 10 Hepes, pH 7.5, 22°C; internal solution: 135 CsAsp, 10 EGTA, 10 Hepes, 5 MgCl$_2$, 4 ATP, pH 7.5., 22°C. Depolarizing pulses were delivered every three to four seconds. Linear leak and capacity currents were subtracted digitally. Cell capacitance was measured from the capacity current elicited by a small depolarization. Currents were measured after leak subtraction at the peak of the current-voltage relation (-20 mV and $+20$ mV for T- and L-currents, respectively). (Part of this figure is reproduced from C. Chen *et al.*[7] by permission of *Science.*)

still expressed T-type channels, and the mean density of T-type current had decreased to less than 15% of that found in control cells. These results imply that there might be a selective growth advantage for cells with few or no T-type Ca channels.

The roles of the two Ca^{2+} channels in the control of cell function and the relation between transformation and functional channel expression remain to be studied. However, our results further demonstrate the widespread occurrence of Ca^{2+} channels and DHP receptors and strengthen the view that the two channel types are separate molecular entities.

FIGURE 2. Densities of Ba currents carried through T-type Ca channels in 3T3 fibroblasts containing a dexamethasone-inducible polyoma middle T oncogene.[11] Recording conditions and measurements of current densities are the same as in FIGURE 1. The mean $+/-$ SEM of the T-current densities (left side) are plotted in the absence of dexamethasone (time 0) and 7 and 28 days after induction of transformation by addition of 5 mM dexamethasone to the culture medium. The bars on the right side indicate the percentage of cells that have a measurable T-type current at the time indicated.

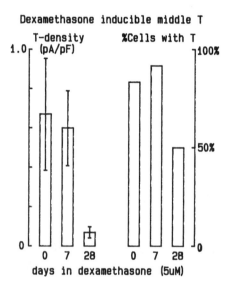

Dexamethasone inducible middle T

REFERENCES

1. MOOLENAAR, W. H., *et al.* 1984. J. Biol. Chem. **259:** 8066-8069.
2. HESKETH, T. *et al.* 1985. Nature **313:** 481-484.
3. BRAVO, R. *et al.* 1985. EMBO J. **4:** 1193-1197.
4. MORGAN, J. I. & T. CURRAN 1986. Nature **322:** 552-555.
5. YIN, H. L. & T. P. STOSSEL. 1979. Nature **281:** 583-586.
6. OKADA, Y. *et al.* 1981. J. Physiol. **313:** 101-119.
7. CHEN, C. *et al.* 1988. Science **239:** 1024-1026.
8. HELDIN, C. & B. WESTMARK. 1984. Cell **37:** 9-20.
9. BISHOP, J. M. *et al.* 1985. Cell **42:** 23-38.
10. WEINBERG, R. A. 1985. Science **230:** 770-776.
11. RAPTIS, L. *et al.* 1985. Mol. Cell. Biol. **5:** 2476-2485.
12. KAPLAN, D. R. *et al.* 1986. P.N.A.S. USA **83:** 3624-3628.
13. HAMILL, O. P. *et al.* 1981. Pfluegers Arch. **391:** 85-100.

Calcium Channels in Freshly Dissociated Rat Cerebellar Purkinje Cells

LAURA J. REGAN

Department of Neurobiology
Harvard Medical School
Boston, Massachusetts

I have used the whole-cell patch-clamp technique to characterize the voltage-dependent calcium currents in freshly dissociated rat cerebellar Purkinje cells. Small sections of cerebellar cortex from one- to three-week postnatal Long-Evans rats were incubated in 20 units/ml papain at 32-36°C for 1.5 hour.[1] Mechanically dissociated cells were used within 10 hours. Patch pipettes contained (in mM): 120 TEA- or Cs-glutamate, 10 EGTA, 5 MgCl$_2$, 3-4 MgATP, 0.3-1 GTP, and 10 HEPES; 12-15 mM phospho-creatine and 42-50 units/ml creatine phosphokinase were sometimes included. The external solution (in mM) was 5 or 50 BaCl$_2$ or CaCl$_2$, 154 TEACl, 0 or 2 MgCl$_2$, 10 glucose, 1-3 TTX, and 10 HEPES. All solutions were pH 7.4 and 22-25°C.

Dissociated Purkinje cells, identified by their large-diameter somata and single stumps of apical dendrite, had two components of calcium current. One component was transient (FIG. 1), having a decay time constant of 20-30 msec at −30 mV. It activated positive to −70 mV, and reached a peak value between −30 and −40 mV. At holding potentials more positive than −60 mV, it was fully inactivated. These properties are very similar to those of the transient ("T"-type) calcium current first described in chick dorsal root ganglion (DRG) cells.[2] The other component was long lasting (FIG. 2), taking seconds to decay fully during a sustained depolarization. It activated at potentials above −60 mV, and reached a peak value between −20 mV and −30 mV. These properties are similar to those of the long-lasting ("L"-type) current first described in rat DRG cells. In some Purkinje cells, the activation range of the long-lasting current appeared to be broader than that of the peripheral L-type current.[4] All but two of more than 100 cells had the long-lasting current, whereas less than half had the transient current.

The pharmacology of the long-lasting current was investigated. It was relatively resistant to block by 1-2 μM ω-conotoxin (mean decrease = 7.2 + 4.4%, n = 22). The dihydropyridine agonist BAY K8644 slowed the tail current kinetics in 34 out of 38 cells, and increased the size of the current elicited by small depolarizations in 13 out of 13 cells. However, it never increased the size of the peak current in Purkinje cells (as it did in 8/8 DRG cells). The long-lasting current appeared to be relatively resistant to block by micromolar concentrations of the dihydropyridine antagonists nitrendipine and nimodipine (mean decrease = 9.7 + 10.2%, n = 9).

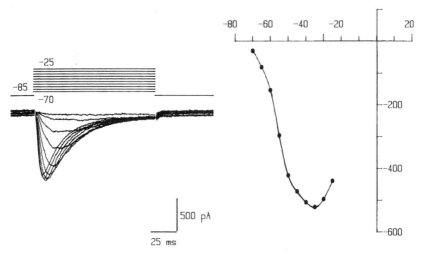

FIGURE 1. Family of currents (**A**) and corresponding peak current-voltage relation (**B**) for a cell having the transient current in relative isolation. Pipette solution was (in mM): 120 Cs-glutamate, 10 EGTA, 5 MgCl$_2$, 14 creatine phosphate, 4 MgATP, 0.3 GTP, 50 units/ml creatine phosphokinase, and 10 HEPES (adjusted to pH 7.4 with CsOH). Charge carrier was 10 mM calcium. Cell P80D.

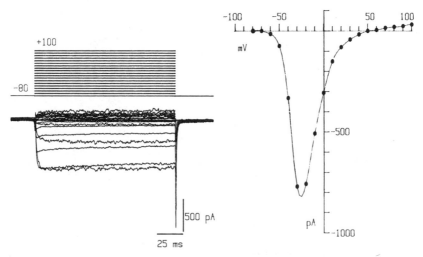

FIGURE 2. Family of currents (**A**) and corresponding peak current-voltage relation (**B**) for a cell having only the long-lasting current. Pipette solution was (in mM): 120 TEA-glutamate, 10 EGTA, 5 MgCl$_2$, 4 MgATP, 0.3 GTP, 3 cAMP, and 10 HEPES (adjusted to pH 7.4 with TEAOH). Charge carrier was 5 mM barium. Cell Q40A.

REFERENCES

1. HUETTNER, J. E. & R. W. BAUGHMAN. 1986. J. Neurosci. **6:** 3044-3060.
2. CARBONE, E. & H. D. LUX. 1984. Biophys. J. **46:** 413-418.
3. KOSTYUK, P. G., N. S. VESELOVSKY & S. A. FEDULOVA. 1981. Neuroscience **6:** 2431-2437.
4. FOX, A. P., M. C. NOWYCKY & R. W. TSIEN. 1987. J. Physiol. **394:** 149-172.
5. NOWYCKY, M. C., A. P. FOX & R. W. TSIEN. 1985. Nature **316:** 440-443.

Two Types of Ca Channels in Mammalian Intestinal Smooth Muscle Cells

H. YABU, M. YOSHINO, T. SOMEYA,
AND M. TOTSUKA

Department of Physiology
Sapporo Medical College
Sapporo 060, Japan

Recent patch-clamp studies of Ca channels have shown the existence of multiple types of voltage-dependent Ca channels with different electrophysiological and pharmacological properties in many excitable cells including neuronal[1,2] and cardiac[3] cells. In this study, single Ca-channel currents were recorded from smooth muscle cells isolated from guinea pig taenia coli by the cell-attached patch-clamp technique in order to examine whether multiple types of Ca channel exist. Two types of unitary Ca-channel currents with large and small conductances were observed in response to depolarizing clamp pulses from a holding potential of -80 mV.[4] The large conductance channel (about 30 pS in 100 mM Ba solution) was activated at more positive potentials (0 mV), and its averaged current decayed much more slowly (FIG. 1A). Dihydropyridine (DHP) Ca antagonist, nifedipine (2 μM), selectively inhibited the large-conductance channel by increasing the proportions of blank sweeps. DHP Ca agonist, BAY K 8644, increased the averaged current by the prolongation of the mean opening times of the large conductance channel.[5] But when the holding potential was set up to -40 mV, BAY K 8644 showed the Ca-channel blocking activity. Thus it was found that the effect of BAY K 8644 was voltage dependent. When Cd was added into the patch pipette solution, the long opening of the unitary currents through the large conductance channel changed into bursts of brief opening times (FIG. 2A), so that the averaged current decreased.

The small conductance channel (about 7 pS in 100 mM Ba solution) activated at relatively more negative potentials (-30 mV), and its averaged current inactivated completely within 100 msec (FIG. 1B). The activity of the small conductance channel disappeared when the steady holding potential was set up to -40 mV from -80 mV. The DHP derivatives, nifedipine and BAY K 8644, had no effect on the small conductance channel and the flickering block by Cd ions was not observed (FIG. 2B). These results suggest that the mammalian intestinal smooth muscle cells contain two types of Ca channels with properties similar to those described for T- and L-type Ca channels in other excitable cells.

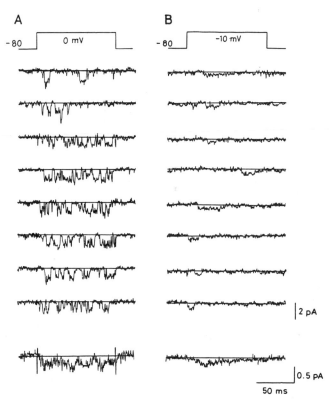

FIGURE 1. Unitary Ca-channel activities of the large (**A**) and small (**B**) conductance channels. The cell resting potential was calibrated to zero by superfusing the high-K solution. The patch membrane was depolarized to the indicated levels from a holding potential of -80 mV. The lower traces in panels A and B showed the averaged current.

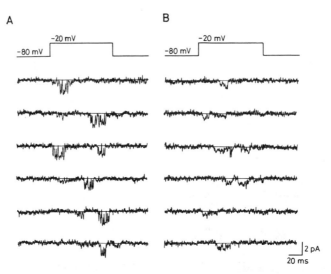

FIGURE 2. Different effects of Cd ions on the large (**A**) and small (**B**) conductance channels. The cell was depolarized to −20 mV from a holding potential of −80 mV. Cd (10 μM) was added in the patch pipette solution. Current traces in A and B are obtained from the same cell. Flickering block by Cd was observed in the large conductance channel (**A**) but not observed in the small conductance channel (**B**).

REFERENCES

1. NOWYCKY, M. C., A. P. FOX & R. W. TSIEN. 1985. Nature **316:** 440-443.
2. FOX, A. P., M. C. NOWYCKY & R. W. TSIEN. 1987. J. Physiol. **394:** 173-200.
3. NILIUS, B., P. HESS, J. B. LANSMAN & R. W. TSIEN. 1985. Nature **316:** 443-446.
4. YOSHINO, M., T. SOMEYA, A. NISHIO & H. YABU. 1988. Pflügers Archiv. **411:** 229-231.
5. YOSHINO, M., A. NISHIO & H. YABU. 1988. Comp. Biochem. Physiol. **89C:** 39-43.

Structure, Function, and Regulation of the Skeletal Muscle Dihydropyridine Receptor[a]

KURT G. BEAM

Department of Physiology
College of Veterinary Medicine and Biomedical Sciences
Colorado State University
Fort Collins, Colorado 80523

TSUTOMU TANABE AND SHOSAKU NUMA

Departments of Medical Chemistry and Molecular Genetics
Kyoto University Faculty of Medicine
Kyoto 606, Japan

INTRODUCTION

Excitation-contraction (E-C) coupling represents an essential but still poorly understood aspect of skeletal muscle function. It is known that depolarization of infoldings of the surface membrane, the transverse tubules (t-tubules), causes calcium to be released from an internal, membrane-bound store, the sarcoplasmic reticulum (SR).[1] This functional arrangement is advantageous because it produces a rapid and nearly simultaneous release of calcium throughout the fiber interior. The mechanism of skeletal muscle E-C coupling is the current focus of intense research efforts. One hypothesis for how this coupling might occur is that calcium entering the muscle fiber across the t-tubular membrane triggers the secondary and much larger release of calcium from the SR.[2,3] Support for this hypothesis was provided by the discovery that skeletal muscle fibers of frogs possess a slowly activated, voltage-dependent calcium current,[4] and that this current flows across the t-tubular membrane.[5] A similar, slow calcium current has been found to be present in skeletal muscle fibers of adult mammals.[6,7] However, blocking this current does not prevent E-C coupling.[8,9] Further, the current activates so slowly[4,6] that an action potential, the normal physiological stimulus for contraction, might be expected to cause very little calcium entry via this pathway.

An alternative hypothesis for E-C coupling in skeletal muscle, suggested by Schneider and Chandler,[10] is that a "voltage sensor" located in the t-tubular membrane is

[a]This research was supported in part by grants from the National Institutes of Health and the Muscular Dystrophy Association (K.G.B.) and by grants from the Ministry of Education, Science and Culture of Japan, the Mitsubishi Foundation, and the Japanese Foundation of Metabolism and Diseases (S.N.).

responsible for controlling the release of calcium from the SR. Specifically, the molecular rearrangement of the voltage sensor in response to changes of t-tubular potential was postulated to gate calcium flow across the SR. Schneider and Chandler demonstrated the presence in frog skeletal muscle of an electrical signal ("charge movement"), which was consistent with the predictions of the voltage-sensor hypothesis.[10] Subsequent work has demonstrated the presence of charge movement in skeletal muscle of other species (cf. Beam & Knudson[11]).

DEVELOPMENTAL REGULATION OF CALCIUM CURRENT

One possible explanation for the presence of the slowly activated calcium current in vertebrate skeletal muscle is that it is vestigial in adult muscle and of importance only in neonatal muscle. Their very small diameter (a few μm compared to ~ 50 μm for the adult) makes it seem plausible that the voltage-dependent entry of extracellular calcium into neonatal muscle fibers could have an important physiological function. For this reason a study[11] was undertaken of calcium-current expression in skeletal muscle fibers acutely isolated from rats of varying postnatal age. The principal result of this study is illustrated in FIGURE 1, which compares calcium currents measured with the whole-cell patch-clamp technique[12] in muscle fibers acutely isolated from a two-day (a) and a 19-day (b) rat. Currents are shown for both a weak depolarization, to -20 mV (A), and a strong depolarization, to $+40$ mV (B). In the two-day fiber, weak depolarization elicits a transient calcium current, "I_{fast}," (A,a) with properties corresponding to T-type[13] calcium currents in other cells. Strong depolarization of the two-day fiber elicits a slowly activating, maintained calcium current, "I_{slow}," (B,a) with properties resembling the slow calcium current of adult skeletal muscle[4,6,7] and L-type[13] calcium currents of other cells. Skeletal muscle myotubes in primary culture possess both I_{fast} and I_{slow} (e.g., FIG. 4a) and thus resemble skeletal muscle acutely isolated from neonatal animals. In the 19-day muscle fiber, I_{fast} is small or absent (A,b) whereas I_{slow} (B,b) is not only present but much larger than in the two-day fiber. The data for the maximum I_{fast} and maximum I_{slow} measured in fibers acutely isolated from animals of varying age are summarized in the lower portion of FIGURE 1. During the first three weeks postnatally I_{fast} is lost, whereas I_{slow} increases substantially. Evidently, I_{fast} must have a function only in the early stages of postnatal development. By contrast, the large postnatal increase in the magnitude of I_{slow} suggests that its expression reflects a function important for adult skeletal muscle. Similar developmental changes in the magnitude of I_{fast} and I_{slow} were reported in a recent study by Gonoi and Hasegawa.[14]

PRIMARY STRUCTURE OF THE DIHYDROPYRIDINE RECEPTOR

I_{slow} is selectively blocked by the dihydropyridine (DHP) calcium-channel blockers[15,16] and the t-tubules of skeletal muscle contain a higher density of DHP binding sites than any other tissue known.[17] Because of this high density, skeletal muscle was used for the biochemical purification of the DHP receptor (reviewed in

Catterall *et al.*[18]). After identification of the 170K polypeptide of the purified rabbit skeletal muscle DHP receptor preparation as the [³H]azidopine[19]-binding component, cDNA encoding this polypeptide was cloned and sequenced.[20] The primary structure of the DHP receptor thus deduced reveals that this protein is homologous with the voltage-dependent sodium channel[21-23] both in amino acid sequence and in proposed

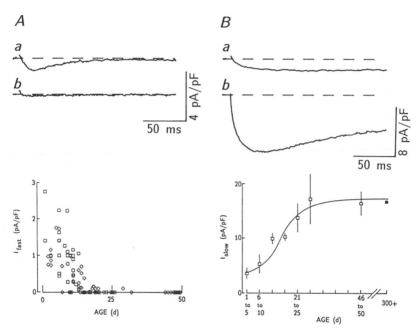

FIGURE 1. Changes in expression of calcium currents during skeletal muscle development. The currents were recorded from muscle fibers of a two-day rat (*a*) and a 19-day rat (*b*) for either a weak (−20 mV) (*A*) or a strong (+40 mV) (*B*) depolarization. The weak depolarization elicits a transient calcium current, I_{fast}, in the two-day fiber (*A,a*) but not in the 19-day fiber (*A,b*). The strong depolarization elicits a maintained calcium current, I_{slow}, in both the two-day (*B,a*) and the 19-day (*B,b*) fiber but the magnitude of I_{slow} is much larger in the 19-day fiber. The maximum values of I_{fast} and I_{slow} recorded in fibers of varying age are plotted beneath the current records of *A* and *B*, respectively. Because there is a large increase in fiber size during postnatal development, the data are presented as current normalized by linear fiber capacitance. The data for I_{fast} were obtained with either 10 m*M* (diamonds) or 50 m*M* (squares) Ca^{2+} in the bath. The data for I_{slow} (open squares, error bars indicate ± SEM) were obtained with 10 m*M* Ca^{2+}; the filled square indicates the average value of peak I_{slow} in omohyoid muscle fibers of adult rats. *a*, cell A76. *b*, cell D48. Figure modified from Beam & Knudson with permission.[11]

transmembrane topology, having four structural units of homology (FIG. 2). Each of the four internal repeats contains a putative voltage sensor segment (S4) with clustered, regularly spaced, positively charged amino acid residues. This structural similarity to the voltage-dependent sodium channel suggests that the DHP receptor in the t-tubular membrane of skeletal muscle may have a dual function, serving both as the voltage

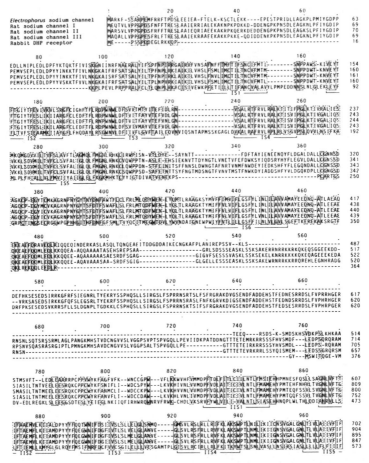

FIGURE 2. Alignment of the amino-acid sequences of the *Electrophorus* electroplax sodium channel (top row), rat brain sodium channels I (second row), II (third row), and III (fourth row), and the rabbit skeletal muscle DHP receptor (bottom row). The one-letter amino acid notation is used. The five sequences from top to bottom have been taken from Refs. 21, 22, 22, 23 & 20, respectively. Sets of five identical amino acid residues at one position are enclosed with solid lines, and sets of five identical or conservative[34] residues at one position with broken lines. Gaps (−) have been inserted to achieve maximum homology. Amino-acid residues are numbered beginning with the initiating methionine, and numbers of the last residues on the individual lines are given on the right-hand side. Positions in the aligned sequences including gaps are numbered beginning with the initiating methionine, and position numbers are given above the sequences. The putative transmembrane segments S1-S6 in each of repeats I-IV are indicated; the termini of these segments have been tentatively assigned. The amino-acid differences resulting from the nucleotide differences found among the individual clones are as follows (position numbers in the aligned sequences are given in parentheses. Sodium channel I: deletion (724-734); Asn (533); Lys (647); Asn (884). Sodium channel II: Lys (552); His (1104); Met (1191); Gly (1328). Sodium channel III: Ile (297); Leu (305); Thr (384); Lys (563); Arg (1196). DHP receptor: Lys (243); Asp (378); Leu (2442).

```
            980        1000        1020        1040        1060
ALVGFQLFGKNYKEYVCKISDDCELPRWHMNDFFHSFLIIVFRALCGEWIETMWDCMEVGQVF--------MCLAVYRMYVIIIGNLVMLNLFLALLLSSFSS  795
AVVGMQLFGKSYNKECVCKIATDCKLPRWHMNDFFHSFLIVFRVLCGEWIETMWDCMEVAGQA--------MCLTVFMMVMVIIGNLVMLNLFLALLLSSFSA  997
AVVGMQLFGKSYNKECVCKISNDCELPRWHMNDFFHSFLIVFRVLCGEWIETMWDCMEVAGQT--------MCLTVFMMVMVIIGNLVMLNLFLALLLSSFSS  988
AVVGMQLFGKSYKECVCKINVDCKLPRWHMNDFFHSFLIVFRVLCGEWIETMWDCMEVAGQT--------MCLIVFMLVMVIIGNLVMLNLFLALLLSSFSS  940
ALLGMQLFGGRYD------FEDTEVRRSNFDNFPQALISVFQVLTGEDWNVSVMYNGIMAYGGPSYPGVLVCIL-YFIIILFVCGNYILLNVFLAIIAVDNL--  664
                                                                       IIS6

           1080       1100       1120       1140       1160
DNLSSIEEDDEVNSLQVASERISRAKNWVK--I--FITGTVQALVLWLVDGKKPPSDDVVGEEGDN-EGKKDTLPL-NYLDGEKIVDGIT------NCVE-  882
DNLAATDDDDENMNNLQIAVDRMHKGVAYVKRKIYEFIQQSFVRKQKILDEIKPL-DDL-NNRKDNCTSNHTT-EIGKDLDCLKDVNGTTSGIGTGSSVEK  1094
DNLAATDDDDENMNNLQIAVGRMQKGIDFVKRKIREFIQKAFVRKQKALDEIKPL-EDL-NNKKDSCISNNHTTIEIGKDLNYLKDGNGTTSGIG--SSVEK  1084
DNLAATDDDDENMNNLQIAVGRMQKGIDFVKKNIRECFRKAFFRKPKVII-EI----QE--GNKIDSCMSSNNTGIEISKELNYLKDGNGTTSGVGTGSSVEK  1033
-----------------AEA-ESLTSAQKA--KAEERKRKKMS--RGL-PDK-TEEEKSVMAKKLE-QKPK----------------------  710

           1180       1200       1220       1240       1260
------------SPTLNLPIKVGES--EIEEEGLVDSSDEEDTNKKRHAL-NDEDSSV--CSTVDYSPSEQDPLAKEEE-EEEEEEPE--ELESKD  958
YIIDESDYMSFINNPSLTVTVPIAVGES--DFENLNTEDFSSESDSLEESKEKL-NESSSSS--EGSTVDIGAPA--------E-EGPVMEPE----ETLE  1176
YVVDESDYMSFINNPSLTVTVPIALGES--DFENLNTEEFSSESDMEESKEKL-N-ATSSS--EGSTVDIGAPA--------EGEOPEAEPE----ESLE  1166
YVIDENDYMSFINNPSLTVTVPIAVGES--DFENLNTEEFSSESELEESKEKL-N-ATSSS--EGSTVDNVAPPR--------EGEOAEIEPE----EDLK  1115
--------------------GEGIPTTAKLKVDEF--ESNVNIVKDPYPS-ADFPGDDEEDEPEIPVSP--------RP-RPLAELQLKEKAVPI  773

           1280       1300       1320       1340       1360
PEACFTEKCIWRFPFLDVDITQGKGKIWWNLRRTCYTIVEHDYFETFIIFMILLSSGVLAFEDIYWRRRVIkVILEYADKVFTYVFIVEMLLKWVAYG  1057
PEACFTEGCVQRFKCCQISVEEGRGKQWWNLRRTCFRIVEHNWFETFIVFMILLSSGALAFEDIYIDQRKTIKTMLEYADKVFTYIFILEMLLKWVAYG  1275
PEACFTEDCVRKFKCCQISIEEGKGKLWWNLRKTCYKIVEHNWFETFIVFMILLSSGALAFEDIYIEQRKTIKTMLEYADKVFTYIFILEMLLKWVAYG  1265
PEACFTEGCIKKFPFCQVSTEEGKGKIWWNLRKTCYSIVEHNWFETFIVFMILLSSGALAFEDIYIEQRKTIKTMLEYADKVFTYIFILEMLLKWVAYG  1214
PEA----------------SSFFIFSPT-NKVRVLCHRIVNATMETNEILEFILLSSAALAEDP-JRAESVRNOILGYFDIAETSVEFTVELVILKMTTYGA  856
                                                      IIS1                          IIS2

           1380       1400       1420       1440       1460
FKRYFT---DAWCWLDFVIIVGAGSIMGITSSLLGYEELGAIKNLRTIRALRPLRALSRFEGMKVVVRALLGAIPSIMNVLLVCLIFWLIFSIMGVNLFAGK  1154
YQTYFT---NAWCWLDFLIVDVSLVSLTANALGYSELGAIKSLRTLRALRPLRALSRFEGMRVVVNALLGAIPSIMNVLLVCLIFWLIFSIMGVNLFAGK  1372
FQMYFT---NAWCWLDFLIVDVSLVSLTANALGYSELGAIKSLRTLRALRPLRALSRFEGMRVVVNALLGAIPSIMNVLLVCLIFWLIFSIMGVNLFAGK  1362
FQTYFT---NAWCWLDFLIVDVSLVLVANALGYSELGAIKSLRTLRALRPLRALSRFEGMRVVVNALLGAIPSIMNVLLVCLIFWLIFSIMGVNLFAGK  1311
FLHKGSFCRNYFNILDLLVVAVSLISMGLES---STISVVKILRVLRVLRPLRAINRAKGLKHVVQCVFVAIRTIGNIVLVTTLLQFMFACIEVQLFKGK  953
            IIS3                                IIS4                       IIS5

           1480       1500       1520       1540       1560
FYKCINT---TTDEILPVEEVNNRSDCMAL-MYTNE-VRWVNLKVNYDNAGMGHGYLSLLQVSTFKGWMDIMYAAVDSREVEDQPINEINVYMYLYFVIFIV  1249
FYHCVNT---TTGDTFEIIEVNNHSDCLKL-IERNETANMKNVKVNFDNVGFGYLSLLQVATFKGWMDIMYAAVDSRNVELQPKYEESLYMYLYFVIFIII  1468
FYHCINY---TTGEMFDVSVVNNYSECQAL-IESNGTARMKNVKVNFDNVGAGYLALLQVATFKGWMDIMYAAVDSRDVKLQPIYEENLYMYLYFVIFIII  1458
FYHCVNT---TTGDMFDIEEVNNHTDCLKL-IGXQ-ARMKNVKVNFDNVGAGYLALLQVATFKGWMDIMYAAVDSRDVKLQPIYEENLYMYLYFVIFIII  1404
FFSCNDLSKMTIEEE--CRGYYYVYKDXGDPTQMELR-PRQWIHNDFHFDNVLSAMMSLFTVSTFEGWPQLLYRAIDSNEDMGPVYNVNARVEMAIFFIIIII  1050
                                                                                         IIS6

           1580       1600       1620       1640       1660
FGAFFTLNLFIGVIIDNFNRQKQKLGGEDLFMTEEQKKYYNAMKKLGSKKAAKCIPRPSNVVQGMVFDIVTDQFDTDIFIMALTCIIMVAMMVVESEDQSQV  1349
FGSFFTLNLFIIGVIIDNFNQQKKKFGGQDIFMTEEQKKYYNAMKKLGSKKPQKPIPRPGNKFQGMVFDIVTRQVFDISIMILICLNMVTMMVETDDQSDY  1568
FGSFFTLNLFIIGVIIDNFNQQKKKFGGQDIFMTEEQKKYYNAMKKLGSKKPQKPIPRPANKFQGMVFDIVTRQVFDISIMILICLNMVTMMVETDDQSKY  1558
FGSFFTLNLFIGVIIDNFNQQKKKFGGQDIFMTEEQKKYYNAMKKLGSKKPQKPIPRPANKFQGMVFDFVTKQVFDISIMILICLNMVTMMVETDDQSKY  1504
LIAFFMMMFIFVGFVINTFQEQGET-EYKNCELDKMDKKQCVQYALKARPLR------CYIPKNPYQYQVMVVTSSYFEYLMFALIMTIICLGMQHYHQSEE  1145
    IIS6                                                               IVS1

           1680       1700       1720       1740       1760
KKDILSQINYIFVIIFTVEELLKLLALR-QYFFTVGWNVFDFVVVIISIIGLIISVIVGMFLAELIEKYFVSPT-LF-------------RVIR-LARIARVLRLI  1431
VTSILSRINLVFIVLFTGECVLKLISLR-HYYFTIGWNIFDFVVVIISIVGMFLAELIEKYFVSPT-LF-------------RVIR-LARIGRILRLI  1650
MTNLLYALSRINLVFIVLFTGECVLKLISLR-YYYFTIGWNIFDFVVVILSIVGMFLAELIEKYFVSPT-LF-------------RVIR-LARIGRILRLI  1640
MTLVLSRINLVFIVLFTGEFLLKLISLR-YYYFTIGWNIFDFVVVILSIVGMFLAELIEKYFVSPT-LF-------------RVIR-LARIGRILRLI  1586
MNHISDILNVAFTIIFTLEMILKLLAFKARGYFGDPMNVFDFLIVIGSIIDVILSE--IDTFLASSGGLYCLGGGCVNVDPDESARILRSSAFFRLFRVMRLI  1244
     IVS2                         IVS3                                          IVS4 -

           1780       1800       1820       1840       1860
RA---AKGIRTLLFALMMSLPALFNIGLLLFLIMFIFSIFGMSNFAYVKKQGG--VDDIFNFETFGNSMICLFEITTSAGWDGLLLPTLNTGPPDCDPDV  1526
XG---AKGIRTLLFALMMSLPALFNIGLLLFLVMFIYAIFGMSNFAYVKREVG--IDDMFNFETFGNSMICLFOITTSAGWDGLLLAPILNSKPPDCDPNK  1745
XG---AKGIRTLLFALMMSLPALFNIGLLLFLVMFIYAIFGMSNFAYVKREVG--IDDMFNFETFGNSMICLFOITTSAGWDGLLLAPILNSAPPDCDPEK  1735
XG---AKGIRTLLFALMMSLPALFNIGLLLFLVMFIYAIFGMSNFAYVKREVG--IDDMFNFETFGNSMICLFOITTSAGWDGLLLAPILNNAPPDCDPDA  1681
KLLSRAEGVRTLLWTFIKSFQALPYVALIIVMLFFIYAVIGMQMFGKIALVDGTQINRNNNFQTFPQAVLLLFRCATGEAWQEILLACSYGKL--CDPES  1342
      IVS5

           1880       1900       1920       1940       1960
EN-PGTDVRGNEGNPGKGIITFFCSYIILISFLVVVNMYIAIILENFGVAQEESSDLLCEDDFVMFDETWEKFDKVKGTQFLDYNDLPRFVNALQEPMRIPN-  1624
VN-PGSSVKGDCGNPSVGIFFFFVSYIIISFLVVVNMYIAVILENFSVATEESAEPLSEDDFEMFYEVWEKFDPDATQFIEFCKLSDFAAALDPPLLIAK-  1843
DH-PGSSVKGDCGNPSVGIFFFFVSYIIISFLVVVNMYIAVILENFSVATEESAEPLSEDDFEMFYEVWEKFDPDATQFIEFCKLSDFAAALDPPLLIAK-  1833
IH-PGSSVKGDCGNPSVGIFFFFVSYIIISFLVVVNMYIAVILENFSVATEESAEPLSEDDFEMFYEVWEKFDPDATQFIEFCKLSDFAAALDPPLLIAK-  1779
DYAPGEEY--TKCG-TNFAYYYFISFYMLCAFLIINLFVAVIMDNFDYLTRDWS--ILGPHHLDEFKAIWAEYDPEAKGRIKHLDVVTLLRRIQPPLGFGKF  1438
     IVS6

           1980       2000       2020       2040       2060
-PNRH--KLAKFDHYVVMEDKISYLDVLLAVTQ---EVLGSGTTEMEA-------------------------------------MRLSII-  1670
-PNKL--OLIAMDLPMVSGDRIHCLDILFAFTK---RVLGESGEMDA-------------------------------------LRIQM-  1889
-PNKV--OLIAMDLPMVSGDRIHCLDILFAFTK---RVLGESGEMDA-------------------------------------LRIQM-  1879
-PNKV--OLIAMDLPMVSGDRIHCLDILFAFTK---RVLGESGEMDA-------------------------------------LRIQM-  1825
CPHRVACKRLVGNMPNSLNSDGTVTFNATLFALVRTALKIXTEGNFEQANEELRAIIKKIWKRTSMKLLDQVIPPIGDDEVTVGKFYATFLIGEHFRRKFMK  1538

           2080       2100       2120       2140       2160
-QAKFXKDNPSPTFFEPVYTTLRRKEEFAESVVTQORAFRQYLLMRAVSHASFLSQIKHMNEGPKGGVGS---QDSLITQKMNALYRGN-PELTMPLEQQI  1765
-EERFMASNPSKVSYQPITTTLKRKQEEVSAVIIQRAYRRHLLKRTVKQASFTYNKNKL----KGGANLLVKED-MIIDRINENSITEKTDLTMS------  1978
-EERFMASNPSKVSYEPITTTLKRKQEEVSAVIIQRAYRRYLLKQKVKKVSSIYKKDKG----KEDEGTPIKED-TIIDKLNENSTPEKTDVTPS------  1968
-EDRFMASNPSKVSYEPITTTLKRKQEEVSAVIIQRNYRCYLLKQRLKNISSKYDKETI----KGRIDLPIKGD-MVIDKLNGNSTPEKTDGSSS------  1914
RQEEYYGYRPKKKDTVQ-IQAGLRTIEEEAAPEIRBTISGDLTAEEELERAMVEAAMEE----RIFRRTGGLFG-QV------DTFLERTNSLPP------  1620

           2180       2200       2220       2240       2260
KPMLDKPR-MPSL-SVPETYPIQIPKEVTNEVI-LHSAPMVRQNYSYSGAIVVR-ESI-V-----------------------------------  1820
-----TAACPPSYDRSV--TKPIVEKHE--QEGK-DEKA-KGK------------------------------------------  2009
-----TTS-PPSYDSV--TKPEKEKFE--KD-K-SEKEDKGKD---------------IR-ESK-K-----------------------  2005
-----TTS-PPSYDSV--TKPDKEKFE--KD-K-PEKEIKGKE---------------VR-ENQ-K-----------------------  1951
-----VMA---NQRPL--QFAEIEMEEL--LEJ-SPVFLEDFPQD---------ARTNPLARANTNNANANVAYGNSNHSNNQMFSSVHCEREFPGEAETPA  1698

           2280       2300       2320       2340       2360
-------------------------------------------------------------------------------------  1820
-------------------------------------------------------------------------------------  2009
-------------------------------------------------------------------------------------  2005
-------------------------------------------------------------------------------------  1951
AGRGALSHSHRALGPHSKPCAGKLNGQLVQPGMPINQAPPAPCQQPSTDPPERGQRRTSLTGSLQDEAPQRRSSEIGSTPRRPAPATALLIQEALVRGGLD  1798

           2380       2400       2420       2440
-------------------------------------------------------------------------------------
-------------------------------------------------------------------------------------
-------------------------------------------------------------------------------------
-------------------------------------------------------------------------------------
TLAADAGFVTATSQALADACQMEPEEVEVAATELLKARESVQGMASVPGSLSRRSSLGSLDQVQGSQETLIPPRP  1873
```

sensor for E-C coupling and as a calcium channel.[20] An involvement of the t-tubular DHP receptor in E-C coupling was suggested previously on the basis of observed effects of DHP on charge movement and myoplasmic calcium transients.[24]

ANALYSIS OF A MOUSE MUTANT WITH DEFECTIVE E-C COUPLING

Muscular dysgenesis (*mdg*) is an autosomal, recessive mutation of mice.[25] Mice homozygous (*mdg/mdg*) for this mutation only survive until birth, at which time they die of respiratory arrest due to complete paralysis of their skeletal muscles. Analysis of skeletal muscle acutely isolated from afflicted animals[26] and of primary cultures of dysgenic myotubes[27,28] shows that this mutation eliminates E-C coupling in skeletal muscle. The sarcolemma of dysgenic muscle is functional, having the ability to generate propagated action potentials.[26–28] The ability of caffeine to produce contraction of dysgenic muscle[26,29] suggests that these cells possess a functional SR and contractile apparatus. Electrophysiological analysis[30] demonstrates that dysgenic myotubes in primary culture possess the I_{fast} calcium current characteristic of normal embryonic skeletal muscle, but that I_{slow} is completely lacking (e.g., see FIG. 4b). Similar results have been obtained for acutely dissociated muscle fibers from muscular dysgenic mice (B. Adams and K. Beam, unpublished). The mutation does not appear to affect calcium currents of cardiac or neuronal cells.[30]

The muscular dysgenesis phenotype is transmitted as a single recessive defect and has been maintained in highly inbred colonies of mice. Thus, it was reasonable to test the possibility that the mutation affects the structural gene for the skeletal muscle DHP receptor. For this purpose, in collaboration with Dr. J. Powell, we carried out restriction-fragment analysis of genomic DNA from homozygous normal (+/+), homozygous mutant (*mdg/mdg*), and heterozygous (+/*mdg*) animals using four different restriction endonucleases and nine different cDNA probes for the rabbit skeletal muscle DHP receptor.[31] FIGURE 3 illustrates the results of this analysis for cDNA probes 2 and 6, which correspond to repeat I and part of repeat III of the rabbit skeletal muscle DHP receptor, respectively. Both probes revealed restriction fragments that differed in size between +/+ and *mdg/mdg* DNA. The +/*mdg* DNA yielded restriction fragments that corresponded in size both to those of +/+ and of *mdg/mdg*, as expected if the heterozygote carries one normal and one mutant copy of the affected gene. The restriction fragment differences revealed by probe 2 were distinct from those revealed by probe 6. None of the other probes revealed differences in restriction fragments between +/+ and *mdg/mdg* DNA that were distinct from those revealed by probes 2 and 6. Probe 1 showed some of the same *Eco*RI restriction fragment differences between +/+ and *mdg/mdg* as revealed by probe 2, but showed common restriction fragments for the other three restriction endonucleases used. Probe 5 showed the same *Apa*I restriction fragment difference between +/+ and *mdg/mdg* as revealed by probe 6, but also showed, for all four endonucleases (including *Apa*I), restriction fragments that were common to +/+ and *mdg/mdg*. Probes 7, 8, and 9 revealed no differences in restriction fragments between +/+ and *mdg/mdg* DNA. Thus, the restriction fragment analysis indicates that the muscular dysgenesis mutation is associated with alterations of at least two distinct regions of the structural gene for the skeletal muscle DHP receptor.

The analysis of poly(A)$^+$ RNA from skeletal muscle of 18-day embryos of +/+ and *mdg/mdg* mice revealed that the mRNA that presumably codes for the mouse skeletal muscle DHP receptor is strongly reduced in dysgenic skeletal muscle.[31] Fur-

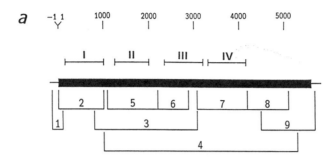

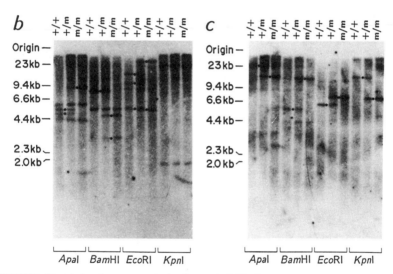

FIGURE 3. Restriction-fragment analysis of genomic DNA from normal and mutant mice using cDNA probes for the rabbit skeletal muscle DHP receptor. *a*: The heavy bar indicates the protein-coding region and the thin lines indicate the 5'- and 3'-noncoding regions of the cDNA for the rabbit skeletal muscle DHP receptor. The Roman numerals indicate the four regions of internal repeat. The top set of numbers indicates the nucleotide number, the first residue of the ATG triplet encoding the initiating methionine being +1. The numbered brackets below indicate the regions spanned by the cDNA probes used. *b,c*: Autoradiograms of blot hybridization analysis using cDNA probes 2 and 6, respectively. The restriction endonucleases are indicated below the autoradiograms. Genomic DNA was obtained from a homozygous normal (+ / +), a homozygous mutant (m/m) and a heterozygous (+ /m) mouse as indicated. Restriction fragments that differ in size between DNAs from the + / + and the *mdg / mdg* mouse are indicated by diamonds. Figure modified from Tanabe *et al.*[31] with permission.

thermore, analysis using antibodies against rabbit skeletal muscle proteins showed that the DHP-binding polypeptide was present in skeletal muscle membranes of normal neonatal mice but undetectable in skeletal muscle membranes of *mdg / mdg* neonatal mice.[32] Examination of several other proteins suggested that this evidently complete deficit is restricted to the DHP-binding polypeptide. Thus, both the genetic and biochemical analyses suggest that the primary defect in muscular dysgenesis is to alter, and to prevent expression of, the skeletal muscle DHP receptor gene.

In primary culture, a significant fraction of normal myotubes are able to contract upon depolarization as a result of spontaneous action potential discharge or of electrical stimulation applied via an extracellular pipette. Dysgenic myotubes, however, do not contract upon depolarization. If this failure of E-C coupling in dysgenic skeletal muscle is a result of mutation of the skeletal muscle DHP receptor gene, then one would predict that it would be possible to repair this defect by supplying normal copies of the gene. Furthermore, if the DHP receptor functions as a calcium channel, any restoration of E-C coupling ought to be accompanied by a restoration of I_{slow}, the DHP-sensitive calcium current that is present in normal myotubes (FIG. 4a) but absent in dysgenic myotubes (FIG. 4b). To test these predictions,[31] we prepared an expression plasmid ("pCAC6") that carries cDNA encoding the rabbit skeletal muscle DHP receptor. This expression plasmid was microinjected[33] into nuclei of dysgenic myotubes *in vitro*. The injected myotubes were examined on subsequent days for the appearance of functional E-C coupling as evidenced by the appearance of spontaneous contractions or contractions elicited by electrical stimulation. To date, of the ~600 myotubes that survived the injection procedure, 115 were found to have restored E-C coupling (41 of these contracted spontaneously and 74 contracted in response to electrical stimulation). Those cells that were observed to contract were examined with the whole-cell patch-clamp technique to determine which calcium currents were present. As shown in FIGURE 4c, injected dysgenic myotubes that had been observed to contract displayed an I_{slow} similar to that of normal myotubes. This restoration of I_{slow} was observed for all 26 of the injected dysgenic myotubes from which patch-clamp recordings were obtained. I_{slow} in injected dysgenic myotubes is blocked by the dihydropyridine (+)PN 200-110 with about the same efficacy as I_{slow} of normal myotubes.[31]

In principle, the restoration of contractile activity in injected dysgenic myotubes might arise from calcium entering via I_{slow}. This would differ from E-C coupling of normal skeletal muscle and myotubes, for which calcium entry is not required.[8,9] To determine whether calcium entry is involved in the E-C coupling restored by injection of dysgenic myotubes, the calcium-channel blocker Cd^{2+} was used. At a concentration of 0.5 mM, Cd^{2+} produces a >90% block of I_{slow} (FIG. 5a). However, the presence of 0.5 mM Cd^{2+} did not prevent electrical stimulation from producing robust contractions in pCAC6-injected dysgenic myotubes. Similar results were obtained when calcium currents were abolished by replacing extracellular Ca^{2+} with Mg^{2+} (data not shown). Thus E-C coupling of injected dysgenic myotubes resembles that of normal skeletal muscle in which depolarization directly causes the release of calcium from the SR without the necessity for voltage-dependent calcium currents to trigger the process.

CONCLUSIONS

The ability of cDNA encoding the skeletal muscle DHP receptor to restore normal phenotype in dysgenic muscle strongly supports the hypothesis that the skeletal muscle

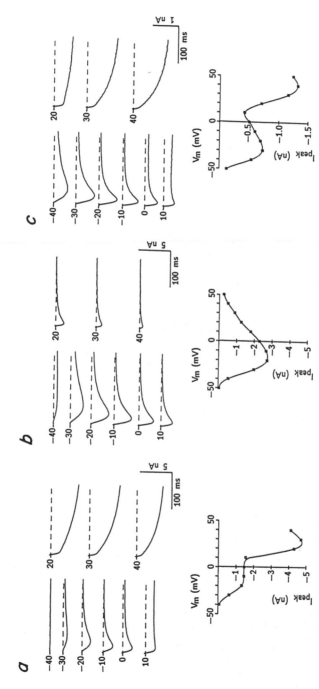

FIGURE 4. Calcium currents in a normal ($+/mdg?$) myotube (*a*), a dysgenic (mdg/mdg) myotube (*b*), and a dysgenic myotube injected with the expression plasmid pCAC6 (*c*). Test potentials are indicated next to each current trace. The peak current (I_{peak}) is plotted against test potential (V_m) beneath the current traces for each cell. (From Tanabe *et al.*[31] Reprinted by permission from *Nature*.)

DHP receptor functions both as an essential component of E-C coupling and as the slow calcium channel. In the former role, it most likely represents the t-tubular voltage sensor[10,20,24] that controls the release of calcium from the SR. With respect to the latter role, one can imagine that it functions to replenish intracellular stores of calcium. Certainly, the substantial developmental increase in the magnitude of I_{slow} argues that this current represents the expression of a vital muscle function. The ability to express and manipulate the cDNA for the skeletal muscle DHP receptor should help to define the nature of the interaction between the DHP receptor and the SR calcium release channel and the physiological role of I_{slow}.

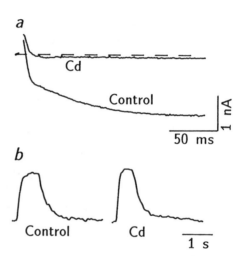

FIGURE 5. Restoration of E-C coupling by injection of pCAC6 into dysgenic myotubes. a: measurement of I_{slow} in a pCAC6-injected dysgenic myotube before (control) and during exposure to 0.5 mM Cd^{2+}. b: Electrically evoked contractions in a pCAC6-injected dysgenic myotube before (control) and during exposure to 0.5 mM Cd^{2+}. The methods for evoking and recording contractions were as described previously.[31] Vertical scale: arbitrary units.

REFERENCES

1. COSTANTIN, L. L. 1975. Prog. Biophys. Molec. Biol. **29:** 197-224.
2. ENDO, M., M. TANAKA & Y. OGAWA. 1970. Nature **228:** 34-36.
3. FORD, L. E. & R. J. PODOLSKY. 1970. Science **167:** 58-59.
4. SANCHEZ, J. A. & E. STEFANI. 1978. J. Physiol. London **283:** 197-209.
5. NICOLA SIRI, L., J. A. SANCHEZ & E. STEFANI. 1980. J. Physiol. London **305:** 87-96.
6. DONALDSON, P. L. & K. G. BEAM. 1983. J. Gen. Physiol. **82:** 449-468.
7. WALSH, K. B., S. H. BRYANT & A. SCHWARTZ. 1986. J. Pharmacol. Exp. Ther. **236:** 403-407.
8. GONZALEZ-SERRATOS, H., R. VALLE-AGUILERA, D. A. LATHROP & M. DEL CARMEN GARCIA. 1982. Nature **298:** 292-294.
9. KNUDSON, C. M., S. D. JAY & K. G. BEAM. 1986. Biophys. J. **49:** 13a.
10. SCHNEIDER, M. F. & W. K. CHANDLER. 1973. Nature **242:** 244-246.
11. BEAM, K. G. & C. M. KNUDSON. 1988. J. Gen. Physiol. **91:** 799-815.
12. HAMILL, O. P., A. MARTY, E. NEHER, B. SAKMANN & F. J. SIGWORTH. 1981. Pflügers Arch. Ges. Physiol. **391:** 85-100.
13. NOWYCKY, M. C., A. P. FOX & R. W. TSIEN. 1985. Nature **316:** 440-443.
14. GONOI, T. & S. HASEGAWA. 1988. J. Physiol. London **401:** 617-637.

15. ALMERS, W., R. FINK & P. T. PALADE. 1981. J. Physiol. London **312:** 177-207.
16. COGNARD, C., M. LAZDUNSKI & G. ROMEY. 1986. Proc. Natl. Acad. Sci. USA **83:** 517-521.
17. JANIS, R. A. & A. SCRIABINE. 1983. Biochem. Pharmacol. **32:** 3499-3507.
18. CATTERALL, W. A., M. J. SEAGAR & M. TAKAHASHI. 1988. J. Biol. Chem. **263:** 3535-3538.
19. FERRY, D. R., M. ROMBUSCH, A. GOLL & H. GLOSSMANN. 1984. FEBS Lett. **169:** 112-118.
20. TANABE, T., H. TAKESHIMA, A. MIKAMI, V. FLOCKERZI, H. TAKAHASHI, K. KANGAWA, M. KOJIMA, H. MATSUO, T. HIROSE & S. NUMA. 1987. Nature **328:** 313-318.
21. NODA, M., S. SHIMIZU, T. TANABE, T. TAKAI, T. KAYANO, T. IKEDA, H. TAKAHASHI, H. NAKAYAMA, Y. KANAOKA, N. MINAMINO, K. KANGAWA, H. MATSUO, M. RAFTERY, T. HIROSE, S. INAYAMA, H. HAYASHIDA, T. MIYATA & S. NUMA. 1984. Nature **312:** 121-127.
22. NODA, M., T. IKEDA, T. KAYANO, H. SUZUKI, H. TAKESHIMA, M. KURASAKI, H. TAKAHASHI & S. NUMA. 1986. Nature **320:** 188-192.
23. KAYANO, T., M. NODA, V. FLOCKERZI, H. TAKAHASHI & S. NUMA. 1988. FEBS Lett. **228:** 187-194.
24. RIOS, E. & G. BRUM. 1987. Nature **325:** 717-720.
25. GLUECKSOHN-WAELSCH, S. 1963. Science **142:** 1269-1276.
26. BOURNAUD, R. & A. MALLART. 1987. Pflügers Arch. Ges. Physiol. **409:** 468-476.
27. POWELL, J. A. & D. M. FAMBROUGH. 1973. J. Cell. Physiol. **82:** 21-38.
28. KLAUS, M. M., S. P. SCORDILIS, J. RAPALUS, R. T. BRIGGS & J. A. POWELL. 1983. Dev. Biol. **99:** 152-165.
29. BOWDEN-ESSIEN, F. 1972. Dev. Biol. **27:** 351-364.
30. BEAM, K. G., C. M. KNUDSON & J. A. POWELL. 1986. Nature **320:** 168-170.
31. TANABE, T., K. G. BEAM, J. A. POWELL & S. NUMA. 1988. Nature **336:** 134-139.
32. KNUDSON, C. M., N. CHAUDHARI, A. H. SHARP, J. A. POWELL, K. G. BEAM & K. P. CAMPBELL. 1989. J. Biol. Chem. **264:** 1345-1348.
33. KUDO, A., F. YAMAMOTO, M. FURUSAWA, A. KUROIWA, S. NATORI & M. OBINATA. 1982. Gene **19:** 11-19.
34. DAYHOFF, M. O., R. M. SCHWARTZ & B. C. ORCUTT. 1978. *In* Atlas of Protein Sequence and Structure. Vol. 5, Suppl. 3. M. O. Dayhoff, Ed.: 345-352. National Biomedical Research Foundation. Silver Spring, MD.

Calcium Channels Reconstituted from the Skeletal Muscle Dihydropyridine Receptor Protein Complex and Its α_1 Peptide Subunit in Lipid Bilayers[a]

DIETER PELZER,[b] AUGUSTUS O. GRANT,[c,d]
ADOLFO CAVALIÉ,[b] SIEGRIED PELZER,[b]
MANFRED SIEBER,[e] FRANZ HOFMANN,[e] AND
WOLFGANG TRAUTWEIN[b]

[b] *II. Physiologisches Institut*
Medizinische Fakultät
Universität des Saarlandes
D-6650 Homburg/Saar, Federal Republic of Germany

[c] *Cardiovascular Division, Department of Medicine*
Duke University Medical Center
Durham, North Carolina 27710

[e] *Physiologische Chemie*
Medizinische Fakultät
Universität des Saarlandes
D-6650 Homburg/Saar, Federal Republic of Germany

INTRODUCTION

The dihydropyridines (DHPs), as exemplified by nifedipine, nitrendipine, PN 200-110, and so on, have been shown to inhibit calcium influx through some voltage-sensitive calcium channels (VSCC) *in vivo* (see Refs. 1-8a for reviews and references). High-affinity binding sites for the tritiated congeners of some of these compounds have been identified in a variety of tissues *in vitro* (see Refs. 5, 6, 9-12 for reviews and references). The high-affinity DHP receptor (DHPR) purified from skeletal muscle transverse (t) tubules or microsomes[13-17] (see Glossman & Striessnig[12] for review) appears to be a multisubunit protein with an approximate molecular mass of 210,000 daltons. Reconstitution of this protein into lipid vesicles[18] or lipid bilayers[19] is associated with the reconstitution of VSCC activity. However, lipid bilayer recordings revealed a heterogeneity of VSCC activity from purified DHPR (pDHPR) prepara-

[a] Supported by the Deutsche Forschungsgemeinschaft, SFB 246, Projects A1 and B2.

[d] Recipient of a research fellowship from the Alexander von Humboldt Foundation.

138

tions.[17,20-23] It is reassuring that functionally and pharmacologically similar classes of VSCC have also been identified after incorporation of solubilized DHPR (sDHPR) samples and DHPR-containing t-tubule membrane vesicles (mDHPR) from skeletal muscle into lipid bilayers.[17,20,23] Because of their apparent persistent coexistence in DHPR preparations after extensive purification, the multiple types of VSCC are likely to be closely related in structure and cannot be separated by DHP (and/or phenyl-alkylamine) binding alone.

Purified skeletal muscle DHPR protein complexes that form functional VSCC after reconstitution into lipid bilayers consist of several tightly, noncovalently associated polypeptides, which are termed α_1, α_2, β, and γ (see Glossman & Striessnig[12] for review and references). Apparent molecular weights for the respective peptide subunits of pDHPR range from 155,000-200,000 Da for α_1 under reducing (R) and nonreducing (NR) conditions, 165,000-175,000 Da (NR) or 130,000-145,000 Da (R) for α_2, 50,000-65,000 Da (R/NR) for β, and 30,000-35,000 Da (R/NR) for γ. A small δ subunit (22,000-30,000 Da [R]) is disulfide linked to α_2; α_2, γ, and δ are heavily glycosylated, whereas α_1 is poorly glycosylated and β appears to have no glycoprotein at all. Hydrophobic regions are predominantly found in α_1 and γ, and to a lesser extent in α_2 and δ. The DHP and phenylalkylamine drug receptor sites are located on α_1. This subunit (together with the β subunit) is the substrate for protein kinases A and C as well as an intrinsic protein kinase from skeletal muscle triads. The α_1 subunit carries the regulatory sites for L-type VSCC drugs and has a primary structure highly homologeous to the α subunit of the voltage-dependent sodium channel. Thus, α_1 should be a major constituent of the L-type VSCC *in vivo* and *in vitro*. However, direct experimental (functional) proof for this "biochemical" assumption is lacking up to now. In fact, the identity of any of the observed peptide subunits of the pDHPR protein complex with L-type VSCC and their relation to functionally and pharmacologically distinct types of VSCC is not yet clear. As suggested by Sieber *et al.*,[24] elucidation of this problem requires work directed toward reconstituting the isolated peptide subunits of the pDHPR protein complex.

In this communication, we first summarize the functional and pharmacological properties of two types of elementary VSCC activity reconstituted from skeletal muscle sDHPR and pDHPR protein samples into lipid bilayers in our laboratory. We then proceed with a description of preliminary experimental evidence from single-channel recordings on the function and modulation of isolated peptide subunits from the skeletal muscle pDHPR protein complex in lipid bilayers, and finally end with a trial to establish the relation of VSCC activity and properties in lipid bilayers to VSCC types, function, and modulation in intact t-tubule membranes and skeletal muscle fibers.

METHODS

Materials and Solutions

Bovine heart phosphatidylethanolamine (PE) and bovine brain phosphatidylserine (PS) were obtained from Avanti Polar Lipids (Birmingham, AL), and cholesterol was purchased from Sigma Chemical (St. Louis, MO). Lipid solutions were prepared from a mixture of 70% PE, 15% PS, and 15% cholesterol dissolved in *n*-hexane (1 mg per ml). (±)-Methoxyverapamil ((±)-D600), (±)-BAY K 8644, (+)-PN 200-

110 and adenosine-5-(γ-thio)-triphosphate (ATP-γ-S) were from Knoll AG (Ludwigshafen, FRG), Bayer AG (Leverkusen, FRG), and Serva (Heidelberg, FRG), respectively. The catalytic (C) subunit of the cAMP-dependent protein kinase (cA kinase) was prepared and purified from bovine heart.[25,26] The bath and pipette filling solutions usually contained 90 mM BaCl$_2$, 5 mM Hepes, and 0.02 mM TTX. When indicated (e.g., FIG. 2), the BaCl$_2$ concentration in the pipette was reduced to 20 mM. The pH was adjusted to 7.4 by Ba(OH)$_2$, and the experiments were carried out at room temperature. All other materials used were of the highest quality and purity available to us.

Purification of the DHP-Binding Site and Separation of the Peptide Subunits of pDHPR

Membranes were prepared from white rabbit skeletal muscle according to Flockerzi et al.[16] Solubilization and purification of the DHP binding site was carried out as described[16] with modifications introduced by Sieber et al.[24] The peptides of pDHPR were separated by molecular size-exclusion chromatography on HPLC columns.[24] Solubilized and purified DHPR preparations as well as α_1 (from the leading edge of peak I of eluted peptides), β (from the tailing part of peak III of eluted peptides), and γ (from the peak fraction of peak IV of eluted peptides) (see Sieber et al.[24] for details) were reconstituted into lipid vesicles. The vesicles were then stored in small aliquots at $-80°$C until they were used for incorporation into lipid bilayer membranes.

Lipid Bilayers

Solvent-free lipid bilayers were formed from monolayers on the tips of thick-walled, unpolished, uncoated, hard borosilicate glass patch pipettes (5-10 MΩ) by the contact method as described elsewhere.[27] For lipid monolayers, 5-15 μl lipid solution was spread at the surface of the salt solution contained in the 1.5-ml bath. Glass-bilayer seal resistances were in the order of 10-50 GΩ and the probability of successful bilayer formation was $\geq$ 0.75. The confidence of our bilayer assembly was periodically tested with gramicidin that regularly formed pores with elementary conductances of 3-7 pS (100 mM NaCl) in our bilayers, in agreement with previous reports in other bilayer set-ups.[28] Before incorporation of DHPR protein- or peptide subunit-containing liposomes into the lipid bilayers, their electrical stability was always assured by applying constant potential gradients between -100 and $+100$ mV across the artificial membranes. Only those bilayers with stable artifact-free current baselines for 15-30 minutes were considered reliable for further experimentation. With this test at the beginning of each experiment, about 20-25% of tightly sealed bilayers had to be discarded because of the spontaneous occurrence of single-channel-like current pulses of variable amplitudes. By contrast to protein- or peptide-related single-channel activity, no sharp peaks could be detected in the current amplitude histograms of such bilayer-related elementary events. Additionally, sequences of single-channel-like bilayer activity were accompanied by irreversible stepwise decrements of the seal resistance. These observations suggest that spontaneous single-channel-like elementary current pulses in pure bilayers may arise from "micro ruptures" of the bilayer itself and/or of the glass-bilayer seal. Support for this idea comes from experiments, which show that monomolecular films of stearic acid properly compressed on the water surface are not completely homogeneous but can include densely scattered small holes.[29]

Incorporation of VSCC into Lipid Bilayers

Reconstituted DHPR protein complexes or isolated pDHPR peptides were incorporated into preformed artifact-free lipid bilayers by adding 10-40 μl of the liposome suspension to the bath underneath the bilayer membrane. The responsiveness of incorporated channels to chemical agents was tested by applying the chemicals at desired concentrations to the bath side of the lipid bilayer.

Sidedness of VSCC in Lipid Bilayers

According to Tanabe *et al.*,[30] all potential cAMP-dependent phosphorylation sites are located on the cytoplasmic side of the DHPR molecule or rather its α_1 subunit. When DHPR protein- or α_1 peptide-containing liposomes were successfully reconstituted as VSCC, the functional consequences of VSCC phosphorylation were observed when the C subunit of the cA kinase was applied to the bath side of the bilayer in the presence of MgATP.[19,22,23] This implies that DHPR and/or α_1 molecules functioning as VSCC are likely to be positioned in our bilayer assembly with their cytoplasmic sides facing the bath resulting in an inside-out configuration of single-channel recording.

Electrical Recording and Data Analysis

Elementary currents were recorded at constant potentials using a conventional patch-clamp amplifier (L/M-EPC 7, List Medical Electronic, Darmstadt, FRG). The potential across the bilayer membrane was controlled by clamping the pipette potential (V_p) with respect to the bath. According to the recording configuration, a positive membrane potential corresponds to a negative voltage signal on the pipette voltage display and vice versa. In the result section of this paper, pipette polarities are given exclusively. Positive current signals at positive V_p (negative membrane potentials) are displayed as upward deflections from the zero current baseline and reflect cationic current flow from the pipette into the bath. Vice versa, negative current signals at negative V_p (positive membrane potentials) are shown as downward deflections from the zero current baseline and represent cationic current flow from the bath into the pipette. Anionic current flow into the opposite direction was excluded by measurements of the current reversal potential (V_r) following a decrease of the $BaCl_2$ concentration in the pipette to 20 mM. Single-channel currents were recorded continuously on an FM tape recorder at 7½ or 15 ips, and then played back through a four-pole Bessel filter (-3 dB) at one-fourth of the digitizing rate (1.5-3.6 kHz) for analysis on a Nicolet MED-80 or PDP 11-73 computer as described elsewhere.[31] The open-state probability (P_o) is defined as the time average current divided by the single-channel current amplitude, and was calculated from 3-35-minute records continuously chopped into two-second segments.

RECONSTITUTION OF MULTISUBUNIT sDHPR AND pDHPR PROTEINS

Success of Reconstitution and Characterization of Elementary Conductance Levels in Successful Trials

The DHP binding site solubilized and purified from rabbit skeletal muscle microsomes appears to be a multisubunit protein with five major peptide bands of apparent molecular masses of 165 (α_1), 130 (α_2), 55 (β), 32 (γ), and 28 (δ) kDa under reducing conditions.[16,19,24] Under nonreducing conditions, the 130-kDa (α_2) and the 28-kDa peptides (δ) migrate as a single peptide of 165 kDa, the latter being unrelated to the 165-kDa peptide (α_1), which does not change its molecular mass upon reduction.[24] In the attempt to study the functional and regulatory properties of this target protein for Ca^{2+} antagonists and cAMP-dependent phosphorylation, sDHPR and pDHPR protein complexes were reconstituted into lipid vesicles, and the vesicles were incorporated into lipid bilayer membranes formed at the tips of patch pipettes. In most experiments, either no reconstitution was obtained within 60 minutes of observation ($136/252 \simeq 54\%$), or reconstitution of doubtful single-channel activity was achieved ($96/252 \simeq 38.1\%$) with individual elementary conductances between 3 and 80 pS. This activity was insensitive to block by ($\pm$)-D600, ($+$)-PN 200-110, and even $CoCl_2$ in concentrations up to 40 mM. Only in a low percentage of trials ($20/252 \simeq 7.9\%$), the reconstitution of sDHPR and pDHPR was unequivocally associated with the reconstitution of VSCC activity, as judged from the responsiveness of single-channel events to organic VSCC blockers (($\pm$)-D600), VSCC activators (($\pm$)-BAY K 8644) and/or cAMP-dependent phosphorylation, or from sensitivity to the inorganic VSCC blocker $CoCl_2$. FIGURE 1 shows typical elementary current records obtained after successful incorporation of sDHPR (FIG. 1A) and pDHPR (FIG. 1B) protein complexes as pharmacologically regulated VSCC. In both DHPR preparations, two single-channel current levels were regularly identified, either singly (FIG. 1Ab, c; Bb, c), or together (FIG. 1Aa; Ba) in the same bilayer. At a V_p of -70 mV, the larger elementary current level had a unitary current amplitude of 1.45 pA, which roughly corresponds to a conductance of 20 pS (FIG. 1Ac; Bc). By contrast, the unitary amplitude of the smaller single-channel current level was 0.65 pA at the same V_p, which gives a conductance of about 9 pS (FIG. 1Ab; Bb). In addition, the 9-pS events exhibit distinctly longer open times than the 20-pS activity. In the example experiments (FIG. 1Ab, c; Bb, c), the 20- and 9-pS conductance levels were observed singly in different bilayers for more than two hours without interconversion of one into the other. This observation, which is typical for all stable experiments with only one conductance level to start with, favors the conclusion that the two conductance levels represent activity from two distinct channel molecules rather than that they arise from the same molecule.

Current-Voltage Relations

FIGURE 2 shows the 20-pS (FIG. 2Aa) and the 9-pS (FIG. 2Ba) condu ....is
of elementary current separately at constant pipette potentials in symmetrical 90 mM
BaCl$_2$ solutions. The records shifted to the left were obtained from sDHPR, whereas
the traces displaced to the right were recorded from pDHPR. The elementary current-
V_p relationships were linear over a wide potential range (FIG. 2Ab, c; Bb, c; filled
circles) and slope conductances were 19.6 (FIG. 2Ab; sDHPR), 20.1 (FIG. 2Ac;
pDHPR), 8.5 (FIG. 2Bb; sDHPR) and 9.5 pS (FIG. 2Bc; pDHPR), respectively. The
zero current potential (V_r) was around 0 mV in each case. With 20 mM BaCl$_2$ in the
pipette and 90 mM BaCl$_2$ in the bath (FIG. 2Ab, c; Bb, c; open circles), the respective
single-channel current-V_p relations remained linear with slope conductances of 20.9

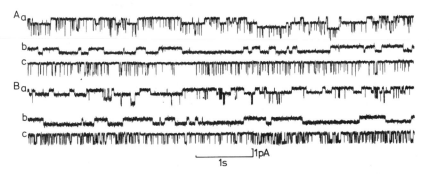

FIGURE 1. Single-channel recordings after incorporation of the solubilized (**A**) or purified (**B**)
DHP receptor into lipid bilayers. The current records were obtained in six different bilayers. In
all cases, activity started within the first 5-7 minutes after addition of DHPR to the bath side
underneath the bilayer membrane and lasted for more than two hours. The pipette potential
(V_p) was -75 mV (Aa), -70 mV (Ab, Ac), -65 mV (Ba, Bb), and -60 mV (Bc). At
negative V_p, downward deflections represent channel openings and reflect negative elementary
cationic current flowing from the bath into the pipette. The pipette and bath contained 90 mM
BaCl$_2$ solutions. The records were digitized at 1.5 kHz.

(FIG. 2Ab; sDHPR), 18.9 (FIG. 2Ac; pDHPR), 9.3 (FIG. 2Bb; sDHPR), and 10 pS
(FIG. 2Bc; pDHPR), respectively. However, V_r ranged between $+18.7$ and $+22.7$
mV, as expected if Ba^{2+} ions are the charge carrier of both types of elementary activity.

Gating Kinetics

The voltage dependence of the kinetic behavior of the 20- (FIG. 3A) and the 9-
pS (FIG. 3B) elementary events was analyzed singly in separate bilayers. In symmetrical 90 mM BaCl$_2$ solutions (FIGS. 3 and 4), the open and closed time distributions
of both types of single-channel activity could be fitted by two exponential functions

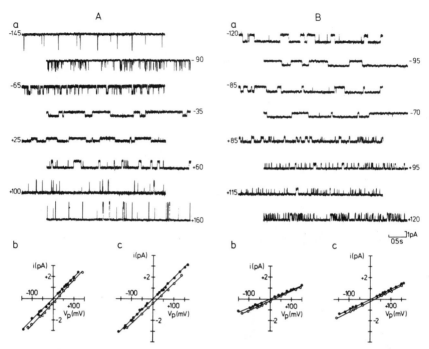

FIGURE 2. Current-voltage relations and selectivity of two types of elementary events. The original traces of single-channel current (Aa, Ba) were recorded at various pipette potentials (V_p), indicated beside each record, in symmetrical 90 mM BaCl$_2$ solutions. In this and all following figures, downward deflections of elementary current at negative V_p (positive membrane potentials) represent channel openings, whereas at positive V_p (negative membrane potentials) channel openings are represented by upward deflections of single-channel current. The records shifted to the left in Aa and Ba were obtained from sDHPR, whereas the traces displaced to the right were recorded from pDHPR, respectively. The sampling frequency was 1.5 kHz. The lower panels show elementary current-voltage relations from mean values ($4 \leq n \leq 8$, SD not shown) of single-channel current amplitudes in symmetrical 90 mM BaCl$_2$ solutions ($\bullet$), and with 20 mM BaCl$_2$ in the pipette and 90 mM BaCl$_2$ in the bath ($\circ$), for solubilized (Ab,Bb) and purified (Ac, Bc) DHP receptor. The straight lines are least-squares fits to the data with the parameters:

A b. ($\bullet$) $\gamma = 19.6$ pS, $V_r = +0.9$ mV; ($\circ$) $\gamma = 20.9$ pS, $V_r = +18.7$ mV,
    c. ($\bullet$) $\gamma = 20.1$ pS, $V_r = +0.9$ mV; ($\circ$) $\gamma = 18.9$ pS, $V_r = +22.7$ mV,
B b. ($\bullet$) $\gamma = 8.5$ pS, $V_r = +0.3$ mV; ($\circ$) $\gamma = 9.3$ pS, $V_r = +18.8$ mV,
    c. ($\bullet$) $\gamma = 9.5$ pS, $V_r = +2.1$ mV; ($\circ$) $\gamma = 10.0$ pS, $V_r = +19.5$ mV.

over a wide potential range. At V_p -80 mV, the time constants τ_{01} and τ_{02} were 1.3 and 8.1 msec for the 20-pS single-channel events (FIG. 3a), and 7.2 and 268.2 msec for the 9-pS elementary activity (FIG. 3Ba). Closed time constants (not shown) were 2.5 and 79 msec, and 1.7 and 32 msec, respectively, at the same potential. Typically, at $V_p \simeq -80$ mV, τ_{02} of the 9-pS events was 30-50 times longer that τ_{02} of the 20-pS events; τ_{01} was at the most 10 times longer for the 9-pS single-channel activity. A potential change to V_p $+62$ mV prolonged both open times of the 20-pS elementary currents (FIG. 3Ab), but a further increase to V_p +165 mV shortened them again (FIG. 3Ac). By contrast, similar changes in V_p produced a continuous decrease in both open times of the 9-pS single-channel events (FIG. 3Bb,c). In conclusion, the τ_{02}-V_p (FIG. 4Aa) and P_o-V_p (FIG. 4Ab) relations were bell-shaped for the 20-pS elementary currents with a maximum around 0 mV, and sigmoidal for the 9-pS single-channel activity with the longest τ_{02}s and highest P_os at negative V_p in symmetrical 90 mM BaCl$_2$ solutions (FIG. 4Ba,b). The voltage dependence of τ_{01} for both conductance levels of elementary current pulses (not shown) was similar to that of τ_{02} (see FIG. 3).

Chemical Responsiveness

FIGURES 5 and 6 show the sensitivity of the two conductance levels of single-channel activity occurring singly in separate bilayers to (±)-D600 (FIG. 5Aa,d; Ba,d), (±)-BAY K 8644 (FIG. 5Ab,d; Bb,d), cAMP-dependent phosphorylation (FIG.

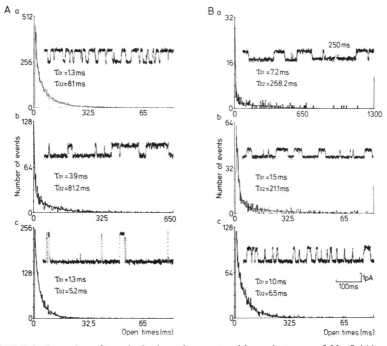

FIGURE 3. Open times from single-channel currents with conductances of 20 pS (**A**) and 9 pS (**B**). V_p was −80 mV (Aa, Ba), +62 mV (Ab), +165 mV (Ac), +85 mV (Bb), and +150 mV (Bc). Open time distributions were compiled from recordings showing only one current level and digitized at 2.0–3.1 kHz. A biexponential function with time constants τ_{01} and τ_{02} was fitted to each open time distribution by a nonlinear, unweighted, least-squares method. The insets show representative example records digitized at 1.5 kHz. The pipette and bath contained 90 mM BaCl$_2$ solutions.

5Ac,d; Bc,d) and CoCl₂ (FIG. 6A,B). Bath application of 25 μM (±)-D600 shortened 20-pS open-channel lifetimes and prolonged closed times between channel openings and/or groups of openings (FIG. 5Aa). With (±)-BAY K 8644 (10 μM) (FIG. 5Ab) or C subunit of cA kinase (1 μM plus 1 mM ATP-γ-S and 0.3 mM MgCl₂) (FIG. 5Ac) being present in the bath, 20-pS open-channel lifetimes became longer and gaps between channel openings and/or groups of openings got shorter. As consequence of these drug-induced changes in gating behavior, P_o of the 20-pS conductance level at V_p between −52 mV and −70 mV was reduced 2.5-6 times by 25 μM (±)-D600 (apparent $IC_{50} \simeq$ 10-14 μM, not shown) and enhanced 2.5-6 times by 10 μM (±)-BAY K 8644 and cAMP-dependent phosphorylation (1 μM C, 1 mM ATP-γ-S, 0.3

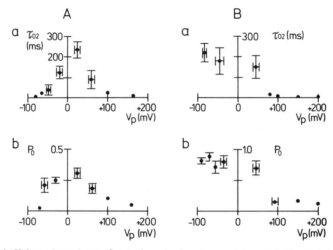

FIGURE 4. Voltage dependence of open times (τ_{o2}) and open-state probability (P_o) from single-channel events with conductances of 20 pS (**A**) and 9 pS (**B**). Each point represents the mean value of τ_{o2}s and P_os determined in 3-10 experiments, where only one conductance level was recorded at at least three potentials. The pipette and bath contained 90 mM BaCl₂ solutions. The sampling frequency was 2.0-3.6 kHz.

mM MgCl₂) (FIG. 5Ad). By contrast, gating behavior and P_o of the 9-pS conductance level were largely insensitive to these chemicals at similar V_p and concentrations (FIG. 5Ba-d). Both conductance levels of channel activity were sensitive to block by 1 mM CoCl₂ (FIG. 6A,B). In the presence of the blocking ion (lower traces), individual openings took the form of bursts of brief openings. Discrete fluctuations between nonconducting and conducting current levels became more frequent. These interruptions of the unitary currents were found in excess of those closings found in the absence of the blocking ion and attributed to gating transitions (upper traces). The number of Co^{2+}-induced closed periods outnumbered the closings of the channels seen in the absence of added CoCl₂ (not shown).

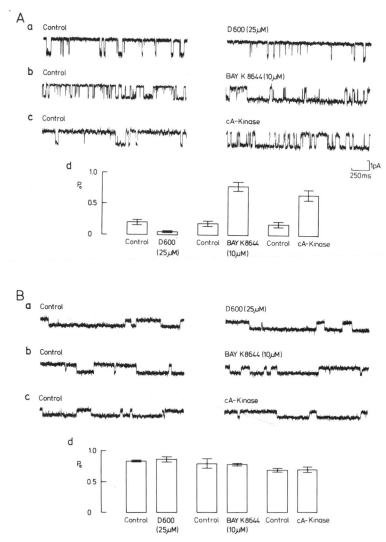

FIGURE 5. Responsiveness of the two conductance levels from DHP receptor to (±)-D600, (±)-BAY K 8644 and cA-kinase in lipid bilayers. Single-channel currents with slope conductances of 20 pS (**A**) and 9 pS (**B**) were recorded singly for 5-10 minutes. (±)-D600 (25 μM), (±)-BAY K 8644 (10 μM) or C subunit (1 μM) of the cA-kinase plus ATP-γ-S (1 mM) and MgCl$_2$ (0.3 mM) were then applied to the bath side of the bilayer, and the recording was continued for 10-20 min. The original records were obtained at V$_p$ of −65 mV (Aa), −52 mV (Ab), −60 mV (Ac), −70 mV (Ba), −60 mV (Bb), and −55 mV (Bc). The sampling frequency was 1.5 kHz. The responsiveness of the DHP receptor to (±)-D600, (±)-BAY K 8644 and cA-kinase (Ad, Bd) is expressed in terms of open-state probability (P$_o$). Each value is the mean P$_o$ from three to five measurements in 5-15 minute-long records. V$_p$ was −57.5 ± 7.5 mV (Ad), −65 ± 5 mV (Bd, control and (±)-D600), −55 ± 5 mV (Bd, control and (±)-BAY K 8644), and −57.5 ± 2.5 mV (Bd, control and cA-kinase).

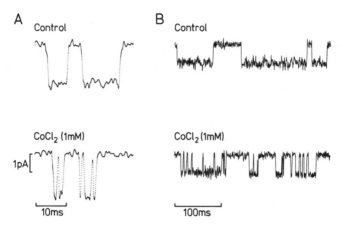

FIGURE 6. Responsiveness of the two conductance levels from DHP receptor to $CoCl_2$ in lipid bilayers. Single-channel currents with slope conductances of 20 pS (**A**) and 9 pS (**B**) were recorded singly for 5-10 minutes. $CoCl_2$ was then applied to the bath side of the bilayer, and the recording was continued for 10-20 minutes. The example records were obtained at V_p of -100 mV.

RECONSTITUTION OF ISOLATED PEPTIDE SUBUNITS OF THE pDHPR PROTEIN COMPLEX

Success of Reconstitution and Characterization of Elementary Conductance Levels in Successful Trials

The isolated α_1, β, and γ peptide subunits of the pDHPR protein complex were investigated in a total of 245 trials of reconstitution. Reconstitution of α_1, β, and γ was attempted 147, 50, and 48 times, respectively, in separate bilayers. With α_1, no functional reconstitution was achieved in 125 trials ($\simeq$ 85%). In 17 experiments ($\simeq$ 11.6%), α_1 gave single-channel activity with elementary conductances between 2.4 and 71.5 pS. Single-channel current pulses of different unitary amplitudes more often occurred together in the same bilayer, rather than singly in separate bilayers, with the accidental indication of the largest conductance state being a multiple of about equally spaced subconductance levels. However, this activity was not unequivocally sensitive to VSCC-blocking agents (($\pm$)-D600, (+)- or ($\pm$)-PN 200-110, $CoCl_2$). Only in five experiments ($\simeq$ 3.4%), was the reconstitution of α_1 associated with the reconstitution of single-channel current fluctuations with elementary conductances between 15 and 26 pS, which were unequivocally sensitive to block by ($\pm$)-D600 and augmentation by ($\pm$)-BAY K 8644 and cAMP-dependent phosphorylation. These unitary conductance events were observed singly in the five separate bilayers for a long time ($>$ 1 hr) of continuous single-channel recording at constant V_p without clear evidence of interconversion to another type of activity or the occupancy of subconductance states. Lipid bilayers, after incorporation of β or γ, did not exhibit any single-channel activity whatever. Thus, in the following sections, we will focus only on the properties of the pharmacologically regulated elementary activity reconstituted from α_1.

Current Voltage and Open-State Probability-Voltage Relations

FIGURE 7A shows original records of the pharmacologically regulated conductance level of elementary current reconstituted from α_1, at three constant pipette potentials in symmetrical 90 mM BaCl$_2$ solutions. The elementary current-V_p relationship was linear over a wide potential range and the slope conductance was 20.5 pS (FIG. 7B). The zero current potential (V_r) was around 0 mV. Open times were short at V_p −80 mV. A potential change to V_p −35 mV prolonged open times of the 20-pS elementary currents, but a further increase to V_p +180 mV shortened them again (FIG. 7A). The P_o-V_p relation was bell-shaped with a maximum around 0 mV (FIG. 7C).

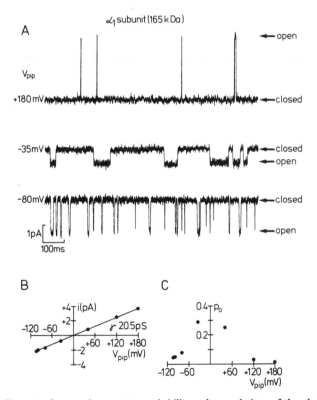

FIGURE 7. Current-voltage and open-state probability-voltage relations of the pharmacologically regulated single-channel activity reconstituted from α_1. (A) The original traces of single-channel current were recorded at various pipette potentials (V_{pip}), indicated beside each record, in symmetrical 90 mM BaCl$_2$ solutions. The sampling frequency was 1.5 kHz. (B) Elementary current-voltage relation constructed from single-channel current amplitude measurements at various V_{pip} in symmetrical 90 mM BaCl$_2$ solutions (●). The straight line is a least-squares fit to the data with the parameters: γ = 20.5 pS, V_r = +0.1 mV. (C) Voltage-dependence of the open-state probability (P_o) calculated as the time average current divided by the single-channel current amplitude.

Chemical Responsiveness

FIGURE 8 shows the sensitivity of the 20-pS elementary currents reconstituted from α_1 to (±)-D600 (FIG. 8A), (±)-BAY K 8644 (FIG. 8B), and cAMP-dependent phosphorylation (FIG. 8C) at V_p −100 mV. Bath application of 25 μM (±)-D600 shortened channel openings and prolonged closed times between channel openings and/or groups of openings (FIG. 8A). With (±)-BAY K 8644 (10 $\mu M;$ FIG. 8B) or C subunit of cA kinase (1 μM plus 1 mM ATP-γ-S and 0.3 mM MgCl$_2$; FIG. 8C) being present in the bath, open-channel lifetimes became longer and gaps between channel openings and/or groups of openings got shorter. As consequence of these drug-induced changes in gating behavior, P_o of the 20-pS single-channel current pulses was reduced about six times by 25 μM (±)-D600 and enhanced between 5.8 and 7.5 times by 10 μM (±)-BAY K 8644 and cAMP-dependent phosphorylation (1 μM C, 1 mM ATP-γ-S, 0.3 mM MgCl$_2$).

SUMMARY AND CONCLUSIONS

In the first part of this study, we show that sDHPR and pDHPR preparations reconstituted into lipid bilayers formed on the tips of patch pipettes exhibit two divalent cation-selective conductance levels of 9 and 20 pS, similar in single-channel conductance to VSCC reported in a variety of intact preparations[32–34] (see Pelzer et al.[8,8a] and Tsien et al.[35] for review). The larger conductance level is similar to the VSCC identified in intact rat t-tubule membranes[36] and described in sDHPR and pDHPR preparations,[19,22] and shares many properties in common with activity from L-type VSCC.[8,8a,35] It is sensitive to augmentation by the DHP agonist (±)-BAY K 8644 and cAMP-dependent phosphorylation, and to block by the phenylalkylamine (±)-D600 and the inorganic blocker CoCl$_2$. Its open-state probability and open times are increased upon depolarization as expected for a voltage-dependent activation process. Upon depolarization beyond the reversal potential, however, open-state probability and open times decline again. A reasonable way to explain the bell-shaped dependence of open times and open-state probability on membrane potential is to assume voltage-dependent ion-pore interactions that produce closing of the channel at strong negative and positive membrane potentials. By contrast, the smaller conductance level may be similar to the 10.6-pS t-tubule VSCC described by Rosenberg et al.[37] and may best be compared with T-type VSCC.[8,8a,35] It is largely resistant to augmentation by (±)-BAY K 8644 and cAMP-dependent phosphorylation or block by (±)-D600, but is sensitive to block by CoCl$_2$. Its open times and open-state probability show a sole dependence on membrane potential where depolarization increases both parameters sigmoidally from close to zero up to a saturating level. Both elementary conductance levels do not exhibit significant inactivation over a wide potential range, which may suggest that skeletal muscle VSCC inactivation is either poorly or not voltage-dependent at all. This possibility seems in agreement with bilayer recordings on reconstituted intact t-tubule membranes[36,37] and voltage-clamp recordings on intact fibers.[38–40] It supports the idea that the decline of Ca^{2+} current in intact skeletal muscle fibers may

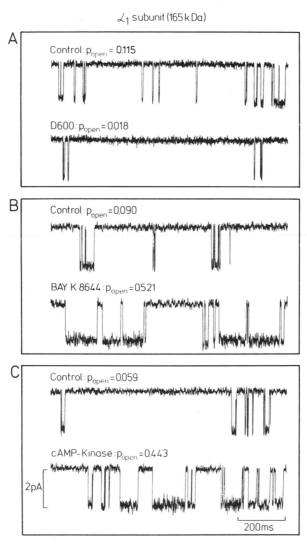

FIGURE 8. Responsiveness of the 20.5-pS conductance level reconstituted from α_1, to (±)-D600 (**A**), (±)-BAY K 8644 (**B**) and cA-kinase (**C**). Single-channel currents were recorded for 5-10 minutes at V_p −100 mV (upper example traces in A-C). (±)-D600 (25 μM, A), (±)-BAY K 8644 (10 μM, B) or C subunit (1 μM) of the cA-kinase plus ATP-γ-S (1 mM) and MgCl$_2$ (0.3 mM) (C) were then applied to the bath side of the bilayer, and the recording was continued at V_p −100 mV for 10-20 minutes (lower example traces in A-C). The sampling frequency was 1.5 kHz. The responsiveness of the α_1 subunit of pDHPR to (±)-D600, (±)-BAY K 8644 and cA-kinase is quantified in terms of open-state probabilities (P$_{open}$).

be due to Ca^{2+} depletion from the t-tubule system[38] and/or to inactivation induced by Ca^{2+} release from the sarcoplasmic reticulum.[40]

We consistently observe two conductance levels of 9 and 20 pS, either singly, or together in the same bilayer from solubilized DHPR samples and even highly purified DHPR preparations. The observations that (1) both conductance levels are observed in intact t-tubule membranes,[17] (2) the two conductance levels clearly exhibit distinct kinetic, pharmacologic and regulatory properties,[20,23] and (3) both conductance levels from sDHPR and pDHPR samples can be observed singly in different bilayers for more than two hours without interconversion[17,23] strongly argue against the conclusion that the two conductance levels arise from the same channel molecule. Rather, they reflect activity from two distinct channel molecules with closely related molecular structures that coexist even after extensive purification and cannot be separated by DHP (and/or phenylalkylamine) binding alone. In this context, amiloride, which selectively blocks T-type VSCC in a variety of cell types,[41,42] may, in concert with DHPs and phenylalkylamines, enable identification and separation of various VSCC types.

Our results[19,20,22,23] and those of Talvenheimo et al.[17] firmly establish the presence of multiple functional VSCC in sDHPR and pDHPR preparations. However, the picture has been complicated by the finding that a large fraction of DHPR in skeletal muscle does not appear to be VSCC.[43,44] The low success in reconstituting skeletal muscle sDHPR and pDHPR as functional VSCC may also indicate that only a few percent of DHPRs in skeletal muscle t-tubule membranes operate as VSCC. The rest may either be unrelated to ionic channels and may serve another function than ion transport[45] or the majority of DHPR in skeletal muscle are silent channels that cannot be activated by voltage and/or cAMP-dependent phosphorylation,[23] but probably need direct coactivation by a G protein[46] or another essential component. Nevertheless, two components of macroscopic Ca^{2+} current have been identified in intact skeletal muscle fibers[47] and in cultured skeletal muscle cells,[48,49] a slow DHP-sensitive component, and a fast, transient DHP-insensitive component. It is possible that the two conductance levels we[20,23] and others[17] have observed from intact t-tubule membranes, sDHPR samples and pDHPR preparations underlie the two components of voltage-dependent Ca^{2+} current in skeletal muscle.

In the second part of this study, we show that the isolated α_1 peptide subunit of the pDHPR protein complex can be reconstituted into lipid bilayers to give volt-age-, ($\pm$)-D600-, ($\pm$)-BAY K8644-, and cA kinase-sensitive 20-pS single-channel activity. Elementary currents from α_1 compare well in conductance, voltage dependence, and pharmacology with the 20-pS single-channel events reconstituted from sDHPR and pDHPR preparations.[17,19,20,22,23] Like the latter, they closely resemble activity from L-type VSCC.[8,8a,35] The evidence presented suggests that the α_1 peptide subunit of the pDHPR protein complex from skeletal muscle is the major functional part of L-type VSCC that contains the conducting pore, a voltage sensor, and the sites for VSCC blockers, VSCC activators and cAMP-dependent phosphorylation through which the L-type VSCC is modulated in vivo and in vitro.[8,8a,35] The β and γ subunits seem to be also part of the same VSCC as inferred by the constant stoichiometric ratio of these peptides to be copurified with α_1,[24] and the finding that antibodies against the β and γ components modulate L-type VSCC function.[50,51] The α_2 subunit disulfide linked to the δ subunit contains no binding sites for DHPs and phenylalkylamines and is unrelated to the α_1 subunit, as suggested by proteolytic digestion of the isolated peptides.[24] If we do not want to assume copurification of an undetectable channel protein X, α_2/δ may be candidates to underlie the T-type-like VSCC activity consistently coreconstituted from skeletal muscle pDHPR preparations.

REFERENCES

1. SCHWARTZ, A. & D. J. TRIGGLE. 1984. Ann. Rev. Med. **35:** 325-339.
2. FLECKENSTEIN, A., C. VAN BREEMEN, R. GROB & F. HOFFMEISTER. 1985. Cardiovascular Effects of Dihydropyridine Type Calcium Antagonists and Agonists. Springer, Berlin/Heidelberg/New York/Tokyo.
3. REUTER, H., H. PORZIG, S. KOKUBUN & B. PROD'HOM. 1985. Trends Neurosci. **8:** 396-400.
4. REUTER, H., H. PORZIG, S. KOKUBUN & B. PROD'HOM. 1986. *In* Proteins in Excitable Membranes. B. Hille & P. M. Fambrough, Eds. John Wiley. New York.
5. JANIS, R. A., J. P. SILVER & D. J. TRIGGLE. 1987. Adv. Drug Res. **16:** 309-591.
6. VENTER, J. C. & D. J. TRIGGLE. 1987. Structure and Physiology of the Slow Inward Calcium Channel. Alan R. Liss. New York.
7. BECHEM, M. & M. SCHRAMM. 1988. *In* Isolated Adult Cardiomyocytes. H. M. Piper & G. Isenberg, Eds. Vol. II. In press. CRC Press Inc. Boca Raton, FL.
8. PELZER, D., A. CAVALIÉ, T. F. MCDONALD & W. TRAUTWEIN. 1988. *In* Isolated Adult Cardiomyocytes. H. M. Piper & G. Isenberg, Eds. Vol. II. In press. CRC Press Inc. Boca Raton, FL.
8a. PELZER, D., S. PELZER & T. F. MCDONALD. 1989. Rev. Physiol. Biochem. Pharmacol. In press.
9. TRIGGLE, D. J. & R. A. JANIS. 1984. *In* Modern Methods in Pharmacology. N. Back & S. Spector, Eds. Vol. **2:** 1-28. Allen R. Liss. New York.
10. CATTERALL, W. A. 1986. *In* Membrane Control of Cellular Activity. H. C. Lüttgau, Ed.: 3-27. Gustav Fischer. Stuttgart/New York.
11. GLOSSMANN, H., D. R. FERRY, J. STRIESSNIG, A. GOLL & K. MOOSBURGER. 1987. TIPS **8:** 95-100.
12. GLOSSMANN, H. & J. STRIESSNIG. 1988. Vitam. Horm. **44:** 155-328.
13. BORSOTTO, M., J. BARHANIN, R. I. NORMAN & M. LAZDUNSKI. 1984. Biochem. Biophys. Res. Commun. **122:** 1357-1366.
14. CURTIS, B. M. & W. A. CATTERALL. 1984. Biochemistry **23:** 2113-2118.
15. BORSOTTO, M., J. BARHANIN, M. FOSSET & M. LAZDUNSKI. 1985. J. Biol. Chem. **260:** 14255-14263.
16. FLOCKERZI, V., H. J. OEKEN & F. HOFMANN. 1986. Eur. J. Biochem. **161:** 217-224.
17. TALVENHEIMO, J. A., J. F. WORLEY III & M. T. NELSON. 1987. Biophys. J. **52:** 891-899.
18. CURTIS, B. M. & W. A. CATTERALL. 1986. Biochemistry **25:** 3077-3083.
19. FLOCKERZI, V., H. J. OEKEN, F. HOFMANN, D. PELZER, A. CAVALIÉ & W. TRAUTWEIN. 1986. Nature (London) **323:** 66-68.
20. CAVALIÉ, A., V. FLOCKERZI, F. HOFMANN, D. PELZER & W. TRAUTWEIN. 1987. J. Physiol. (London) **390:** 82P.
21. MCKENNA, E. J., J. S. SMITH, J. MA, J. VILVEN, P. VAGHY, A. SCHWARTZ & R. CORONADO. 1987. Biophys. J. **51:** 2.
22. TRAUTWEIN, W., A. CAVALIÉ, V. FLOCKERZI, F. HOFMANN & D. PELZER. 1987. Circ. Res. **61**(Suppl. 1): 17-23.
23. PELZER, D., A. CAVALIÉ, V. FLOCKERZI, F. HOFMANN & W. TRAUTWEIN. 1988. *In* The Calcium Channel: Structure, Function, and Implications. M. Morad, W. Nayler, S. Kazda & M. Schramm, Eds.: 217-230. Springer. Berlin/Heidelberg/New York/Tokyo.
24. SIEBER, M., W. NASTAINCZYK, V. ZUBOR, W. WERNET & F. HOFMANN. 1987. Eur. J. Biochem. **167:** 117-122.
25. BEAVO, J. A., P. J. BECHTEL & E. G. KREBS. 1974. Methods Enzymol. **38:** 299-308.
26. HOFMANN, F. 1980. J. Biol. Chem. **255:** 1559-1564.
27. HANKE, W., C. METHFESSEL, U. WILMSEN & G. BOHEIM. 1984. Bioelectrochem. Bioenerget. **12:** 329-339.
28. FINKELSTEIN, A. & O. S. ANDERSEN. 1981. J. Membr. Biol. **59:** 155-171.
29. UYEDA, N., T. TAKENAKA, K. AOYAMA, M. MATSUMOTO & Y. FUJIYOSHI. 1987. Nature (London) **327:** 319-321.
30. TANABE, T., H. TAKESHIMA, A. MIKAMI, V. FLOCKERZI, H. TAKAHASHI, K. KANGAWA, M. KOJIMA, H. MATSUO, T. HIROSE & S. NUMA. 1987. Nature (London) **328:** 313-318.

31. CAVALIÉ, A., D. PELZER & W. TRAUTWEIN. 1986. Pflügers Arch. **406:** 241-258.
32. NILIUS, B. P., P. HESS, J. B. LANSMAN & R. W. TSIEN. 1985. Nature (London) **316:** 443-446.
33. NOWYCKY, M. C., A. P. FOX & R. W. TSIEN. 1985. Nature (London) **316:** 440-443.
34. WORLEY III, J. F., J. W. DEITMER & M. T. NELSON. 1986. Proc. Natl. Acad. Sci. USA **83:** 5746-5750.
35. TSIEN, R. W., P. HESS, E. W. MCCLESKEY & R. L. ROSENBERG. 1987. Ann. Rev. Biophys. Biophys. Chem. **16:** 265-290.
36. AFFOLTER, H. & R. CORONADO. 1985. Biophys. J. **48:** 341-347.
37. ROSENBERG, R. L., P. HESS, J. P. REEVES, H. SMILOWITZ & R. W. TSIEN. 1986. Science **231:** 1564-1566.
38. ALMERS, W., R. FINK & P. T. PALADE. 1981. J. Physiol. (London) **312:** 177-207.
39. ALMERS, W., E. W. MCCLESKEY & P. T. PALADE. 1984. J. Physiol. (London) **353:** 565-583.
40. ALMERS, W., E. W. MCCLESKEY & P. T. PALADE. 1985. *In* Calcium in Biological Systems. R. P. Rubin, G. B. Weiss & J. W. Patney, Jr., Eds.: 312-330. Plenum. New York.
41. TANG, C. M. & M. MORAD. 1988. Biophys. J. **53:** 22a.
42. TANG, C. M., F. PRESSER & M. MORAD. 1988. Science **240:** 213-215.
43. SCHWARTZ, L. M., E. W. MCCLESKEY & W. ALMERS. 1985. Nature (London) **314:** 747-751.
44. RIOS, E. & G. BRUM. 1987. Nature (London) **325:** 717-720.
45. AGNEW, W. S. 1987. Nature (London) **328:** 297.
46. YATANI, A., J. CODINA, Y. IMOTO, J. P. REEVES, L. BIRNBAUMER & A. M. BROWN. 1987. Science **238:** 1288-1292.
47. COTA, G. & E. STEFANI. 1986. J. Physiol. (London) **370:** 151-163.
48. BEAM, K. G., C. M. KNUDSON & J. A. POWELL. 1986. Nature (London) **320:** 168-170.
49. COGNARD, C., M. LAZDUNSKI & G. ROMEY. 1986. Proc. Natl. Acad. Sci. USA **83:** 517-521.
50. VILVEN, J., A. T. LEUNG, T. IMAGAWA, A. H. SHARP, K. P. CAMPBELL & R. CORONADO. 1988. Biophys. J. **53:** 556a.
51. REUTER, H. & H. PORZIG. 1988. Nature (London) **336:** 113-114.

Role of the Ryanodine Receptor of Skeletal Muscle in Excitation-Contraction Coupling

MICHAEL FILL,[a] JIANJIE MA,[a] C. MICHAEL KNUDSON,[b]
TOSHIAKI IMAGAWA,[b] KEVIN P. CAMPBELL,[b]
AND ROBERTO CORONADO[a]

[a] *Department of Physiology and Molecular Biophysics*
Baylor College of Medicine
Houston, Texas 77030

[b] *Department of Physiology and Biophysics*
The University of Iowa School of Medicine
Iowa City, Iowa 52242

In skeletal muscle, contraction is initiated by a depolarization of the transverse tubular membrane (t-tubule), which in turn signals the release of Ca from the sarcoplasmic reticulum (SR). A key protein involved in this process is the ryanodine receptor, an SR membrane protein of MW 450,000 that binds the alkaloid ryanodine with nanomolar affinity and is present exclusively at the junction between t-tubule and SR membranes.[1-4] The ryanodine receptor plays a dual role: Functionally, it is the putative Ca-release channel of the SR,[4] and structurally, it is the major protein responsible for forming "bridges" or "feet" that anatomically connect t-tubule and SR.[2,5] Here we demonstrate that the ryanodine receptor is steeply gated by both voltage and protons, and for the first time *in vitro,* we measured nonlinear capacitance (charge movement) that may be involved in the gating of this channel protein.

We recently identified the 450,000-Da ryanodine receptor-feet protein (FIG. 1A) as the Ca-release channel of native SR.[4,9,10] This was achieved using the planar bilayer recording technique and by comparing ligand-dependent gating, ionic selectivity, and pharmacology of purified ryanodine receptors to that of native Ca-release channels.[11,22] Voltage dependence was a gating property notoriously absent in our study and in studies by others that followed.[12,13] Its inconspicuousness in our earlier work is related to the effect of protons. At pH 7.4 (FIG. 1B) the channel dwells in a fast gating mode ($p = 0.38$). A drop to pH 7.2 drives the channel into an almost closed condition ($p = 0.08$) and at pH 7.0 the channel never opens ($p \ll 0.01$). Reversibility is shown in the last record of FIGURE 1 where alkalinization from pH 7.0 to pH 7.6 reopens the channel, resulting in a higher level of activity ($p = 0.72$) than seen at pH 7.4. Over this narrow range of pH, slope conductance is not affected (FIG. 2C, inset), and the kinetics remain fast, with a mean open event duration of approximately 100 μsec. Thus for all practical purposes, a change of 0.6 unit in solution pH from 7.6 to 7.0 units is sufficient to make the channel switch from an almost all-open to an almost all-closed conformation. The fitted Hill coefficient for data in FIGURE 1 was $n = 6.8$ and the apparent pK_a was 7.5. FIGURE 2 describes the ensuing changes in voltage

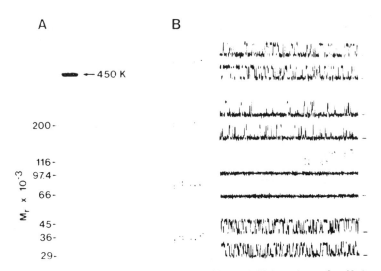

FIGURE 1. Ryanodine receptor polypeptide composition and pH dependence of purified receptor channels. (**A**) SDS-PAGE analysis of 3 μg of purified ryanodine receptor on 3-12% gradient gel and stained with Coomassie blue. The 450,000-Da receptor, indicated by the arrow (450K), migrated as a single band. Standards are indicated at the left. Purity of the 450,000-Da protein determined by gel scans was greater than 95%. Scatchard analysis of [³H]ryanodine binding to the purified receptor (detailed in Smith *et al.*[10]) yielded a straight line with an apparent B_{max} and K_d of 490 pmoles/mg and 7.0 nM, respectively. (**B**) Single-channel activity mediated by the purified receptor in solutions of different pH. All records from the same experiment, HP +60 mV (380 pS open-channel conductance), 3-kHz cutoff frequency. Baseline current is indicated next to each record. Solutions in both chambers were 0.25 M KCl, 0.5 mM EGTA (pCa 7), 25 mM MOPS (3-[*n*-morpholino]propanesulfonic acid) pH 7.4. pH (indicated next to each record) was adjusted with calibrated aliquots of MOPS (free acid) or KOH. To increase frequency response, K ions (instead of Ca or Ba) were used as current carriers. At low Ca (pCa > 6), K conductance is 6-10-fold larger than Ca conductance (detailed in Smith *et al.*[10]).

Ryanodine receptor was purified from isolated adult rabbit skeletal muscle triads by immunoaffinity chromatography (as in Imagawa *et al.*[4] with modifications introduced in Smith *et al.*[10]). Isolated triads were solubilized in 1% CHAPS in 0.5 M NaCl and buffer A (0.5 M sucrose, 0.75 mM benzamidine, 0.1 mM PMSF, and 50 mM Tris-HCl at pH 7.4) at the protein concentration 1 mg/ml in the presence of several protease inhibitors. Solubilized triads were applied to a MAb-XA7-Sepharose column (20 ml) and recycled overnight. Receptor was eluted in 0.3% CHAPS, 0.15% asolecithin, and 0.5 M KSCN in buffer A. KSCN was exchanged for KCl using a Pharmacia PD-10 column. Planar bilayers were cast from 20 mg/ml brain phospholipids in decane (PE/PS = 1/1 wt/wt). The two chambers in contact with the bilayer were defined as *cis* (side of receptor addition, connected to head-stage amplifier) and *trans* (protein-free side, connected to ground potential). Solutions in *cis* and *trans* chambers were the same, 250 mM KCl, 1 mM EGTA (pCa 7), 25 mM MOPS pH 7.0, 7.2, 7.4, or 7.6. For capacitance measurements we used 50 mM NaCl and 10 mM HEPES-TRIS, pH 7.2. Receptor concentration in solutions was 0.5-1 μg/ml for single-channel measurements and 80-100 μg/ml from capacitance measurements. An equal amount of receptor buffer (0.3% CHAPS, 0.15% asolecithin, 0.5 M KCl; 50 mM Tris pH 7.2) was always present in the *trans* chamber. At the concentrations used, CHAPS detergent (3-[(3-cholamidopropyl)dimethylammonio]-1-propanesulfonate) had no effect on conductance or mechanical stability. Head-stage amplifier was a List EPC7 (List-Electronic, DA-Eberstadt, West-Germany) and pulse protocols were constructed using Basic-Fastlab software (Indec System, Sunnyvale, CA, USA). Current records were filtered at 3-10-kHz corner frequency on an eight-pole Bessel (Frequency Devices, Haverhill, MA, USA) and digitized at 14-30 μsec/point.

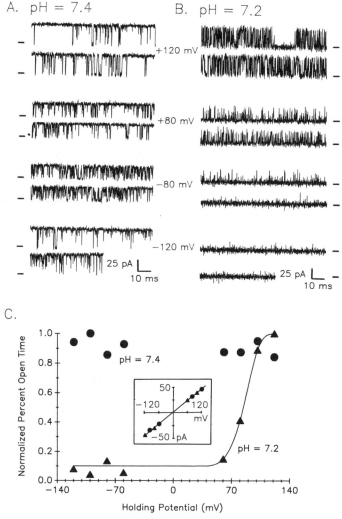

FIGURE 2. Voltage-dependence of ryanodine receptor channel. (**A,B**) Channel activity in *cis* solution pH 7.4 and 7.2 at the indicated holding potentials. Openings at every voltage are shown as deflecting the trace upward. Baseline is given for each trace. (**C**) The fraction of time spent open during a representative 10-sec interval (p, open probability) is plotted against holding potential at pH 7.2 (triangles) and pH 7.4 (circles). p at HP $+100$ mV was given a value of unity at both pHs. Actual values are $p = 0.82$ (pH 7.4) and $p = 0.41$ (pH 7.2). The current voltage curve (inset) at each pH was adequately fitted by the same line with a slope of 380 pS.

dependence. At pH 7.4 (FIG. 2A), the open probability remains fairly constant over the complete range of test potential, $+120$ mV to -120 mV. In this experiment open probability averaged 0.87 (FIG. 2C, circles). At pH 7.2 (FIG. 2B), a similar level of activity was present at positive potentials, and it dropped to 0.02 and less at potentials below $+50$ mV (FIG. 2C, triangles). Transition from open to closed was steep, with a midpoint at $+85.7$ mV (9.3 mV standard deviation, $n = 8$) and an e-fold increase in open probability every 3.5 mV. At pH 7.0 no openings were observed at any test potential. Thus, the receptor channel is steeply dependent on both membrane potential and pH with the switching from closed to open occurring over an unusually narrow range.

A demonstration of voltage dependence prompted us to search for charge movement associated with this protein. In skeletal muscle, nonlinear charge movements are argued to be a manifestation of some molecular event that is involved in transduction of t-tubule depolarization to SR calcium release. Experimentally, charge movement was looked for as a small voltage-dependent component of the total membrane capacitance. To ensure a large density of receptors in the planar membrane, we used 100-200 times more protein (80 μg/ml) than used in single-channel experiments. To avoid trivial artifacts related to adding receptor to only one chamber, we added protein-free receptor buffer (see legend FIG. 1) to the opposite chamber. Hence, the only asymmetry in the system was the protein itself. We measured the voltage-dependent component of membrane capacitance, $C(V)$, as a function of holding potential, V. Capacitance was determined in the range of -150 mV to $+150$ mV by integrating of the "on" or "off" charging current in response to a constant test pulse. The test pulse was applied midway during a long holding pulse. In order to compare the shape of the voltage-dependent component, we subtracted the capacitance at V $= 0$ mV from each curve and plotted $C(V) - C(V = \emptyset)$ as a function of V^2 (FIG. 3). Bare bilayers and bilayers with protein-free buffer displayed a characteristic minimum at V $= 0$ mV and a quadratic increase in $C(V)$ as a function of V, symmetric with respect to zero. This phenomenon has been well described and is due to electrostriction.[14,15] Capacitance of bare bilayers was typically 12% higher at $+150$ mV (or -150 mV) than at 0 mV. A plot of $C(V)$ versus V^2 (FIG. 3, circles) shows that in the absence of receptor each arm of the curve, the one at positive and the one at negative potentials, varies linearly with V^2 with a slope of approximately 4.9 pF/V^2. Receptor protein (FIG. 3, triangles) had two effects on this relationship. It significantly increased the steepness at positive potentials, and it decreased the steepness at negative potentials. The effects are complex with no less than three slopes clearly different from the slope expected on the basis of electrostriction alone. One component at negative potentials is weakly voltage dependent; a second component around 0 mV is steeply voltage dependent, and a third component at large positive potentials tends to saturate with a limiting slope of about 20 pF/V^2, six times larger than without receptor. Thus clearly, the ryanodine receptor *in vitro* mediates nonlinear charge movement of several kinds.

Although we could not specify components responsible for gating, the bulk of the charge correlated well with open-channel probabilities. FIGURE 4A shows receptor charge movement (protein minus protein-free capacitance) as a function of holding voltage. More charge is moved at progressively positive potentials with the largest change occurring between -60 mV and $+60$ mV. Under similar conditions, records of single channels are shown in FIGURE 4B. Activation occurs at positive potentials with few openings below $+60$ mV. Thus the two processes have voltage dependencies that increase in the same direction. This is plotted in FIGURE 4C where charge movement during either the "on" transient (squares) or "off" transient (circles) was found to be superimposable with open-channel probabilities (diamonds). The latter measured independently from single-channel records obtained at a low concentration of receptor. The equivalence of charge moved during the "on" and "off" of the test pulse is given in the inset of FIGURE 4. One followed the other in the protein-free control (lower curves) and in the presence of receptor (upper curves).

Four hallmarks of Ca release studied in skinned fibers and SR vesicles,[16–25] namely, activation of release by micromolar Ca and millimolar adenine nucleotide, and inhibition by micromolar ruthenium red and millimolar Mg, can be traced to the channel formed by the 450,000-Da ryanodine receptor.[10] The purified polypeptide alone contains the binding site for ryanodine and the regulatory sites for Ca, ATP, and ruthenium

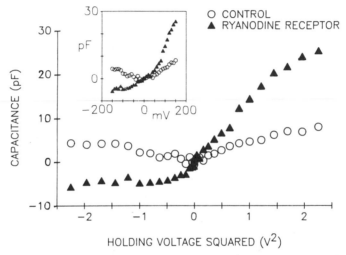

FIGURE 3. Voltage-dependent capacitance mediated by ryanodine receptor. Capacitance as a function of voltage $C(V)$, was measured using a two-pulse protocol. A square voltage pulse of 10-msec duration and constant +10 mV amplitude (test pulse) was added to square voltage pulse of 200-msec duration and variable amplitude (holding voltage pulse). Test pulse was applied 100 msec into the holding pulse. Holding voltage was varied between −150 mV to +150 mV. Capacitance was measured by integrating the first 8 msec of the "on" transient of the test pulse at each holding voltage. Curves are the average of four separate experiments. Circles correspond to bilayers with receptor protein and receptor buffer on the *cis* and *trans* sides, respectively. Capacitance at 0 mV holding voltage was subtracted from each curve (control $C(V = 0) = 314$ pF; ryanodine receptor $C(V = 0) = 347$ pF. $C(V) - C(V = 0)$ is plotted against holding voltage, mV (inset) or V^2 (center).

red.[4] The identified receptor channel is novel in several respects. It has a large conductance for K and Ca ions and a fast kinetics composed of open times in the range of 60-100 μsec.[10] These properties are expected of a channel that mediates SR Ca release. The ryanodine receptor *in vivo* forms part of a large oligomeric structure that spans the 100-Å gap between t-tubules and SR membranes.[2,5,12,13] Recent evidence suggests that the attachment to the t-membrane may contain the dihydropyridine (DHP) receptor.[26] The DHP receptor has been identified as the t-tubule voltage sensor of excitation-contraction coupling.[8] Thus, our observation of nonlinear capacitance may lead to the possibility that the ryanodine receptor protein may sense and convey

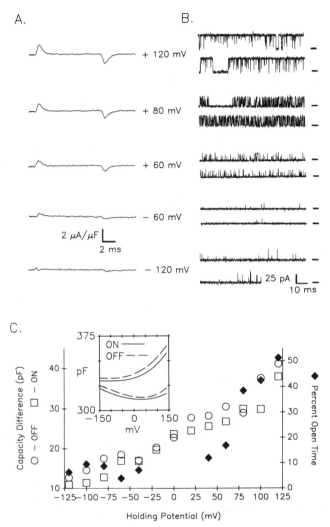

FIGURE 4. Correlation charge movement and channel opening in the ryanodine receptor. (**A**) Charge movement mediated by the receptor protein obtained by subtracting pairs of test pulses (protein minus protein free) at each indicated holding potential. (**B**) Single-channel activity at the corresponding holding voltages measured in a separate experiment. Channel opening is always shown as upward deflections from the baseline which is indicated for each record. (**C**) Capacitance difference after subtraction of protein minus protein-free bilayers (open symbols) and open-channel probability (filled diamonds) are plotted against holding voltage. Inset shows C(V) versus V separately in protein-free (lower curves in inset) and protein-containing (upper curves in inset) bilayers. The "on" and "off" transients have the same voltage dependence.

t-tubule voltage changes to the SR. Further experiments, however, emphasizing the pharmacology and specificity of the nonlinear capacitance are required to assess its significance *in vivo*. (

REFERENCES

1. PESSAH, I. N., A. O. FRANCINI, D. J. SCALES, A. L. WATERHOUSE & J. E. CASIDA. 1986. Calcium ryanodine receptor complex solubilization and partial characterization from skeletal muscle functional sarcoplasmic reticulum vesicles. J. Biol. Chem. **261:** 8643-8648.
2. INUI, M., A. SAITO & S. FLEISCHER. 1987. Purification of the ryanodine receptor and identity with feet structures of functional terminal cisternae of sarcoplasmic reticulum from fast skeletal muscle. J. Biol. Chem. **262:** 1740-1747.
3. CAMPBELL, K. P., C. M. KNUDSON, T. IMAGAWA, A. T. LEUNG, J. L. SUTKO, S. D. KAHL, C. R. RAAB & L. MADSON. 1987. Identification and characterization of the high affinity [^{3}H]ryanodine receptor of the junctional sarcoplasmic reticulum Ca^{2+} release channel. J. Biol. Chem. **262:** 6460-6463.
4. IMAGAWA, T., J. S. SMITH, R. CORONADO & K. P. CAMPBELL. 1987. Purified ryanodine receptor from skeletal muscle sarcoplasmic reticulum is the Ca^{2+} permeable pore of the calcium release channel. J. Biol. Chem. **262:** 16636-16643.
5. FERGUSON, D. G., H. W. SCHWARTZ & C. FRANZINI-ARMSTRONG. 1984. Subunit structure of junctional feet in triads of skeletal muscle: A freeze-drying, rotatory-shadowing study. J. Cell Biol. **99:** 1735-1742.
6. MELZER, W., M. F. SCHNEIDER, B. J. SIMON & G. SZUCS. 1986. Intramembrane charge movement and calcium release in frog skeletal muscle. J. Physiol. **373:** 481-511.
7. BERWE, D., G. GOTTSCHALK & H. CH. LUTTGAU. 1987. Effects of the calcium antagonist gallopamil (D600) upon excitation-contraction coupling in toe muscle fibers of the frog. J. Physiol. **385:** 693-707.
8. RIOS, E. & G. BRUM. 1987. Involvement of dihydropyridine receptors in excitation-contraction coupling in skeletal muscle. Nature **325:** 717-720.
9. SMITH, J. S., T. IMAGAWA, K. P. CAMPBELL & R. CORONADO. 1988. Permeability and gating properties of the purified ryanodine receptor-calcium release channel from skeletal muscle sarcoplasmic reticulum. Biophys. J. **53:** 422a.
10. SMITH, J. S., T. IMAGAWA, J. MA, M. FILL, P. K. CAMPBELL & R. CORONADO. 1988. Purified ryanodine receptor from rabbit skeletal muscle is the calcium release channel of sarcoplasmic reticulum. J. Gen. Physiol. **92(1):** 1-27.
11. SMITH, J. S., R. CORONADO & G. MEISSNER. 1986. Single-channel measurements of the calcium release channel from skeletal muscle sarcoplasmic reticulum. Activation by calcium ATP and modulation by magnesium. J. Gen. Physiol. **88:** 573-588.
12. LAI, F. A., H. P. ERIKSON, E. ROUSSEAU & G. MEISSNER. 1988. Purification and reconstitution of the calcium release channel from skeletal muscle. Nature **331:** 315-319.
13. HYMEL, L., M. INUI, S. FLEISCHER & H. G. SCHINDLER. 1988. Purified ryanodine receptor of skeletal muscle sarcoplasmic reticulum forms Ca^{2+}-activated oligomeric Ca^{2+} channels in planar bilayers. Proc. Natl. Acad. Sci. USA, in press.
14. ALVAREZ, O. & R. LATORRE. 1978. Voltage-dependent capacitance in lipid bilayers made from monolayers. Biophys. J. **21:** 1-17.
15. WHITE, S. 1980. How electric fields modify alkane solubility in lipid bilayers. Science **207:** 1075-1077.
16. MEISSNER, G., E. DARLING & J. EVELETH. 1986. Kinetics of rapid Ca^{2+} release by sarcoplasmic reticulum effects of Ca^{2+}, Mg^{2+}, and adenine nucleotides. Biochemistry **25:** 236-243.
17. OGAWA, Y. & S. EBASHI. 1983. J. Biochem. Tokyo **93:** 1271-1285.
18. ENDO, M. & O. KITAZAWA. 1976. Proc. Japan Acad. **52:** 599.
19. MORII, H. & Y. TONOMURA. 1983. The gating behavior of a channel for Ca^{2+} induced release in fragmented sarcoplasmic reticulum. J. Biochem. Tokyo **93:** 1271-1285.

20. NAGASAKI, K. & M. KASAI. 1983. Fast release of calcium from sarcoplasmic reticulum vesicles monitored by chlortetracycline fluorescence. J. Biochem. Tokyo **94:** 1101-1109.
21. VOLPE, P., G. SALVATI & A. CHU. 1986. Calcium-gated calcium channels in sarcoplasmic reticulum of rabbit skinned skeletal muscle fibers. J. Gen. Physiol. **87:** 289-303.
22. SMITH, J. S., R. CORONADO & G. MEISSNER. 1985. Sarcoplasmic reticulum contains adenine nucleotide-activated calcium channels. Nature **361:** 446-449.
23. FABIATO, A. 1985. Time and calcium dependence of activation and inactivation of calcium-induced release of calcium from the sarcoplasmic reticulum of a skinned canine cardiac purkinje cell. J. Gen. Physiol. **85:** 247-289.
24. STEPHENSON, E. W. & R. J. PODOLSKY. 1977. Influence of magnesium on chloride-induced calcium release in skinned muscle fibers. J. Gen. Physiol. **69:** 17-35.
25. STEPHENSON, E. W. 1981. Calcium dependence of stimulated ^{45}Ca efflux from skinned muscle fibers. J. Gen. Physiol. **77:** 419-443.
26. KNUDSON, C. M., T. IMAGAWA, S. D. KAHL, M. G. GAVER, A. T. LEUNG, A. H. SHARP, S. D. JAY & K. P. CAMPBELL. 1988. Evidence for physical association between junctional sarcoplasmic reticulum ryanodine receptor and junctional transverse tubular dihydropyridine receptor. Biophys. J. **53:** 605a.

Reconstitution of Cardiac Sarcoplasmic Reticulum Calcium Channels[a]

ALAN J. WILLIAMS AND RICHARD H. ASHLEY

Department of Cardiac Medicine
Cardiothoracic Institute
University of London
London SW3 6LY, United Kingdom

INTRODUCTION

The sarcoplasmic reticulum (SR) membrane network of mammalian cardiac and skeletal muscle forms an entirely intracellular membrane system that surrounds the cell's contractile apparatus. The system also forms specialized junctional regions with the sarcolemma at the cell surface and within the cell interior at the level of the transverse tubules. The location and regional specialization of the SR reflects its major role in excitation-contraction coupling.

It has long been accepted that the SR of striated muscle cells is responsible for lowering cytosolic Ca^{2+} and hence initiating relaxation. Ca^{2+} is accumulated into the lumen of the SR via abundant ATP-dependent pumps. The part played by the SR in the elevation of cytosolic Ca^{2+} and hence in the initiation of contraction has been more controversial, particularly in mammalian cardiac muscle. However, the work of Fabiato[1-3] with mechanically skinned cardiac myocytes has now established the SR network as a major source of Ca^{2+} for excitation-contraction coupling in this tissue.

A variety of mechanisms have been suggested for Ca^{2+} release from the SR in both cardiac and skeletal muscle. Of these, the most plausible involves a triggering of release by the relatively small amounts of Ca^{2+} entering the cell through voltage-dependent channels during the action potential. So-called Ca^{2+}-induced Ca^{2+} release has now been demonstrated in skinned preparations and isolated SR membrane vesicles of both skeletal and cardiac muscle. $^{45}Ca^{2+}$ flux measurements have revealed that the pathway for Ca^{2+}-induced Ca^{2+} release is localized in the junctional regions of both skeletal and cardiac SR membrane systems.[4-7] This work has also established that release may be stimulated by ATP, its nonhydrolyzable analogues, and caffeine and may be inhibited by Mg^{2+}, ruthenium red, and calmodulin.[4,5,7] The high rates of release demonstrated in these experiments are consistent with a role for this process in striated muscle excitation-contraction coupling and require that the process is channel mediated.

[a]This work was supported by grants from the British Heart Foundation and a Wellcome Trust Training Fellowship in Mental Health to RHA. Sulmazole was a gift of Boehringer Ingelheim.

163

A scheme representing the various pathways for ion transport revealed by isotope flux experiments and their location within the SR membrane network, is given in FIGURE 1. In addition to the routes of Ca^{2+} accumulation and release outlined above, the SR contains structures conferring K^+ and Cl^- permeabilities on the membrane.[8,9] These systems could allow charge compensation during the electrogenic Ca^{2+} fluxes associated with excitation–contraction coupling.

Details of the mechanisms underlying these conductances have been revealed by experiments in which purified skeletal and cardiac SR membrane vesicles have been incorporated into reconstituted membrane systems and conductance studied at the single-channel level under voltage-clamp conditions. Using this approach K^+,[10-12] Cl^-,[13,14] and Ca^{2+}[6,13,15] channels have been resolved at the single-channel level. Particularly dramatic progress has been made with the SR Ca^{2+} channel. Meissner and his colleagues have demonstrated relatively large conductance Ca^{2+} and ATP-activated

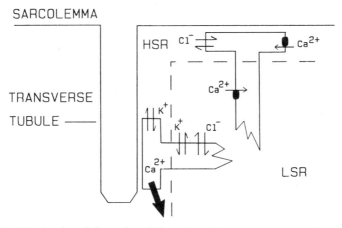

FIGURE 1. The location of the various SR membrane ion conductances revealed by isotope flux studies. Ca^{2+} pumps ($\rightarrow$), K^+ channels, and Cl^- channels occur throughout the SR network. Ca^{2+} channels are restricted to the heavy SR (HSR), which is equivalent to the junctional regions of the system and may be separated from the longitudinal or light SR (LSR) by density gradient centrifugation.

single channels derived from the junctional regions of both skeletal[13,15] and cardiac[6] SR membranes. These studies have recently been extended with the identification of the SR Ca^{2+} channel as a component of the ryanodine receptor and the purification and functional reconstitution of the channel from rabbit skeletal[16,17] and canine cardiac[18] SR membranes.

In this report we will limit our discussion to the properties of the "native" sheep cardiac SR Ca^{2+} release channel. That is the channel revealed by the incorporation of isolated junctional cardiac SR membrane vesicles into planar phospholipid bilayers without further purification. The channel observed under these conditions shares a number of properties with those previously described in rabbit skeletal and canine cardiac SR membranes. The channel is activated by Ca^{2+} and ATP and "inhibited" by Mg^{2+} from the cytosolic side of the membrane. In addition to modulation by these physiologically important agents, the channel may also be activated by sulmazole, a caffeine-like cardiotonic drug.

METHODS

SR Membrane Isolation

Sheep hearts were obtained from a local abattoir and transported to the laboratory bathed in cold cardioplegic solution.[19] Following tissue homogenization in the presence of 1 mM PMSF, ventricular muscle SR membrane vesicles were isolated by differential centrifugation essentially as described by Meissner.[20] Membrane vesicles originating from the junctional and longitudinal regions of the SR network were separated by centrifugation through discontinuous sucrose density gradients. Junctional SR vesicles were collected at the 30-40% (wt/wt) interface after sedimentation at 150,000 g_{max} for two hours. After pelleting, the membrane vesicles were suspended in 0.4 M sucrose, rapidly frozen in liquid N_2, and stored at $-80°C$.

Incorporation of Vesicles into Planar Bilayers

Isolated membrane vesicles were incorporated into planar phospholipid bilayers formed across either a 300- or 500-μm hole in a polystyrene partition separating two solution-filled chambers designated cis and trans. Bilayers were formed across the hole using dispersions of phosphatidylethanolamine (Avanti Polar Lipids, Birmingham, Alabama) in n-decane (30 mg/ml). The trans chamber was held at ground and the cis chamber could be clamped at a range of holding potentials relative to ground. Current flow was monitored using an operational amplifier as a current-voltage converter.[21] SR membrane vesicle fusion was achieved using the conditions described by Smith et al.[13] Initially, both cis and trans chambers contained 50 mM choline chloride, 5 mM $CaCl_2$ and 10 mM Hepes titrated to pH 7.4 with Tris. After vesicle addition to the cis chamber, the choline chloride concentration in this chamber was raised to 250 mM by adding aliquots of a 3 M stock solution. The formation of an osmotic gradient across the bilayer generally led to vesicle fusion and the appearance of Cl^--selective channels.[6]

Once a fusion event had occurred, both the cis and trans chambers were perfused with the solutions required for the resolution of Ca^{2+}-release channels. As has been previously demonstrated with the K^+-selective channels of both skeletal[22] and cardiac[23] muscle SR membranes, SR channels almost invariably incorporate into planar phospholipid bilayers with a fixed orientation, the solution on the cis side of the bilayer corresponding to the cytosolic medium and that on the trans to the SR lumen. Consequently, the cis chamber was perfused with 20 volumes of a solution containing 250 mM Hepes and 125 mM Tris (pH 7.4) plus 1 mM EGTA and varying amounts of $CaCl_2$ to give the desired free calcium concentration (monitored with a Ca^{2+}-sensitive electrode[24]). The trans chamber was perfused with between three and nine volumes of either 250 mM Hepes titrated to pH 7.4 with $CaOH_2$ (free Ca^{2+} concentration 53 mM) or 250 mM glutamic acid titrated to pH 7.4 with $CaOH_2$ (free Ca^{2+} concentration 65 mM).

If, as was generally the case, Ca^{2+} release channels had been incorporated into the bilayer along with Cl^- channels, Ca^{2+} current fluctuations were now observed. Using the approach outlined above, it was possible to stop vesicle fusion after a single fusion event by perfusing unfused vesicles out of the cis chamber. Single fusion events usually produced bilayers containing between one and three Ca^{2+} channels. Alternatively, large numbers of Ca^{2+} channels could be incorporated into the bilayer by allowing several fusion events to occur. In this case single-channel current fluctuations were obscured and the observed "macroscopic" currents represented mean values of the entire population of channels present in the bilayer.

Data Acquisition and Analysis

Channel data was displayed on an oscilloscope and recorded on FM tape. For analysis, data were filtered using a four-pole RC-mode filter at a front panel setting of 300 to 400 Hz (Krohn-Hite 3200R, Avon, MA). Filtered data was digitized at 2 kHz using either an Indec PDP 11/73 based lab system (Indec, Sunnyvale, CA) or an AT-based system (Intracel, Cambridge, U.K.). Single-channel open probabilities (Po) were obtained by integrating amplitude histograms constructed from three-minute steady-state recordings. Channel open and closed lifetimes were determined by 50% amplitude threshold analysis of the digitized data. Lifetimes were stored in sequential files and displayed in noncumulative histograms. Individual times were fitted to a probability density function (pdf) using the method of maximum likelihood.[25] Lifetimes of less than 4 msec were not included in the fitting procedure and a missed events correction was employed.[25]

RESULTS AND DISCUSSION

Identification of Calcium-Activated Calcium Release Channels

With μM activating Ca^{2+} on the *cis* side of the membrane and high (approximately 50 mM) Ca^{2+} on the *trans* side of the membrane, divalent cation currents flow from the *trans* to the *cis* chamber at negative holding potentials and reverse at approximately +30 mV (FIG. 2).

These currents were almost always abolished by reducing the *cis* (activating) Ca^{2+} to subnanomolar levels. Occasional channels were apparently Ca^{2+} insensitive. These are not discussed here. Current flow through a single sheep cardiac SR Ca^{2+} release channel is proportional to applied potential with either Ca^{2+} or Mg^{2+} as the charge carrier (In both cases the *cis* chamber contained 250 mM Hepes, 125 mM Tris, pH 7.4, 1 mM EGTA with 100 μM free Ca^{2+}. The *trans* chamber contained 250 mM Hepes titrated to pH 7.4 with either $CaOH_2$ or $MgOH_2$). Under these conditions the slope conductance for Ca^{2+} was 89 pS and that for Mg^{2+} was 61 pS, yielding a conductance ratio, Mg^{2+}/Ca^{2+} of 0.7. Other unit conductances or substates were occasionally seen. As with the skeletal SR Ca^{2+} release channel, Ba^{2+} is also an effective carrier of current from *trans* to *cis*. The slope conductance obtained in the presence of 50 mM *trans* Ba^{2+} was 140 pS.

The reversal potentials obtained from the data in FIGURE 2 and similar plots, produced permeability ratios[26] for $Ca^{2+}/Tris^+$ of approximately 8.5, which is in good agreement with the value reported for the rabbit skeletal muscle SR Ca^{2+} release channel.[13] Taken together, these results indicate that the SR Ca^{2+} release channels are not particularly selective for Ca^{2+}. However, in the prevailing intracellular ionic environment, Ca^{2+} will undoubtedly be the sole current-carrying species in the channel.

Modulation of Channel Gating by Voltage and Ligands

Voltage Dependence

"Macroscopic" current-voltage relationships obtained from bilayers containing many cardiac SR Ca^{2+} release channels were not linear (FIG. 3: *cis*—10 μM activating Ca^{2+}, *trans*—65 mM Ca^{2+}); current increased disproportionately as the applied potential was made more positive. As single-channel current was directly proportional to applied potential, the most likely explanation of this observation is that the probability of the channel being in the open state is dependent upon voltage, Po increasing

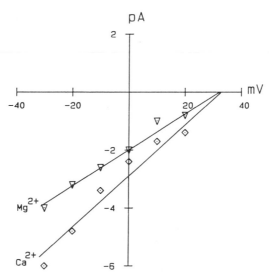

FIGURE 2. Current-voltage relationships for a single sheep cardiac SR Ca^{2+} release channel with either Ca^{2+} or Mg^{2+} as the charge-carrying species (see text for details). Ca^{2+} slope conductance is 89 pS, Mg^{2+} slope conductance is 61 pS.

as the applied potential is made more positive. This conclusion is supported by observations at the single-channel level (not shown), the channel tending to dwell in the open state for long periods at holding potentials approaching the reversal potential. Similar voltage dependence has been reported for "macroscopic" Ba^{2+} currents in the skeletal SR Ca^{2+} channel.[13]

The K^+-selective channel of both skeletal and cardiac SR membrane networks shows very marked voltage dependence. As with the Ca^{2+} release channel, the K^+ channel opening is favored at positive holding potentials,[22,23] a situation that is equivalent to the cytosol being at a positive potential with respect to the SR lumen. However, it is unlikely that the voltage dependence of either of these SR channels is of any great physiological significance. Under physiological conditions the high K^+ conductance of the membrane would prevent the generation of a potential of more than a few millivolts across the SR membrane.[27]

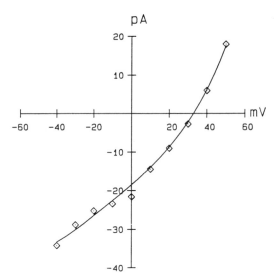

FIGURE 3. "Macroscopic" current-voltage relationship for the sheep cardiac SR Ca^{2+} release channel. The *cis* chamber contained 250 m*M* Hepes, 125 m*M* Tris pH 7.4, 1 m*M* EGTA, 10 μ*M* free Ca^{2+}. The *trans* chamber contained 250 m*M* glutamic acid titrated to pH 7.4 with $CaOH_2$ (free Ca^{2+}; 65 m*M*). Points are steady-state currents monitored one minute after clamping the membrane to the appropriate holding potential.

Physiological Ligands—Calcium, ATP, and Magnesium

$^{45}Ca^{2+}$ release from isolated canine cardiac SR membrane vesicles is stimulated by Ca^{2+} and ATP.[6,7] Similar activation of the sheep cardiac SR Ca^{2+} channel may be seen at the single-channel level. The experiment shown in FIGURE 4 demonstrates the activation of a typical single sheep cardiac SR Ca^{2+} channel by Ca^{2+} and ATP added to the solution on the cytosolic side of the membrane. In all cases the applied potential was held constant at 0 mV. At a cytosolic Ca^{2+} concentration of 1 μ*M,* the channel opened only very rarely. In fact, the openings were too sparse to register a measurable Po. Elevation of the free-Ca^{2+} concentration to 10, 100, and 1000 μ*M* produced a progressive increase in the measured Po. Although we see no evidence of channel "run-down," it is clear from these traces that at a given free-Ca^{2+} concentration, some variation in Po with time was observed. In an effort to nullify these variations, the Po values quoted in the figure were obtained from amplitude histograms accumulated over the long (3-min) periods shown in the figure. Po is markedly dependent on the free-Ca^{2+} concentration at the cytosolic face of the channel; however, it is clear that at an applied potential of 0 mV, Ca^{2+} alone cannot fully activate the sheep cardiac SR Ca^{2+} channel. In the experiment shown in FIGURE 4, 1 m*M* Ca^{2+} produced a Po of only 0.38. Ca^{2+} concentrations in excess of 1 m*M* tended to reduce Po as has previously been reported for the skeletal muscle SR Ca^{2+} release channel.[15] It should be noted that the Po values quoted here were obtained in essentially neutral bilayers (100% phosphatidylethanolamine) and are therefore not directly comparable with the values obtained by Meissner and his colleagues, who have studied the SR Ca^{2+} release channels in bilayers containing a proportion of negatively charged phospholipids. The net negative surface charge present on such bilayers will obviously lead to an elevation of the divalent cation concentration sensed by the channel to a level above that of the bulk free concentration.

Although the sheep cardiac SR Ca^{2+} release channel cannot be fully activated with Ca^{2+} as the sole ligand, this may be achieved by combinations of Ca^{2+} and ATP. This is demonstrated in FIGURE 4 by the addition of 1 mM ATP to the cytosolic face of the channel in the presence of 1 mM Ca^{2+}. Under these conditions Po increased from 0.38 to 1.0. Channel activation by ATP does not appear to involve hydrolysis of the nucleotide. Similar activation may be obtained with nonhydrolyzable analogues such as AMP-PCP.[13]

A comparison of the rabbit skeletal and canine cardiac SR Ca^{2+} release channels suggests that the cardiac channel is the more Ca^{2+} dependent. At very low (nanomolar) Ca^{2+} concentrations in cytosolic solution, the cardiac channel could not be activated

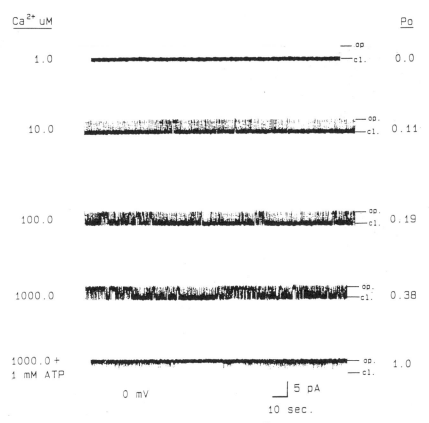

FIGURE 4. Fluctuations of a single sheep cardiac SR Ca^{2+} channel recorded in the presence of increasing levels of activating Ca^{2+}, or Ca^{2+} plus ATP, in the *cis* chamber. The *cis* chamber contained 250 mM Hepes, 125 mM Tris pH 7.4, 1 mM EGTA. The free-Ca^{2+} concentration was raised by sequential additions of a stock $CaCl_2$ solution. The *trans* chamber contained 250 mM Hepes titrated to pH 7.4 with $CaOH_2$ (free Ca^{2+}; 53 mM). Open probability (Po) was determined as described in the methods section. The applied potential was 0 mV.

by millimolar concentrations of ATP, while it was possible to partially activate the skeletal muscle channel under such conditions.[6] It is not clear if this reflects a physiologically significant difference between the two species of channel.

Millimolar concentrations of Mg^{2+} are known to inhibit Ca^{2+}-induced $^{45}Ca^{2+}$ efflux from both skeletal and cardiac SR membrane vesicle preparations.[5-7] Similarly,

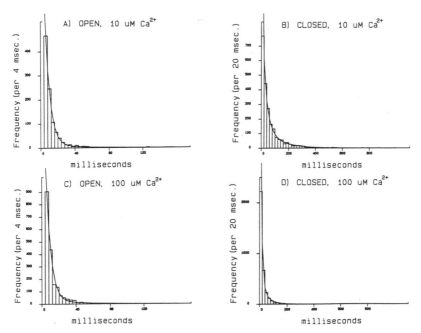

FIGURE 5. Single-channel lifetimes in the presences of 10 μM (**A, B**) or 100 μM (**C, D**) activating Ca^{2+}. Lifetimes were determined as described in the METHODS section. Lifetime histograms are shown together with probability density functions (pdfs) fitted by the method of maximum likelihood. The pdfs are as follows:

(**A**) Open times with 10 μM Ca^{2+}; $f(t) = 0.93 \exp(-t/5.79) + 0.07\exp(-t/45.29)$.

(**B**) Closed times with 10 μM Ca^{2+}; $f(t) = 0.17\exp(-t/6.96) + 0.49\exp(-t/30.74) + 0.34\exp(-t/124.0)$.

(**C**) Open times with 100 μM Ca^{2+}; $f(t) = 0.92\exp(-t/5.8) + 0.08 \exp(-t/44.87)$.

(**D**) Closed times with 100 μM Ca^{2+}; $f(t) = 0.17\exp(-t/5.56) + 0.59\exp(-t/10.4) + 0.24\exp(-t/47.32)$. The unit for all time constants is msec. The applied potential was 0 mV.

Mg^{2+} added in millimolar concentrations to the cytosolic face of either the reconstituted skeletal or cardiac SR Ca^{2+}-release channels inhibits channel opening,[6,15] probably by competing for a Ca^{2+} binding site (Ashley & Williams, submitted for publication). It has been proposed that Mg^{2+} may act as an important regulatory factor of the SR Ca^{2+} release channels *in vivo*.[15]

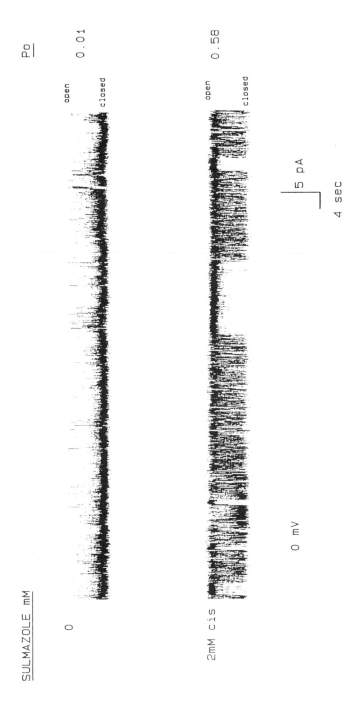

FIGURE 6. Activation of the sheep cardiac SR Ca^{2+} release channel by sulmazole. The top trace is a representative portion of a recording made in the presence of 10 μM activating Ca^{2+} in the *cis* chamber. The bottom trace is a representative portion of a recording of the same channel following the addition of 2 mM sulmazole to the *cis* chamber. Open probability values were obtained as described in the METHODS section from three-minute steady-state recordings. The applied potential was 0 mV.

Lifetime Analysis after Calcium Activation

Open and closed lifetime distributions of a single sheep cardiac SR Ca^{2+} release channel in the presence of either 10 or 100 μM activating Ca^{2+} as the sole ligand are shown in FIGURE 5. At both Ca^{2+} concentrations the most likely fits were obtained with double exponentials for the open times and triple exponentials for the closed times.[28] The probability density functions fitted by the method of maximum likelihood to the individual lifetimes are given in the legend to FIGURE 5.

Channel open times were unaffected by the elevation of activating Ca^{2+}. Similarly, the distribution of events remained unchanged, with approximately 90% of openings occurring to the short lifetime distribution at both 10 and 100 μM Ca^{2+}.

The increase in Po observed with elevation of the Ca^{2+} concentration on the cytosolic side of the channel must be brought about by the interaction of Ca^{2+} with one or more of the three closed conformations of the channel protein. All three mean closed times were shortened as activating Ca^{2+} was raised and this resulted in an increased rate of channel opening. Thus, as with all ligand-activated channels characterized so far, the SR Ca^{2+} release channel, even with Ca^{2+} as the sole ligand, has multiple open states. Recent work suggests that a branched or cyclic gating scheme will be necessary (Ashley and Williams, submitted for publication).

Pharmacological Modulation by Sulmazole (AR-L 115 BS)

Sulmazole, a benzimidazole derivative, is a cardiotonic drug that, in common with caffeine, is known to produce contractures in isolated cardiac muscle preparations[29] and a direct activation of cardiac muscle myofibrils via an increase in their affinity for Ca^{2+}.[30] We have recently demonstrated activation of the sheep cardiac muscle SR Ca^{2+} channel by this compound.[31] FIGURE 6 demonstrates the activation of a single sheep cardiac SR Ca^{2+} channel, in the presence of 10 μM cytosolic Ca^{2+}, by the addition of 2 mM sulmazole to the solution bathing the cytosolic face of the channel. Before the addition of sulmazole, channel open probability was 0.01. This rose to 0.58 after sulmazole addition.

Sulmazole is capable of activating the channel when added to either the cytosolic or lumenal sides of the channel, and its action is freely reversible.[31] It seems probable that this direct activation of the cardiac SR Ca^{2+} channel and the subsequent elevation of the cytosolic Ca^{2+} concentration contributes significantly to the contractures induced by this compound in mammalian cardiac muscle preparations.

REFERENCES

1. FABIATO, A. 1985. J. Gen. Physiol. **85:** 189-246.
2. FABIATO, A. 1985. J. Gen. Physiol. **85:** 247-289.
3. FABIATO, A. 1985. J. Gen. Physiol. **85:** 291-320.
4. MEISSNER, G. 1984. J. Biol. Chem. **259:** 2365-2374.
5. MEISSNER, G., E. DARLING & J. EVELETH. 1986. Biochemistry **25:** 236-244.
6. ROUSSEAU, E., J. S. SMITH, J. S. HENDERSON & G. MEISSNER. 1986. Biophys. J. **50:** 1009-1014.

7. MEISSNER, G. & J. HENDERSON. 1987. J. Biol. Chem. **262:** 3065-3073.
8. MCKINLEY, D. & G. MEISSNER. 1977. Febs Lett. **82:** 47-50.
9. MEISSNER, G. & D. MCKINLEY. 1982. J. Biol. Chem. **257:** 7704-7711.
10. MILLER, C. 1978. J. Memb. Biol. **40:** 1-23.
11. MILLER, C. 1982. *In* Transport in Biomembranes: Model Systems and Reconstitution. Antolini, R., A. Gliozzi & A. Gorio, Eds.: 99-108. Raven Press. New York.
12. TOMLINS, B., A. J. WILLIAMS & R. A. P. MONTGOMERY. 1984. J. Memb. Biol. **80:** 191-199.
13. SMITH, J. S., R. CORONADO & G. MEISSNER. 1985. Nature **316:** 446-449.
14. TANIFUJI, M., M. SOKABE & M. KASAI. 1987. J. Memb. Biol. **99:** 103-111.
15. SMITH, J. S., R. CORONADO & G. MEISSNER. 1986. J. Gen. Physiol. **88:** 573-588.
16. IMAGAWA, T., J. S. SMITH, R. CORONADO & K. P. CAMPBELL. 1987. J. Biol. Chem. **263:** 16636-16643.
17. LAI, F. A., H. P. ERICKSON, E. ROUSSEAU, Q. Y. LIU & G. MEISSNER. 1988. Nature **331:** 315-319.
18. LAI, F. A., K. ANDERSON, E. ROUSSEAU, Q. Y. LIU & G. MEISSNER. 1988. Biochem. Biophys. Res. Commun. **151:** 441-449.
19. TOMLINS, B., S. E. HARDING, M. S. KIRBY, P. A. POOLE-WILSON & A. J. WILLIAMS. 1986. Biochim. Biophys. Acta **856:** 137-143.
20. MEISSNER, G. 1986. J. Biol. Chem. **261:** 6300-6306.
21. MILLER, C. 1982. Philos. Trans. R. Soc. London B **299:** 401-411.
22. LABARCA, P., R. CORONADO & C. MILLER. 1980. J. Gen. Physiol. **76:** 397-424.
23. GRAY, M. A., R. A. P. MONTGOMERY & A. J. WILLIAMS. 1985. J. Memb. Biol. **88:** 85-95.
24. OEHME, M., M. KESSLER & W. SIMON. 1976. Chimia **30:** 204-206.
25. COLQUHOUN, D. & F. SIGWORTH. 1983. *In* Single-Channel Recording. B. Sakmann & E. Neher, Eds.: 191-263. Plenum. New York.
26. LEE, K. S. & R. W. TSIEN. 1984. J. Physiol. London **354:** 253-272.
27. GARCIA, A. M. & C. MILLER. 1984. J. Gen. Physiol. **83:** 819-839.
28. ASHLEY, R. H. & A. J. WILLIAMS. 1988. J. Physiol. London **406:** 89p.
29. TRUBE, G. & W. TRAUTWEIN. 1981. Drug Res. **31:** 185-188.
30. SOLARO, R. J. & J. C. RUEGG. 1982. Circ. Res. **51:** 290-294.
31. ASHLEY, R. H. & A. J. WILLIAMS. 1988. J. Physiol. London **406:** 213p.

Expression of mRNA Encoding Voltage-Dependent Ca Channels in *Xenopus* Oocytes

Review and Progress Report[a]

HENRY A. LESTER, TERRY P. SNUTCH, JOHN P.
LEONARD,[b] JOEL NARGEOT,[c] NATHAN DASCAL,[d]
BENSON M. CURTIS,[e] AND NORMAN DAVIDSON

*Division of Biology and
Division of Chemistry and Chemical Engineering
California Institute of Technology
Pasadena, California 91125*

INTRODUCTION

We are studying electrically excitable Ca channels with an approach that combines nucleic acid molecular biology and membrane biophysics. Gurdon *et al.*[1] first employed *Xenopus* oocytes as a system for heterologous expression of RNA. This system was later adapted for electrophysiological studies of ion channels by Barnard, Miledi, and their colleagues.[2-5] Although other expression systems are being developed for membrane channels [Claudio *et al.*[6] and Beam *et al.*, this volume], the oocyte is now the most common, robust, and best-characterized system. Experiments in which channels are expressed by RNA injections in oocytes have at least three major goals.

[a] This research was sponsored by grants from the National Institutes of Health (GM-10991 and GM-29836), from the United States—Israel Binational Science Foundation, from the CNRS (LP 8402), and from INSERM (U249), by postdoctoral fellowships from the American Heart Association to T. S. and J. P. L., from the NIH to B. M. C., and by a Bantrell Fellowship to N. D. J. N. thanks Laboratoire Servier for a travel grant.

[b] Present address: Department of Biological Sciences, University of Illinois at Chicago, Chicago IL 60680.

[c] Present address: Centre de biochimie macromoleculaire, C. N. R. S., 34033 Montpellier, France.

[d] Present address: Department of Physiology and Pharmacology, Sackler School of Medicine, Tel Aviv University, Ramat Aviv, Israel.

[e] Present address: Immunex Corporation, Seattle, WA 98101.

1. *Reconstitution:* Chemical and electrical excitability can be reconstituted in oocytes via RNA injection. This heterologous expression system can be exploited (**a**) to confirm the cloning of all subunits of a channel or a receptor[7–12]; (**b**) to employ electrophysiological or biochemical advantages of the oocyte[13]; (**c**) to determine the number and size of the RNA components necessary for a response[14,15]; (**d**) to obtain information about hormonal regulation of mRNA levels.[16,17]

2. *cDNA Cloning:* If an assay has been established in oocytes for a reconstituted excitability phenomenon, one can use this assay for screening a cDNA library in order to isolate a clone that encodes a channel, receptor, or transport molecule. Relevant recent examples concern the serotonin $5HT_{1C}$ receptor from choroid plexus,[18] the substance K receptor from bovine stomach,[19] and the Na-glucose transporter from intestine.[20]

3. *Structure-function relations:* After obtaining complete clones for all subunits of the desired protein, one can employ bacteriophage promoters to synthesize RNA *in vitro*. Subunits are mixed to form interspecific hybrid proteins; and the clones are mutated to test specific hypotheses about amino acids important in the function. Studies of this sort are revealing facts about agonist binding, drug blockade, and permeation at acetylcholine receptor proteins.[7,8,21–24]

Our experiments on Ca^{2+} channels have not yet progressed past Stage 1. Full-length cDNA clones have been described that encode the dihydropyridine receptor of skeletal muscle (see paper by Tanabe *et al.*), which is presumably a component of a voltage-dependent Ca^{2+} channel. However, no oocyte expression studies have been reported with RNA synthesized *in vitro* from these clones. Perhaps other subunits are required for which clones are not yet available. This hypothesis is likely in view of the fact that the only successful expression experiment reported to date has employed a cell line that expresses other peptide components of the Ca^{2+} channel (Beam *et al.*, this volume). It is also possible that the oocyte cannot perform posttranslational modifications appropriate to produce a functional channel.

This paper describes further progress toward understanding the properties of voltage-dependent Ca channels that can be reconstituted with the oocyte assay.

EXPERIMENTAL PROCEDURES

mRNA Size Fractionation

Total cellular RNA was isolated from the hearts of 2- to 16-day-old rats or from brains of 14- to 16-day-old rats using a lithium chloride/urea procedure. Poly (A)+ RNA was isolated by a single binding to either oligo(dT)-cellulose (Collaborative Research, type 3) or poly(U)-Sepharose (Pharmacia, type 4B). A sucrose-gradient method was employed to fractionate RNA according to molecular length. Poly(A)+ RNA was heat denatured at 70°C for three minutes cooled on ice and then layered onto a 6-20% sucrose gradient containing 15 mM PIPES, pH 6.5/5 mM EDTA/ 0.25% Sarkosyl. Gradients were centrifuged in an SW 27.1 rotor at 24K rpm for 18 hours at 4°C. Fractions of approximately 0.45 ml were collected and the RNA recovered by ethanol precipitation. Individual fractions were pooled into sets of three adjacent fractions and the RNA reprecipitated. The RNA from each pool was dissolved in H_2O to give a final concentration equivalent to 4.5 $\mu g/\mu l$ of unfractionated poly(A)+ RNA. These samples were used for RNA blots and oocyte injections.

Oocyte Isolation, Injection, and Electrophysiology

Xenopus laevis were anesthetized in 0.15% tricaine (Sigma) and several ovarian lobes were excised. Ovarian tissue and follicular cells were removed by treatment with 2 mg/ml collagenase (Sigma, type 1A) in Ca-free saline (82.5 mM NaCl/2 mM KCl/1 mM MgCl$_2$/5 mM HEPES, pH 7.4). Stage V and VI oocytes[25] were selected and allowed to recover overnight in ND96 (96 mM NaCl/2 mM KCl/1.8 mM CaCl$_2$/1 mM MgCl$_2$/ 5 mM HEPES, pH 7.4/2.5 mM pyruvate/100 U/ml penicillin/ 100 μg/ml streptomycin). Oocytes were injected with 50-70 nl of RNA solution by positive displacement using a 10-μl micropipette (Drummond Scientific Co., Broomall, PA).

RNA Transfer and Hybridization

RNA was denatured with 2.2 M formaldehyde/50% formamide/5 mM sodium acetate/1 mM EDTA/20 mM MOPS, pH 7.0 at 65°C for 15 minutes and then cooled on ice. Samples were separated through a 1.1% agarose gel containing 2.2 M formaldehyde for 8 hours at 3.5 V/cm. After electrophoresis, RNA was transferred overnight by capillary blot in 20 × SSPE (3.6 M NaCl/0.2 M NaH$_2$PO$_4$/20 mM EDTA) to Hybond-N (Amersham). The size distribution and recovery of mRNA from sucrose gradient pools was determined by hybridization of Northern blot filters to a ^{32}P-labeled poly(dT) probe. Preparation of the probe and hybridization were as previously described.[10]

Electrophysiology

Macroscopic currents were measured with a two-microelectrode voltage-clamp circuit (Dagan Instruments, Minneapolis, MN) on oocytes incubated in ND96[26] for 2-5 days after injection. Microelectrodes were filled with 3 M KCl and had resistances of 0.5-2 MΩ. Ba currents were detected in 40 mM Ba(OH)$_2$/50 mM NaOH/2 mM KOH/5 mM HEPES titrated to pH 7.4 with methanesulphonic acid. To block Na currents, tetrodotoxin (TTX) was added to a final concentration of 1 μM. Other currents were detected in ND96 saline without added antibiotics and pyruvate. Pulse protocols and data analysis were as previously described.[27]

RESULTS

Ca channels are typically monitored with Ba as the permeant ion. This procedure is followed with oocytes as well, in order to eliminate two complicating Ca-mediated phenomena: (1) activation of the Cl current and (2) inactivation of the Ca channels themselves.

Heart RNA

The best-characterized current induced by heart RNA resembles the L-type Ca currents observed in heart muscle.[26] This current was activated at membrane potentials more positive than about −30 mV and was maintained for several seconds. It was blocked by 20 μM Cd and by micromolar concentrations of the dihydropyridine Ca blocker nifedipine. It was enhanced by the dihydropyridine "Ca agonist" Bay K 8644.[28] The single-channel conductance was 18-25 pS.[29]

Neurotransmitters modulate the Ca channels expressed from heart RNA; and these actions resemble those for the Ca channels of intact heart cells.[26] Thus, β-adrenergic agonists increased the current, and this action was mimicked by forskolin and by cAMP injection. Muscarinic agonists decreased the current; it is not known whether this action proceeds via the same intracellular messenger as for the muscarinic decrease of Ca currents in heart cells.

Fractionation of heart mRNA showed that Ca channels were induced best by RNA in a rather narrow size range (5-7 kb, TABLE 1).

TABLE 1. Size Fractionation of Heart RNA-Inducing Ba Currents

Size Class (kb)	I_{Ba} (nA)
7-11	10
6-8	18
5-7	60
4-6	12
1-3	10
Non-injected	10

NOTE: Each datum is the average of three to eight oocytes.

Brain mRNA

Brain RNA induces Ca channels with different characteristics than does heart RNA.[27,30] Observable currents were present at test potentials more positive than about −30 mV. The current-voltage relation peaked at about +15 mV. The currents inactivated rather slowly, by about 50% after several seconds. The currents were insensitive to dihydropyridines, both antagonists such as nifedipine and agonists such as Bay K 8644, even at concentrations up to 20 μM. They were also insensitive to verapamil (10 μM). They were blocked by Cd^{2+} with an apparent dissociation constant of 6 μM and by Ni at about 500 μM. They were not blocked by ω-conotoxin. In single-channel experiments, the Ba currents persisted in recordings from outside-out patches. The current-voltage relation revealed a conductance of 12-15 pS.

The characteristics of these Ba currents induced by brain RNA are in general agreement with those reported for Ca fluxes in brain synaptosomes by Nachshen and Blaustein[31] and Nachshen.[32] They also match the characteristics reported for "N"-type Ca channels[33] in several respects. There is pleasing agreement with the idea that "N"-type channels are primarily responsible for the Ca influx that leads to transmitter release at brain synapses.[34] Perhaps the most serious point of disagreement is the insensitivity of our currents to ω-conotoxin.[35]

In contrast to the fractionation results reported for heart RNA, Ca channels are induced by brain RNA in two rather broad size ranges: relatively large size (6-10 kb) and relatively small size (1-3 kb) (TABLE 2 and FIGS. 1, 2). Furthermore, the characteristics of the current induced by fractionated RNA were similar to those induced by total RNA, with regard to voltage dependence, inactivation, phorbol ester stimulation, and Cd blockade. FIGURES 1 and 2 give examples of currents induced by each faction.

DISCUSSION

There are clearly major differences between the expression of brain and heart Ca channels in oocytes. We shall comment on the fact that the distribution of brain mRNA inducing Ca channels is much broader than that for heart mRNA Na channels. All known Na channels[36,37] and the putative Ca channel subunit[38] are so large that

TABLE 2. Size Fractionation of Brain RNA Inducing Ba and Na Currents

Size Fraction (kb)	I_{Na} (nA)	I_{Ba} (nA)	Number of Oocytes
6-10	479 ± 219	361 ± 60	5
1-3	0	132 ± 13	15
Noninjected	0	22 ± 2	5

NOTE: Amplitudes (mean ± SEM) of voltage-dependent currents in oocytes injected with the fractions given.

their mRNAs have coding regions in excess of 5.5 kb. Such a coding region could in theory be accommodated by any mRNA in excess of this size, such as (a) the active fractions from the heart mRNA or (b) the higher MW active fractions from the brain mRNA. Clearly a shorter coding region is inducing currents in oocytes injected with the lighter fraction of rat brain mRNA. Although other interpretations are possible, the simplest is that some Ca channels can be homo- or hetero-oligomers of subunits that correspond to only one or two of the four homology units in the known Na or Ca channels. Thus, the novel Ca-channel proteins might have the same organization as the K-channel proteins recently described from the *Shaker* locus of *Drosophila*[12,39] and expressed in oocytes.[11,12]

A point of difference between brain Ca and Na channels expressed in oocytes concerns the apparent sufficiency of a single mRNA size fraction. That is, the pharmacology and kinetics of the induced Ba currents do not differ between total mRNA, the high-MW fraction, and the low-MW fraction. In contrast, normal Na currents are not induced in oocytes by RNA encoding the ~200 kDa α subunit alone; other proteins—encoded by RNA in the size range 1-2.5 kb—must also be present for normal kinetic and voltage-sensitive properties of the channel. Krafte *et al.*[14] and Auld *et al.*[37] consider several possibilities about the nature of these protein(s). Briefly, (a)

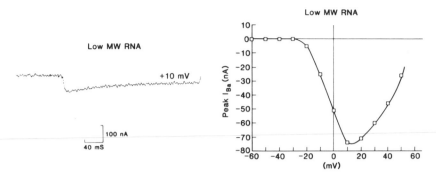

FIGURE 1. Left, a Ba current waveform for a voltage step to +10 mV from a holding potential of −80 mV, in an oocyte injected several days previously with low-MW brain poly (A) RNA. The inactivation time course is quite similar to that observed with unfractionated RNA. Right, the current-voltage relation is also indistinguishable from that with unfractionated RNA.

mammalian brain and skeletal muscle Na channels contain one or two small subunits that could be necessary for normal function; (b) the additional protein(s) could be required for the very extensive and unusual glycosylations; and/or (c) the additional protein(s) could play another role in assembly. A similar but less completely described situation may apply to the expression of brain K currents in oocytes.[41]

It is ironic that mRNA from skeletal muscle, the only tissue from which a putative Ca channel has been cloned,[38] induces little or no detectable Ca channels in oocytes. This situation has thus far limited attempts to use the cloned DHP receptor in functional studies of the muscle Ca channel. The papers by Tanabe *et al.* and by Beam *et al.* (this volume) describe an alternative expression system that will lead to new insights about Ca-channel function.

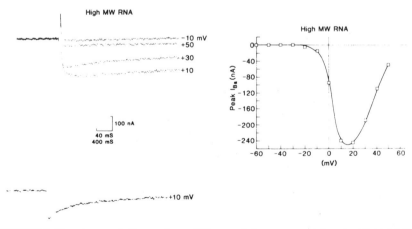

FIGURE 2. Experiments similar to those of FIGURE 1, in an oocyte injected with high-MW brain poly (A) RNA. The waveforms and current-voltage relation resemble those seen with low-MW RNA (FIG. 1) and with unfractionated RNA.

There are many additional questions about the molecular biology of Ca channels and their expression in oocytes. For instance, dihydropyridine-sensitive high-threshold ("L"-type) and low-threshold transient ("T"-type) Ca channels have often been described in the brain; yet these are not detected in the brain RNA-injected oocytes, despite the fact that heart RNA induces dihydropyridine-sensitive channels. Further progress will call for cloning and expression of Ca channels from brain and heart.

REFERENCES

1. GURDON, J. B., C. D. LANE, H. R. WOODLAND & G. MARBAIX. 1971. Use of frog eggs and oocytes for the study of messenger RNA and its translation in living cells. Nature **233:** 177-182.
2. BARNARD, E. A., R. MILEDI & K. SUMIKAWA. 1982. Translation of exogenous messenger RNA coding for nicotinic acetylcholine receptors produces functional receptors in *Xenopus* oocytes. Proc. R. Soc. London B. **215:** 241.
3. MILEDI, R., I. PARKER & K. SUMIKAWA. 1982. Synthesis of chick brain GABA receptors by frog oocytes. Proc. R. Soc. London B. **216:** 509-515.
4. MILEDI, R. & K. SUMIKAWA. 1982. Synthesis of cat muscle acetylcholine receptors by *Xenopus* oocytes. Biomed. Res. **3:** 390-399.
5. GUNDERSEN, C. B., R. MILEDI & I. PARKER. 1983. Voltage-operated channels induced by foreign messenger RNA in *Xenopus* oocytes. Proc. R. Soc. London B. **220:** 131-140.
6. CLAUDIO, T., W. N. GREEN, D. S. HARTMAN, D. HAYDEN, H. L. PAULSON, F. J. SIGWORTH, S. M. SINE & A. SWEDLUND. 1987. Genetic reconstitution of functional acetylcholine receptor channels in mouse fibroblasts. Science **238:** 1688-1694.
7. MISHINA, M., T. TOBIMATSU, K. IMOTO, K. I. TANAKA, Y. FUJITA, K. FUKUDA, M. KURASAKI, H. TAKAHASHI, Y. MORIMOTO, T. HIROSE, S. INAYAMA, T. TAKAHASHI, M. KUNO & S. NUMA. 1985. Localization of functional regions of acetylcholine receptor α-subunit by site-directed mutagenesis. Nature **313:** 364-369.
8. WHITE, M. M., K. MIXTER-MAYNE, H. A. LESTER & N. DAVIDSON. 1985. Mouse-*Torpedo* hybrid acetylcholine receptors: Functional homology does not equal sequence homology. Proc. Natl. Acad. Sci. USA **82:** 4852-4856.
9. NODA, M., T. IKEDA, T. KAYANO, H. SUZUKI, H. TAKESHIMA, H. TAKAHASHI, H. M. KUNO & S. NUMA. 1986. Expression of function sodium channels from cloned cDNA. Nature **322:** 826-828.
10. GOLDIN, A. L., A. DOWSETT, H. LUBBERT, T. SNUTCH, R. DUNN, H. A. LESTER, W. A. CATTERALL & N. DAVIDSON. 1986. Messenger RNA coding for only the α subunit of the rat brain Na channel is sufficient for expression of functional channels in *Xenopus* oocytes. Proc. Natl. Acad. Sci. USA **83:** 7503-7507.
11. TIMPE, L. C., T. L. SCHWARZ, B. L. TEMPEL, D. M. PAPAZIAN, Y. N. JAN & L. Y. JAN. 1988. Expression of functional potassium channels from *Shaker* cDNA in *Xenopus* oocytes. Nature **331:** 143-145.
12. IVERSON, L., M. A. TANOUYE, H. A. LESTER, N. DAVIDSON & B. RUDY. 1988. Expression of A-type potassium channels from *Shaker* cDNAs. PNAS **85:** 5723-5727.
13. DASCAL, N., C. IFUNE, R. HOPKINS, T. P. SNUTCH, H. LUBBERT, N. DAVIDSON & H. A. LESTER. 1986. Involvement of a GTP-binding protein in mediation of serotonin and acetylcholine responses in *Xenopus* oocytes injected with rat brain messenger RNA. Mol. Brain Res. **1:** 201-209.
14. KRAFTE, D. A., T. P. SNUTCH, J. P. LEONARD, N. DAVIDSON & H. A. LESTER. 1988. Evidence for the involvement of more than one mRNA species in controlling the inactivation process of rat and rabbit brain Na channels expressed in *Xenopus* oocytes. J. Neurosci. **8:** 2859-2868.
15. LUBBERT, H., T. P. SNUTCH, N. DASCAL, H. A. LESTER & N. DAVIDSON. 1987. Rat brain 5HT$_{1C}$ receptors are encoded by a 5-6 kbase mRNA size class and are functionally expressed injected *Xenopus* oocytes. J. Neurosci. **7:** 1159-1165.

16. BOYLE, M. B., N. J. AZHDERIAN, N. J. MacLUSKY, F. NAFTOLIN & L. K. KACZMAREK. 1987. Xenopus oocytes injected with rat uterine RNA express very slowly activating potassium currents. Science **235:** 1221-1224.

17. ORON, Y., R. E. STRAUB, P. TRAKTMAN & M. C. GERSHENGORN. 1987. Decreased TRH receptor mRNA activity precedes homologous downregulation: Assay in oocytes. Science **238:** 1406-1408.

18. LUBBERT, H., B. HOFFMAN, T. P. SNUTCH, T. VAN DYKE, A. J. LEVINE, P. R. HARTIG, H. A. LESTER & N. DAVIDSON. 1987. cDNA cloning of a serotonin 5HT $_{1C}$ receptor by using electrophysiological assays of mRNA injected *Xenopus* oocytes. Proc. Natl. Acad. Sci. **84:** 4332-4336.

19. MASU, Y., K. NAKAYAMA, H. TAMAKI, Y. HARADA, M. KUNO & S. NAKANISHI. 1987. cDNA cloning of bovine substance-K receptor through oocyte expression system. Nature **329:** 836-838.

20. HEDIGER, M., M. J. COADY, T. S. IKEDA & E. M. WRIGHT. 1987. Expression cloning and cDNA sequencing of the Na$^+$/glucose cotransporter. Nature **330:** 379-381.

21. SAKMANN, B., C. METHFESSEL, M. MISHINA, T. TAKAHASHI, T. TAKAI, M. KURASAKI, K. FUKUDA & S. NUMA. 1985. Role of acetylcholine receptor subunits in gating of the channel. Nature **318:** 538-543.

22. MISHINA, M., T. TAKAI, K. IMOTO, M. NODA, T. TAKAHASHI, S. NUMA, C. METHFESSEL & B. SAKMANN. 1986. Molecular distinction between fetal and adult forms of muscle acetylcholine receptor. Nature **321:** 406-411.

23. IMOTO, K., C. METHFESSEL, B. SAKMANN, M. MISHINA, Y. MORI, T. KONNO, K. FUKUDA, M. KURASAKI, H. BUJO, Y. FUJITA & S. NUMA. 1986. Location of a delta-subunit region determining ion-transport through the acetylcholine-receptor channel. Nature **324:** 670-674.

24. YOSHII, K., L. YU, K. MIXTER-MAYNE, N. DAVIDSON & H. A. LESTER. 1987. Equilibrium properties of mouse-*Torpedo* acetylcholine receptor hybrids expressed in *Xenopus* oocytes. J. Gen. Physiol. **90:** 553-573.

25. DUMONT, J. N. 1972. Oogenesis in *Xenopus laevis* (Daudin). I. Stages of oocyte development in laboratory maintained animals. J. Morphol. **136:** 153-180.

26. DASCAL, N., T. P. SNUTCH, H. LUBBERT, N. DAVIDSON & H. A. LESTER. 1986. Expression and modulation of voltage-gated calcium channels after RNA injection in *Xenopus* oocytes. Science **231:** 1147-1150.

27. LEONARD, J. P., J. NARGEOT, T. SNUTCH, N. DAVIDSON & H. A. LESTER. 1987. Ca channels induced in *Xenopus* oocytes by rat brain mRNA. J. Neurosci. **7:** 875-881.

28. SNUTCH, T. P., J. P. LEONARD, J. NARGEOT, H. LUBBERT, N. DAVIDSON & H. A. LESTER. 1987. Characterization of voltage-gated calcium channels in *Xenopus* oocytes after injection of RNA from electrically excitable tissues. *In* Cell Calcium and Membrane Transport. D. Eaton & L. Mandel, Eds.: 153-166. Society of General Physiologists Series, Vol. 42. Rockefeller Univ. Press. New York.

29. MOORMAN, J. R., Z. ZHOU, G. E. KIRSCH, A. E. LACERDA, J. M. CAFFREY, D. M-K. LAM, R. H. JOHO & A. M. BROWN. 1987. Expression of single calcium channels in *Xenopus* oocytes after injection of mRNA from rat heart. Am. J. Physiol. **253:** H985-991.

30. LEONARD, J. P., T. P. SNUTCH, N. DAVIDSON & H. A. LESTER. 1987. Ca channels induced in *Xenopus* oocytes by rat brain mRNA: Size fractionation and physiological studies. Soc. Neurosci. Abstr. **13:** 795.

31. NACHSHEN, D. A. & M. P. BLAUSTEIN. 1979. The effects of some organic "calcium antagonists" on calcium influx in presynaptic nerve terminals. Mol. Pharmacol. **16:** 579-586.

32. NACHSHEN, D. A. 1985. The early time course of potassium stimulated calcium uptake in presynaptic nerve terminals isolated from the rat brain. J. Physiol. **361:** 251-268.

33. NOWYCKY, M. C., A. P. FOX & R. W. TSIEN. 1985. Long-opening mode of gating of neuronal calcium channels and its promotion by the dihydropyridine calcium agonist Bay K 8644. Proc. Natl. Acad. Sci. USA **82:** 2178-2182.

34. HIRNING, L. D., A. P. FOX, E. W. McCLESKEY, B. M. OLIVERA, S. A. THAYER, R. J. MILLER & R. W. TSIEN. 1988. Dominant role of N-type Ca^{2+} channels in evoked release of norepinephrine from sympathetic neurons. Science **239:** 57-61.

35. McCleskey, E. W., A. P. Fox, D. H. Feldman, L. J. Cruz, B. M. Olivera & D. Yoshikami. 1987. ω-Conotoxin: Direct and persistent blockade of specific types of calcium channels in neurons but not muscle. Proc. Natl. Acad. Sci. USA **84:** 4327-4331.
36. Noda, M., T. Ikeda, T. Kayano, H. Suzuki, H. Takeshima, M. Kurasaki, H. Takahashi & S. Numa. 1986. Existence of distinct sodium channel messenger RNAs in rat brain. Nature **320:** 188-192.
37. Auld, V., A. L. Goldin, D. S. Krafte, J. Marshall, J. M. Dunn, W. A. Catterall, N. Davidson, H. A. Lester & R. J. Dunn. 1988. A rat brain Na channel α subunit with novel gating properties. Neuron **1:** 449-461.
38. Tanabe, T., H. Takeshima, A. Mikami, V. Flockerzi, H. Takahashi, K. Kangawa, M. Kojima, H. Matsuo, T. Hirose & S. Numa. 1987. Primary structure of the receptor for calcium channel blockers from skeletal muscle. Nature **328:** 313-318.
39. Schwarz, T. L., B. L. Tempel, D. M. Papazian, Y. N. Jan & L. Y. Jan. 1988. Multiple potassium-channel components are produced by alternative splicing at the *Shaker* locus in *Drosophila.* Nature **331:** 137-342.
40. Kamb, A., J. Tseng-Crank & M. A. Tanouye. 1988. Multiple products of the drosophila *Shaker* gene contribute to potassium channel diversity. Neuron **1:** 421-430.
41. Rudy, B., J. H. Hoger, N. Davidson & H. A. Lester. 1987. At least two mRNA species contribute to the properties of rat brain "A"-type potassium channels expressed in *Xenopus* oocytes. Neuron **1:** 649-658.

Expression of Voltage-Dependent Ca Channels from Skeletal Muscle in *Xenopus* Oocytes

ILANA LOTAN, ARIELA GIGI, AND
NATHAN DASCAL

Division of Physiology and Pharmacology
Sackler Faculty of Medicine
Tel Aviv University
Ramat Aviv 69978, Israel

Oocytes were injected with RNA from hearts and hind-limb skeletal muscles of rats (7-17 days old) or rabbits (2-10 days old) and incubated for three to seven days as described previously.[1] Currents through Ca^{2+} channels were recorded using the two-electrode voltage clamp technique, in a solution containing 40 mM Ba^{2+}, 2 mM K^+, 5 mM HEPES (pH 7.5), and 40 to 50 mM Na^+ or TEA^+. The anion was either methanesulfonate,[1] acetate, or Cl^-. In many instances, and always when the Cl^--containing solution was used, oocytes were injected with 200-800 pmol EGTA, to prevent activation of Ca-dependent Cl^- currents.

As we reported previously, heart RNA directed a transient (20-40 nA), dihydropyridine (DHP)-insensitive Ba current (I_{tr}), and a dominant (50-400 nA), slow L-type current (I_{sl}); both were activated at voltages above -30 mV. I_{tr} was fully inactivated at about -10 mV, when I_{sl} was only marginally inactivated.[1] BAY K8644 (1 μM) enhanced I_{sl} by about 200%. I_{tr} was more susceptible to block by Ni^{2+} than I_{sl} (ID_{50} of 40 and 350 μM, respectively).

Despite the abundance of DHP receptor (DHPR) in the skeletal muscle, muscle RNA directed small (5-30 nA), slow current (I_{sl}), whereas the dominant (and sometimes the only) current was DHP-insensitive, transient, and inhibited by Ni^{2+} (ID_{50} = 40 μM). I_{sl} was much less sensitive to Ni^{2+}. Voltages at which the steady-state inactivation was half-maximal were about -45 mV for I_{tr}, and about -10 mV for I_{sl}. BAY K8644 (1 μM) enhanced I_{sl} by 30%.

It has been proposed that only a few percent of DHPR in the muscle correspond to active Ca^{2+} channels,[2] while the rest may be the voltage sensors of excitation-contraction coupling.[3] On the other hand, the primary structure of DHPR deduced from the cDNA sequence suggests that DHPR is the channel-forming subunit of the Ca^{2+} channel.[4] To test whether DHPR is indeed related to a Ca channel, we prepared two oligonucleotides, cDHPR-O_1 (57 bases) and cDHPR-O_2 (81 bases), complementary to two transmembrane stretches of cDNA of DHPR cloned by Tanabe *et al.*[4] Both oligonucleotides selectively hybrid-arrested the expression of the L channel (I_{sl}) from rabbit and rat heart, leaving I_{tr} unaffected. Control experiments verified the specificity of these effects: These oligonucleotides did not affect the expression of voltage-dependent Na^+ and transient K^+ channels from rat and chick brain. Moreover, oligonucleotides complementary to cDNA of transmembrane stretches of the voltage-dependent Na^+ channel did not affect the expression of the heart Ca^{2+} channels. We conclude that DHPR of skeletal muscle is an essential component of a Ca^{2+} channel, and that fast and slow Ca^{2+} channels are separate molecular entities.

REFERENCES

1. DASCAL, N., et al. 1986. Science **231:** 1147-1150.
2. SCHWARTZ, L. M., E. W. MCCLESKEY & W. ALMERS. 1985. Nature **314:** 747-751.
3. RIOS, E. & G. BRUM. 1987. Nature **325:** 717-720.
4. TANABE, T. et al. 1987. Nature **328:** 313-318.

A Molecular Model of Excitation-Contraction Coupling at the Skeletal Muscle Triad Junction via Coassociated Oligomeric Calcium Channels[a]

LIN HYMEL,[b] HANSGEORG SCHINDLER,[b] MAKOTO INUI,[c] SIDNEY FLEISCHER,[c] JÖRG STREISSNIG,[d] AND HARTMUT GLOSSMANN[d]

[b] Institute for Biophysics
University of Linz
Linz, Austria

[c] Department of Molecular Biology
Vanderbilt University
Nashville, Tennessee

[d] Institute for Biochemical Pharmacology
University of Innsbruck
Innsbruck, Austria

We have recently shown that both the ryanodine[1] and dihydropyridine (DHP)[2] receptors purified from skeletal muscle form Ca^{2+}-specific oligomeric Ca^{2+} channels in planar lipid membranes. The purified ryanodine receptor revealed several of the properties expected for the sarcoplasmic reticulum Ca^{2+}-release channel, including activation by submicromolar Ca^{2+}, while the purified DHP receptor showed many of the characteristics of a phosphorylation- and voltage-dependent L-type Ca^{2+} channel. These observations, together with mounting evidence that both channels are localized at the triad junction[3,4] would support a "Ca^{2+}-induced Ca^{2+} release" mechanism. On the other hand, the discovery that DHPs block the rapid, depolarization-induced charge movement in the transverse tubule[5] suggests more direct, conformational coupling between the two channel complexes. This study represents the first attempt to characterize and compare biophysical properties of the purified channel proteins in consideration of the molecular mechanism of excitation-contraction coupling.

An important biophysical characteristic of both channels is their organization into oligomeric complexes (oligochannels), evidenced by the multitude of time-correlated subconductance states observed[1,2] and the size and composition of purified ryanodine receptor.[3] Conductance characteristics for the purified Ca^{2+}-release channel appear

[a] Supported by grants from the Fonds zur Förderung der Wissenschaftlichen Forschung (Project Grant S45-1 to HG and S45-3 to HS) and the National Institutes of Health (DK14632 and HL32711 to SF).

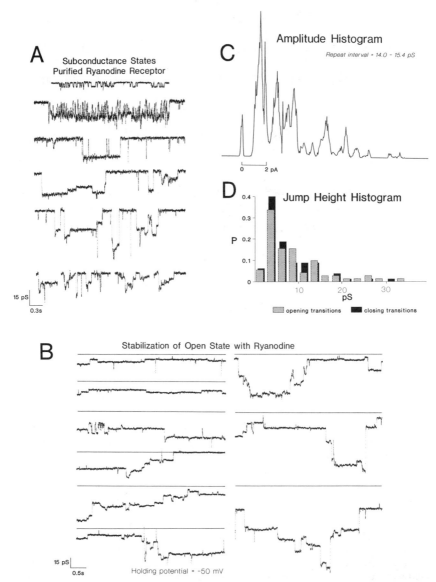

FIGURE 1. (**A**) Selected channel activity showing stabilized subconductance levels. Channel openings shown as downward deflections. Solutions: *cis,* 100 m*M* KCl, 10 m*M* Hepes/Tris (pH 7.4), 1 m*M* ATP, 2 μ*M* CaCl$_2$; *trans,* 50 m*M* CaCl$_2$, 100 m*M* KCl, 10 m*M* Hepes/Tris (pH 7.4). Bilayer consisted of soybean phospholipid/cholesterol (6/1 wt/wt), with approximately 20-50 ryanodine receptor polypeptides per membrane, added from the *cis* side. Holding potential = −100 mV at the *cis* side. The reversal potential under these conditions was +47 mV. (**B**) Same as in A, but holding potential = −50 mV and 10 μ*M* ryanodine has been added to the *cis* side. (**C**) Amplitude histogram for the data in B. (**D**) Jump height histogram for the data in B. Conductance transitions were grouped (openings and closings treated separately) into bins at 2.5-pS intervals. Gaps are for clarity of presentation only.

in FIGURE 1A (for similar data on DHP receptor *cf.* Glossman *et al.*[6]). The smallest well-defined conductance state was 3.8 pS (upper trace), while larger conductances and gating transitions were often integer multiplies of this value (lower traces). Ryanodine stabilized open-channel states, thus simplifying analysis of their distribution (FIG. 1B). Amplitude histogram analysis (FIG. 1C) revealed a series of major peaks

A

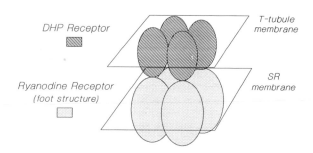

B

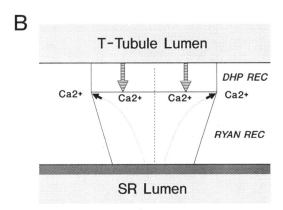

FIGURE 2. Model of the structural (**A**) and functional (**B**) organization of the coassociated calcium channels at the triad junction. In B the large arrows indicate the flow of information (Ca^{2+} ions, protein conformational transitions, or electrostatic redistribution) from the DHP-sensitive Ca^{2+} channel complex to the ryanodine-sensitive Ca^{2+} channel complex (foot structure or Ca^{2+} release channel). The smaller arrows represent the exit of Ca^{2+} ions from the sarcoplasmic reticulum terminal cisternae through the foot structure.

at intervals of 14.0-15.3 pS, or about four times the elementary conductance. The peak heights were Poisson-distributed, implying that the open states of oligomeric ryanodine receptor complexes are determined by independent 15-pS complexes, or, on the average, about four elementary units. We then analyzed the size distribution of opening and closing transitions (FIG. 1D), and found for both a distinct maximum at approximately 4 pS, again obeying a Poisson distribution. The distribution of closing transitions was slightly shifted to smaller values than that of openings, and the overall number of closings exceeded that of openings by 13%. That is, the tendency was for closings to be more often resolvable into smaller discrete steps, whereas openings were more often synchronized, larger transitions. Both opening and closing transitions of about 14-15 pS showed a slightly higher frequency than expected for the Poisson distribution fitted to the remaining data. Because the Poisson distribution implies statistically independent events, these data would indicate that, although the decision to *remain* open or closed is made in 15 pS units, single gating events occur as randomly determined multiples of the elementary unit of 3.8 pS, with 15 pS transitions slightly favored. Moreover, opening transitions tend to be somewhat more cooperative than channel closings, similar to the behavior of DHP receptor[6] and porin[7] channels.

Thus, the functional cooperativity witnessed in the gating properties of both DHP receptor and ryanodine receptor Ca^{2+} channels would support morphological indications that the ryanodine receptor[8] and perhaps also the DHP receptor[4] *in situ* have fourfold symmetry and consist of multiples of an elementary unit. A simple structural model would require the interaction, in orthogonal symmetry, of four structural elements across the triad junction (FIG. 2A). It should be noted that, given the low conductance values that we have tentatively assigned to "monochannels,"[1,2] the four structural elements depicted in FIGURE 2A could themselves be composed of as many as four still smaller functional domains. In addition, the basic complex consisting of the two coassociated oligochannel complexes might well be structurally stabilized or functionally modified via further associated proteins.[9] Functional coupling between the two channel complexes remains to be elucidated. Given the proposed close proximity of the channel complexes, the flux of Ca^{2+} from the DHP-sensitive channel might be sufficient to activate the Ca^{2+}-release channel within milliseconds, even before a measurable Ca^{2+} current develops (FIG. 2B). Alternatively, direct, voltage-sensitive, conformational coupling may turn on the Ca^{2+}-release channel. Another interesting possibility is suggested by the tendency, discussed above, for the opening transitions of associated channels to be synchronized. Because we have observed this property for several different apparently unrelated channels, it may be related to the one property common to all ion channels, namely ion flux through the channel. Perhaps, then, merely the intimate association of the two oligochannel complexes suffices to couple the opening of both.

REFERENCES

1. HYMEL, L., M. INUI, S. FLEISCHER & H. SCHINDLER. 1988. Proc. Natl. Acad. Sci. USA 85: 441-445.
2. HYMEL, L., J. STRIESSNIG, H. GLOSSMANN & H. SCHINDLER. 1988. Proc. Natl. Acad. Sci. USA 85: 4290-4294.
3. INUI, M., A. SAITO & S. FLEISCHER. 1987. J. Biol. Chem. 262: 1740-1747.
4. BLOCK, A. B., C. FRANZINI-ARMSTRONG, F. A. LAI & G. MEISSNER. 1988. Biophys. J. 53: 470a.
5. RIOS, E. & G. BRUM. 1987. Nature 325: 717-720.
6. GLOSSMANN, H., et al. 1989. Structure of calcium channels. Ann. N.Y. Acad. Sci. This volume.
7. SCHINDLER, H. & J. P. ROSENBUSCH. 1981. PNAS 78: 2302-2306.
8. SAITO, A., M. INUI, M. RADERMACHER, J. FRANK & S. FLEISCHER. 1988. J. Cell Biol. 107: 211-219.
9. CHADWICK, C., M. INUI & S. FLEISCHER. 1988. J. Biol. Chem. 263: 10872-10877.

Modulation of Calcium Channels by Charged and Neutral Dihydropyridines

R. S. KASS, J. P. ARENA, AND S. CHIN

Department of Physiology
University of Rochester Medical Center
Rochester, New York 14642

INTRODUCTION

It is well known that dihydropyridines (DHPs) are the most potent and selective compounds that have been used to probe (L-type) calcium channels in a variety of cell types (reviewed by Triggle and Venter[1]). These compounds modulate channel gating[2-5] and may parallel the activity of endogenous ligands.[6] Because of their selectivity and potency, the DHPs have been used as ligands to isolate and purify the calcium-channel protein.[7-10]

The purpose of the work described briefly in this paper and more completely in Kass and Arena[11] was to investigate the modulation of calcium channels by DHPs under conditions in which the ionization of the drug molecule could be modified. The work was designed to compare calcium-channel block by neutral and charged DHPs in order to gain insight into the manner by which these molecules gain access to the receptor that is closely associated with the calcium-channel protein. The results indicate that the drug-bound DHP receptor is accessible to external hydrogen ions. This observation places restrictions on the location of the DHP receptor in the cell membrane.

METHODS

The experiments summarized in this paper were carried out in single ventricular myocytes that had been isolated following the procedure of Mitra and Morad.[12] Experimental details are provided in Kass and Arena[11] but are briefly summarized here. Membrane currents were recorded using a whole-cell arrangement of the patch-clamp procedure,[13] and membrane currents were recorded in solutions that eliminated Na- and K-channel currents.[14] Solutions were changed via gravity-fed chambers that could be selected with an electronic valve. Drugs were dissolved in millimolar stock solutions and diluted down appropriately for each experiment. Solutions were buffered as described in Krafte and Kass[15] over a range of external pH 6.0 to 10.0

In this study, two DHP derivatives were used. Nisoldipine, a gift of Miles Laboratories, was dissolved in polyethylene glycol 400. This drug is entirely in the neutral form over the pH range of the study. Amlodipine, a gift from Pfizer Central Research (Sandwich, U.K.) was dissolved in water. Amlodipine has a pK_a of 8.6 and is thus 94% charged at pH 7.4 and is 99% neutral at pH 10.0.

Voltage protocols were designed to investigate the kinetics of the development of and recovery from block. We were thus interested in the DHP receptor associated with "antagonism" of calcium-channel currents.[16-19] Trains of pulses were applied to minimize experimental time. In order to study onset of block, the cell membrane was held at a negative voltage under stimulation-free conditions for at least two minutes. Then, the holding potential was changed and a series of pulses was applied to measure I_{Ca}. The pulse frequency and duration, which were varied, are noted in each figure caption. Recovery from block was studied with a similar approach, but in this case the cell was held at a positive conditioning voltage for a given duration. Then, the membrane was stepped to a negative voltage, and pulses were applied to measure I_{Ca}.

RESULTS

The voltage dependence of neutral DHP antagonists is well established. These compounds inhibit channels at nanomolar concentrations if cell membrane potentials are depolarized to voltages more positive than -50 mV and have little inhibitory activity if the cell resting potential is more negative than -70 mV.[3,4] When membrane potential is changed, steady-state effects of these derivatives (drugs such as nifedipine, nitrendipine, nisoldipine, and PN 200-110) redevelop within tens of seconds. Because the pK_as of these compounds are less than three, these drugs are neutral over a pH range of 6.0 to 10.0, and over this pH range the kinetics of drug-induced block and recovery from block are constant.[11]

Like the neutral DHPs, amlodipine has little inhibitory activity on membrane currents if cell resting potential is negative to -70mV. FIGURE 1 illustrates this point by showing currents measured from a -80 mV holding potential in the absence and presence of amlodipine. In order to emphasize the fact that this drug had little effect on I_{Ca} at these voltages, the cell was held at -80 mV without pulsing for five minutes in the presence of amlodipine before currents were measured. In this experiment, there was no change in current compared to the drug-free conditions. In other experiments slight enhancement of I_{Ca} was induced by amlodipine.[11] The data shown in FIGURE 1 were obtained in pH$_o$ 7.4, but similar results were found in pH$_o$ 10.0. Thus both neutral and ionized amlodipine molecules are not effective inhibitors of I_{Ca} at negative cell membrane potentials.

Effects of Neutral Amlodipine

We next investigated the development of and recovery from block of I_{Ca} by amlodipine under conditions in which the drug molecule was predominantly neutral (pH$_o$ 10.0). We found that the neutral amlodipine molecule was characterized by a voltage dependence that closely resembled previously investigated neutral DHPs. In solutions buffered to pH 10.0, I_{Ca} block by amlodipine develops at voltages positive to -50 mV and is only slightly accelerated by the addition of trains of voltage pulses. Furthermore, as is the case for other neutral DHPs, I_{Ca} block by neutral amlodipine is rapidly and completely relieved when the cell membrane is returned to negative voltages. These two properties of neutral amlodipine are illustrated in FIGURES 2 and 3.

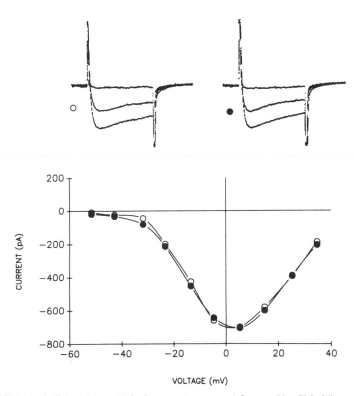

FIGURE 1. Amlodipine does not block currents measured from -80 mV holding potential. Currents measured in response to 40-msec voltage pulses applied from a -80 mV holding potential are plotted against pulse voltage. Pulses were applied at 0.5 Hz and currents were measured in the absence ($\bigcirc$) and presence ($\bullet$) of 3 μM amlodipine. The cell was exposed to the drug at a -80 mV holding potential and held without stimulation for five minutes in the presence of drug before currents were measured. The inset shows current traces in response to pulses to -30, -15, and $+10$ mV in the absence (left) and presence (right) of amlodipine. Conditions: 2 mM Ca; pH$_o$ = 7.4. (From Kass and Arena.[11] Reprinted by permission.)

FIGURE 2 shows an experiment in which the membrane potential was changed from -80 mV to -40 mV. Trains of pulses were imposed in the absence and presence of drug to determine the kinetics of drug block. The figure shows that amlodipine block develops at -40 mV even when brief (20-msec) pulses are applied infrequently (0.2 Hz). After 20 seconds at -40 mV, approximately 45% of the available current had been blocked.

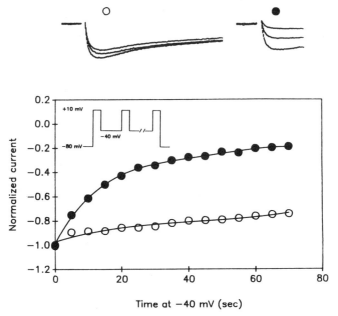

FIGURE 2. Onset of amlodipine block in pH_o 10.0. Currents were measured in response to pulse train protocols in the absence (open symbols) and presence (filled symbols) of amlodipine (10 μM) in pH_o 10.0. In the pulse train protocol HP was changed from -80 to -40 mV, and pulses were applied once every five seconds from the -40 mV holding potential. Pulse widths were 200 msec in control and 20 msec in the presence of amlodipine. Currents are plotted against time after change from -80 mV to the -40 mV holding potential. The smooth curves through the data are drawn through the points as visual aids. Insets show the 1st, 4th, and 15th current traces for control (first 66 msec) and amlodipine (20 msec).

FIGURE 3 shows the recovery of current after applying conditioning pulses of 20 sec to 0 mV in the absence and presence of drug. Neutral amlodipine promotes a slow component to the recovering current, which is consistent with the recovery from block of drug-bound channels and is typical of other neutral DHPs.[3,4] FIGURES 2 and 3 together make an additional point about the action of neutral amlodipine. The same fraction of current is blocked after a 20-second conditioning pulse to 0 mV (FIG. 3) as is blocked by 20 seconds at -40 mV (FIG. 2) indicating that differences in the state of the calcium channel at these two voltages are not very important to the development of neutral amlodipine block.

Effects of Ionized Amlodipine

Although neutral amlodipine closely resembled other previously characterized DHPs, the properties of the ionized drug were much different. Block of I_{Ca} by ionized amlodipine (pH_o 7.4) was promoted by depolarized cell voltages, but the development of block was very slow if trains of brief pulses were applied (FIG. 4). The development of block was speeded if pulse width was increased for each pulse of the train (FIG. 5), but recovery from block was always incomplete and slow. Poor recovery from block is evident in FIGURE 5 in which the cell was held at -80 mV between the two runs carried out in the presence of drug. Block that had developed during the train protocol was irreversible during this time period, whereas over comparable times, block by neutral drug is fully reversible (FIG. 3).

Protonation of the Drug-Bound Receptor by External Hydrogen Ions

Thus we concluded that the degree of ionization of the amlodipine molecule affected the time course of the development and recovery from block. Furthermore, we reasoned that the marked difference in the time course of recovery from block by neutral and ionized amlodipine could be used as an assay for the accessibility of the drug-bound receptor to external H^+. We designed an experiment to test for this possibility. In this experiment we exposed a cell to ionized amlodipine and promoted block by

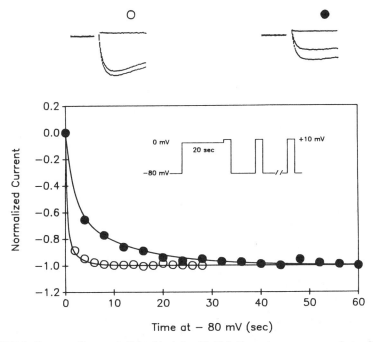

FIGURE 3. Recovery from amlodipine block in pH_o 10.0. Currents were measured at -80 mV after applying 20-second conditioning pulses to 0 mV in the absence ($\bigcirc$) and presence ($\bullet$) of 3 μM amlodipine. Currents were measured in response to 20-msec pulses to 0 mV and plotted against time after termination of the 20-second conditioning pulse to 0 mV. The smooth curves are exponential functions with the following time constants: 0.4, 2.3 sec (control); 1.9, 12 sec (drug). The inset shows the 1st, 2nd, and 15th current traces in the absence (left) and presence (right) of drug.

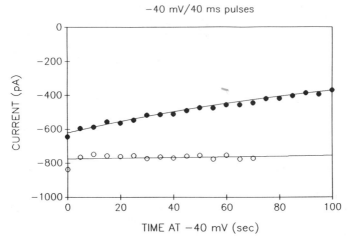

FIGURE 4. Slow development of block in pH_o 7.4. Currents measured in the absence (open) and presence (filled) of 3 μM amlodipine in response to 40-msec pulses applied once every five seconds in a train protocol. The holding potential was changed from -80 mV to -40 mV during the onset of the train.

imposing a train of 200-msec pulses from a -40 mV holding potential. As expected, block developed slowly and was not reversible at -80 mV (FIG. 6). We then changed to an external solution buffered to pH 10.0, and measured currents in response to brief pulses applied from -80 mV. Amlodipine block was relieved within 30 seconds of the solution change (FIG. 6). We repeated this experiment over a pH range of 6.0 to 10.0 and consistently found the same result: Irreversible I_{Ca} block by ionized amlodipine was fully relieved in pH_o 10.0. Because we had previously found that block and recovery from block by other neutral DHPs was unchanged over this pH range, we concluded that, in fact, the degree of ionization of the amlodipine molecule affects its access to the DHP receptor and the drug-bound receptor can be influenced by changes in external pH. The simplest interpretation of our results is that extracellular protons must be able to interact with the amlodipine molecule whether or not it is bound to the DHP receptor.

CONCLUSIONS AND IMPLICATIONS

Our experiments have shown that external pH modifies the block of I_{Ca} by amlodipine over a pH range where the degree of ionization of the drug molecule is expected to change. Neutral drug block is enhanced by depolarization and is reversibly relieved at negative voltages. Voltage protocols that do not require channel openings promote block by the neutral amlodipine molecule. Ionized amlodipine also inhibits I_{Ca} at positive voltages, but block is enhanced by pulses that open channels and promote inactivation. Block by ionized amlodipine is only partially relieved upon hyperpolarization and the time course of the recovery that is observed is 10 times slower than that for neutral drug (see Kass and Arena[11] for details).

In many ways our results resembled the effects of pH_o on local anesthetic block of sodium channels that has been well characterized by the modulated receptor formalism.[20-22] A possible parallel between calcium and sodium channels would be particularly interesting because of the homology between the two channel types already documented from molecular studies (reviewed by Catterall[23]).

Our experimental results place limitations on the location of the DHP receptor responsible for antagonistic activity. The drug-bound receptor is accessible to external hydrogen ions. Thus the receptor either lies within the channel pore but near the external surface of the cell membrane, or the receptor lies within the lipid bilayer adjacent to the channel pore but also near the external surface of the membrane. Ongoing experiments are designed to distinguish between these two possibilities.

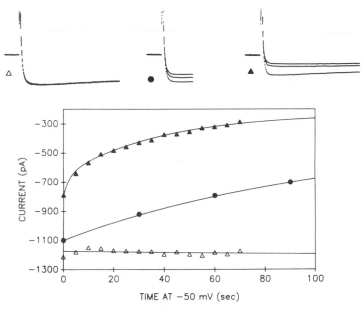

FIGURE 5. Development of amlodipine block in pH_o 7.4: influence of holding potential. A pulse train protocol similar to that described in FIGURE 2 was applied in the absence (△) and presence of 3 μM amlodipine (▲). HP was changed from -80 to -50 mV, and 200-msec pulses were applied at 0.2 Hz. In the presence of drug, a single-pulse protocol (●) was also applied to test for the role of holding potential in amlodipine block. Here, 20-msec pulses were applied once every 30 seconds. In the presence of drug, the single-pulse protocol was applied first, followed by the train protocol. The cell was held for three minutes without stimulation at the -80 mV holding between runs. Measured currents are plotted against the time after changing HP from -80 to -50 mV. Line through control data is intended only as a visual aid. The curves in the presence of drug are fitted exponential functions with the following time constants: pulse train, 2.5, 38 sec; single pulse, 100 sec. The insets show currents measured 0, 30, and 60 seconds after changing HP to -50 mV. The first 66 msec of the 200-msec pulses are shown. (From Kass and Arena.[11] Reprinted by permission.)

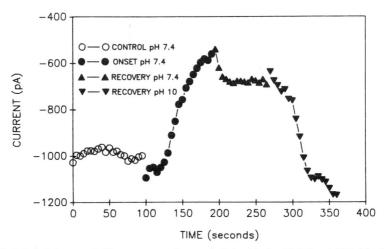

FIGURE 6. Influence of pH_o on recovery from amlodipine block: pH_o 7.4 and 10.0. Membrane currents were measured using voltage protocols that favor development of and recovery from block in solutions buffered to pH_o 7.4 and 10.0. Control currents were measured in the absence of drug in response to 200-msec pulses applied from a -40 mV holding potential ($\bigcirc$) in pH_o 7.4. The cell was then exposed to amlodipine (3 μM) at pH_o 7.4, and currents were measured in response to the same voltage protocol ($\bullet$) during the solution change. After development of block, the holding potential was changed to -80 mV and currents were measured in response to 20-msec pulses with pH_o still 7.4. These currents ($\blacktriangle$) indicate little recovery from block. With the holding potential fixed at -80 mV, pH_o was then changed to 10.0, and currents were measured in response to 20-msec pulses recorded during the solution change ($\blacktriangledown$). Recovery was rapidly apparent. Pulse rate was 0.2 Hz and pulse voltage was $+10$ mV in all runs. Currents are plotted against time after application of first pulse in control. Conditions: 2 mM Ca; Cell 12162. (From Kass and Arena.[11] Reprinted by permission.)

REFERENCES

1. TRIGGLE, D. J. & J. C. VENTER. 1987. Structure and Physiology of the Slow Inward Calcium Channel.: 1-281. Alan R. Liss. New York.
2. HESCHELER, J., D. PELZER, G. TRUBE & W. TRAUTWEIN. 1982. Does the organic calcium channel blocker D600 act from the inside or outside on the cardiac cell membrane? Pflugers Arch. **393:** 287-291.
3. SANGUINETTI, M. C. & R. S. KASS. 1984. Voltage-dependent block of calcium channel current in the calf cardiac Purkinje fiber by dihydropyridine calcium channel antagonists. Circ. Res. **55:** 336-348.
4. BEAN, B. P. 1984. Nitrendipine block of cardiac calcium channels: High-affinity binding to the inactivated state. Proc. Natl. Acad. Sci. USA **81:** 6388-6392.
5. KASS, R. S. 1987. Voltage-dependent modulation of cardiac calcium channel current by optical isomers of Bay K8644: Implications for channel gating. Circ. Res. **61**(suppl I): I1-I15.
6. TRIGGLE, D. J., D. A. LANGS & R. A. JANIS. 1989. Ca^{2+} channel ligands: Structure-function relationships of the 1,4-dihydropyridines. Med. Res. Rev. In press.
7. CURTIS, B. M. & W. A. CATTERALL. 1984. Purification of the calcium antagonist receptor of the voltage-sensitive calcium channel from skeletal muscle transverse tubules. Biochemistry **23:** 2113-2118.

8. BORSOTTO, M., J. BARHANIN, R. I. NORMAN & M. LAZDUNSKI. 1984. Purification of the dihydropyridine receptor of the voltage-dependent calcium channel from skeletal muscle transverse tubule using (+)[3H]PN200-110. Biochem. Biophys. Res. Commun. **122:** 1357-1366.

9. FLOCKERZI, V., H. J. OEKEN, F. HOFMANN, D. PELZER, A. CAVALIE & W. TRAUTWEIN. 1986. Purified dihydropyridine-binding site from skeletal muscle T-tubules is a functional calcium channel. Nature (London) **323:** 66-68.

10. TANABE, T., H. TAKESHIMA, A. MIKAMI, V. FLOCKERZI, H. TAKAHASHI, K. KANGAWA, M. KOJIMA, H. MATSUO, T. HIOSE & S. NUMA. 1987. Primary structure of the receptor for calcium channel blockers from skeletal muscle. Nature (London) **328:** 313-318.

11. KASS, R. S. & J. P. ARENA. 1989. Influence of pH_o on calcium channel block by amlodipine, a charged dihydropyridine compound: Implications for location of the dihydropyridine receptor. J. Gen. Physiol. In press.

12. MITRA, R. & M. MORAD. 1985. A uniform enzymatic method for dissociation of myocytes from hearts and stomachs of vertebrates. Am. J. Physiol. **249:** H1056-H1060.

13. HAMILL, O. P., A. MARTY, E. NEHER, B. SAKMANN & F. J. SIGWORTH. 1981. Improved patch-clamp techniques for high-resolution current recording from cells and cell-free membrane patches. Pflugers Arch. **391:** 85-100.

14. ARENA, J. P. & R. S. KASS. 1988. Block of heart potassium channels by clofilium and its tertiary analogs: Relationship between drug structure and type of channel blocked. Mol. Pharmacol. **34:** 60-66.

15. KRAFTE, D. S. & R. S. KASS. 1988. Hydrogen ion modulation of Ca channel current in cardiac ventricular cells: Evidence for multiple mechanisms. J. Gen. Physiol. **91:** 641-657.

16. KOKUBUN, S., B. PROD'HOM, C. BECKER, H. PORZIG & H. REUTER. 1987. Studies on Ca channels in intact cardiac cells: Voltage-dependent effects and cooperative interactions of dihydropyridine enantiomers. Mol. Pharmacol. **30:** 571-584.

17. WILLIAMS, J. S., I. L. GRUPP, G. GRUPP, P. L. VAGHY, L. DUMONT, A. SCHWARTZ, A. YATANI, S. HAMILTON & A. M. BROWN. 1985. Profile of the oppositely acting enantiomers of the dihydropyridine 202-791 in cardiac preparations: Receptor binding, electrophysiological and pharmacological studies. Biochem. Biophys. Res. Commun. **131:** 13-21.

18. HAMILTON, S. L., A. YATANI, K. BRUSH, A. SCHWARTZ & A. M. BROWN. 1987. A comparison between the binding and electrophysiological effects of dihydropyridines on cardiac membranes. Mol. Pharmacol. **31:** 221-231.

19. BROWN, A. M., D. L. KUNZE & A. YATANI. 1986. Dual effects of dihydropyridines on whole cell and unitary calcium currents in single ventricular cells of guinea-pig. J. Physiol. (London) **379:** 495-514.

20. HILLE, B. 1977. Local anesthetics: Hydrophilic and hydrophobic pathways for the drug-receptor reaction. J. Gen. Physiol. **69:** 497-515.

21. HILLE, B. 1977. The pH-dependent rate of action of local anesthetics on the Node of Ranvier. J. Gen. Physiol. **69:** 475-496.

22. HONDEGHEM, L. M. & B. G. KATZUNG. 1977. Time and voltage dependent interaction of antiarrhythmic drugs with cardiac sodium channels. Biochim. Biophys. Acta **472:** 373-398.

23. CATTERALL, W. A. 1988. Structure and function of voltage-sensitive ion channels. Science **242:** 50-61.

Structure of Calcium Channels[a]

H. GLOSSMANN,[b] J. STRIESSNIG,[b] H.-G. KNAUS,[b]
J. MÜLLER,[b] A. GRASSEGGER,[b] H.-D. HÖLTJE,[c]
S. MARRER,[c] L. HYMEL,[d] AND H. G. SCHINDLER [d]

[b] Institute of Biochemical Pharmacology
University of Innsbruck
A-6020 Innsbruck, Austria

[c] Department of Pharmacy
Free University of Berlin
D-1000 Berlin 33, Federal Republic of Germany

[d] Institute of Biophysics
University of Linz
A-4040 Linz, Austria

INTRODUCTION

L-type Ca^{2+} channels have distinct, allostericly communicating drug receptor domains that interact with cation binding sites.[1,2] Ca^{2+}-channel drugs, radioactively labeled, are valuable tools in characterizing, solubilizing, and purifying the channel polypeptides.[3] Photoaffinity labels were necessary to identify the drug-receptor carrying $alpha_1$ polypeptide in skeletal muscle[4-6] and $alpha_1$-type polypeptides in heart[8,9] or brain membranes.[10] High-affinity, irreversible photolabels are also useful in controlling preparations employed for reconstitution studies as being structurally intact and in probing for the amino acids within the primary sequence of $alpha_1$[11] that are essential for drug binding. An alternative approach to find receptor domains within the (deduced) primary amino acid sequence is presented, based on a theoretically derived receptor model.[12,13] The model accurately predicted functional properties of the sadopine diastereomers.[14] The focus of this paper is on the development of new radiolabeled Ca^{2+}-channel drugs, including an arylazide photoligand for the benzothiazepine-selective drug-receptor domain, novel [^{35}S]-labeled 1,4-dihydropyridines, receptor modeling, and on the properties of reconstituted Ca^{2+} channels from skeletal muscle with intact $alpha_1$ subunits, as shown by photolabeling.

[a] This work was supported by Fonds zur Förderung der wissenschaftlichen Forschung (Schwerpunktprogramm), Bundesministerium für Wissenschaft und Forschung und Deutsche Forschungsgemeinschaft.

NOVEL HIGH-AFFINITY LIGANDS FOR THE C$_A^{2+}$ CHANNEL

HOE-166

HOE-166 is the enantomerically pure benzothiazinone (R)-$(+)$-3,4-dihydro-2-isopropyl-4-methyl-2-[2-[4-[4-[2-(3,4,5-trimethoxyphenyl)-ethyl]-piperazinyl]-butoxyl-phenyl]-2H-1,4-benzothiazine-3-one-dihydrochloride (HOE-166). This compound is a potent C$_A^{2+}$-channel blocker in a variety of experimental systems, whereas the respective (S)-$(-)$-enantiomer was approximately 10-fold less potent.[15] To evaluate whether HOE-166 was interacting in a simple competitive manner with one of the well-defined drug-receptor binding domains of L-type calcium channels, a detailed analysis of the equilibrium binding as well as kinetic properties of the benzothiazepine (BT), the phenylalkylamine (PA), and the 1,4-dihydropyridine (DHP) selective site was performed in guinea-pig brain, heart, and skeletal muscle membranes, each representing a different "isochannel," characterized by its "isoreceptor" [see Glossmann & Striessnig[1] for further details].

Our experiments indicated that HOE-166 was an allosteric regulator for the BT, PA, and DHP receptor domains. In every case binding of a domain-selective ligand was inhibited by a mechanism not explainable by simple competitive interaction. Specifically, HOE-166 accelerated $(+)$-[^{3}H]PN 200-110 dissociation from the DHP receptor in skeletal muscle transverse (T) tubule membranes, although the main effect was to increase the K_D (dissociation constant) in equilibrium saturation assays. A recent paper claims, in contrast, that HOE-166 binds competitively to the DHP receptor and does not accelerate 1,4-DHP ligand dissociation.[16] When equilibrium binding saturation studies are performed at 37°C with [^{3}H]HOE-166 (specific activity between 27-52 Ci/mmole) and skeletal muscle partially purified T-tubule membranes (as described by Glossmann & Ferry[17]), the K_D is 239 ± 36 pM and the B$_{max}$ (maximum density of binding sites) 15.4 ± 68 pmoles per mg of protein ($n = 4$). HOE-166 and its (S)-$(-)$-enantiomer compete with apparent Hill slopes of 1.08 ± 0.1 and 1.09 ± 0.1 and K_i values of 0.46 ± 0.1 nM and 12.6 ± 3.8 nM, respectively. The dissociation kinetics at 15° and 37°C is monoexponential but the K_{-1} (dissociation rate constant) is increased in dilution chase experiments in a concentration-dependent manner by unlabeled HOE 166—a phenomenon previously (albeit in a more complex fashion) found for the $(-)$-[^{3}H]desmethoxyverapamil[18,19] or $(\pm)$-[^{3}H]verapamil-labeled PA receptors.[20] $(+)$-PN 200-110 inhibited [^{3}H]HOE-166 binding in skeletal muscle with an IC$_{50}$ value of 11.7 ± 1.9 nM and an apparent Hill coefficient of 1.44 (FIG. 1). The inhibition was 100%, indicating that the allosterism was complete. In guinea-pig brain membranes, $(-)$-desmethoxyverapamil inhibits [^{3}H]HOE-166 binding 100% with an apparent Hill coefficient of 0.54 ± 0.1; but $(+)$-PN 200-110 or $(\pm)$-BAY K 8644 inhibit only by 44% and 53%, respectively. The inhibition by the DHPs is much less pronounced when brain membranes are depleted for divalent cations as previously described.[17] Thus, there is no evidence that HOE-166 binds competitively to the 1,4-DHP receptor. Interestingly, the stereoselectivity of HOE-166 and its (S)-(−)-enantiomer is reversed for the drosophila (head membrane) PA receptor. When these membranes (prepared as described by Pauron *et al.*[21]) are labeled with $(-)$-[^{3}H]desmethoxyverapamil or [N-methyl-^{3}H]LU 49888, we cannot confirm the extremely high eudismic ratios for PAs as reported by Pauron *et al.*[21] Instead, we find for all chiral PAs eudismic ratios of 10-20, for example, the IC$_{50}$ values for $(-)$-desmethoxyverapamil and $(+)$-desmethoxyverapamil are 0.69 and 5.7 nM, respectively.[22] HOE-166 has an IC$_{50}$ value of 88.4 ± 5.9 nM in this system, but the (S)-(−)-

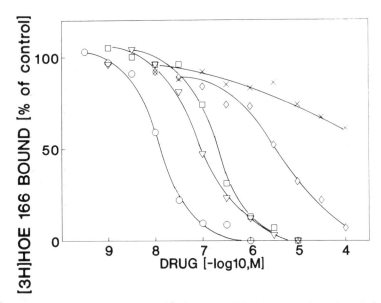

FIGURE 1. Pharmacological profile of [³H]HOE 166 binding to membrane-bound skeletal muscle Ca^{2+} channels.

Partially purified guinea pig skeletal muscle T-tubule membranes (0.02 mg) were incubated with 0.1-0.8 nM of [³H]HOE 166 for 60 minutes at 37°C in the absence and presence of increasing concentrations of (+)-PN200-110 (○), (−)-D888 (▽), (+)-D888 (□), (+)-*cis*-diltiazem (◇) and (−)-*cis*-diltiazem (×). Nonspecific binding (defined as the binding in the presence of 1 μM unlabeled HOE-166) was subtracted from total binding to yield specific binding. Binding-inhibition data were computer-fitted to the general dose-response equation. The following binding parameters (± asymptotic SD from two to three pooled experiments) were obtained (IC$_{50}$ = concentration causing 50% inhibition of specific binding; n$_H$ = slope factor): (+)-PN200-110: IC$_{50}$ = 11.7 ± 1.9 nM, n$_H$ = 1.44; (−)-D888: IC$_{50}$ = 92 ± 12 nM, n$_H$ = 1.17; (+)-D888: IC$_{50}$ = 203 ± 11 nM; n$_H$ > 1.3; (+)-*cis*-diltiazem: IC$_{50}$ = 4.6 ± 1.6 μM, n$_H$ = 0.60 (maximal inhibition 90% of control binding at 100 μM); (−)-*cis*-diltiazem: IC$_{50}$ > 50 μM.

enantiomer has 30.1 ± 2.5 nM. As the drosophila PA receptor is devoid of 1,4-DHP binding domains,[21,22] this stereoselective interaction (albeit reversed compared to the L-type channel) strongly supports our claim that the novel class of benzothiazinone Ca^{2+}-channel drugs (for which HOE-166 is one example) does not bind to the DHP receptor in a competitive manner.

Sadopine—A Novel, High-Affinity, High-Specific-Activity Probe for the 1,4-DHP Receptors

The diastereomers of the ³⁵S-labeled, high-specific-activity (> 1000 Ci/mmol) 1,4-dihydropyridine, sadopine (1,4-dihydro-2,6-dimethyl-4-(2-trifluormethylphenyl)-pyridine-3,5-dicarboxyl-3-[2-(N-tertbutyloxycarbonyl-L-methionyl)-aminoethyl]-ester-5-ethylester) were synthesized.[23] This novel 1,4-DHP has two chiral centers. ($-$)-Sadopine stands for the (R)-(S)-diastereomer, whereas ($+$)-sadopine is the (S)-(S)-diastereomer. Based on the allosteric interaction with [^{3}H]($+$)-*cis*-diltiazem-labeled BT receptors in skeletal muscle, 1,4-DHPs can be differentiated in two different classes.[24] At temperatures > 20° or 25°C, 1,4-DHPs that allostericly stimulate binding are always antagonists and this includes the antagonistic enantiomers (R)-($-$)-202-791 and (R)-($+$)-Bay K 8644, whereas agonistic 1,4-dihydropyridines, for example, CGP 28392 (most likely (S)-CGP 28392, see Refs. 12 and 13), (S)-($+$)-202-791, or (S)-($-$)-Bay K 8644 are always inhibitory.[25]

Interestingly, the same phenomen was found for the PA receptor regulation by 1,4-DHPs in digitonin-purified calcium channels from guinea-pig skeletal muscle.[26] Unlabeled ($-$)-sadopine and ($+$)-sadopine (synthesized from the respective optically pure precursors and purified by HPLC) behave in both systems exactly as previously described for agonistic or antagonistic 1,4-DHPs. ($-$)-Sadopine stimulates ($+$)-*cis*-[^{3}H]diltiazem binding, ($+$)-sadopine inhibits (FIG. 2). The unlabeled diastereomers also change the dissociation kinetics of the labeled BT: ($+$)-Sadopine accelerates and ($-$)-sadopine inhibits. An example is shown in FIG. 3. ($-$)-Sadopine stimulates ($-$)-

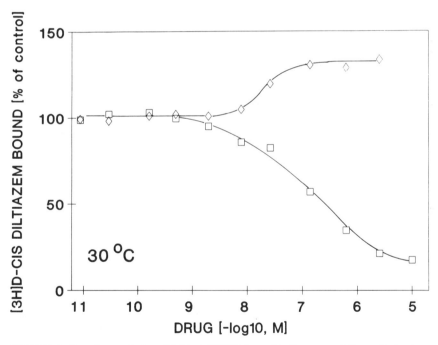

FIGURE 2. Opposite regulation of ($+$)-*cis*-[^{3}H]diltiazem binding to partially purified guinea pig skeletal muscle T-tubule membranes by ($-$)sadopine and ($+$)sadopine. The experiments were performed by incubating 0.154-0.264 μg/ml protein with 2.97-3.34 nM radioligand in the absence or presence of the sadopine diastereomers (at the indicated concentrations) at 30°C for 60 minutes. Nonspecific binding was determined in the presence of 10 μM ($+$)-*cis*-diltiazem. Data [$\pm$ asymptotic standard deviation] are means from three independent experiments. The IC$_{50}$ value for ($+$)-sadopine (□) was 150.4 $\pm$ 33.8 nM, the EC$_{50}$ value for ($-$)-sadopine (◇) was 23.9 $\pm$ 3.9 nM.

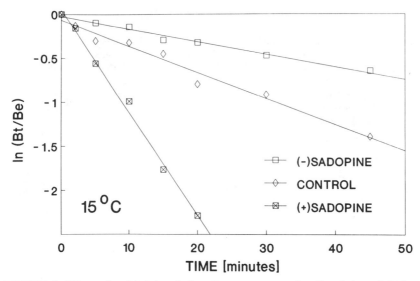

FIGURE 3. Effects of unlabeled sadopine diastereomers on the dissociation of $(+)$-*cis*-[³H]diltiazem. Guinea-pig skeletal muscle T-tubule membrane protein (0.155 mg/ml) was labeled with 1.55 nM $(+)$-*cis*-[³H]diltiazem at 15°C. Dissociation was started by addition of 10 μM $(+)$-*cis*-diltiazem with or without (control) of 1 μM $(-)$ or $(+)$sadopine. Data were fitted by linear regression analysis. Control: r = 0.97 k_{-1} = 0.028 min⁻¹; 1 μM $(-)$sadopine present: r = 0.981, k_{-1} = 0.015 min⁻¹; 1 μM $(+)$sadopine present: r = 0.989, k_{-1} = 0.114 min⁻¹.

[³H]desmethoxyverapamil binding to the purified skeletal muscle channel, whereas the $(+)$-diastereomer is a potent negative heterotropic allosteric regulator, exactly as described for the agonistic 1,4-DHPs.[26]

To test whether the allosterism (between the DHP and the BT domain) is reciprocal, we have [³⁵S]-radiolabeled $(+)$, $(-)$, and $(\pm)$-sadopine, each with a specific radioactivity of > 1000 Ci/mmole. These novel radiolabels are perfect for autoradiographic studies, detect minute amounts of DHP receptor sites, and are also expected to interact with the multiple drug resistance (p = 170) glycoprotein as does azidopine.[27] It was found that the allosterism was indeed reciprocal. Binding of $(+)$-[³⁵S]sadopine is stimulated by $(+)$-*cis*-diltiazem, whereas binding of the [³⁵S]-labeled $(+)$-diastereomer is inhibited. There was no effect on the equilibrium binding of the [³⁵S]-labeled $(\pm)$-sadopine. Thus, the lack of allosteric interaction of $(+)$-*cis*-diltiazem with the latter ligand is more apparent than real. It reflects the sum of the individual effects. Being completely opposite and symmetrical, they cancel out—a rather unique finding in Ca²⁺ channel research. Despite their opposite regulation by $(+)$ *cis*-diltiazem (and different kinetic constants see FIG. 4) both enantiomers have the same K_D for the membrane-bound Ca²⁺ channel in skeletal muscle T-tubule membranes. It seemed logical that the different orientation of the bulky sadopine side chain within the 1,4-DHP binding domain determined whether $(+)$-*cis*-diltiazem acted as a positive heterotropic allosteric regulator or as a negative one. In order to approach this problem at the structural level, the sadopine enantiomers were tested on the hypothetical DHP receptor model described below. Both enantiomers were predicted to be antagonistic compounds, which was confirmed by electrophysiological and classical pharmacological experiments in both heart and smooth-muscle preparations.[23]

Azidodiltiazem—A Novel Photoaffinity Label for the Benzothiazepine Receptor

From the three major chemical classes of Ca^{2+}-channel drugs (1,4-DHPs, PAs, BTs), mainly 1,4-DHPs have been used to monitor channel isolation. Employing radiolabeled phenylalkylamines (e.g., (−)-[³H]desmethoxyverapamil or [*N*-methyl-³H]LU49888) and 1,4-DHPs ((+)-[³H]PN200-110 or (−)-[³H]azidopine) the co-purification of the binding activities for the three main chemical classes of Ca^{2+}-channel drugs and preservation of their allosteric coupling mechanisms was demonstrated.[26] Indirect evidence for the presence of the BT binding domain in the purified preparation was the allosteric stimulation of 1,4-DHP binding and inhibition of PA binding by (+)-*cis*-diltiazem. Due to the low affinity of BTs for the detergent-solubilized purified channels, direct labeling with (+)-*cis*-[³H]diltiazem in reversible binding studies has so far been unsuccessful. We therefore developed an optically pure arylazide BT photoaffinity probe, (+)-*cis*-(2*S*,3*S*)-5-[2-(4-azidobenzoyl)aminoethyl]-2,3,4,5-tetrahydro-3-hydroxy-2-(4-methoxyphenyl)-4-oxo-1,5-benzothiazepine (azidodiltiazem) for irreversible labeling. The specific activity was 41.0 Ci/mmol.

Kinetic studies revealed that unlabeled azidodiltiazem does not affect the dissociation rate of [³H](+)-*cis*-diltiazem from its receptor in skeletal muscle T-tubule membranes and therefore binds to the same site. Unlabeled azidodiltiazem inhibited (+)-*cis*-[³H]diltiazem binding with an IC_{50} (± SD) value of 2.2 ± 0.3 μM (slope factor = 1.12 ± 0.12) being about 25-fold less potent than (+)-*d-cis*-diltiazem itself

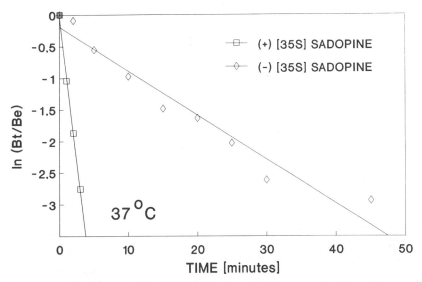

FIGURE 4. Dissociation kinetics of ³⁵S-labeled sadopine diastereomers. Partially purified guinea pig skeletal muscle T-tubule membranes (62.4–71.2 μg/ml) were preincubated for 45 minutes with 34.3–41.2 pM [(+)sadopine] or 75 minutes [(−)sadopine] at 37°C. The dissociation reaction was initiated by addition of 1 μM (+)PN200-110. Transformation of the data by plotting ln(Bt/Be) yield a k_{-1} of 0.0841 min^{-1} for labeled (−)sadopine and 0.911 min^{-1} for (+)sadopine, respectively. The specific activities of both ligands were 1100 Ci/mmol.

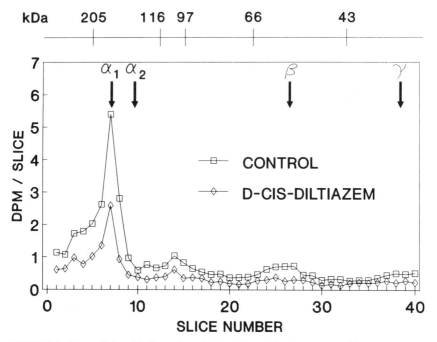

FIGURE 5. Photoaffinity labeling of purified rabbit skeletal muscle Ca^{2+} channels with [^{3}H]azidodiltiazem. Purified rabbit skeletal muscle Ca^{2+} channel protein (0.005 mg) was incubated (20 min at 25°C, 90 min at 2°C) with [^{3}H]azidodiltiazem (270 nM) in a final volume of 1 ml in the absence (□) and presence (◇) of 100 μM (+)-*cis*-diltiazem. The incubation mixture was transferred to Petridishes and irradiated on ice for 60 minutes with a Philips TL30W/08 blacklight lamp. After irradiation the samples were dialyzed against 0.05% (wt/vol) SDS and lyophilized. After resuspension in 0.1 ml of water followed by 0.1 ml of electrophoresis sample buffer (125 mM Tris-HCl, pH 6.8, 10% (wt/vol) SDS, 20% glycerol (vol/vol), containing bromphenol blue) the proteins were resolved on a 8% SDS-polyacrylamide gel. Standard proteins were run on a separate lane on the same gel. The proteins were stained with Coomassie blue and the incorporated activity was determined by gel slicing.[28] The migration of the subunits (arrows) of the purified Ca^{2+} channel protein and of the standard proteins is indicated.

(IC_{50} = 0.089 ± 0.013 nM, slope factor = 1.04 ± 0.14). The affinity of both benzothiazepines was also estimated indirectly from their potency to displace (−)-[^{3}H]desmethoxyverapamil binding. The respective IC_{50} values were 8.54 ± 2.1 and 3.31 ± 0.67 μM. For photolabeling experiments the Ca^{2+} channel was purified as described.[28] Five to ten μg protein from fractions containing the (+)-[^{3}H]PN200-110 and (−)-[^{3}H]desmethoxyverapamil binding activity were incubated with 100-270 nM of [^{3}H]azidodiltiazem in the absence and presence of unlabeled drugs and irradiated with UV light (FIG. 5). Coomassie staining of the SDS-PAGE-separated proteins revealed the usual polypeptide composition of the purified Ca^{2+}-channel preparation (alpha$_1$, alpha$_2$, beta, and gamma polypeptides with apparent molecular weights of 175, 140, 59, and 34 kDa under reducing conditions). Gel slicing (FIG. 5) or autoradiography (not shown) demonstrates that incorporation of [^{3}H]azidodiltiazem occurred mainly into the alpha$_1$ polypeptide, whereas incorporation into the other polypeptides was around the background levels. Labeling of the alpha$_1$ subunit was proportional to the protein concentration employed. Incorporation of [^{3}H]azidodiltiazem could be blocked by the inclusion of 100 μM *d-cis*-diltiazem (FIG. 5) or ($-$)-desmethoxyverapamil (not shown). Although the purified Ca^{2+} channel has low affinity for [^{3}H]azidodiltiazem, our experiments are first to show the colocalization of the BT receptor with the 1,4-DHP and PA domains on the purified alpha$_1$ subunit.

A HYPOTHETICAL RECEPTOR MODEL FOR CALCIUM CHANNEL-MODULATING 1,4-DIHYDROPYRIDINES

Here we present a theoretically derived receptor model for Ca^{2+} channel-modulating DHPs that leads to greater understanding of the molecular mechanism of action of Ca^{2+} agonists as well as antagonists. As mentioned above, we could predict the antagonistic properties of the sadopine diastereomers within the framework of this model. The theoretical methods used include force field and semiempirical quantum chemical methods as well as molecular graphics and have been described previously.[12,13] A systematic investigation of the molecular electrostatic potentials (MEPs) of the DHPs has led to the discovery of a defined segment of space where antagonists and agonists possess MEPs with opposite signs.[12,13] In this area the MEP of antagonistic DHPs is positive whereas the MEP of agonistic DHPs is negative. It can be assumed that the opposite electrostatic fields will influence the potential of the receptor protein in a different manner. In order to analyze how this difference can be transferred to the receptor, interaction complexes between DHPs and a tryptophan molecule (serving as a simplified MEP sensor model) were constructed and analyzed (FIG. 6). The

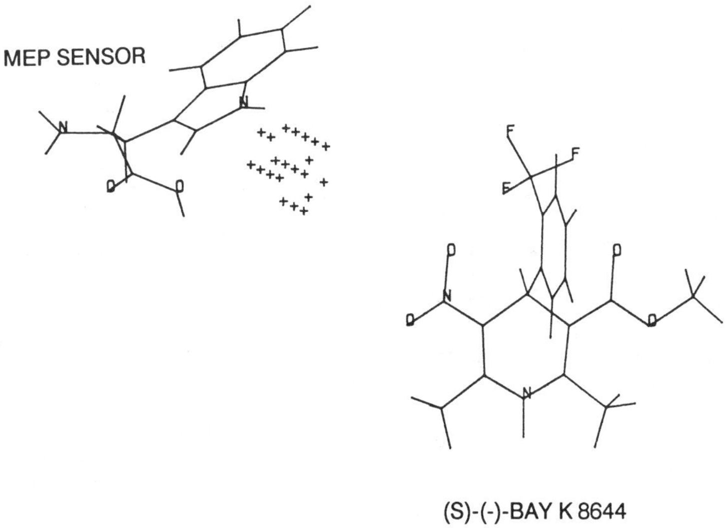

(S)-(-)-BAY K 8644

FIGURE 6. Interaction complex between the postulated MEP sensor (represented by tryptophan) and the calcium agonist (*S*)-($-$)-BAY K 8644. The crosses indicate the space where the molecular electrostatic potential of Ca^{2+} agonists and antagonists differ.

calculated effective MEP changes induced by DHPs are listed in TABLE 1. The data clearly demonstrate that Ca^{2+} agonists shift the molecular electrostatic potential of the MEP sensor model in a negative direction. On the contrary Ca^{2+} antagonists enhance the potential of the MEP sensor. In our view these effects may be the driving forces for the opposite biological activities of Ca^{2+} agonists and antagonists. The MEP sensor converts the potential changes to secondary effects that finally result in an enhanced or decreased transmembrane Ca^{2+} flux, respectively.

Because the MEP changes correlate only qualitatively with the biological activity (agonism versus antagonism) of the DHPs, we tried to extend the MEP sensor model to describe the interaction of Ca^{2+} antagonists with their receptor in a more quantitative manner. To this end we added two additional binding sites to the already identified MEP sensor. A basic amino acid served as counterpart of negatively charged substituents of the DHP phenyl moiety. At this binding site electrostatic forces dominate. On the other hand, an aromatic amino acid was predominantly involved in van der Waals interactions with alkyl ester groups at the second binding site (FIG. 7).

TABLE 1. Calculated Changes of the Molecular Electrostatic Receptor Potential Induced by 1,4-Dihydropyridine Derivatives

1,4-Dihydropyridine Derivative	ΔMEP[a]
Agonists	
(S)-(−)-BAY K 8644	−316.1
(R)-H 160/51[b]	− 34.8
(S)-(+)-Sandoz 202 791	−320.3
Antagonists	
(R)-(−)-BAY K 8644	90.0
(S)-H 160/51*	44.8
(R)-(−)-Sandoz 202 791	72.4
Nifedipine	84.6

[a] ΔMEP: change in molecular electrostatic receptor potential expressed in kJ.
[b] Predicted configuration.

Using lysine and phenylalanine simulating the two binding sites, interaction energies with the DHP ligands have been calculated (TABLE 2, column 1). According to the correlation coefficient, 88% of the variation of the experimental data can be explained by the interaction energies (alpha < 0.001). Based on the correlation equation, $\log(IC_{50})$ values were calculated (TABLE 2, column 3). The calculated data agree well with the experimentally derived constants.

This hypothetical binding site model (which was derived exclusively on the basis of theoretical methods) is nicely supported by the (deduced) primary structure of the receptor for Ca^{2+}-channel drugs (alpha$_1$ subunit) from skeletal muscle.[11] Using the similarity matrix analysis technique, four homologous sequence repeats were identified within the 1873 amino acids of the whole alpha$_1$ subunit.[11] Each repeat consists of five hydrophobic segments (S1, S2, S3, S5, and S6) and one positively charged segment (S4). The S4 segments contain five or six Lys or Arg residues at every third position and at least one aromatic amino acid is found in three of the S4 segments. Presumably, segments S1-S6 form an alpha helix. Tanabe et al.[11] suggest that the positively charged

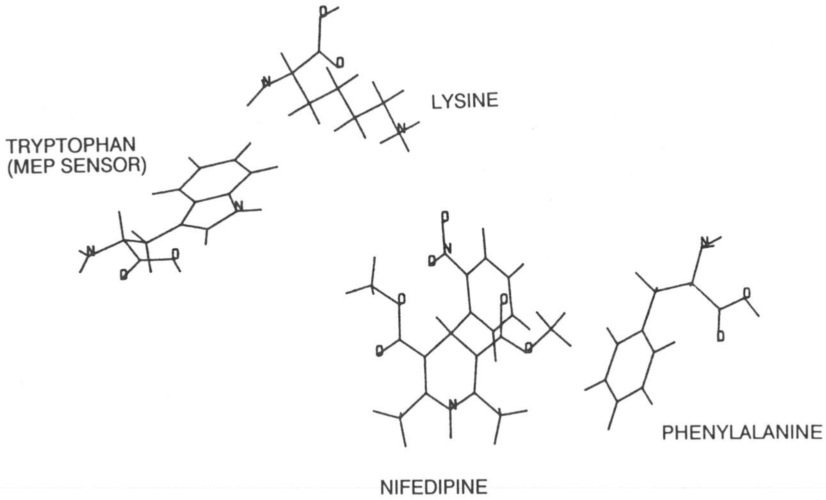

FIGURE 7. Extended binding site model for Ca^{2+} channel modulating 1,4-dihydropyridines. As an example of a DHP ligand, nifedipine is displayed.

S4 segments could act as a voltage sensor of the excitation-contraction (E-C) coupling. Recently Rios *et al.*[29] have postulated that the voltage sensor may be the alpha$_1$ subunit itself or that the drug receptor complex closely interacts with the voltage sensor. It is also well known that binding (and action) of the DHPs are highly voltage-dependent. It can be deduced from these arguments that our binding site model should have some structural similarities with the S4 segments.

In order to prove this we constructed a computer model of the supposed alpha-helical structure of the I-S4 segment using standard geometries. An optimization of the side-chain torsional angles was performed with a scan routine to avoid van der

TABLE 2. Calculated Interaction Energies of the Postulated Binding Site Model with Calcium Antagonists (Column 1), Biological Activities That Were Correlated with the Interaction Energies (Column 2), and Expected $-\log(IC_{50})$ Values Calculated on the Basis of the Correlation Equation (Column 3)

Model[a]	1[b]	2[c]	3
R-(+)-BAY E 6927	−59.9	7.1	7.3
R-(+)-Nitrendipine	−59.8	7.7	7.3
R-(+)-Nimodipine	−55.8	7.9	8.1
Nifedipine	−55.8	8.0	8.1
S-(−)-Nimodipine	−52.3	8.6	8.8
S-(−)-Nitrendipine	−52.5	8.7	8.7
S-(−)-BAY E 6927	−51.4	9.3	9.0

[a] $R = 0.94$; $S = 0.28$.
[b] Expressed in kJ/mole.
[c] Biological activities: $-\log(IC_{50})$ for the K^+-stimulated contraction of strips of rabbit aorta, from Bellemann *et al.*[30]

Waals overlap. Then a search for similarities between the postulated receptor model and the standard helix was carried out.

The result of this manipulation shows that the spatial orientation of [Phe][167] and [Arg][174] agrees well with the geometry of the amino acids tryptophan and lysine of the binding site model (FIG. 8). No corresponding I-S4 segment residue could be found for the third amino acid (phenylalanine) of the binding site model. This might be due to the fact that the I-S4 segment consists only of 19 amino acids and therefore phenylalanine is located too far away and resides on an other part of alpha$_1$. From our point of view it is most important that the MEP-sensitive volume of the receptor model is completely surrounded by the I-S4 segment, whereas the van der Waals

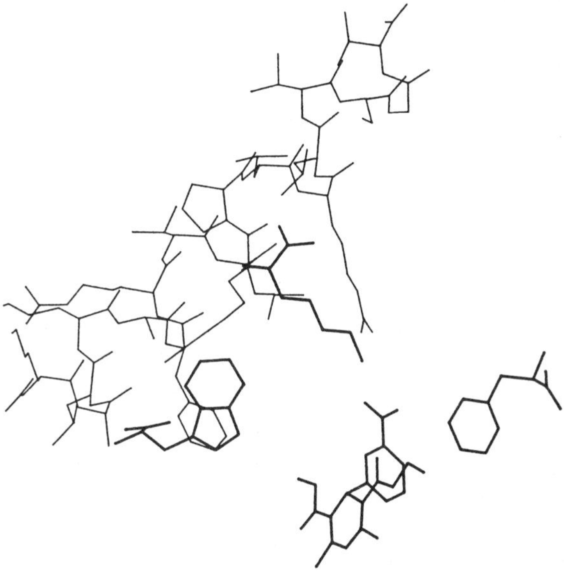

FIGURE 8. The thin structure represents the I-S4 segment that fit well to our postulated binding site model (bold). Nitrendipine represents an Ca^{2+} antagonistic-acting DHP (bold). All hydrogen atoms are not shown. The primary sequence of this segment is (residues 161-179): Val-Lys-Ala-Leu-Arg-Ala-Phe-Arg-Val-Leu-Arg-Pro-Leu-Arg-Leu-Val-Ser-Gly-Val (see Tanabe[11]).

interaction at the phenylalanine binding site can be carried out by other protein segments. In conclusion, our results provide an attractive explanation of how structurally similar DHPs can exert opposite effects at the voltage-dependent calcium channel. It could be shown that Ca^{2+} agonists reduce the molecular electrostatic receptor potential while it is increased by the antagonists. The potential changes must be converted by the alpha$_1$ subunit to a cascade of secondary effects pharmacologically recognizable as an enhanced or reduced Ca^{2+} flux into the cell. In addition the receptor model can explain quantitative structure-activity relationships of DHP Ca^{2+} antagonists, for example, for the sadopine diastereomers. Calculated interaction energies show a very good correlation with corresponding biological activities. The postulated receptor model shows geometrical and physicochemical properties that are comparable to those of a selected segment of the (deduced) primary structure of the alpha$_1$ subunit from rabbit skeletal muscle. Irreversible labeling experiments (for which we have developed the tools) must confirm or refute our hypothesis.

ILLUMINATING THE STRUCTURE OF CALCIUM CHANNELS VIA FUNCTIONAL RECONSTITUTION

Skeletal Muscle Ca^{2+}-Antagonist Receptors

The value of the purification and reconstitution approach for structural and functional characterization of membrane components has been excellently demonstrated using the Ca^{2+}-antagonist receptors from the skeletal muscle transverse tubule (see Glossmann & Striessnig[1] for a review). Without the reconstitution data, a role for this controversial receptor protein as a functioning channel would remain in doubt. It is now clear that, at least in isolated form, the receptor complex is a functioning Ca^{2+} channel, although the precise characteristics seem to vary a great deal from one laboratory to another. Nonetheless, in order to proceed with physiological, pharmacological, and biophysical characterization of the molecular complex, the functional qualities of the biochemical preparation must be delineated. We have shown that our preparation of the skeletal muscle Ca^{2+}-antagonist receptor is biochemically and functionally intact.[33,34] The alpha$_1$ polypeptide, which is known to be quite labile,[35] in our preparation retains its full size as shown by photolabeling.[28,32] When the preparation is incorporated into planar lipid bilayers, it forms Ba^{2+}-selective channels (P_{Ba}/P_{Na} = 30) with a recovery of at least 50%—calculated for 1,4-DHP binding sites. The channel activity was found to be entirely dependent on phosphorylation of the alpha$_1$ subunit by the cAMP-dependent protein kinase[32,33] and is inhibitable by phenylalkylamines from the appropriate side of the membrane at micromolar concentrations. In the absence of divalent cations, the channel is highly permeable to monovalent cations and can be blocked by low concentrations of Ca^{2+}; Cd^{2+} blocks the Ba^{2+} current effectively at 100 μM. These characteristics would qualify the channel as a typical "L-type" calcium channel.

The "single-channel" conductance of the skeletal muscle L-type calcium channel in 90-100 mM Ba^{2+}, after direct incorporation of T-tubule membranes without detergent solubilization and purification, has been reported as either 9-12 pS[36,37] or 20-25 pS.[38,39] The lower value of about 12 pS contrasts with the universally found higher value of 20-25 pS in cardiac muscle and other cell types.[40-43] We observed (using reconstituted, purified receptors) a variety of multilevel conductance events, the smallest and simplest of which were only 0.9-1.0 pS, and larger events that sometimes yielded stabile bursts of about 12, 25, or even 50-60 pS. We interpret the complex appearance of "single-channel" events to represent associates ("oligochannels") of single-receptor complexes, whose open-close transitions are governed by allosteric interactions. We have shown that the voltage dependence of reconstituted receptors depends on association state: Freshly incorporated, singly distributed receptors yield a macroscopic membrane current that depends linearly on voltage, similar to the typical voltage dependence of single-channel current in patch-clamp experiments, while associates formed later in time show a distinctly nonlinear macroscopic current.[32] This result implies that the voltage dependence so characteristic for L-type Ca^{2+} channels is rooted in allosteric interactions.

Here, we further characterize the subconductance levels of reconstituted Ca^{2+} channels. The results provide new insight into their structure and gating mechanism.

Methods

To assess the functionality of a biochemical preparation, it is essential to choose an appropriate and unbiased method. For the purified Ca^{2+}-antagonist receptor, the methods that have been applied include: (1) fusion of reconstituted vesicles containing the protein with a Muller-Rudin "black lipid" film,[44,45] (2) dilution of a detergent suspension of the channel protein in the presence of a black film,[44] (3) adding reconstituted vesicles in the vicinity of a planar lipid bilayer supported in a patch-clamp pipette,[38] (4) isotope flux studies using suspensions of reconstituted vesicles,[46] and (5) spreading reconstituted vesicles at the air-water interface to form protein-containing lipid monolayers, followed by folding of two such monolayers to form a bilayer.[31-34] Of the aforementioned methods, only (4) and (5) can give quantitative information on the extent of functionality of biochemical preparations. The other methods are purely qualitative, since the incorporation of channel proteins is based on random events that are not understood and cannot be controlled or observed. Methods 1-3 are therefore limited to confirming expected functional properties for an unknown fraction of the molecules in an isolated preparation.

Isotope flux studies are limited in time resolution and require relatively large amounts of isolated material. The vesicle-derived planar bilayer method, however, has the advantages of electrical and biochemical accessibility, sensitivity, and is a semiquantitative ("VSB" = "vesicle-derived, septum-supported bilayer") method.[47,48] Spreading of a protein-containing vesicle suspension yields a monolayer with similar protein content to that of the vesicles (within a factor of two), enabling estimation of the number of protein molecules incorporated into each bilayer. Under proper conditions of spreading kinetics and lateral pressure,[48,49] transformation of vesicles into a monolayer does not lead to protein denaturation. VSB membranes have extremely stable baseline properties and can withstand much higher holding potentials for long periods than black films without electrical breakdown. Furthermore, the VSB method is the only one that allows both single-channel and macroscopic currents to be measured using the same reconstituted vesicle preparation, a property that is very useful when measuring the effects of activators and inhibitors. For these reasons, we have chosen the VSB method for our reconstitution studies of purified Ca^{2+} antagonist-receptor complex.

Significance of Subconductance Levels for Structure and Gating of the Calcium Channel Isolated from Skeletal Muscle

In order to more clearly evaluate the biophysical nature of subconductance states, it was necessary to choose conditions where such states become stabilized, since the usual activity was characterized by rapid transitions among various conductance levels. We obtained significant prolongation of open conductance states by adding 10 μM

(+)-PN200-110 to channels phosphorylated with protein kinase A.[32,34] Under these conditions, the DHP antagonist (+)-PN200-110 exhibits strongly agonistic properties. The channel activity was not only strongly increased by (+)-PN200-110, but the conductance transitions became more clearly resolvable due to prolonged dwell time in each open state (range of hundreds of msecs to secs), now orders of magnitude longer than the time response of the amplifier (0.3 msec). The conductance state

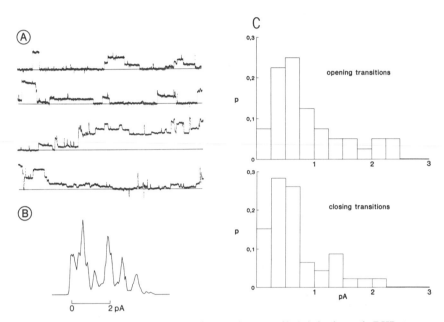

FIGURE 9. (A) Subconductance states of reconstituted purified skeletal muscle DHP receptor in a vesicle-derived planar bilayer containing approximately 100 PN200-110 receptor sites. Solutions: *cis* 100 mM BaCl$_2$ 100 mM NaCl, 9 mM Hepes/Tris (pH 7.4); *trans* 110 mM NaCl, 10 mM Hepes/Tris (pH 7.4). In addition, the *cis* side also contained (for phosphorylation) 1 mM ATP-γ-S, cAMP-dependent protein kinase (catalytic subunit), 1 mM MgCl$_2$ and (for stabilization of conductance states) 10 μM (+)-PN 200-110. Holding potential was 0 mV. E$_{..}$ = −55 mV (*cis* side). Ba^{2+}-selective current is shown as upward deflection from the baseline position (drawn in).

(B) Amplitude histogram from the data depicted in A. The histogram was constructed on a point for point basis from the digitized current recording and normalized to the most frequently occupied bin.

(C) Histograms of current transition amplitudes ("jump-height" histogram) from the data in A. Opening (upward) and closing (downward) transitions were analyzed separately.

transitions, clearly resolvable as single events, ranged from at least 3 to 60 pS (FIG. 9A). We then constructed a simple amplitude histogram of the data to determine which conductance levels predominate. As seen in FIGURE 9B, the histogram is characterized by several peaks that are about equally spaced at a current interval corresponding to 11.3-13.6 pS. Interestingly, the heights of the major peaks approximate a Poisson distribution, implying that the frequently populated multiples of about

12 pS overlap as the result of independently functioning 12-pS units. The conductance level transitions themselves (conductance "jumps") contain a different kind of information. Therefore, the opening and closing transitions were themselves subjected to histogram analysis (FIG. 9C). The results were again Poisson distributed, with the opening transitions peaking at about 11 pS and the closing transitions at about 7 pS. We interpret the conductance jumps as indicating the size of cooperatively gating oligomeric channels associated with them. In this light, the data in FIGURE 9C would mean that gating occurs under these conditions in randomly determined oligomeric units. Moreover, the most frequent "gating unit" for channel opening (11 pS) coincides well with the most frequently occupied conductance states (multiples of 12 pS), whereas channel closing more frequently occurred in somewhat smaller units (7 pS). Indeed, the number of closing events in FIGURE 9A is 15% greater than the number of openings. This would imply that channel opening is more highly cooperative than closing, and is reminiscent of the behavior of porin channel triplets.[50]

Physiological Consequences of the Oligochannel Structure of the Skeletal Muscle Ca^{2+}-Antagonist Receptor—an Integral Component of the Excitation-Contraction Coupling Mechanism

These results raise several interesting questions. First, although the grouping of reconstituted, purified Ca^{2+}-antagonist receptors into 12 pS "gating units" corresponds well with the major conductance state observed for native T-tubule Ca^{2+} channels,[36,37] why do we need special conditions to observe these structures after purification and reconstitution? Without the agonistic effect of (+)-PN200-110, no single conductance state predominates in our hands, implying a different organization of the channel proteins in the native membrane, perhaps involving a component lost during purification. Second, what is the nature of the conductance jumps smaller than 12 pS? Our current data would support a true single-channel conductance of 0.9-1.0 pS,[32] although we cannot, at present, exclude the possibility that this represents a true sublevel of a larger structural unit. More data are required to clarify this issue. We are also investigating the effects of pharmacological agents on the various structural and gating elements of the reconstituted Ca^{2+} channel.

Finally, the role of the skeletal muscle Ca^{2+}-antagonist receptor as an oligomeric Ca^{2+} channel must be considered in terms of its function in excitation-contraction coupling. Recently, it has come to light that the "feet structures" bridging the junctional space at the triad in both skeletal and cardiac muscle are referable to the ryanodine receptor and are identical with the Ca^{2+}-release channel (Hymel et al.[51] and references cited therein). In addition, anatomical evidence suggests that the feet structures may be attached to the T-tubule membrane via proteins organized with orthogonal symmetry, while charge-coupling experiments would suggest that the Ca^{2+}-antagonist receptor serves as voltage sensor directly coupled to (at least some of) the Ca^{2+}-release channels of the sarcoplasmic reticulum.[29] We find it highly provocative that both structures evidently involved in the junctional association are oligomeric Ca^{2+} channels and are currently involved in investigating whether co-association of these two oligomeric channel arrays might not be crucial in determining both their structural and functional organization.

ACKNOWLEDGMENTS

The authors would like to thank the chemists and pharmacologists of the following companies for generous supply of compounds: Bayer AG (Wuppertal, F.R.G.), Goedecke AG (Freiburg, F.R.G.), Knoll AG (Ludwigshafen, F.R.G.), and Sandoz AG (Basel, Switzerland).

REFERENCES

1. GLOSSMANN, H. & J. STRIESSNIG. 1988. Vitam. Horm. **44:** 155-328.
2. JANIS, R. A. & D. J. TRIGGLE. 1987. Adv. Drug. Res. **16:** 309-591.
3. GLOSSMANN, H. & J. STRIESSNIG. 1988. ISI Atlas Pharmacol. **2:** 202-210.
4. FERRY, D. R., M. ROMBUSCH, A. GOLL & H. GLOSSMANN. 1984. FEBS. Lett. **169:** 112-118.
5. STRIESSNIG, J., H. G. KNAUS, M. GRABNER, K. MOOSBURGER, W. SEITZ, H. LIETZ & H. GLOSSMANN. 1987. FEBS. Lett.: 247-253.
6. CATTERALL, W. A., M. J. SEAGAR & M. TAKAHASHI. 1988. J. Biol. Chem. **263:** 3535-3538.
7. GLOSSMANN, H., D. R. FERRY, J. STRIESSNIG, A. GOLL & K. MOOSBURGER. 1987. Trends Pharm. Sci. **8:** 95-100.
8. FERRY, D. R., A. GOLL & H. GLOSSMANN. 1987. Biochem. J. **243:** 127-135.
9. SCHNEIDER, T. & F. HOFMANN. 1988. Eur. J. Biochem. **174:** 369-375.
10. STRIESSNIG, J., H. G. KNAUS & H. GLOSSMANN. 1988. Biochem. J. **253:** 39-47.
11. TANABE, T., H. TAKESHIMA, A. MIKAMI, V. FLOCKERZI, H. TAKAHASHI, K. KANGAWA, M. KOJIMA, H. MATUSO, T. HIROSE & S. NUMA. 1987. Nature **238:** 313-318.
12. HÖLTJE, H.-D. & S. MARRER. 1987. J-Camd. **1:** 23-30.
13. HÖLTJE, H.-D. & S. MARRER. 1988. Quant. Struct.-Act. Relat. **7:** 174-177.
14. GLOSSMANN, H., H.-G. KNAUS, J. STRIESSNIG, S. MARRER & H.-D. HÖLTJE. 1988. Naunyn-Schmiedebergs Arch. Pharmacol. **338:** R20.
15. STRIESSNIG, J., E. MEUSBURGER, M. GRABNER, H. G. KNAUS, H. GLOSSMANN, J. KAISER, B. SCHÖLKENS, R. BECKER, W. LINZ & R. HENNING. 1988. Naunyn-Schmiedebergs Arch. Pharmacol. **337:** 331-340.
16. QAR, J., J. BARHANIN, G. ROMEY, R. HENNING, U. LERCH, R. OEKONOMOPULOS, H. URBACH & M. LAZDUNSKI. 1988. Mol. Pharmacol. **33:** 363-369.
17. GLOSSMANN, H. & D. R. FERRY. 1985. Methods Enzymol. **109:** 513-550.
18. GLOSSMANN, H., D. R. FERRY, A. GOLL, J. STRIESSNIG & M. SCHOBER. 1985. J. Cardiovasc. Pharmacol. **7:** 6-6.
19. REYNOLDS, I. J., A. D. SNOWMAN & S. H. SNYDER. 1986. J. Pharmacol. Exp. Ther. **237:** 731-738.
20. GOLL, A., D. R. FERRY & H. GLOSSMANN. 1984. Eur. J. Biochem. **141:** 177-186.
21. PAURON, D., J. QAR, J. BARHANIN, D. FOURNIER, A. CUANY, M. PRALAVORIO, J. B. BERGE & M. LAZDUNSKI. 1987. Biochemistry **26:** 6311-6315.
22. GREENBERG, R., J. STRIESSNIG, A. KOZA, P. DEVAY, H. GLOSSMANN & L. M. HALL. J. Insect. Biochem. In press.
23. KNAUS, H. G., J. STRIESSNIG, H. GLOSSMANN, S. HERING, E. SCHWENNER, G. KINAST, R. GROSSER & M. MARSMANN. 1988. Naunyn-Schmiedebergs Arch. Pharmacol. **338:** R36.
24. GLOSSMANN, H., T. LINN, M. ROMBUSCH & D. R. FERRY. 1983. FEBS. Lett. **160:** 226-232.
25. GLOSSMANN, H., D. R. FERRY, A. GOLL, J. STRIESSNIG & G. ZERNIG. 1985. Arzneimittelforschung **35:** 1917-1935.
26. STRIESSNIG, J., A. GOLL, K. MOOSBURGER & H. GLOSSMANN. 1986. FEBS. Lett. **197:** 204-210.

27. YANG, C.-P.H., W. MELLADO & S. B. HORWITZ. 1988. Biochem. Pharmacol. **37:** 1417-1421.
28. STRIESSNIG, J., K. MOOSBURGER, A. GOLL, D. R. FERRY & H. GLOSSMANN. 1986. Eur. J. Biochem. **161:** 603-609.
29. RIOS, E. & G. BRUM. 1987. Nature **325:** 717-720.
30. BELLEMANN, P., A. SCHADE & R. TOWART. 1983. Proc. Natl. Acad. Sci. USA **80:** 2356-2360.
31. GLOSSMANN, H., J. STRIESSNIG, L. HYMEL & H. SCHINDLER. 1988. Ann. N.Y. Acad. Sci. USA **522:** 150-161.
32. HYMEL, L., J. STRIESSNIG, H. GLOSSMANN & H. SCHINDLER. 1988. Proc. Natl. Acad. Sci. USA **85:** 4290-4294.
33. GLOSSMANN, H., J. STRIESSNIG, L. HYMEL & H. SCHINDLER. 1987. Biomed. Biochim. Acta **46:** 351-356.
34. GLOSSMANN, H., J. STRIESSNIG, L. HYMEL, G. ZERNIG, H.-G. KNAUS & H. SCHINDLER. 1988. *In* The Calcium Channel: Structure, Function and Implications. M. Morad *et al.,* Eds.: 168-192. Springer Verlag. Berlin.
35. VAGHY, P. L., J. STRIESSNIG, K. MIWA, H. G. KNAUS, K. ITAGAKI, E. MCKENNA, H. GLOSSMANN & A. SCHWARTZ. 1987. J. Biol. Chem. **262:** 14337-14342.
36. MA, J. & R. CORONADO. 1988. Biophys. J. **53:** 387-395.
37. ROSENBERG, R. L., P. HESS, J. P. REEVES, H. SMILOWITZ & R. W. TSIEN. 1986. Science **231:** 1564-1566.
38. FLOCKERZI, V., H. J. OEKEN, F. HOFMANN, D. PELZER, A. CAVALIE, & W. TRAUTWEIN. 1986. Nature **323:** 66-68.
39. AFFOLTER, H. & R. CORONADO. 1985. Biophys. J. **48:** 341-347.
40. BRUM, G., W. OSTERRIEDER & W. TRAUTWEIN. 1984. Pflüger's Arch. **401:** 111-118.
41. REUTER, H. 1983. Nature **301:** 569-574.
42. NILIUS, B., P. HESS, J. B. LANSMAN & R. W. TSIEN. Nature **316:** 443-446.
43. CHEN, C. F., M. J. CORBLEY, T. M. ROBERTS & P. HESS. 1988. Science **239:** 1024-1026.
44. SMITH, J. S., E. J. MCKENNA, J. J. MA, J. VILVEN, P. L. VAGHY, A. SCHWARTZ & R. CORONADO. 1987. Biochemistry **26:** 7182-7188.
45. TALVENHEIMO, J. A. III, J. F. WORLEY & M. T. NELSON. 1987. Biophys. J. **52:** 891-899.
46. CURTIS, B. M. & W. A. CATTERALL. 1986. Biochemistry **25:** 3077-3083.
47. SCHINDLER, H. 1980. FEBS Lett. **122:** 77-79.
48. SCHINDLER, H. 1988. Methods Enzymol. In press.
49. SCHURHOLZ, W. & H. SCHINDLER. 1988. Eur. Biophys. J. In press.
50. SCHINDLER, H. & J. P. ROSENBUSCH. 1981. Proc. Natl. Acad. Sci. USA **78:** 2302-2306.
51. HYMEL, L., M. INUI, S. FLEISCHER & H. SCHINDLER. 1988. Proc. Natl. Acad. Sci. USA **85:** 441-445.

Calcium Channels in Smooth Muscle

Properties and Regulation[a]

D. J. TRIGGLE,[b] W. ZHENG, M. HAWTHORN,
Y. W. KWON, X.-Y. WEI, A. JOSLYN,
J. FERRANTE, AND A. M. TRIGGLE

School of Pharmacy
State University of New York
Buffalo, New York 14260

INTRODUCTION

Smooth muscle is homogenous in neither function nor physical characteristics, and it has been remarked, with some asperity, "that the most constant property of smooth muscle is its variability." It is agreed, however, that a key component of tension development in smooth muscle is an elevation of intracellular Ca^{2+} subsequent to chemical, electrical, or mechanical stimulation. In this respect smooth muscle behaves similarly to cardiac and skeletal muscle and to motile systems generally. However, important differences do exist between smooth and other muscle systems in the control processes by which elevation of intracellular Ca^{2+} from a resting level of approximately $10^{-8} M$ to a stimulated level of approximately $5 \times 10^{-7} M$ is linked to tension development. Additionally, smooth muscle is capable both of generating rapid responses and of sustaining prolonged periods of tension (tone). Determination of the control processes regulating Ca^{2+} mobilization in smooth muscle, both visceral and vascular, must accommodate these characteristics of tension development.

A widely accepted hypothesis holds that Ca^{2+}-calmodulin (CM) activation of myosin light chain kinase leads to the phosphorylation of myosin light chain which, in its phosphorylated form, interacts with actin. A cycling actin-myosin cross-bridge process is presumed to underlie tension development (for review see Kamm and Stull[1]). A more recently developed alternative hypothesis holds that the processes underlying the generation and maintenance of tension in smooth muscle may be quite different.[2] The initial component of response is assumed to involve the CM-dependent activation of myosin light chain kinase and the sustained component of response is postulated to occur through Ca^{2+} activation of membrane-associated protein kinase C by the diacylglycerol formed in the receptor-initiated, phospholipase C mediated breakdown

[a] This work was supported by grants from the National Institutes of Health (HL 10003, HL/AI 31178, GM 07145, and 2507 05454). Additional support from the Miles Institute for Preclinical Pharmacology is gratefully acknowledged.

[b] Address for correspondence: D. J. Triggle, Office of The Dean, 126 Cooke, School of Pharmacy, State University of New York, Buffalo, NY, 14260.

of phosphatidylinositol diphosphate (PIP2). There are attractive features to this model. It accommodates the well-known ability of smooth muscle to undergo brief phasic and prolonged tonic responses, it incorporates an important Ca^{2+} signaling pathway, it permits the muscle to exhibit energy-efficient tension responses, it prevents the persistent overloading of cells with Ca^{2+} during prolonged responses, and it is consistent with recent observations that the distribution of Ca^{2+} is both temporally and kinetically heterogeneous.[3,4] Finally, and of particular importance, the model draws attention to the important roles that multiple Ca^{2+}-mobilizing pathways may play in the differential regulation of smooth muscle tone.

In principle, Ca^{2+} mobilization to satisfy the requirements of excitation-contraction coupling in smooth muscle can occur through several distinct pathways (reviewed in Refs. 5-8; FIG. 1). The release of Ca^{2+} from the sarcoplasmic reticulum by the second messenger IP3 is a pathway of major significance for intracellular mobilization.[2,8] Additionally, a ryanodine-sensitive Ca^{2+} release process, known to be of importance in cardiac and skeletal muscle, may also operate in smooth muscle.[9] Sarcoplasmic reticulum of smooth muscle is adjacent to the plasma membrane and hence Ca^{2+} movements across these two cellular membranes may be intimately linked. Ca^{2+} movements across plasma membranes are frequently considered to be three major pathways—leak, receptor-operated channels, and voltage-dependent channels.[5-10] The former categories, particularly the leak channel, are not well-defined entities in smooth muscle, either vascular or nonvascular.[11] Receptor-operated channels may be viewed in terms of two limiting models in which the receptor and the channel are a single protein or association of proteins or where the channel and the receptor are physically discrete but are linked through one or more cytosolic or membrane messengers. The actions of inositol 1,3,4,5-tetraphosphate in promoting plasma membrane Ca^{2+} influx[2,12] provide a biochemical basis to one class of receptor-operated channel. Electrophysiological studies are now providing a direct approach to the definition of receptor-operated Ca^{2+} channels.[13]

In contrast, Ca^{2+} movements in smooth muscle through voltage-dependent Ca^{2+} channels in response to receptor-initiated and depolarizing signals have been relatively

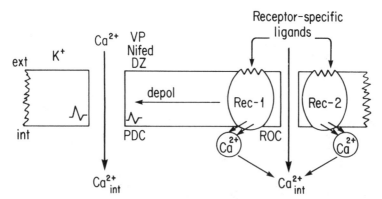

FIGURE 1. Representation of cellular pathways of Ca^{2+} mobilization. Depicted are Ca^{2+} entry through voltage-dependent Ca^{2+} channels, entry through receptor-operated Ca^{2+} channels, and mobilization from intracellular stores. No attempt has been made to indicate the several distinct categories (at least three) of voltage-dependent Ca^{2+} channels or to indicate possible biochemical intermediates (IP3, IP4, PKC, etc.) that modulate receptor-operated and potential-dependent Ca^{2+} channels.

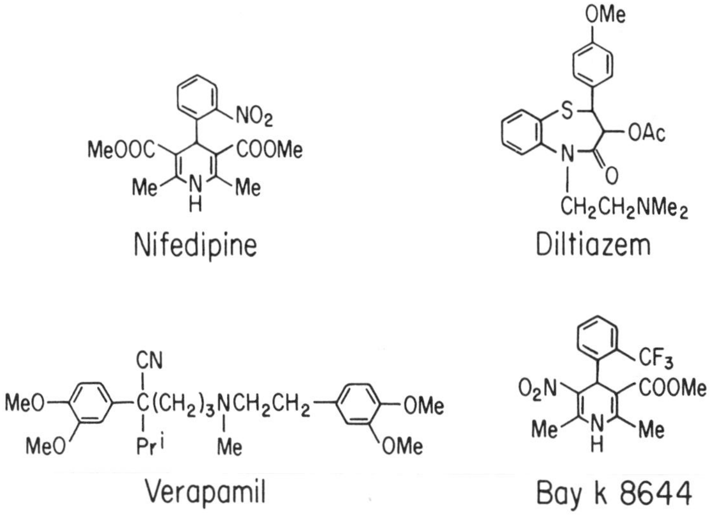

FIGURE 2. Structural formulas of the major chemical categories of Ca^{2+} channel antagonists and activators.

well studied. The availability, first of the Ca^{2+}-channel antagonists including the clinically available verapamil, nifedipine, and diltiazem and subsequently of the 1,4-dihydropyridine Ca^{2+}-channel activators including Bay K 8644 (FIG. 2), has proved to be important to the pharmacological classification of tension responses and $^{45}Ca^{2+}$ uptake processes and to the electrophysiologic characterization of the channels involved.[14] The 1,4-dihydropyridine-sensitive or L-type Ca^{2+} channel, of large conductance and slow inactivation, is of particular functional importance in the cardiovascular system, although similar or identical channels are found in other excitable cells including brain and skeletal muscle.[15–17] Furthermore, the L channel is not the only Ca^{2+} channel present in smooth muscle: At least one other major channel type, the T channel, characterized by activation at more negative membrane potentials, rapid inactivation and insensitivity or relative insensitivity to 1,4-dihydropyridines, is also present[15,16,18–20] and may have a role in the generation of pacemaker activity.

Smooth muscle thus presents a variety of Ca^{2+} mobilization processes that will contribute, in proportions varying according to stimulus and time, to the total tension response (FIG. 1). Our approach has been to characterize the voltage-dependent Ca^{2+} channels in smooth muscle through structure–activity studies of Ca^{2+}-channel ligands and to compare the derived structure–activity relationships with those observed in other excitable tissues.

CA^{2+} CHANNELS IN VISCERAL SMOOTH MUSCLE

The longitudinal smooth muscle of guinea pig ileum offers a number of advantages to the study of Ca^{2+} channels in smooth muscle. Our previous work in this preparation has demonstrated the following properties:

1. It is sensitive to several modes of stimulation including K^+ depolarization and activation of muscarinic, histamine, 5-hydroxytryptamine, and polypeptide hormone receptors.[21,22]

2. The tissue exhibits both specific and nonspecific desensitization after receptor activation.[22,23]

3. The responses to these several stimulants are dependent upon extracellular Ca^{2+}.[21,22]

4. The responses consist of phasic (fast) and tonic (slow) components of response, the latter exhibiting an apparently greater sensitivity to extracellular Ca^{2+}.[22,24]

5. Tension responses to both K^+ depolarization and muscarinic receptor activation are accompanied by an uptake of $^{45}Ca^{2+}$.[24]

6. Both tension response and $^{45}Ca^{2+}$ uptake are sensitive to inorganic (Co^{2+}, Ln^{3+}) and organic Ca^{2+}-channel antagonists.[24,25]

7. $^{45}Ca^{2+}$ uptake in response to a series of muscarinic agonists is dependent upon the intrinsic activity of the agonist.[26]

8. The high sensitivity of response to cations of the Ln^{3+} series is dependent upon the ionic radius of the cation, Tm^{3+} being the most potent, and is characterized by a saturable high-affinity binding of Tm^{3+}.[25]

9. Radioligand binding studies with $[^3H]$1,4-dihydropyridines reveal a high binding density, approximately 1000 fmoles/mg protein,[27] that is associated with the plasmalemmal fraction,[28] and whose properties correlate with the pharmacological activities of these agents.

Collectively, these observations have characterized excitation-contraction coupling in guinea pig ileal longitudinal smooth muscle to be mediated dominantly by Ca^{2+} mobilization through 1,4-dihydropyridine-sensitive Ca^{2+} channels activated and modulated by depolarization and receptor-initiated events, respectively.

An extensive series of 1,4-dihydropyridine antagonists related to nifedipine has been examined through pharmacologic and radioligand binding techniques.[28,29] For the antagonism of the tonic component of response to muscarinic receptor activation in a series of achiral 1,4-dihydropyridine analogues of nifedipine-bearing ortho, meta, para, or multiple substituents in the 4-phenyl ring, the following quantitative structure-activity relationship [QSAR] was obtained:[29]

$$\log 1/IC_{50} = 0.62\ \pi\ +\ 1.96\ \sigma_m\ -\ 0.44 L_m\ -\ 3.26 B_{1[p]}\ -\ 1.51 L_m' + 14.23$$

$$(n\ =\ 46,\ r\ =\ 0.90,\ s\ =\ 0.67,\ F\ =\ 33.93)$$

showing that activity is dependent upon the hydrophobic character of the substituent (π), the electron withdrawing characteristics (σ_m) and steric effects (L, B_1) exerted at the meta (m) and para (p) positions of the phenyl ring. The QSAR relationship confirms that steric interactions are most unfavorable at the para position and more unfavorable at the meta' position than at the meta position and is consistent with solid-state structures that show that the preferred conformation bears the substituent of the pseudo axially oriented 4-phenyl ring oriented away from the flattened boat conformation adopted by the 1,4-dihydropyridine ring (for reviews of geometry of 1,4-dihydropyridines see Refs. 14 and 30-32).

There are no sufficiently comprehensive studies to permit the elucidation of similar QSAR for other substituent changes. However, in several series of compounds bearing different ester groups, the effects of phenyl ring substitution are expressed independently of the ester substitution pattern[32] (FIG. 3). Thus, the contributions of these two sets of structural change to the linear free energy relationship are expressed independently.

The greater sensitivity of the tonic or sustained component of response of ileal smooth muscle to depolarization or receptor activation is well documented and a variety of explanations offered from discrete channel types mediating phasic and tonic responses to state-dependent interactions.[5,7,14,21,28] However, comparisons of the sequences of activities for a series of substituted 1,4-dihydropyridines against phasic and tonic responses in ileal and other smooth muscle reveals that the same rank order is followed regardless of absolute activity.[28,34–36] This is consistent with ligand interaction at a common receptor that is expressed in variable affinity states.

Comparisons of pharmacologic and binding affinities for an extensive series of 1,4-dihydropyridines (FIG. 4) reveal correlations ranging from 1 : 1 to 100 : 1 according to the stimulant, K^+ depolarization or muscarinic receptor activation, and the phasic or tonic component of response. An explanation of these findings, couched in terms

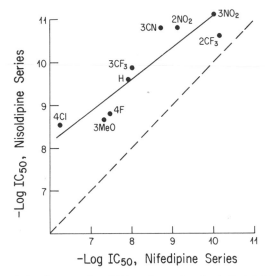

FIGURE 3. Correlation between the abilities of two phenyl-substituted 1,4-dihydropyridines bearing different ester group substitution patterns to bind to Ca^{2+} channels in ileal longitudinal smooth muscle membranes. Nifedipine, 3,5-dicarbomethoxy; nisoldipine, 3-carbomethoxy,5-carboisobutoxy. (Data from Fossheim *et al.*[33])

of the modulated receptor hypothesis, indicates that, consistent with much electrophysiological evidence,[37–39] high-affinity interactions of 1,4-dihydropyridines are associated with the inactivated state of the channel favored by prolonged depolarization and likely the dominant state in membrane fragments. Accordingly, the 1 : 1 correlation between binding and pharmacologic affinities for the tonic component of K^+ depolarization-induced responses reflects 1,4-dihydropyridine interactions at this depolarization-favored state. The approximately 100-fold lower sensitivity measured against the phasic component of the K^+ depolarization-induced response may reflect a preferential interaction with the resting state of the channel, an equilibrium that dominates during the brief time needed for the generation of this component of response. It is of interest that the phasic and tonic components of response to the muscarinic agonist methylfurmethide (MF; FIG. 4) show a sensitivity to the 1,4-dihydropyridines inter-

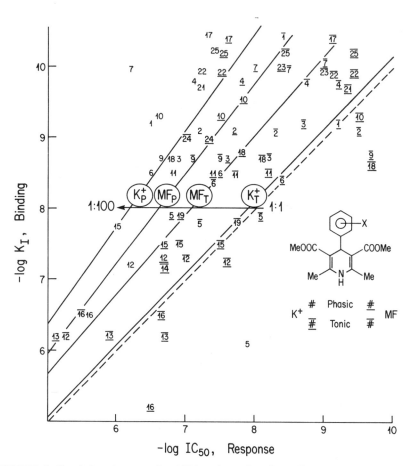

FIGURE 4. Correlations between the abilities of a series of 1,4-dihydropyridines to inhibit specific [³H]nitrendipine binding to guinea pig ileal longitudinal smooth muscle membranes and to inhibit the phasic and tonic components of tension response to K⁺ depolarization and methylfurmethide (muscarinic) receptor activation in the same tissue. The 1,4-dihydropyridines are 2,6-dimethyl-3,5-dicarbomethoxy-4-substituted phenyl-1,4-dihydropyridine where the substituents are: 1, 2-CN; 2, 2-NO2; 3, 2-Me; 4, 2-Cl; 5, 2-OMe; 6, 2-F; 7, 3-NO2; 8, 3-OMe; 9, 3-CN; 10, 3-Cl; 11, 3-F; 12, 3-Me; 13, 4-Cl; 14, 4-Me; 15, 4-F; 16, 4-NO2; 17, F5; 18, 2-F,6-Cl; 19,H ; 21, nitrendipine; 22, nimodipine; 23 (−)3628; 24, (+)3629; 25, nisoldipine.

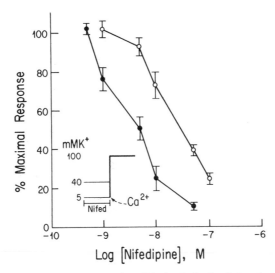

FIGURE 5. The activity of nifedipine against K^+ depolarization-induced response (total 100 mM K^+) after preincubation in physiologic saline containing 5 mM K^+ ($\bigcirc$) or 40 mM K^+ ($\bullet$) with or without various concentrations of nifedipine for 30 minutes before the addition of Ca^{2+} and K^+ to total 100 mM K^+. The inset figure shows the addition protocol.

mediate between those observed for the corresponding components of response to K^+ depolarization. This may reflect the extent to which muscarinic receptor activation changes membrane potential, contributions from intracellular Ca^{2+} release mediated by the polyphosphatidylinositol or other pathway or modulation of the Ca^{2+} channel through biochemical intermediate control (c-AMP, IP3, G protein).

Although the voltage dependence of 1,4-dihydropyridine interactions with the L class of Ca^{2+} channels is best documented from electrophysiological observations,[37–39] it can also be observed in functional smooth muscle experiments.[40–42] The data of FIGURE 5 and TABLE 1 show for ileal longitudinal muscle that preincubation of the preparation with elevated K^+ before generation of response enhances the sensitivity of the phasic, but not the tonic, component of response to nifedipine and other Ca^{2+} channel antagonists. The design of the experiment, although not the time scale, is comparable to the voltage-clamp experiments employed in electrophysiologic studies.

TABLE 1. Activities of 1,4-Dihydropyridines against K^+-Induced (100 mM) Responses of Guinea Pig Ileal Longitudinal Smooth Muscle under Preincubation Conditions of 5 and 40 mM K^+

| | $IC_{50} \times 10^{-9}$ M | | |
1,4-Dihydropyridine	K^+ 40 mM	K^+ 5 mM	Ratio
Nifedipine	3.9	31.1	7.9
Nitrendipine	0.53	5.36	10.1
Nisoldipine	1.3	11.2	8.6
Nicardipine	1.3	11.9	9.0

These observations can be interpreted in terms of an enhanced affinity of the antagonists for a state of the channel favored by depolarization.

By comparison with the 1,4-dihydropyridine antagonists, there are few quantitative data available to characterize a QSAR for the 1,4-dihydropyridine activators. It is known, however, that the most potent activators possess the 5-nitro group in the 1,4-dihydropyridine ring (reviewed in Refs. 14,15,32) and that enantiomeric discrimination between the S-activator and the R-antagonist forms is important.[43,44] In a small series of activators the effects of the phenyl ring substitution pattern parallel those observed in the antagonist series, ortho $\geq$ meta > para.[45] Additionally, there is also a close correlation between affinities derived in radioligand and pharmacological experiments (FIG. 6a). Differences do exist, however, between the activator and antagonist series and the effects of substituents are larger in antagonist than in activator ligands: This difference is clearly apparent in the nitro-substituted series (FIG. 6b). This indicates that activators and antagonists likely probe different environments or conformations of the 1,4-dihydropyridine receptor.

The ileal longitudinal smooth muscle of the guinea pig is the best characterized preparation in terms of structure-activity relationships derived from pharmacologic and radioligand binding data. However, sufficient data have been accumulated for bladder and uterine smooth muscle to determine that similar or identical conclusions operate for these preparations also, whereby tonic responses are more sensitive than phasic responses and where the same rank order of activity holds regardless of the absolute activity expressed.[34,35] Additionally, the absolute values of the pharmacologic and radioligand binding affinities determined in ileal, uterine, and bladder smooth muscles exhibit only minor differences, consistent with the thesis that the receptor

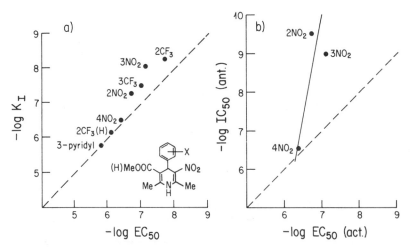

FIGURE 6. (a) The abilities of a series of 1,4-dihydropyridine activators to activate tension response (EC_{50}) in ileal longitudinal smooth muscle and to inhibit specific [^{3}H]nitrendipine binding in membrane preparation. (b) Comparison of the effects of the nitro substitution pattern in the phenyl ring of 1,4-dihydropyridine antagonists (2,6-dimethyl-3,5-dicarbomethoxy-4-[X-nitrophenyl]1,4--dihydropyridines) and activators (2,6-dimethyl-3-carbomethoxy-5-nitro-4-[X-nitrophenyl]-1,4-dihydropyridines) in tension responses in guinea pig ileal longitudinal smooth muscle. (Data from Kwon et al.[45])

TABLE 2. Binding Properties of [³H]1,4-Dihydropyridines in Smooth Muscle

Tissue	Ligand	Temp. (°C)	K_D (nM)	B_{max} (fmoles/mg)	Reference
Vascular Canine:					
Aorta	[³H]Nitr	25	0.25	20	46
Mesenteric	[³H]Nitr	25	0.31	25	46
Rat:					
aorta	[³H]Nitr	25	0.92	33	47
mesenteric	[³H]Nitr	25	0.82	10	47
tail	[³H]Nitr	25	0.36	550	48
cerebral	[³H]PN 200-110	25	0.06	45	49
Human:					
umbilical	[³H]Nitr	25	0.21	139	50
Bovine:					
aorta	[³H]PN 200 110	37	0.49	152	51
mesenteric	[³H]PN 200 110	37	0.62	77	51
renal	[³H]PN 200 110	37	0.33	26	51
Visceral Guinea Pig:					
Ileum	[³H]Nitr	25	0.16	1100	27
Ileum	[³H]Nimod	25	0.12	670	52
Ileum	[³H]PN 200 110	25	0.04	820	52
Bovine:					
trachea	[³H]Nitr	25	0.15	15	53
Rat:					
bladder	[³H]Nitr	25	0.20	85	34
vas deferens	[³H]Nitr	25	0.20	230	54
lung	[³H]Nitr	25	0.22	77	34
uterus	[³H]Nitr	25	0.10	180	35

sites associated with the voltage-dependent Ca^{2+} channel in these preparations are probably identical.

Less data are available for other smooth muscles, particularly vascular tissue, but equilibrium binding studies for both vascular and nonvascular smooth muscle do not suggest major differences in the binding sites (TABLE 2). There do exist, however, substantial differences in binding site density. With the assumption that all of the binding sites represent functional Ca^{2+} channels, the significance of the differences in density is not clear, and they need to be evaluated in terms of the cell volumes of the several smooth muscles. In principle, however, binding site densities may indicate the relative importance to excitation-contraction coupling of Ca^{2+} mobilization through this pathway.

The direct studies of drug receptors associated with Ca^{2+} channels through radioligand binding approaches have not demonstrated major apparent differences, other than binding site density, between different smooth muscles. There are, however, serious limitations to the majority of ligand binding studies thus far available. For the most part these studies have been carried out in microsomal or other membrane preparations. These conditions almost certainly reflect the properties of the channel in a single noninterconvertible state, depolarized and inactivated, and which may not reflect precisely the states available in intact functional smooth muscle. Binding studies on intact, functional cells under voltage control will be very helpful in elucidating such properties.

It is clear, however, that smooth muscle does show very large differences in sensitivity to 1,4-dihydropyridines and other ligand classes and that these differences in sensitivity underlie the pharmacologic and therapeutic uses and differentiation between the different classes of Ca^{2+}-channel drugs. Thus, nifedipine and other 1,4-dihydropyridines are vascular smooth muscle selective agents with few cardiodepressant effects, exhibit apparent regional vascular bed selectivity, and are more effective in hypertensive than in normotensive situations.[14,15,55,56]

In principle, selectivity of action may arise from a variety of factors alone or, more frequently, in combination. These factors include:

(a) Source of Ca^{2+} mobilized;
(b) Pharmacokinetics of drug action;
(c) Class of Ca^{2+} channel activated;
(d) Activation of cardiovascular reflex-compensating pathways;
(e) State-dependence [frequency and voltage] of drug action;
(f) Activator-antagonist-partial activator properties; and
(g) Pathological state (hypertension, etc.).

The significance of all of these factors in determining the selectivity of action of Ca^{2+}-channel ligands has not been fully established. Clearly, stimuli that mobilize

TABLE 3. Voltage-Dependent Parameters of [³H]PN 200-110 Binding to Polarized and Depolarized Neonatal Cultured Rat Cardiac Cells[62]

$[K^+]$, mM	k_1 min$^{-1}M^{-1}$	k_{-1} min^{-1}	k_{-1}/k_1
50	2.3×10^8	0.018	7.9×10^{-11}
5	2.2×10^8	0.53	2.4×10^{-9}

Ca^{2+} from an intracellular store or through non-1,4-dihydropyridine-sensitive Ca^{2+} channels will be major determinants of all-or-none selectivity patterns. More subtle patterns of selectivity are exerted through state-dependent interactions, whereby the affinity of a drug is determined by its preferential interaction with or access to a particular channel state or states.

Voltage-dependent interactions of 1,4-dihydropyridine antagonists with cardiac and smooth muscle systems have been demonstrated electrophysiologically[14,15,37-39,57] and are consistent with enhanced activity at an inactivated state of the channel. Limited functional studies in smooth muscles also indicate that the apparent affinity of 1,4-dihydropyridines and other antagonists are increased under conditions of increased depolarization.[40-42,58] Very limited radioligand binding studies in vascular smooth muscle suggest that in rat mesenteric artery the affinity of [³H]PN 200 110 is increased by an approximately fivefold factor from 2×10^{-10} M to 0.4×10^{-10} M by depolarization.[58] However, a much smaller difference was found with [³H]nifedipine binding to isolated cells from pig coronary artery.[59] Whether preferential association with inactivated channel states occurs in all smooth muscles remains to be determined. In rabbit ear artery a preferential association of nifedipine with the resting channel state has been proposed.[60]

More detailed analyses of voltage-dependent [³H]1,4-dihydropyridine binding are available from cardiac cell preparations.[61,62] The affinity of [³H]PN 200 110 is increased

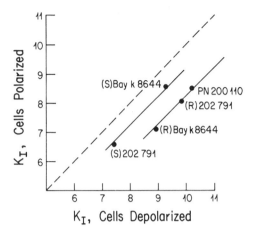

FIGURE 7. Correlation between the binding affinities of a series of 1,4-dihydropyridine activators (S BAY K 8644 and S 202 791) and antagonists (PN 200-110, R BAY K 8644 and R 202-791) measured against [³H]PN 200-110 binding in cultured neonatal cardiac cells in polarized (5 m*M* KCl) and depolarized (50 m*M* KCl) media. The dashed line represents 1 : 1 equivalency. (Data from Wei *et al.*[62])

by approximately 100-fold in the depolarized preparation, and this increase is mediated by a decrease in the dissociation rate constant (TABLE 3). The increased binding energy associated with interaction with the depolarized state corresponds approximately to the formation of one new hydrogen bond. Competition studies show that other antagonists also demonstrate similar voltage-dependent interactions, but that 1,4-dihydropyridine Ca^{2+} channel activators demonstrate little voltage dependence (FIG. 7).

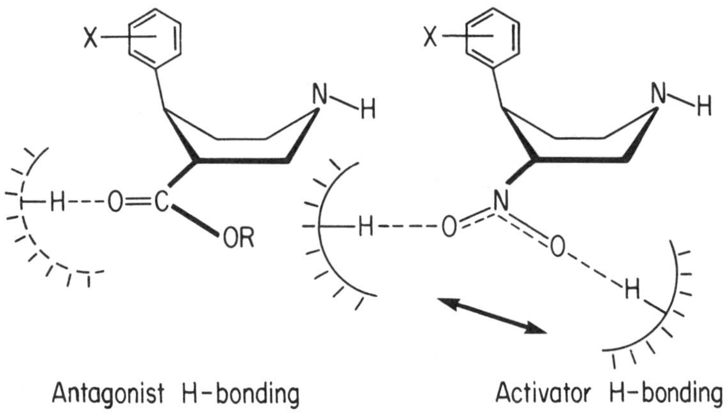

Antagonist H-bonding   Activator H-bonding

FIGURE 8. Schematic representation of modes of 1,4-dihydropyridine antagonist and activator binding in polarized and depolarized states. The high affinity of an antagonist is attributed to its formation of a hydrogen bond preferentially in the depolarized state. The lack of state-dependent interaction of the activator molecule is attributed to its ability to form hydrogen bonds with approximately equal efficacy in the polarized and depolarized states.

The absence of significant voltage dependence of interaction of the activators so far studied indicates that these agents are relatively state-independent in their interactions. An interpretation in terms of binding orientations of antagonists and activators is depicted in FIGURE 8. The relative absence of voltage-dependent binding of activator 1,4-dihydropyridines is consistent with previous studies indicating that S Bay K 8644 behaves as an activator or antagonist at polarized and depolarized membrane potentials, respectively.[48,63] There are important implications to this conclusion for it suggests that 1,4-dihydropyridines should exist that will be both cardiotonic and vasodilatory—an important combination of properties.

REFERENCES

1. KAMM, K. E. & J. T. STULL. 1985. The function of myosin and myosin light chain kinase phosphorylation in smooth muscle. Ann. Rev. Pharmacol. Toxicol. **25:** 593-620.
2. RASMUSSEN, H., Y. TAKUWA & S. PARK. 1987. Protein kinase C in the regulation of smooth muscle contraction. FASEB J. **1:** 177-185.
3. YADA, T., S. OIKI, S. UEDA & Y. OKADA. 1986. Synchronous oscillation of the cytoplasmic Ca^{2+} concentration and membrane potential in cultured epithelial cells (Intestine 407). Biochim. Biophys. Acta **887:** 105-112.
4. LIPSCOMBE, D., D. V. MADISON, M. POENIE, H. REUTER, R. Y. TSIEN & R. W. TSIEN. 1988. Spatial distribution of calcium channels and cytosolic calcium transients in growth cones and cell bodies of sympathetic neurons. Proc. Natl. Acad. Sci. USA **85:** 2398-2402.
5. BOLTON, T. 1979. Mechanisms of action of transmitters and other substances on smooth muscle. Physiol. Rev. **3:** 606-718.
6. CAUVIN, C., R. LOUTZENHISER & C. VAN BREEMEN. 1983. Mechanisms of calcium antagonist-induced vasodilation. Ann. Rev. Pharmacol. Toxicol. **23:** 373-396.
7. HURWITZ, L. 1986. Pharmacology of calcium channels and smooth muscle. Ann. Rev. Pharmacol. Toxicol. **26:** 225-258.
8. JOHNS, A., P. LEIJTEN, H. YAMAMOTO, K. HWANG & C. VAN BREEMEN. 1987. Calcium regulation in vascular smooth muscle contractility. Am. J. Cardiol. **59:** 18A-23A.
9. IMAGAWA, T., J. S. SMITH, R. CORONADO & K. P. CAMPBELL. 1987. Purified ryanodine receptor for skeletal muscle sarcoplasmic reticulum is the Ca^{2+} permeable pore of the Ca^{2+} release channel. J. Biol. Chem. **262:** 16636-16643.
10. VAN BREEMEN, C., P. AARONSON, R. LOUTZENHISER & K. MEISHERI. 1982. Ca fluxes in isolated rabbit aorta and guinea pig taenia coli. Fed. Proc. **41:** 2891-2897.
11. LOUTZENHISER, R., P. LEITJEN, K. SAIDA & C. VAN BREEMEN. 1985. Calcium compartments and mobilization during contraction of smooth muscle. *In* Calcium and Contractility. A. K. Grover & E. E. Daniel, Eds. Humana Press. Clifton, NJ.
12. IRVINE, F. & R. MOOR. 1986. Microinjection of inositol 1,3,4,5-tetrakisphosphate activates sea urchin eggs by a mechanism dependent on external Ca^{2+}. Biochem. J. **240:** 917-920.
13. BENTHAM, C. D. & R. W. TSIEN. 1987. A novel receptor-operated Ca^{2+} permeant channel activated by ATP in smooth muscle. Nature **328:** 275-278.
14. JANIS, R. A., P. SILVER & D. J. TRIGGLE. 1987. Drug action and cellular Ca^{2+} regulation. Adv. Drug. Res. **16:** 309-591.
15. TRIGGLE, D. J. & R. A. JANIS. 1987. Calcium channel ligands. Ann. Rev. Pharmacol. Toxicol. **27:** 346-369.
16. HESS, P., A. P. FOX, J. B. LANSMAN, B. NILIUS, M. C. NOWYCKY & R. W. TSIEN. 1986. Calcium channel types in cardiac, neuronal and smooth muscle-derived cells: Differences in gating, permeation and pharmacology. *In* Ion Channels in Neural Membranes. J. M. Ritchie, R. D. Keynes & L. Bolis, Eds.: 227-252. Alan R. Liss. New York.
17. MILLER, R. J. 1987. Multiple Ca^{2+} channels and neuronal function. Science **235:** 46-52.
18. CAFFREY, J. M., I. R. JOSEPHSON & A. M. BROWN. 1986. Calcium channels of amphibian stomach and mammalian aorta smooth muscle. Biophys. J. **49:** 1237-1242.

19. STUREK, M. & K. HERMSMEYER. 1986. Calcium and sodium channels in spontaneously contracting vascular muscle cells. Science **233**: 475-478.
20. NAKAZAWA, K., H. SAITO & N. MATSUKI. 1988. Fast and slowly inactivating components of Ca channel current and their sensitivities to nicardipine in isolated smooth muscle cells from rat vas deferens. Pflug. Arch. **411**: 289-295.
21. CHANG, K.-J. & D. J. TRIGGLE. 1983. Quantitative aspects of drug-receptor interactions. I. Ca^{2+} and cholinergic receptor activation in smooth muscle: A basic model for drug-receptor interactions. J. Theor. Biol. **40**: 125-154.
22. SIEGEL, H., K. JIM, G. T. BOLGER, P. GENGO & D. J. TRIGGLE. 1984. Specific and nonspecific desensitization of guinea pig ileal smooth muscle. J. Auton. Pharmacol. **4**: 109-126.
23. CHANG, K.-J. & D. J. TRIGGLE. 1973. Quantitative aspects of drug receptor interactions. II. The role of Ca^{2+} in desensitization and spasmolytic activity. J. Theor. Biol. **40**: 155-172.
24. ROSENBERGER, L. B., M. K. TICKU & D. J. TRIGGLE. 1979. The effects of Ca^{2+} antagonists on mechanical responses and Ca^{2+} movements in guinea pig ileal longitudinal smooth muscle. Can. J. Physiol. Pharmacol. **57**: 333-347.
25. TRIGGLE, C. R. & D. J. TRIGGLE. 1976. An analysis of the actions of cations of the lanthanide series on the mechanical responses of guinea pig ileal longitudinal smooth muscle. J. Physiol. (London) **254**: 39-54.
26. ROSENBERGER, L. B. & D. J. TRIGGLE. 1979. Ca^{2+} utilization in guinea pig ileal longitudinal smooth muscle in response to a series of muscarinic agonists. Can. J. Physiol. Pharmacol. **57**: 1375-1380.
27. BOLGER, G. T., P. GENGO, R. KLOCKOWSKI, E. LUCHOWSKI, H. SIEGEL, R. A. JANIS, A. M. TRIGGLE & D. J. TRIGGLE. 1983. Characterization of binding of the Ca^{2+} channel antagonist, [^{3}H]nitrendipine, to guinea pig ileal smooth muscle. J. Pharmacol. Exp. Ther. **225**: 291-309.
28. GROVER, A. K., C.-Y. KWAN, E. LUCHOWSKI, E. E. DANIEL & D. J. TRIGGLE. 1984. Subcellular localization of [^{3}H]nitrendipine binding in smooth muscle. J. Biol. Chem. **259**: 2223-2226.
29. COBURN, R. A., M. WIERZBA, M. SUTO, A. J. SOLO, A. M. TRIGGLE & D. J. TRIGGLE. 1988. 1,4-Dihydropyridine antagonist interactions at the calcium channel: A QSAR approach. J. Med. Chem. **31**: 2103-2107.
30. TRIGGLE, A. M., E. SHEFTER & D. J. TRIGGLE. 1980. Crystal structures of calcium channel antagonists. J. Med. Chem. **23**: 1442-1445.
31. FOSSHEIM, R., K. SVARTENG, A. MOSTAD, C. RØMMING, E. SHEFTER & D. J. TRIGGLE. 1982. Crystal structures and pharmacological activities of calcium channel antagonists: 2,6-Dimethyl-3,5-dicarbomethoxy-4-(unsubstituted, 3-methyl, 4-methyl, 3-nitro, 4-nitro and 2,4-dinitrophenyl)-1,4-dihydropyridine. J. Med. Chem. **25**: 126-131.
32. TRIGGLE, D. J., D. A. LANGS & R. A. JANIS. 1989. Ca^{2+} channel ligands: Structure-activity relationships of the 1,4-dihydropyridines. Med. Res. Revs. In press.
33. FOSSHEIM, R., A. JOSLYN, A. J. SOLO, E. M. LUCHOWSKI, A. RUTLEDGE & D. J. TRIGGLE. 1987. Crystal structures and pharmacologic activities of 1,4-dihydropyridine calcium channel antagonists of the isobutyl methyl 2,6-dimethyl-4-(substituted phenyl)-1,4-dihydropyridine-3,5-dicarboxylate (nisoldipine) series. J. Med. Chem. **31**: 300-305.
34. YOUSIF, F. B., G. T. BOLGER, A. RUZYCKY & D. J. TRIGGLE. 1985. Ca^{2+} channel antagonist actions in bladder smooth muscle: Comparative pharmacologic and [^{3}H]nitrendipine binding studies. Can. J. Physiol. Pharmacol. **63**: 453-462.
35. RUZYCKY, A. L., D. J. CRANKSHAW & D. J. TRIGGLE. 1987. Ca^{2+} channel ligand activities in uterine smooth muscle: Influence of hormonal status. Can. J. Physiol. Pharmacol. **65**: 2085-2092.
36. FLECKENSTEIN, A., C. VAN BREEMEN, R. GROSS & F. HOFFMEISTER, Eds. 1986. Cardiovascular effects of dihydropyridine-type calcium antagonists and agonists. Springer-Verlag. Berlin and New York.
37. SANGUINETTI, M. C. & R. S. KASS. 1984. Voltage-dependent block of calcium channel current in calf cardiac Purkinje fibers by dihydropyridine calcium channel antagonists. Circ. Res. **55**: 336-348.

38. BEAN, B. P. 1984. Nitrendipine block of cardiac calcium channels: High affinity binding to the inactivated state. Proc. Natl. Acad. Sci. USA **81:** 6388-6392.
39. BEAN, B. P., M. STUREK, A. PUGA & K. HERMSMEYER. 1986. Calcium channels in muscle cells isolated from rat mesenteric arteries: Modulation by dihydropyridine drugs. Circ. Res. **59:** 229-235.
40. BURGES, R. A., D. G. GARDINER, M. GWILT, J. A. HIGGINS, K. J. BLACKBURN, S. F. CAMPBELL, P. E. CROSS & J. K. STUBBS. 1987. Calcium channel blocking properties of amlodipine in vascular smooth muscle and cardiac muscle in vitro: Evidence for voltage modulation of vascular dihydropyridine receptors. J. Cardiovasc. Pharmacol. **9:** 110-119.
41. NELSON, M. T. & J. F. WORLEY, III. 1988. Dihydropyridine inhibition of single calcium channels and contraction in rabbit mesenteric artery depends on voltage. J. Physiol. (London), in press.
42. TRIGGLE, D. J., M. HAWTHORN & W. ZHENG. 1988. Potential-dependent interactions of nitrendipine and related 1,4-dihydropyridines in functional smooth muscle preparations. J. Cardiovasc. Pharmacol. **12**(Suppl. 4): 591–593.
43. HOF, P. R., U. T. RUEGG, A. HOF & A. VOGEL. 1985. Stereoselectivity at the calcium channel: Opposite actions of the enantiomers of a 1,4-dihydropyridine. J. Cardiovasc. Pharmacol. **7:** 689-693.
44. FRANCKOWIAK, G., M. BECHEM, M. SCHRAMM & G. THOMAS. 1985. The optical isomers of the 1,4-dihydropyridine Bay K 8644 show opposite effects on Ca channels. Eur. J. Pharmacol. **114:** 223-226.
45. KWON, Y. W., G. FRANCKOWIAK, D. A. LANGS, M. HAWTHORN, A. JOSLYN & D. J. TRIGGLE. 1989. Pharmacologic and radioligand binding analysis of the actions of 1,4-dihydropyridine activators related to Bay K 8644 in smooth muscle, cardiac muscle and neuronal preparations. Naunyn. Schmied. Arch. Pharmacol. In press.
46. TRIGGLE, C. R., D. K. AGRAWAL, G. T. BOLGER, E. E. DANIEL, D. Y. KWAN, E. M. LUCHOWSKI & D. J. TRIGGLE. 1982. Calcium channel antagonist binding to isolated vascular smooth muscle membranes. Can. J. Physiol. Pharmacol. **60:** 1738-1741.
47. SCHIEBINGER, R. J. & K. KONTRIMUS. 1985. Dietary intake of sodium chloride in the rat influences [³H]nitrendipine binding to adrenal glomerulosa cell membranes but does not alter binding to vascular smooth muscle membranes. J. Clin. Invest. **76:** 2165-2170.
48. WEI, X.-Y., E. M. LUCHOWSKI, A. RUTLEDGE, C. M. SU & D. J. TRIGGLE. 1986. Pharmacologic and radioligand binding analysis of the actions of 1,4-dihydropyridine activator-antagonist pairs in smooth muscle. J. Pharmacol. Exp. Ther. **70:** 209-212.
49. GODFRAIND, T. & N. MOREL. 1986. Identification of Ca channels in microvessels isolated from rat brain. Br. J. Pharmacol. **89:** 507P.
50. GOPALAKRISHNAN, V., L. E. PARK & C. R. TRIGGLE. 1985. The effect of the calcium channel agonist, Bay K 8644, on human vascular smooth muscle. Eur. J. Pharmacol. **113:** 447-451.
51. PINQUIER, J.-L., S. URIEN, P. CHAUMET-RIFFAUD, A. COMPTE & J.-P. TILLEMENT. 1988. Binding of [³H]isradipine [PN 200-110] on smooth muscle cell membranes from different bovine arteries. J. Cardiovasc. Pharmacol. **11:** 402-406.
52. RAMPE, D., E. M. LUCHOWSKI, A. RUTLEDGE, R. A. JANIS & D. J. TRIGGLE. 1987. Comparative aspects and temperature-dependence of [³H]1,4-dihydropyridine Ca^{2+} channel antagonist and activator binding to neuronal and muscle membranes. Can. J. Physiol. Pharmacol. **65:** 1452-1460.
53. CHENG, J. B., A. BEWTRA & R. G. TOWNLEY. 1984. Identification of calcium antagonist receptor sites using [³H]nitrendipine in bovine tracheal smooth muscle membranes. Experientia **40:** 207-208.
54. TRIGGLE, D. J. & R. A. JANIS. 1985. Nitrendipine: Binding sites and mechanism of action. *In* Nitrendipine. A. Scriabine, S. Vanov & K. Deck, Eds.: 33-52. Urban and Schwarzenberg. Baltimore and Munich.
55. GODFRAIND, T., N. MOREL & M. WIBO. 1988. Tissue specificity of dihydropyridine-type calcium antagonists in human isolated tissues. Trends Pharmacol. Sci. **9:** 37-39.
56. RESNICK, L. M. 1987. Uniformity and diversity of calcium metabolism in hypertension. Am. J. Med. **82**(Suppl. 1B): 16-26.
57. YATANI, A., C. L. SEIDEL, J. ALLEN & A. M. BROWN. 1987. Whole-cell and single-

channel calcium currents of isolated smooth muscle cells from saphenous vein. Circ. Res. **60:** 523-533.

58. MOREL, N. & T. GODFRAIND. 1987. Prolonged depolarization increases the pharmacological effects of dihydropyridines and their binding affinities for calcium channels of vascular smooth muscle. J. Pharmacol. Exp. Ther. **243:** 711-715.

59. SUMIMOTO, K., M. HIRATA & H. KURIYAMA. 1988. Characterization of [^{3}H]nifedipine binding to intact vascular smooth muscle cells. Am. J. Physiol. **254:** C45-52.

60. HERING, S., D. J. BEECH, T. B. BOLTON & S. P. LIM. 1988. Actions of nifedipine and Bay K 8644 is dependent on calcium channel state in single smooth muscle cells from rabbit ear artery. Pflug. Arch. **411:** 590-592.

61. KOKUBUN, S., B. PRODH'HOM, C. BECKER, H. PORZIG & H. REUTER. 1986. Studies on Ca channels in intact cardiac cells: Voltage-dependent effects and cooperative interactions of dihydropyridine enantiomers. Mol. Pharmacol. **30:** 571-584.

62. WEI, X.-Y., A. RUTLEDGE & D. J. TRIGGLE. 1989. Voltage-dependent binding of 1,4-dihydropyridine Ca^{2+} channel antagonists and activators in cultured neonatal rat ventricular cells. Mol. Pharmacol. In press.

63. KASS, R. S. 1987. Voltage-dependent modification of cardiac calcium channel current by optical isomers of Bay K 8644: Implications for channel gating. Cir. Res. **61**(Suppl. 1): 1-5.

The Inhibitory Effects of Omega-Conotoxins on Ca Channels and Synapses[a]

D. YOSHIKAMI, Z. BAGABALDO,[b] AND

B. M. OLIVERA

Department of Biology
University of Utah
Salt Lake City, Utah, 84112

INTRODUCTION

The omega toxin from the venom of the marine snail *Conus geographus,* ωCgTX (subclass GVIa), is a 27-amino-acid peptide[1] that was originally described as a potent inhibitor of synaptic transmission in the frog[2] and of Ca channels in cultured dorsal root ganglion (DRG) neurons from chick.[3-5] We have examined ωCgTX's ability to block Ca channels and synaptic transmission in a wide variety of tissues from different organisms (TABLE 1). We have also studied the omega toxin from *C. magus,* ωCmTX (subclass MVIIa), a 25-amino-acid peptide with a structure homologous to that of ωCgTX[6] (see FIG. 5).

In Part I of this report we review the efficacy of ωCgTX and ωCmTX in tissues from different animals and document the effect of ωCmTX on neuromuscular transmission in frog and chick. Rapidly advancing biochemical studies on the receptors for the toxins have been recently reviewed[7] and will not be discussed here.

In Part II we scrutinize results from the frog neuromuscular junction to glean insight into the functional organization of Ca channels at the synapse. Analysis of experimental data in the light of simple theoretical models suggests that synaptic Ca channels at the frog neuromuscular junction do not act cooperatively. Instead, release at a given site appears to be mediated by the activity of a single Ca channel.

[a] This research was supported by National Science Foundation Grant BNS-8316076, National Institutes of Health Grant GM 38919, and a grant from the Research Committee of the University of Utah.

[b] On leave from the Department of Physiology, College of Medicine University of the Phillipines, Manila, Phillipines.

MATERIALS AND METHODS

Omega Toxins

ωCmTX and ωCgTX (subclass MVIIa and GVIa, respectively; see sequences in FIG. 5), purified to homogeneity by HPLC,[1] were used.

Frog Muscle

Synaptic currents were recorded extracellularly for many hours from a stable "minimuscle" system we specifically devised to obtain dose-response curves of toxins and other agents. The medial three-quarters of a cutaneus pectoris muscle from a 2.5-inch *R. pipiens* was cut away, leaving the lateral quarter of the muscle with its fibers (except for those at the cut edge) intact and innervated. Such a minimuscle had two advantages over the intact muscle. First of all, its diminutive size allowed it to be placed in a relatively small bath ($\leq 40 \ \mu$l) to conserve toxin. Second, the minimuscle consisted of a more homogeneous population of muscle fibers with respect to their diameters and locations of endplates, and their synapses were activated more synchronously upon nerve stimulation.

The minimuscle was placed in a recording chamber of dimensions $\sim 1 \times 3 \times 15$ mm fabricated from Sylgard, a silicone elastomer. The chamber could be covered with a removable Sylgard cover, which served two functions. It minimized evaporation when the bath fluid was static (during measurements in toxin), and it maintained a fixed extracellular fluid volume so that reproducible synaptic current measurements were obtained. For recording, one of a pair of bare Pt wire (250 μm dia.) electrodes was placed at the endplate region, and the other was placed at a myotendenous end of the minimuscle. Leads from these electrodes were fed to a high-gain differential AC preamplifier. The motor nerve was electrically stimulated about once every 30-60 sec with a supramaximal, 0.1-msec pulse, and data were captured and stored on an Apple II computer with an A/D board (APL-HR14 from RC Electronics, Santa Barbara, CA) or on a MacIntosh II computer with a MacAdios II board (GW Instruments, Cambridge, MA).

In experiments to obtain dose-response curves, synaptically evoked action potentials of the muscle were abolished by pretreating the preparation briefly with α-bungarotoxin (αBuTX, 0.2 μM, ~ 15 min), which permanently suppressed and maintained the endplate potential at a subthreshold level. The time necessary to achieve this was determined by monitoring currents in the muscle. αBuTX was washed out when the fast currents due to action potentials in the muscle had been abolished, leaving only the slower synaptic current, which will be referred to as the compound endplate potential, EPP (*cf.* Weakly[8]) because the entire population of muscle endplates contribute to the response. Compound EPPs with fairly constant amplitudes could be measured for hours (see e.g. FIG. 3). Intracellular recordings from sample preparations revealed that the intracellular EPP was reduced to a level < 10 mV by the αBuTX-treatment (not illustrated).

The bath was perfused at a rate of ~ 100 ml/min. with Ringer's containing 0.2 mg/ml lysozyme. Lysozyme served as a carrier protein to minimize nonspecific binding of peptide toxin to tissue. The preparation was exposed to ωCmTX by halting the perfusion and replacing the bath solution with ωCmTX-Ringer's. All measurements in ωCmTX-containing solutions were done in a static bath. In several instances, the toxin recovered from the bath was subjected to HPLC analysis (cf. Olivera et al.[1]), which showed that $\geq \frac{2}{3}$ of the toxin applied to the preparation could be recovered. This demonstrated that the majority of the toxin was free in solution.

TABLE 1. Effects of ωCgTX in Various Tissues and Species

Animal	Preparation	Effect
Pond snail	CNS neuron	None on Ca current[a]
Aplysia	Bag cell	None on Ca current[b]
Drosophila	Neuromuscular synapse	None on transmission[c]
Elasmobranch	Electroplax synaptosomes	Blocks depolarization-induced ATP release[d]
Frog	Neuromuscular synapse	Blocks transmitter release[e]
	Sympath. ganglion synapse	Blocks transmitter release[f]
	Spinal cord synapse	Blocks transmission[f]
	Cardiac muscle	Weak, if any, block of Ca current[b]
Snake	Neuromuscular synapse	Blocks transmission[c]
Chick	Cultured DRG neuron	Blocks N- & L- Ca currents irreversibly Blocks T-Ca current reversibly[b,g]
	Cultured myotube	None on Ca current[b]
	Parasympath. gangl. synapse	Blocks chemical, not electrical, transmission[c]
Adult mouse	Neuromuscular synapse	None on transmission[f]
	Symp. ganglion synapse	None on transmission[f]
Cultured rat	Cultured hippocampal neuron	Blocks some, but not all, Ca currents[b]
Guinea pig	Cardiac muscle	None on Ca current[b]

NOTE: A nominal toxin concentration of 1 μM was used in these experiments.

[a] Whole-cell voltage clamp of Lymnaea neurons (B. Yazejian & D. Yoshikami, unpublished; cf. Byerly & Hagiwara[32]).

[b] Whole-cell voltage clamp.[5]

[c] Muscle contraction evoked by motor nerve stimulation (cf. Ian & Jan[33] for Drosophila, Burden et al.[34] for snake); extracellularly recorded postganglionic nerve responses in chick ciliary ganglia evoked by preganglionic nerve stimulation (Yoshikami, unpublished; cf. Martin & Pilar[35]).

[d] High [K+]-evoked release of ATP.[11]

[e] Intracellular recording.[2]

[f] Sympathetic ganglia; extracellularly recorded postganglionic nerve responses evoked by preganglionic nerve stimulation; spinal cord: ventral root potentials in response to dorsal root stimulation; neuromuscular junction: nerve stimulation-evoked muscle contraction[12] (also Yoshikami, unpublished).

[g] Whole-cell voltage clamp.[4]

In one preparation (circles in FIG. 4 & 9), ωCgTX, instead of αBuTX, was used to irreversibly attenuate the EPPs to subthreshold levels. Because the untreated endplates in these muscle preparations have a safety factor > 10 (D. Y., unpublished), this ωCgTX-treatment "preblocked" release > 90%.

Frog Ringer's solution consisted of (in mM): 111 NaCl, 2KCl, 1.8 CaCl₂, 10 NaHEPES, pH 7.2.

Chick Muscle

The biventer cervicis muscle (*cf.* Ginsborg & Warriner[9]) from 19-day chick embryos was used. The muscle was placed in a chamber constructed from a 250-μl polyethylene (Eppendorf) microfuge tube that had been split longitudinally in half to form an elongated trough. Septa constructed of Mylar or Sylgard sheets were glued with cyanoacrylate across the trough to partition it into three serial compartments—two small ones followed by a larger one. The muscle was pinned in the third (large) compartment. The motor nerve was draped across the septa, and its cut end was pinned in the first small compartment. One of a pair of stimulating Pt wire electrodes were placed in each of the small compartments. These compartments, which contained portions of the nerve, were filled with avian Ringer's and covered with Vaseline to prevent evaporation. The composition of the avian Ringer's was (in mM) 150 NaCl, 3 KCl, 3 CaCl$_2$, 1 MgCl$_2$, 6.1 glucose, 10 NaHEPES, pH 7.3. The large compartment, in which the muscle was bathed, had a volume of ~100 μl and was perfused at a rate of 600 μl/min. Compound EPPs were measured with a Pt wire bath reference electrode and an extracellular electrode (consisting of a 250 μm diameter Pt wire coated with Sylgard except for its very tip) mounted on a micromanipulator. The ventral surface of the muscle was "scanned" with the electrode while stimulating the motor nerve in order to locate a "hot" spot where a relatively large EPP could be measured. Stimulation of the motor nerve does not evoke any action potentials in the muscle, and EPPs could be observed directly. However, there were fluctuations when synaptic responses were large (see FIG. 2) presumably due to slight movements from muscle contractures evoked by nerve stimulation, which altered the distance between the endplates and the recording electrode. All experiments, with both frog and chick preparations, were performed at room temperature.

RESULTS AND DISCUSSION

Part I: The Effects of ωCgTX and ωCmTX in Different Tissues from Different Organisms Vary, Consonant with the Notion That the Structures of Synaptic Ca Channels Have Changed in the Course of Evolution

The Range of Animals and Tissues in Which ωCgTX Is Effective

TABLE 1 catalogues the preparations from different animals in which ωCgTX has been tested physiologically. The list represents work performed by us and our collaborators, and it is not intended to be exhaustive. The susceptibilities of synapses and Ca channels to ωCgTX vary depending on the tissue and species of animal tested. Although the toxin appears to be ineffective in invertebrates, it is effective in all five families of vertebrates. The electric organ synapse in an elasmobranch is reversibly blocked by ωCgTX,[11] whereas synapses in amphibia,[2,12] reptiles, and birds (D.Y., unpublished) are irreversibly blocked by ωCgTX.

The irreversibility of ωCgTX's block at the frog neuromuscular junction is graphically illustrated in an experiment that made use of extracellular recording to monitor synaptic efficacy over many hours (FIG. 1). Intracellular recordings reveal that the

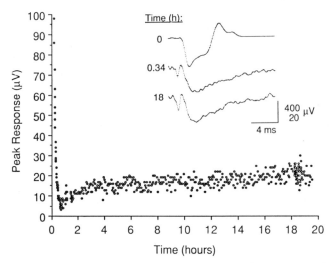

FIGURE 1. ωCgTX blocks synaptic transmission irreversibly at the frog neuromuscular junction. Responses of a cutaneus pectoris muscle to nerve stimulation were monitored by extracellular recording. The peaks of compound action potentials and compound EPPs are shown as a function of time. At time 0 the Ringer's solution in the bath was replaced with a Ringer's solution containing 1 μM ωCgTX. Before the application of toxin, nerve stimulation evoked a suprathreshold EPP which sets up an action potential in the muscle (see inset, response at 0 time). The responses at very early times after toxin application are too large to fit on the graph. The EPP gradually declined until it was subthreshold so that only the EPP without a superimposed muscle action potential was recorded (see inset, response at 0.34 hr). After about 50 minutes, the toxin was washed out. The EPP remained highly attenuated and recovered only very little, if at all (see inset: response at 18 hr). It is unclear whether the apparent slow recovery is real, because other control experiments (not illustrated) show that the EPP frequently increases with time at a rate ~5%/hour, possibly a consequence of motor nerve degeneration in the isolated preparation. **Inset:** Extracellularly recorded responses at the indicated times. The voltage calibration with the higher gain corresponds with the lower two traces. The top trace is that of a muscle action potential superimposed on a suprathreshold EPP. The lower traces show just the EPP. In each trace the stimulus was applied at the beginning of the trace, and in the lower traces (at higher gain) the "glitch" before the downward deflection of the EPP represents the extracellularly recorded action potential of the motor nerve.

ωCgTX's effect is specific—it affects neither the amplitude nor frequency of spontaneous miniature endplate potentials, nor does the toxin affect release evoked by hypertonic treatment, which increases release without involving Ca^{2+} entry.[2] Thus, ωCgTX has no postsynaptic effects, nor does it affect release *per se*. In light of the experiments discussed below that have demonstrated that ωCgTX blocks Ca channels, we conclude that ωCgTX inhibits synaptic transmission by blocking synaptic Ca channels.

Although ωCgTX has been found to be quite potent in blocking synapses in all lower vertebrates tested thus far, it is not uniformly effective in mammals. For example, it is not particularly effective, if at all, in blocking the skeletal neuromuscular junction in mice (see also Sano *et al.*[13]). In addition, the toxin is generally relatively ineffective in blocking Ca channels in muscle.

The Range of Animals and Tissues in Which ωCmTX Is Effective

The omega toxin from *C. magus*, ωCmTX, has been tested on a more limited basis. Similarities as well as differences are seen in the efficacies of ωCmTX and ωCgTX, and TABLE 2 provides a summary comparison of their functional effects. Not shown in TABLE 2 are the observations that micromolar concentrations of ωCmTX, like ωCgTX, do not block synaptic Ca currents in the pond snail *Lymnaea* (Yazejian and Yoshikami, unpublished) and synaptic transmission at the neuromuscular junction in larval *Drosophila* (Yoshikami, unpublished). Thus, neither toxin appears to be effective in invertebrates.

It should be noted that neither ωCgTX nor ωCmTX have effects when injected intraperitoneally into mice; however, they do cause a characteristic shaking behavior when injected intracranially, suggesting that they share common targets in the mammalian CNS.[6]

Comparisons of the Omega Toxin Susceptibilities of Synaptic Ca Channels and Ca Channels in Cultured Neurons

Cultured DRG neurons from chick express three classes of Ca channels: N, T, and L.[14] ωCgTX blocks N and L currents irreversibly, but the toxin's effect on T channels is reversible.[4,5] We presume that ωCgTX's block of transmitter release at synapses is explicable in terms of the toxin acting solely upon Ca channels. It should be noted that ωCgTX-block of synaptic transmission in lower vertebrates, like its block of N and L channels, is irreversible (TABLES 1 & 2). We have previously argued[5] that synaptic Ca channels are likely to be of the N type since both are insensitive to dihydropyridines unlike L channels.

TABLE 2. Comparison of the Blocking Actions of ωCmTX and ωCgTX in Different Tissues from Different Species of Vertebrates

Animal	ωCgTX	ωCmTX
Elasmobranch		
Electric organ	Reversible	Reversible
Frog		
Neuromuscular junction	Irreversible	Reversible
Sympathetic ganglion	Irreversible	Reversible
Snake		
Neuromuscular junction	Irreversible	Not tested
Chick		
Neuromuscular junction	Not tested	Irreversible
Ciliary ganglion	Irreversible	Not tested
DRG Ca Currents		
T	Reversible	Reversible
N & L	Irreversible	Partially reversible

The block of Ca currents with different ωCgTX concentrations in the micromolar range obeys pseudo-first-order kinetics.[4] This is consistent with the notion that it takes a single toxin molecule to block a Ca channel. In addition, whole-cell and patch clamp experiments indicate that the toxin's receptor is either the Ca channel itself or a closely associated component.[5]

Although the effects of ωCmTX are, by and large, similar to those of ωCgTX, they are also distinct in certain instances. ωCmTX irreversibly blocks neuromuscular transmission in chick (FIG. 2). However, whole-cell clamp experiments with chick DRG neurons reveal what appears to be two populations of N and L channels with regard to ωCmTX susceptibility. Whereas $\sim$70% of the N (and L) currents are irreversibly blocked by ωCmTX, 30% is reversibly blocked with a half-life of recovery from block of about 20 minutes.[10] Because ωCmTX-block of synaptic transmission at the chick neuromuscular junction is irreversible, we surmise that synaptic Ca channels there are of the class of N channels that are irreversibly blocked by ωCmTX.

In contrast to ωCmTX's irreversible effect at the chick neuromuscular junction, its block of the neuromuscular junction in frog is readily reversible with a half-life of block of $\sim$8 min following removal of toxin (FIG. 3). This rate is probably slower than the true rate of dissociation of the toxin from the presynaptic terminal, because wash-out of the toxin is expected to be impeded by tissue.

The steady-state dose–response curve of ωCmTX follows the curve for a rectangular hyperbola, and the toxin concentration necessary to block the EPP by a half ($[Tox]_{0.5}$) is 2 μM (FIG. 4). The amplitude of the compound EPP is expected to be a linear

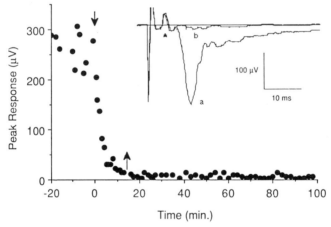

Time (min.)

FIGURE 2. ωCmTX blocks synaptic transmission irreversibly at the chick neuromuscular junction. Compound EPPs were recorded extracellularly from a cluster of endplates in a 19-day chick biventer cervicis muscle. The peaks of the EPPs are plotted as a function of time. At time 0, the bath Ringer's solution was replaced with a solution containing 5 μM ωCmTX. After 14 minutes, the toxin was washed out. The EPP was irreversibly reduced 97% by exposure to toxin [from a mean value of 263 $\pm$ 32 μV (10 responses between t = -20 and -1 min) to 8 $\pm$ 4 μV (10 responses between t = 20 and 35 min)]. The EPPs remained at a constant, low level even after washing for 84 minutes. After washing for 15 hours, responses averaging only 27 $\pm$ 9 μ V (N = 8) were recorded, suggesting that block by toxin had reversed only slightly, if at all. **Inset:** Representative responses. Trace **a,** EPP before addition of toxin (at t = -4). Trace **b,** after toxin (at t = 14 min). Asterisk: Action potential of motor nerve. The EPP was completely abolished by a Ringer's solution containing low $[Ca^{2+}]$, high $[Mg^{2+}]$ (not shown).

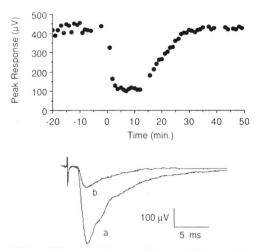

FIGURE 3. ωCmTX blocks synaptic transmission reversibly at the frog neuromuscular junction. The preparation had been pretreated briefly with α-bungarotoxin to permanently reduce the EPP to a subthreshold level (see METHODS). **Top:** The peaks of the compound EPPs are plotted as a function of time. At time 0, the bath Ringer's was replaced with Ringer's containing 7 μM ωCmTX. After 12 minutes, the toxin was washed out with plain Ringer's. During exposure to toxin, 74% of the EPP was blocked. When ωCmTX was washed out, the EPP recovered completely with a half-time for recovery of about 8 minutes. **Bottom:** Representative responses. Trace **a**, compound EPP before exposure to toxin. Trace **b**, EPP in 7 μM ωCmTX. The EPP was completely abolished by Ringer's containing high $[Mg^{2+}]$, low $[Ca^{2+}]$ (not illustrated). When traces **a** and **b** are normalized, they are superimposable (not illustrated) indicating that the time course of the EPP is not altered by toxin treatment and that there are no apparent nonlinearities in the system.

function of transmitter release, so this dose-response curve represents the block of transmitter release as a function of toxin concentration.

A frog muscle with at least 90% of release irreversibly blocked by pretreatment with ωCgTX was unaltered in its sensitivity to ωCmTX (three open circles in FIG. 4). This suggests that synaptic Ca channels are homogenous with respect to their susceptibility to ω toxin. These results are analyzed further in the light of theoretical models in Part II below.

Structure-Function Comparison of ωCmTX and ωCgTX

FIGURE 5 shows the structures of ωCmTX and ωCgTX with their sequences aligned to maximize homology. Twelve out of the 25 residues in ωCmTX are in identical locations to those in ωCgTX, which has a total of 27 residues. Of these twelve residues, six are cysteines that are disulfide bonded in the native peptides. The arrangements of the disulfide bonds that have been determined for ωCgTX[15] is assumed to apply for ωCmTX as well in FIGURE 5. Although the two peptides share an overall 50% sequence homology and presumably a similar secondary structure, they are

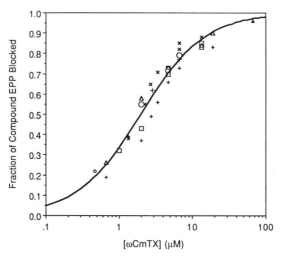

FIGURE 4. Steady-state dose-response curve of ωCmTX at the frog neuromuscular junction. The fractional reduction in the peaks of compound EPPs recorded as in the previous figure are plotted as a function of ωCmTX concentration. The different symbols represent data from different muscle preparations. The three open circles represent data from an experiment in which release was irreversibly attenuated > 90% by pretreatment with ωCgTX (see METHODS). The scatter in the data points is thought to arise in part from inaccuracies in diluting the toxin, because small volumes (in the μl range) were used in order to conserve toxin. The data points follow the curve (solid line) of a rectangular hyperbola: Fraction Blocked = $1 / (1 + [Tox]_{0.5}$ $/ [Toxin])$, with $[Tox]_{0.5}$, the [Toxin] necessary to block release by a half, = 2 μM. Overall, the results suggest that the toxin acts on a homogenous target.

otherwise quite different. It is not known at this time which residues provide the active site(s) of the toxins.

Summary

It is clear that ωCgTX and ωCmTX are effective, to varying degrees, in blocking synaptic Ca channels in a wide range of vertebrates. The variability in the toxins' potency from species to species is thought to be a reflection of the changes that the structures of the channels have undergone over the course of evolution.

Part II: Ca Channels at Synapses Do Not Appear to Act Cooperatively to Effect Transmitter Release

If we assume that ωCmTX blocks synaptic Ca channels in the frog at the rate at which it blocks N and L channels in chick DRG neurons, then the forward rate constant for block, k_1, would be $\sim 5{,}000\ M^{-1}\ sec^{-1}$ (ref. 10). If the equilibrium constant for the blocking reaction, K_i, is considered to be equivalent to $[Tox]_{0.5}$, then $K_i = 2\ \mu M$. From these values the off-rate of the toxin, k_{-1}, is calculated to be 0.01 sec^{-1}, which yields a time constant of recovery of 100 sec. Thus, ωCmTX would behave as a "slow blocker" (*cf.* Hille[16]) in that its lifetime of inhibition is orders of magnitude longer than the lifetime of the open state of the channel, which is in the msec range. Traditionally inorganic Ca channel antagonists, that is, divalent cations (Me^{2+}) such as Mg^{2+}, have been used to examine the effects of inhibiting Ca channels at synapses (e.g. Dodge & Rahamimoff[17]; for review see Silinsky[18]). In contrast to the ω toxins, Me^{2+} antagonists are fast blockers. Thus, the consequences of blocking synaptic transmission with ω toxins versus Me^{2+} antagonists may be quite different. The anticipated effects of Ca channel blockers that act rapidly and slowly with respect to the open time of the channel are qualitatively illustrated in FIGURE 6. This figure provides hypothetical sketches of the distribution of the intracellular $[Ca^{2+}]_i$, in the immediate vicinity of six Ca channels at a fixed instant in time following their activation. FIGURE 6A represents the $[Ca^{2+}]_i$ profile normally. FIGURE 6B depicts the profile when the channels are exposed to a fast Ca-channel antagonist (such as Me^{2+}) at a concentration that blocks the total Ca current (I_{Ca}) by one-half. The influx through each channel would be halved. In contrast, when the total I_{Ca} is blocked by one-half with a slow blocker of Ca channels (such as an ω toxin), half of the channels would be expected to be totally blocked or disabled, while the other half would function normally. The resulting $[Ca^{2+}]_i$ profile depicted in FIGURE 6C is, of course, distinct from that in FIGURE 6B. The differences in $[Ca^{2+}]_i$ profiles could have significant functional consequences (*cf.* Chad & Eckert[19]). Thus, as discussed below, the dose-response curves of ω toxins may be expected to differ qualitatively from those of Ca-channel antagonists such as Me^{2+}.

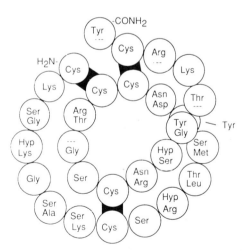

FIGURE 5. Amino acid sequences of ωCgTX and ωCmTX. The upper residue in each circle is for ωCgTX (subclass GVIa) and the lower residue for ωCmTX (subclass MVIIa). The primary structures were determined by Olivera *et al.*[6] The sequences of the peptides have been aligned to maximize their homology. The disulfide arrangement for ωCgTX was determined by Nishiuchi *et al.*[15] and that for ωCmTX is assumed to be similar. As drawn, the peptides appear to be coiled like a garden hose.

In this analysis, our working assumptions are the following:

1. The toxins block release solely by blocking Ca channels. This is a reasonable assumption because we have already demonstrated that ωCgTX action does not have any apparent effects on extracellularly recorded action potentials in the presynaptic terminal nor does it affect release *per se.*[2]

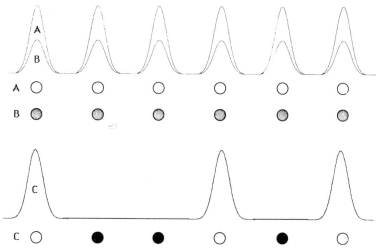

FIGURE 6. Hypothetical sketches depicting the qualitative distribution of the intracellular $[Ca^{2+}]$, $[Ca^{2+}]_i$, in the immediate vicinity of six Ca channels (represented by circles) at a given instant in time following their activation. For purposes of illustration the channels are aligned in rows at evenly spaced intervals and the amount of Ca^{2+} allowed in by each channel is the same. **A** shows the normal $[Ca^{2+}]_i$ profile. The row of open circles indicate that all six channels are active. **B** illustrates the profile in the presence of a fast channel blocker at a concentration that blocks the total Ca current (I_{Ca}) by one-half. Although all of the Ca channels are active, as indicated by the corresponding circles, the influx through each channel is halved. In contrast as shown in **C**, when the total I_{Ca} is blocked by one-half with a slow blocker of Ca channels such as an ω toxin, half of the channels would be expected to be disabled, while the other half would function normally. In effect, ω toxins would reduce the density of Ca channels, and consequently the resulting $[Ca^{2+}]_i$ profile would be quite distinct from that illustrated by **B**.

2. The toxins (TX) block by the reaction scheme:

$$\text{TX} + \text{CHANNEL}_{\text{unblocked}} \underset{k_{-1}}{\overset{k_1}{\rightleftharpoons}} \text{TX:CHANNEL}_{\text{blocked}}$$

where $K_i = k_{-1}/k_1 = ([\text{TX}]\,[\text{CHANNEL}_{\text{unblocked}}])/[\text{TX: CHANNEL}_{\text{blocked}}]$. Thus, the fraction of channels blocked, which equals the probability that a channel is blocked, is given by the equation for a rectangular hyperbola:

$$p = 1/(1 + K_i/[\text{TX}])$$

It might be noted that this is simply a Michaelis-Menten type of relationship.

This simple scheme is supported by the electrophysiological studies that suggest that ωCgTX, and presumably ωCmTX as well, blocks by acting directly on the Ca channel,[5] and, as discussed above, a single toxin molecule appears to be sufficient to block a given channel. Note that in the case of ωCgTX, which acts irreversibly, k_{-1} is zero. It might also be noted that, in practice, [TX] is constant and equals the toxin concentration in the bath, because the number of toxin molecules in the bath far exceeds the number toxin receptors (i.e., presumably channels) in the preparation.

3. All of those Ca channels that affect a common transmitter release site will be considered as part of an *ensemble*. It is assumed that all of the channels in an ensemble are gated independently.

To further our analysis, we define the following:

A. The number of channels in the ensemble is designated by N, and M is the minimum number of channels in the ensemble that must be activated (open) to effect release from a given site.

B. An ensemble is identified by M:N. For example, the 2:3 ensemble consists of three channels wherein two or more of the channels must be active to evoke release.

C. In order to block release at a site, a minimum of M' channels must be blocked, where M' = N − M + 1.

Given the above assumptions and definitions, the block of release by toxin can be related to toxin concentration by a binomial analysis. The fraction of Ca channels blocked by toxin is given by p, and the fraction not blocked is given by q = 1 − p. The fraction of release that is blocked as a consequence of the block of a given fraction of Ca channels depends on the ensemble. The fraction of release that is blocked is signified by $F_{M':N}(p)$ for the ensemble consisting of N channels wherein a minimum of M' of them have to be blocked for release to be blocked. The fraction of release that is blocked can be calculated by summing the appropriate terms in the binomial expansion. Namely,

$$F_{M':N}(p) = \sum_{i=M'}^{N} \frac{N!}{(N-i)!\,i!}\, p^i q^{(N-i)}$$

The fraction of Ca channels blocked, p, as a function of toxin concentration is given by the equation for a rectangular hyperbola:

$$p = \frac{1}{1 + 1/[\text{Toxin}]}$$

These equations have been used to plot dose-response curves of toxin for different ensembles. As shown by samples in FIGURE 7, the curves differ depending on N and M.

The different ensembles can be further distinguished by "preblocking" Ca channels with an irreversible blocker. For certain ensembles the new dose-response curve, obtained after the majority of the Ca channels have been permanently preblocked, will be shifted to the left of the curve obtained before preblocking. FIGURE 8 depicts the expected changes in the $[\text{Tox}]_{0.5}$ for various ensembles when release is preblocked by ~90%. When ~90% of release is irreversibly blocked, the remaining response would yield a dose-response curve with a $[\text{Tox}]_{0.5}$ which differs from that of the control (with no preblock) by a factor of ≥ 1/M' for ensembles where M ≠ N.

FIGURE 9 displays the steady-state dose-response curves for both ωCmTX and Cd^{2+} at the frog neuromuscular junction. As already noted in the discussion of FIGURE 4, the dose-response curve for ωCmTX is not shifted by preblocking release with ωCgTX. Thus, release does not appear to be mediated by ensembles with N ≠ M. The experimental data points in FIGURE 9 are overlayed on normalized dose-response curves for M = N ensembles with N ranging from 1 to 4. It appears that the dose-response curve for ωCmTX corresponds best with that for 1 : 1 ensemble. This leads us to propose the hypothesis that the activity of a single Ca channel mediates transmitter release at a given site.

In contrast to the curve of ωCmTX, the curve of Cd^{2+} fits best the curves for 3:3 or 4:4 ensembles. Note that Cd^{2+} blocked release by a half at a concentration of 2.5 μM, a concentration similar to $[Tox]_{0.5}$ for ωCmTX. Similarly shaped curves have been obtained with Mg^{2+}, which is much less potent (not shown). These results confirm the more extensive studies with Mg^{2+} by Dodge and Rahamimoff[17] and with Cd^{2+} by Cooper and Manalis[20] who showed that these ions behave as competitive antagonist with Ca^{2+} and affect release according to the equation (see Dodge & Rahamimoff[17]):

$$\text{release} = A \left\{ [Ca^{2+}]/(1 + [Ca^{2+}]/K_1 + [Me^{2+}]/K_2) \right\}^N$$

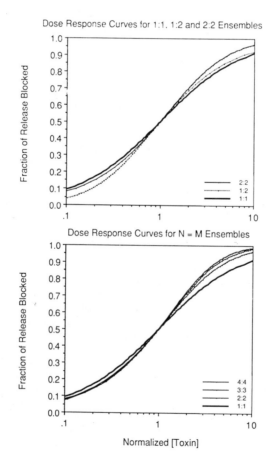

Dose Response Curves for 1:1, 1:2 and 2:2 Ensembles

Dose Response Curves for N = M Ensembles

Normalized [Toxin]

where A is a proportionality constant, K_1 and K_2 are constants, and N has a value between 3 and 4. It is shown in the paragraph below that this equation produces the same dose-response curves as that for N = M ensembles.

When [Ca^{2+}] is maintained constant, and]Me^{2+}] varied, the equation immediately above can be rewritten:

$$F = 1 - \{1/(1 + [Me^{2+}]/K_3)^N\} = 1 - \{1 - 1/(1 + K_3/[Me^{2+}])\}^N$$

where F = fraction of release blocked, and K_3 is a constant. Substitution of K_i/[Tox] for K_3/[Mg^{2+}] produces

$$F = 1 - \{1 - 1/(1 + K_i/[Tox])\}^N$$

However, p = 1/(1 + K_i/[Tox]) and q = 1 - 1/(1 + K_i/[Tox]), so

$$F = 1 - q^N$$

FIGURE 7. Normalized dose-response curves for selected ensembles. **Top:** The dose-response curves for ensembles with M = 1 or 2 and N = 2 are compared with that for the 1 : 1 ensemble. The curves were generated by summing appropriate terms in the binomial expansion and by assuming that the block of Ca channels as a function of [Toxin] follows a simple rectangular hyperbola as described in the text. The fraction of release blocked, F, as a function of the fraction of channels blocked, p, is: F = p for the 1 : 1 ensemble, F = p^2 for the 1 : 2 ensemble, and F = 1 - $(1 - p)^2$ for the 2 : 2 ensemble. The dose-response curves have been normalized so that the [Toxin] necessary to block release by a half, [Tox]$_{0.5}$, for each ensemble coincide. (The relative values of [Tox]$_{0.5}$ are 1, 1.4, and 0.4 for the 1 : 1, 1 : 2, and 2 : 2 ensembles, respectively.) At low relative [Toxin] the curve for the 2 : 2 ensemble more closely resembles that for the 1 : 1 ensemble than the 1 : 2 ensemble, because for both the 1 : 1 and 2 : 2 ensembles only one channel has to be blocked to disable the ensemble. The 1 : 2 ensemble is less readily disabled than the other two ensembles since both channels in the ensemble have to be blocked for the release site associated with the 1 : 2 ensemble to be blocked. At higher relative [Toxin] the reverse is seen, that is, the curve for the 1 : 2 ensemble resembles the 1 : 1 ensemble more closely than does the 2 : 2 ensemble. This can be appreciated by noting that if one goes "down" the dose-response curve, that is, proceed from a saturating [Toxin] to lower ones, in the case of the 2 : 2 ensemble both channels of the ensemble have to become unblocked for the release function to be restored; whereas both the 1 : 2 and 1 : 1 ensembles require that only a single channel in the ensemble be unblocked. Thus, there is a reciprocal symmetry in the curves for the 1 : 2 and 2 : 2 ensembles; that is, by inverting the curve for the 1 : 2 ensemble it can be superimposed on that for the 2 : 2 ensemble. **Bottom:** The dose-response curves for ensembles where M = N are plotted for N = 1 through 4. The fraction of release blocked as a function of the fraction of channels block is F = 1 - $(1 - p)^N$. The curves have been normalized as above. (The relative values of [Tox]$_{0.5}$ are 1, 0.4, 0.26, and 0.19 for the 1 : 1, 2 : 2, 3 : 3, and 4 : 4 ensembles, respectively.). At low relative [Toxin] the curves lie close to each other since in each case only one channel in the ensemble has to be blocked for function to be blocked. At saturating [Toxin] working down, for release function to be restored 1, 2, 3, or 4 channels in the ensemble have to become unblocked for the 1 : 1, 2 : 2, 3 : 3 and 4 : 4 ensembles, respectively. Thus, recovery of the function of the release site associated with each ensemble follows the Nth power of the recovery of Ca channel function. It might be noted that the curve for the 1 : N ensemble (where F = p^N) can be visualized by inverting the curve for the M = N ensemble. At [Toxin] > [Tox]$_{0.5}$ the normalized curve for 2 : 3 ensemble lies close to, if not above, the curve for the 3 : 3 ensemble, and the curves for the 2 : 4 and 3 : 4 ensembles lie close to, if not above, that for the 4 : 4 ensemble (not illustrated). It is evident from these curves that significant differences exist among the dose-response curves for different ensembles.

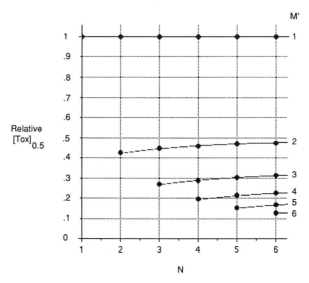

FIGURE 8. Preblocking Ca channels with an irreversible blocker can produce large changes in the dose-response curves of a reversible blocker. Plotted in the graph are the shifts in $[Tox]_{0.5}$, produced by irreversibly preblocking 90% of release, for several different ensembles. The magnitude of the shift is represented in the ordinate as the relative $[Tox]_{0.5}$, which is the ratio of $[Tox]_{0.5}$ before preblock to $[Tox]_{0.5}$ after preblock. The number of channels in the ensemble, N, is represented in the abscissa. Along the right margin of the graph are shown the values of M', the minimum number of channels in the ensemble that must be blocked to block release. The points corresponding to a given M' are connected. In every case except when M' = 1 (i.e, M = N), the $[Tox]_{0.5}$ is reduced by preblocking release. In the case where M' = 1 only a single channel has to be blocked in order for release at the site associated with the ensemble to be blocked, and so preblocking Ca channels cannot affect the likelihood that such an ensemble would become blocked when an additional channel is blocked. However, in the cases where M ≠ N, preblocking of channels would increase the likelihood that remaining active ensembles would become blocked. When 90% of release is preblocked, the effective minimal number of Ca channels that must be blocked to block release shifts, for the most part, from M' to one. Thus, following preblock, $[Tox]_{0.5}$ is reduced to approximately $1/M'$.

which is the same equation as that which describes the dose-response curves for the N = M ensembles.

Dodge and Rahamimoff[17] originally posited the notion that the dose-response curves of Mg^{2+} and Ca^{2+} reflected the cooperative action of Ca^{2+} in effecting release. That notion has been substantiated and elaborated upon with the demonstration that at the squid synapse release is dependent on presynaptic Ca current, I_{Ca}, raised to the third power.[21] The dose-response curves for Me^{2+} antagonists of Ca channels such as that in FIGURE 9, in conjunction with the expected effect of fast blockers as depicted in the sketch in FIGURE 6B, provide a picture consistent with the notion that intracellular Ca^{2+} ions cooperate to effect release.

In contrast, the data with ωCmTX suggests that the channels responsible for Ca^{2+} entry do not normally cooperate with each other in the release process.

It is important to keep in mind the salient physiological and ultrastructural features of the synapse. Classic electrophysiological studies[22] reveal that neurotransmitter is released from the presynaptic terminal in discrete quantities from discrete sites. Release requires the influx of Ca^{2+}, which enters the terminal through channels that are located very close to the sites of release, because release occurs within a fraction of a millisecond following Ca^{2+} entry,[23] and theoretical calculations suggest that Ca^{2+} has time to diffuse only a fraction of a micron in this brief time[24,25] (see Augustine *et al.*[26] for review). Ultrastructural studies of the presynaptic nerve terminal at the frog neuromuscular junction reveal that exocytosis of synaptic vesicles occur along highly organized "active zones."[27–29] A feature of the active zone that has attracted considerable attention is the presence of large intramembrane particles that, at the frog neuromuscular junction, are arrayed in sets of two parallel double rows that are located very close to sites of exocytosis.[30] It has been suggested that the large intramembrane particles represent the Ca channels responsible for mediating the entry of the calcium that triggers transmitter release (e.g. Pumplin *et al.*[31]). If this were so, then about a half-dozen channels would be located along a length equivalent to the diameter of a synaptic vesicle. Thus, conceivably several Ca channels could participate in release at a given site. Our results, however, suggest that the activity of only one channel is sufficient to effect release at a given site.

Further tests, with an improved recording chamber and improved techniques in handling small volumes of toxins, will be pursued in order to reduce the scatter of the data points in the dose-response curve with toxin. In addition, the effects of the toxins on other characteristics of synaptic transmission, such as facilitation, should be examined in light of the proposed hypothesis. Preliminary results indicate that ωCgTX treatment reduces facilitation, as might be expected of an inhibitor that reduces

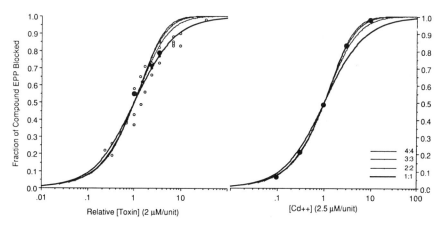

FIGURE 9. Comparison of the dose-response curves of ωCmTX and Cd^{2+} at the frog neuromuscular junction. **Left:** The data points presented in FIGURE 4 are replotted superimposed on curves for ensembles with M = N, where N = 1 to 4. The three points obtained following preblock of release are indicated by filled circles, and they show that the dose-response curve is not appreciably affected by preblock. This suggests that the M ≠ N ensembles are not likely candidates. The data points at high [Toxin] do not fit the curves well for ensembles with M = N where N > 1. However, the data do align well with the curve for the 1 : 1 ensemble, suggesting that a release site is most likely associated with a single Ca channel. **Right:** The dose-response curve for Cd^{2+}. The data are from an experiment on a single muscle preparation and fit the curve for the 3 : 3 or 4 : 4 ensemble which, as described in the text, is the same curve as that expected of a Me^{2+} inhibitor of Ca channels.

the density of active Ca channels (Yoshikami, unpublished). Finally, the generality of the observations should be tested by examining synapses with different morphologies of the presynaptic terminal.

In closing, it might be noted that the analysis presented above may also be useful in studies of other systems dependent on Ca^{2+} entry regulated by Ca channels, for example, Ca^{2+}-activated potassium and chloride channels.

SUMMARY

Omega conotoxins are peptides from snail venom. Two variants, ωCgTX and ωCmTX derived from two species of *Conus,* are the subjects of this report. PART I of this report reviews and discusses the ability of these toxins to inhibit Ca channels and synapses in different tissues from various species of animals. The potencies of these toxins vary depending on the target tissue, consonant with the notion that synaptic Ca channels have changed in the course of evolution. PART II introduces the notion that in contrast to inorganic Ca channel blockers, which act by reducing the amount of Ca^{2+} ions that can permeate an open channel, ω toxins act by reducing the availability of functional Ca channels. Thus, Ca channel-inhibition by ω toxins and that by inorganic blockers are expected to produce qualitatively different alterations in the distribution of intracellular Ca^{2+}. Consistent with this expectation, the dose-response curves of inorganic blockers and ωCmTX differ. The dose-response curves of inorganic blockers are thought to reflect the cooperativity of Ca^{2+} ions in mediating transmitter release. In contrast, comparison of experimental and theoretical dose-response curves of ωCmTX leads us to propose the hypothesis that Ca channels normally do not act cooperatively to effect transmitter release.

ACKNOWLEDGMENTS

We thank Dr. L. M. Okun for providing valuable insight and advice and Dr. D. H. Feldman for helpful discussions.

REFERENCES

1. OLIVERA, B. M., J. M. MCINTOSH, L. J. CRUZ, F. A. LUQUE & W. R. GRAY. 1984. Purification and sequence of a presynaptic peptide toxin from *Conus geographus* venom. Biochemistry **23:** 5087-5090.
2. KERR, L. M. & D. YOSHIKAMI. 1984. A venom peptide with a novel presynaptic blocking action. Nature **308:** 282-284.
3. FELDMAN, D. H. & D. YOSHIKAMI. 1985. A peptide toxin from *Conus geographus* blocks voltage-gated calcium channels. Soc. Neurosci. Abstr. **15:** 517.
4. FELDMAN, D. H., B. M. OLIVERA & D. YOSHIKAMI. 1987. Omega *Conus geographus* toxin: A peptide that blocks calcium channels. FEBS Lett. **214:** 295-300.
5. MCCLESKEY, E. W., A. D. FOX, D. H. FELDMAN, L. CRUZ, B. M. OLIVERA, R. W. TSIEN & D. YOSHIKAMI. 1987. ω-Conotoxin: Direct and persistent block of specific types of calcium channels in neurons but not muscle. Proc. Natl. Acad. Sci. **84:** 4327-4331.

6. OLIVERA, B. M., W. R. GRAY, R. ZEIKUS, J. M. MCINTOSH, J. VARGA, J. RIVIER, V. DE SANTOS & L. J. CRUZ. 1985. Peptide neurotoxins from fish-hunting cone snails. Science **230:** 1338-1343.

7. GRAY, R. W., L. CRUZ & B. M. OLIVERA. 1988. Peptide toxins from venomous *Conus* snails. Ann. Rev. Biochemistry. **57:** 665-700.

8. WEAKLEY, J. N. 1973. The action of cobalt ions on neuromuscular transmission in the frog. J. Physiol. **234:** 597-612.

9. GINSBORG, B. L. & J. WARRINER. 1960. The isolated chick biventer cervicis nerve-muscle preparation. Br. J. Pharmacol. **15:** 410-411.

10. FELDMAN, D. H., Z. BAGABALDO, B. M. OLIVERA & D. YOSHIKAMI. 1987. Omega toxin peptide from *Conus magus* blocks neuromuscular transmission and voltage-gated calcium channels in neurons. Soc. Neurosci. Abstr. **13:** 100.

11. YEAGER, R., D. YOSHIKAMI, J. RIVIER, L. J. CRUZ & G. P. MILJANICH. 1987. Transmitter release from electric organ nerve terminals: Blockade by the calcium channel antagonist, omega *Conus* toxin. J. Neurosci. **7:** 2390-2396.

12. YOSHIKAMI, D., L. M. KERR & K. S. ELMSLIE. 1983. A presynaptically targeted neurotoxin that blocks various chemical synapses in frog. Soc. Neurosci. Abstr. **9:** 882.

13. SANO, K., K. ENOMOTO & T. MAENO. 1987. Effects of synthetic ω-conotoxin, a new type of Ca^{2+} antagonist, on frog and mouse neuromuscular transmission. Eur. J. Pharmacol. **141:** 235-241.

14. NOWYCKY, M. C., A. P. FOX & R. W. TSIEN. 1985. Three types of neuronal calcium channels with different calcium agonist sensitivity. Nature **316:** 440-443.

15. NISHIUCHI, Y., K. KUMAGAYE, Y. NODA, T. X. WATANABE & S. SAKAKIBARA. 1986. Synthesis and secondary-structure determination of ω-conotoxin GVIA: A 27-peptide with three intramolecular disulfide bonds. Biopolymers **25:** S61-S68.

16. HILLE, B. 1984. Chapter 12. *In* Ionic Channels of Excitable Membranes. Sinauer Associates Inc. Sunderland, MA.

17. DODGE, F. A. & R. RAHAMIMOFF. 1967. Cooperative action of calcium ions in transmitter release at the neuromuscular junction. J. Physiol. **193:** 419-432.

18. SILINSKY, E. M. 1985. The biophysical pharmacology of calcium-dependent acetylcholine secretion. Pharmacol. Rev. **37:** 81-132.

19. CHAD, J. E. & R. ECKERT. 1984. Calcium domains associated with individual channels can account for anamolous relations of Ca-dependent responses. Biophys. J. **45:** 993-999.

20. COOPER, G. P. & R. S. MANALIS. 1984. Cadmium: Effects on transmitter release at the frog neuromuscular junction. Eur. J. Pharmacol. **99:** 251-256.

21. AUGUSTINE, G. J. & M. P. CHARLTON. 1986. Calcium dependence of presynaptic calcium current and post-synaptic response at the squid giant synapse. J. Physiol. **381:** 619-640.

22. KATZ, B. 1969. The Release of Neural Transmitter Substances. Liverpool University Press. Liverpool, England.

23. LLINAS, R. 1980. A model of presynaptic Ca^{2+} current and its role in transmitter release. *In* Molluscan Nerve Cells: from Biophysics to Behavior. J. Koester & J. H. Byrne, Eds. Cold Spring Harbor, NY.

24. SIMON, S. & R. R. LLINAS. 1985. Compartmentalization of submembrane calcium activity during calcium influx and its significance in transmitter release. Biophys. J. **48:** 485-489.

25. ZUCKER, R. S. & A. L. FOGELSON. 1986. Relationship between transmitter release and presynaptic calcium influx when calcium enters through discrete channels. Proc. Natl. Acad. Sci. **83:** 3032-3036.

26. AUGUSTINE, G. J., M. P. CHARLTON & S. J. SMITH. 1987. Calcium action in synaptic transmitter release. Ann. Rev. Neurosci. **10:** 633-693.

27. COUTEAUX, R. & M. PECOT-DECHAVASSINE. 1970. Vesicules synaptiques et poches au niveau des zones actives de la jonction neuromaculaire. C.R. Acad. Sci. (Paris) **271:** 2346-2349.

28. HEUSER, J. E., T. S. REESE & D. M. D. LANDIS. 1974. Functional changes in frog neuromuscular junctions studied with freeze-fracture. J. Neurocytol. **3:** 109-131.

29. HEUSER, J. E., T. S. REESE, M. J. DENNIS, Y. JAN, L. Y. JAN & L. EVANS. 1979. Synaptic vesicle exocytosis captured by quick freezing and correlated with quantal transmitter release. J. Cell Biol. **81:** 275-300.

30. HEUSER, J. E. & T. S. REESE. 1981. Structural changes after transmitter release at the frog neuromuscular junction. J. Cell. Biol. **88:** 564-580.
31. PUMPLIN, D. W., T. S. REESE & R. LLINAS. 1981. Are the presynaptic active zone particles the calcium channels? Proc. Natl. Acad. Sci. USA **78:** 7210-7214.
32. BYERLY, L. & S. HAGIWARA. 1982. Calcium currents in internally perfused nerve cell bodies of *Limnea stagnalis.* J. Physiol. **322:** 503-528.
33. JAN, L. Y. & Y. N. JAN. 1976. Properties of the larval neuromuscular junction in *Drosophila melanogaster.* J. Physiol. **252:** 189-214.
34. BURDEN, S. J., H. C. HARTZELL & D. YOSHIKAMI. 1975. Acetylcholine receptors at neuromuscular synapses: Phylogenetic differences detected by snake alpha-neurotoxins. Proc. Natl. Acad. Sci. USA **72:** 3245-3249.
35. MARTIN, A. R. & G. PILAR. 1963. Dual mode of synaptic transmission in the avian ciliary ganglion. J. Physiol. **168:** 433-463.

Voltage-Operated Calcium-Channel Subtypes in Human Neuroblastoma and Rat Pheochromocytoma Cells

E. SHER,[a] A. PANDIELLA,[a,b] R. M. MORESCO,[c] AND
F. CLEMENTI[a]

[a] CNR Center of Cytopharmacology and
Department of Medical Pharmacology
[b] Scientific Institute S. Raffaele
[c] School of Pharmacy
University of Milano
Milano, Italy

Plasma membrane voltage-operated calcium channels (VOCCs) are involved in such important cellular activities as excitability, contraction, and secretion.[1] Different drugs acting on these channels are currently used as therapeutic agents, particularly for peripheral disorders.[2] Much less is known, however, on the presence, characteristics, and drug susceptibility of VOCCs on neuronal cells.[3]

Most of the previous studies were carried out on mixed preparations of neuronal tissues, and a few on cultured cell lines, but rarely those of human origin. We studied VOCC characteristics on IMR32 human neuroblastoma and PC12 rat pheochromocytoma cell lines.[4]

PC12 and IMR32 cells express high-affinity binding sites for [^{125}I]ω-conotoxin (ωCTx), a potent peptide neurotoxin acting specifically on VOCCs.[5] The binding is specific and saturable (TABLE 1). IMR32 cells express binding sites for both

TABLE 1. Characteristics of ω-Conotoxin and PN200-110 Binding to Neuronal Cell Lines

A				
	PC12		IMR32	
B_{max} (fmoles/mg protein)	K_D (pM)		B_{max} (fmoles/mg protein)	K_D (pM)
7.4 ± 2	7.2 ± 1		10.7 ± 0.8	4.7 ± 1

B		
	[^{125}I]ω-Conotoxin	[^{3}H]PN200-110
	(fmoles/mg of protein)	
Control IMR32	17.0 ± 3	24.1
Differentiated IMR32	19.0 ± 0.8	24.0

249

[^{125}I]ωCTx and for the dihydropiridine antagonist [^{3}H]PN200-110 (TABLE 1), and the absolute number of binding sites for both types of ligands is unchanged in cells differentiated *in vitro* with 5-bromodeoxyuridine to obtain a more "mature" neuronal phenotype (TABLE 1).

The binding of [^{125}I]ωCTx is not antagonized by nitrendipine, verapamil, or BAY 8644 (up to 10 μM), although it is prevented by $CaCl_2$ (10 mM). Furthermore the binding of [^{125}I]ωCTx is practically irreversible. The functional properties of the VOCCS in these two cell lines was studied measuring the effect of ωCTx on [Ca^{2+}]$_i$ after KCl-induced depolarization using the fura-2 fluorimetric technique.[4] K^+ depolarization induced a biphasic rise in [Ca^{2+}]$_i$ consisting in an initial, larger [Ca^{2+}]$_i$ spike, followed by a second, sustained plateau phase. The drug sensitivity of these two components was found to be different: The initial phase can be blocked in part by verapamil, dihydropiridines, and diltiazem (TABLE 2), while the second is completely blocked by these compounds. ωCTx was found to preferentially affect the initial spike phase of the [Ca^{2+}]$_i$ transient, but had only a weak effect on the plateau phase. These results suggest a more complex heterogeneity in the gating properties of VOCCs in these neuronal cell lines whose different functioning can be studied by means of their drug sensitivity; in particular they reveal the presence of dihydropiridine-sensitive, ωCTX-insensitive VOCCs, which were thought, until now,[6-7] to be present only on muscle cells.

TABLE 2. Drug Susceptibility of VOCC in PC12 Cells

	Percent Inhibition of [Ca^{2+}]$_i$ rise, nM	
Drug	Peak[a]	Plateau[b]
KCl	—	—
KCl + Ver (50 μM)	75 ± 18	100
KCl + Dil (50 μM)	72 ± 10	100
KCl + Nit (10 μM)	79 ± 24	100
KCl + ωCTx[c] (11 μM)	61 ± 16	27 ± 12

[a] [Ca^{2+}]$_i$ (nM) was measured 15 seconds after the addition of 50 mM KCl to the cell suspension.
[b] [Ca^{2+}]$_i$ was measured four minutes after the addition of 50 mM KCl.
[c] PC12 cells were preincubated with ωCTx for 30 minutes.

REFERENCES

1. TSIEN, R. W. 1983. Annu. Rev. Physiol. **45:** 341-358.
2. URTHALER, F. 1986. Am. J. Med. Sci. **292:** 217-230.
3. MILLER, R. J. 1987. Science **235:** 46-52.
4. SHER, E., A. PANDIELLA & F. CLEMENTI. 1988. FEBS Lett. **235:** 178-182.
5. CRUZ, L. J. & B. M. OLIVERA. 1986. J. Biol. Chem. **261:** 6230-6233.
6. MCCLESKEY, E. W., A. P. FOX, D. H. FELDMAN, L. J. CRUZ, B. M. OLIVERA, R. W. TSIEN & D. YOSHIKAMI. 1987. Proc. Natl. Acad. Sci. USA **84:** 4327-4331.
7. CRUZ, L. J., D. S. JOHNSON & B. M. OLIVERA. 1987. Biochemistry **26:** 820-824.

32,000-Dalton Subunit of the 1,4-Dihydropyridine Receptor[a]

KEVIN P. CAMPBELL, ALAN H. SHARP, AND
ALBERT T. LEUNG

*Department of Physiology and Biophysics
The University of Iowa College of Medicine
Iowa City, Iowa 52242*

INTRODUCTION

The dihydropyridine receptor of the Ca^{2+} channel from skeletal muscle has been purified and characterized by several laboratories (for reviews, see Refs. 1-4). The purified dihydropyridine receptor from rabbit skeletal muscle triads prepared in our laboratory[5] consists of four polypeptides: 175,000 Da, 170,000 Da, 52,000 Da, and 32,000 Da, which have been termed α_2, α_1, β, and γ, respectively.[1] The 170,000-Da polypeptide (α_1 subunit) contains binding sites for the dihydropyridine and phenylalkylamine classes of Ca^{2+}-channel blockers.[6-9] The 175,000-Da polypeptide (α_2 subunit) is the major glycoprotein of the dihydropyridine receptor and has an apparent molecular mass of 150,000 Da on SDS-PAGE after reduction of disulfide bonds.[5] The 52,000-Da polypeptide (β subunit) is tightly associated with the dihydropyridine receptor and is not a proteolytic fragment of a larger subunit.[10] Some evidence exists for one or more small subunits of 24,000-32,000 Da, collectively termed the δ subunit,[6] that are disulfide linked to the α_2 subunit[6,11] and may account for the increase in mobility of the α_2 subunit observed on SDS-PAGE with reduction. The association of the 32,000-Da polypeptide with the receptor has been previously demonstrated only by SDS-PAGE analysis of the purified receptor. However, the 32,000-Da polypeptide has not been identified in preparations of receptor from all laboratories and its relationship to the α_1 and α_2 subunits has not been determined.

RESULTS

Subunit-Specific Polyclonal Antibodies

Dihydropyridine receptor was purified from skeletal muscle triads by WGA-Sepharose affinity chromatography followed by DEAE-cellulose ion-exchange chro-

[a]KPC is an Established Investigator of the American Heart Association and recipient of National Institutes of Health Grants HL-37187, HL-14388, and HL-39265.

matography,[5] and individual components of the receptor were separated by SDS-polyacrylamide gel electrophoresis on 5-16% gradient gels under nonreducing or reducing conditions. Individual bands were cut from the gel, homogenized in Freund's adjuvant, and used for immunization of guinea pigs for the production of ascites containing the polyclonal antibodies.[12] Interfering antibodies to keratin, a common contaminant of SDS-PAGE, were removed by passing ascites fluid through an epidermal protein-Sepharose column. Alternatively, specific anti-dihydropyridine receptor antibodies were prepared by affinity purification.[12]

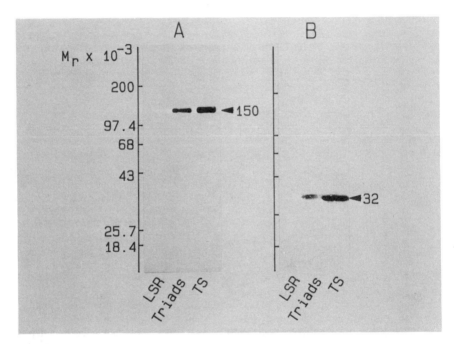

FIGURE 1. Immunoblot staining of α_2 and γ subunits in rabbit skeletal muscle membrane fractions. Isolated rabbit skeletal muscle membrane fractions, light sarcoplasmic reticulum (LSR), triads (Triads), and transverse tubular system (TS) (50 μg each) were subjected to SDS-PAGE under reducing conditions and transferred to nitrocellulose membranes. Indirect immunoperoxidase staining of the nitrocellulose blots was performed using anti-α_2 subunit antiserum (GP5SA, 1:1000 dilution) (**A**) or affinity-purified anti-γ subunit antibody (GP16AP, 1:30) (**B**). Arrowheads indicate the positions of the 150,000-Da α_2 subunit (150) (**A**) and the 32,000-Da γ subunit (32) (**B**). (From Sharp & Campbell[12] with permission of the *Journal of Biological Chemistry.*)

Immunoblot Analysis of Fractions from the Purification of the Dihydropyridine Receptor

Characterization of the 32,000-Da protein of the dihydropyridine receptor in skeletal muscle membranes was performed using affinity-purified antibodies against the nonreduced 32,000-Da protein. The anti-32,000-Da antibodies reacted with a single band of the expected molecular weight in triads (FIG. 1). For comparison, the α_2

subunit was also examined using antiserum against the reduced 150,000-Da protein. Staining for both proteins was completely absent in light sarcoplasmic reticulum membranes and was more intense in transverse tubular membranes consistent with the distribution of [³H]PN200-110 binding and our previous results on the α_1 subunit of the dihydropyridine receptor.[5] These results show the specificity of the polyclonal antibodies and demonstrate that the 32,000-Da protein is distinct from other protein components of the triad and transverse tubular membrane.

Fractions collected during a typical purification of the dihydropyridine receptor were subjected to immunoblot analysis to determine whether the 32,000-Da protein copurifies at each step with the 170,000-Da protein (α_1 subunit) and with the 150,000-Da (α_2 subunit) protein of the receptor. The 170,000-Da protein, which is known to be the dihydropyridine binding component of the receptor, was identified by staining with monoclonal antibody IIC12, which has been previously described.[5] FIGURE 2 demonstrates that the 170,000-, 150,000-, and 32,000-Da proteins are each present in triads and are enriched in the GlcNAc-eluted fraction from the WGA-Sepharose. The elution profile of the 32,000-Da polypeptide from the DEAE-cellulose column was identical to and matched the elution profile of the 170,000-Da and 150,000-Da proteins and the elution profile of dihydropyridine receptor prelabeled with [³H]PN200-110 (not shown). Thus, the 32,000-Da protein copurifies with the dihydropyridine receptor at each step of the purification.

Immunoprecipitation of the 32,000-Da Protein of the Dihydropyridine Receptor by Anti-α_1 and Anti-β Subunit Monoclonal Antibodies

To demonstrate a tight association between the 170,000-, 150,000-, 52,000-, and 32,000-Da proteins, we examined the ability of antibodies directed against the 170,000- and 52,000-Da subunits to immunoprecipitate the 150,000- and 32,000-Da proteins (FIG. 3). Digitonin-solubilized triads containing approximately 20 pmol [³H]PN200-110 binding sites per milligram protein were incubated with monoclonal antibody beads preformed as described.[5] After washing, the beads were extracted with SDS gel sample buffer and the immunoprecipitates were analyzed by immunoblot analysis (FIG. 3B). The 150,000-Da and 32,000-Da polypeptides were both present in the anti-170,000-Da and anti-52,000-Da immunoprecipitates (lanes 2 and 3, respectively). A negative control with an unrelated antibody (lane 1) and a positive control with WGA-Sepharose (lane 4) were also included. This data confirms that the 170,000-, 150,000-, 52,000-, and 32,000-Da polypeptides are tightly associated and suggests that the 150,000- and 32,000-Da proteins are subunits of the dihydropyridine receptor.

The 175,000/150,000-Da and the 32,000-Da Proteins Are Immunologically Distinct

A 32,000-Da protein that appears only after reduction of the dihydropyridine receptor and that is apparently disulfide linked to the α_2 subunit has been reported by Schmid *et al.*[11] To investigate the relationship of the 32,000-Da protein of our preparation to the protein reported by Schmid *et al.*[11] as well as to the α_1 and α_2 subunits of the dihydropyridine receptor, we have produced polyclonal antibodies

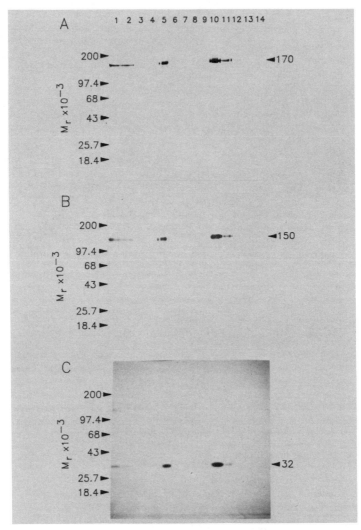

FIGURE 2. Copurification of α_1 (170,000 Da), α_2 (150,000 Da), and γ (32,000 Da) subunits of the dihydropyridine receptor. Dihydropyridine receptor was purified from rabbit skeletal muscle triads as described previously[5] by wheat germ agglutinin affinity chromatography followed by DEAE-cellulose ion-exchange chromatography. Aliquots of various fractions from the purification (50 μl each) were subjected to SDS-PAGE under reducing conditions and transferred to nitrocellulose membranes. The blots were stained by the indirect immunoperoxidase method using monoclonal antibody IIC12, anti-α_1 subunit (**A**); GP5SA, polyclonal anti-α_2 subunit antiserum (**B**); or GP16AP, affinity-purified polyclonal anti-γ subunit antibody (**C**). The samples on the transfers are: lane 1, triads; lane 2, digitonin-solubilized triads; lane 3, void fraction from WGA-Sepharose column; lane 4, wash of WGA-Sepharose column; lane 5, GlcNAc eluate from WGA-Sepharose; lane 6, void fraction from DEAE-cellulose column; lane 7, wash of DEAE-cellulose column; lanes 8-14, fractions 3, 5, 7, 9, 11, 13, and 15, respectively, from a linear 0 to 300 mM NaCl gradient elution of DEAE-cellulose column (fraction 1, no NaCl, fraction 22, 300 mM NaCl). Arrowheads indicate the positions of the 170,000-Da α_1 subunit (170) (**A**), the 150,000-Da α_2 subunit (150) (**B**), and the 32,000-Da γ subunit (32) (**C**). (From Sharp & Campbell[12] with permission of the *Journal of Biological Chemistry*.)

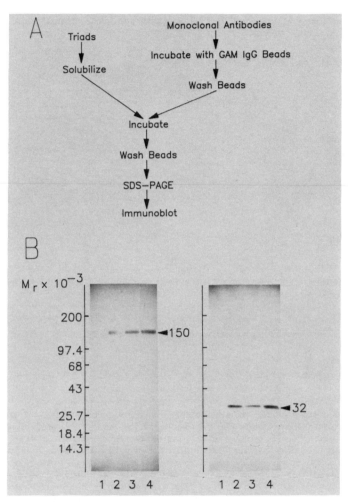

FIGURE 3. Coimmunoprecipitation of the α_2 and γ subunits by anti-α_1 and anti-β subunit antibodies. Isolated triads were solubilized with 1% digitonin and incubated with mAb-goat anti-mouse-IgG-Sepharose or WGA-Sepharose as described.[5] (**A**) Schematic diagram of immunoprecipitation procedure. (**B**) Immunoprecipitates or WGA-Sepharose precipitates were subjected to SDS-PAGE under reducing conditions and transferred to nitrocellulose membranes. Immunoprecipitation was performed using: lane 1, mAb IID8 (anti-cardiac Ca^{2+} + Mg^{2+} ATPase); lane 2, a mixture of mAbs IIID5 and IIC12 (anti-α_1 dihydropyridine receptor subunit); and lane 3, mAb VD2$_1$ (anti-β subunit). As a positive control (lane 4) precipitation was also performed using WGA-Sepharose 6MB. The nitrocellulose blots were stained using polyclonal antiserum GP5SA against the α_2 subunit (left panel) or GP16AP affinity-purified antibody against the γ subunit (right panel). Arrowheads on far right show the positions of the 150,000-Da α_2 subunit and the 32,000-Da γ subunit. (From Sharp & Campbell[12] with permission of the *Journal of Biological Chemistry.*)

against both the nonreduced and reduced forms of the 175,000/150,000-Da protein and the 32,000-Da protein. FIGURE 4 shows immunoblot analysis of the dihydropyridine receptor using these antibodies. Purified dihydropyridine receptor was subjected to SDS-PAGE under either nonreducing (20 mM N-ethylmaleimide) or reducing (10 mM dithiothreitol) conditions and transferred to Immobilon-P transfer membranes. The transfer membranes were then probed with antibodies against the nonreduced

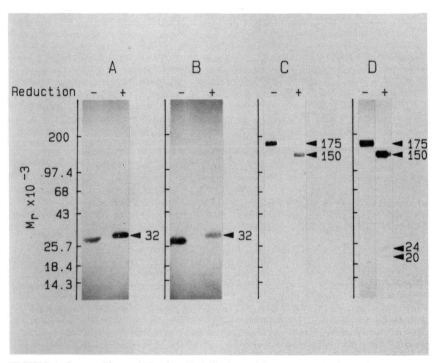

FIGURE 4. Immunoblot analysis of purified dihydropyridine receptor. Purified dihydropyridine receptor was subjected to SDS-PAGE under nonreducing (10 mM NEM) or reducing (10 mM DTT) conditions. Each lane contained 3 μg dihydropyridine receptor except B (6 μg/lane) and D (2 μg/lane). Proteins were then transferred to Immobilon-P membranes and stained by the indirect immunoperoxidase method. Blots were stained with (A) GP16AP, affinity-purified antibodies against the nonreduced 32,000-Da γ subunit; (B) GP11AP affinity-purified antibodies against the reduced 32,000-Da γ subunit; (C) GP5SA epidermal protein-column-adsorbed antiserum against the reduced 150,000-Da α_2 subunit; (D) GP13AP affinity-purified antibody against the nonreduced 175,000-Da form of the α_2 subunit. (From Sharp & Campbell[12] with permission of the *Journal of Biological Chemistry*.)

32,000-Da γ subunit, the reduced 32,000-Da γ subunit, the reduced 150,000-Da α_2 subunit, and the nonreduced 175,000-Da form of the α_2 subunit (FIG. 4, A-D, respectively). Antibodies against either the nonreduced or reduced form of the 32,000-Da protein did not stain other components of the dihydropyridine receptor on immunoblots prepared under either nonreducing or reducing conditions (FIG. 4A, B). Antibodies against the reduced α_2 subunit (150,000-Da protein) stained only the nonreduced 175,000-Da form and the reduced 150,000-Da form of the α_2 subunit (FIG. 4C) and no other components of the receptor. Antibodies against the nonreduced 175,000-Da α_2 subunit stained the nonreduced 175,000-Da and the reduced 150,000-Da forms of the α_2 subunit, but did not stain the 32,000-Da protein even after reduction of the receptor (FIG. 4D). These results show that the 32,000-Da γ subunit and 175,000/150,000-Da α_2 subunit are distinct components of the dihydropyridine receptor and that the 32,000-Da protein is not a proteolytic fragment of another component of the dihydropyridine receptor or linked by disulfide bonds to another component of the dihydropyridine receptor.

SUMMARY

Polyclonal antibodies to the 32,000-Da polypeptide of the 1,4-dihydropyridine receptor of the voltage-dependent Ca^{2+} channel have been produced and used to characterize the association of the 32,000-Da polypeptide (γ subunit) with other subunits of the dihydropyridine receptor. The 32,000-Da polypeptide was found to copurify with α_1 and α_2 subunits at each step of the purification of the dihydropyridine receptor. Monoclonal antibodies against the α_1 and β subunits immunoprecipitate the digitonin-solubilized dihydropyridine receptor as a multisubunit complex that includes the 32,000-Da polypeptide. Polyclonal antibodies generated against both the nonreduced and reduced forms of the α_2 subunit and the γ subunit have been used to show that the 32,000-Da polypeptide is not a proteolytic fragment of a larger component of the dihydropyridine receptor and not disulfide linked to the α_2 subunit. Our results demonstrate that the 32,000-Da polypeptide (γ subunit) is an integral and distinct component of the dihydropyridine receptor.

REFERENCES

1. CAMPBELL, K. P., A. T. LEUNG & A. H. SHARP. 1988. TINS **11**: 425-430.
2. CATTERALL, W. A., M. J. SEAGER & M. TAKAHASHI. 1988. J. Biol. Chem. **263**: 3535-3538.
3. GLOSSMANN, H. & J. STRIESSNIG. 1988. ISI Atlas Pharmacol. **2**: 202-210.
4. HOSEY, M. M. & M. LAZDUNSKI. 1988. J. Membr. Biol. **104**: 81-105.
5. LEUNG, A. T., T. IMAGAWA & K. P. CAMPBELL. 1987. J. Biol. Chem. **262**: 7943-7946.
6. TAKAHASHI, M., M. J. SEAGAR, J. F. JONES, B. F. X. REBER & W. A. CATTERALL. 1987. Proc. Natl. Acad. Sci. USA **84**: 5478-5482.
7. SHARP, A. H., T. IMAGAWA, A. T. LEUNG & K. P. CAMPBELL. 1987. J. Biol. Chem. **262**: 12309-12315.
8. STRIESSNIG, J., H.-G. KNAUS, M. GRABNER, K. MOOSBURGER, W. SEITZ, H. LIETZ & H. GLOSSMAN. 1987 FEBS Lett. **212**: 247-253.
9. SIEBER, M., W. NASTAINCZYK, V. ZUBOR, W. WERNET & F. HOFMANN. 1987. **167**: 117-122.
10. LEUNG, A. T., T. IMAGAWA, B. BLOCK, C. FRANZINI-ARMSTRONG & K. P. CAMPBELL. 1988. J. Biol. Chem. **263**: 994-1001.
11. SCHMID, A., J. BARHANIN, T. COPPOLA, M. BORSOTTO & M. LAZDUNSKI. 1986. Biochemistry **25**: 3492-3495.
12. SHARP, A. H. & K. P. CAMPBELL. 1989. J. Biol. Chem. **264**: 2816-2825.

Monoclonal Antibodies against Calcium Channels

ROBERT I. NORMAN,[a] ALISON J. BURGESS,[a] AND
TIMOTHY M. HARRISON[b]

[a] Department of Medicine and
[b] Department of Biochemistry
University of Leicester
Leicester LE1 7RH, United Kingdom

INTRODUCTION

The influx of calcium through voltage-sensitive calcium channels has been shown to be of central importance to the coupling of excitation to secretion and contractile processes.[1,2] These channels have been subdivided according to their electrophysiological and pharmacological properties into three classes, L, T, and N.[3] The major distinguishing property of the L-type channels is their unique sensitivity to a large group of drugs known collectively as "the calcium-channel antagonists."[3,4] L-type channels are distributed widely in excitable tissues as evidenced by both electrophysiological measurement and binding studies with radiolabeled derivatives of the calcium-channel antagonists.[5] Indeed, vital roles for these channels have been established in the stimulation of contraction in cardiac myocytes and smooth muscle,[6] and a role in the coupling of excitation and contraction in the transverse tubule-sarcoplasmic reticulum system of skeletal muscle has also been implicated.[7]

The availability of radiolabeled derivatives of the calcium-channel antagonists has greatly facilitated the characterization of this class of calcium channels.[5] Antagonist binding occurs to three allosterically coupled binding sites identified as sites for 1,4-dihydropyridines (site I, e.g., nifedipine, PN 200-110), phenylalkylamines (site II, e.g. verapamil), and benzothiazepines (site III, e.g. diltiazem).[5] The identification of radiolabeled antagonists that bind with high affinity and specificity to membrane preparations containing L-type channels, particularly members of the 1,4-dihydropyridine family, has provided useful probes for the antagonist receptors associated with voltage-sensitive calcium-channel activity.[5] Using such radiolabeled 1,4-dihydropyridine derivatives to follow receptor activity, several laboratories have reported on the purification of the L-type voltage-sensitive calcium-channel complex from skeletal muscle.[8-14] Depending on the choice of detergent and conditions employed, five putative component polypeptides have been identified. These have been designated α_1 (M_r 170,000), α_2 (M_r 140,000), β (M_r 54,000), γ (M_r 30,000), and δ (M_r 27,000).[14]

Primary sequence data for the α_1 subunit are already published.[15] This subunit shares high sequence homology with the α-subunit of the voltage-sensitive sodium channel displaying the characteristic internal repeat structure of the sodium-channel protein, with each repeat containing the six putative transmembrane segments S1-S6. Furthermore, the S4 transmembrane segment regions of the sodium-channel protein, which have been suggested to be the voltage sensors of the channel due to their high positive charge, are also highly conserved in the calcium-channel protein sequence. Such similarities between the α_1 calcium-channel protein and the α-subunit of the voltage-sensitive sodium channel have led to suggestions that the α_1 subunit constitutes both the calcium-specific pore or channel and the voltage sensor in excitation-contraction coupling, although this has yet to be established unequivocably.[15] Photoreactive derivatives of calcium-channel antagonists for all three binding sites have been shown to label specifically the α_1 polypeptide of the calcium channel.[16-20] Moreover cyclic AMP-dependent phosphorylation sites have also been identified on this component, indicating that it may also contain sites necessary for channel regulation.[20,21]

While it is clear from this data that the α_1 subunit plays a central role in calcium-channel formation, important questions regarding the functional significance of the other putative calcium-channel subunits are thereby raised. This report concentrates on characterization of the α_2 subunit and presents evidence that it is implicated in influencing antagonist binding to the calcium-channel complex.

PURIFICATION OF THE α_2 CALCIUM-CHANNEL SUBUNIT

Indications that the α_1 subunit may not constitute the entire 1,4-dihydropyridine-sensitive calcium-channel complex are suggested from reports of purifications of 1,4-dihydropyridine-binding protein preparations in which the α_1 polypeptide is apparently not present.[8-10,22-25] Attempts to purify the calcium channel in our laboratory were based on the method of Borsotto *et al.*[8] in which 1,4-dihydropyridine receptor solubilized with CHAPS detergent was purified by sequential gel filtration on Ultrogel A2 and wheat germ agglutinin affinity chromatography. In our hands receptor prepared by this method contained contaminating protein bands when analyzed by sodium dodecyl sulfate polyacrylamide gel electrophoresis. Therefore, fractions containing partially purified receptor eluting from the Ultrogel A2 column were treated with Triton X-100 before wheat germ agglutinin affinity chromatography, in order to "resolubilize" the receptor.[25] The resulting receptor preparations (0.5-1.0 nmol $(+)[^3H]PN$ 200-110 binding sites per mg protein) eluting from the wheat germ agglutinin column revealed a major staining band of M_r 140,000 (α_2) with minor components of M_r 100,000 and M_r 33,000 when analyzed by sodium dodecyl sulfate gel electrophoresis (FIG. 1). These preparations, in which the α_1 subunit was apparently absent, still retained the ability to bind 1,4-dihydropyridines specifically. Similar reports have been made by other laboratories.[8-10,22-25] Indeed, preparations containing only M_r 140,000 and M_r 30,000 components have been reported to contain the binding sites for the three most commonly used calcium channel markers[24] and even to reconstitute functional voltage-sensitive calcium channels when reincorporated into artificial phospholipid bilayers.[26]

MONOCLONAL ANTIBODIES AGAINST THE α_2 CALCIUM-CHANNEL SUBUNIT

We have raised a panel of monoclonal antibodies against the highly purified α_2 component of the 1,4-dihydropyridine receptor associated with the voltage-sensitive calcium channel of skeletal muscle.[25] Of the 31 individual clones isolated, four antibodies produced specific staining of bands in immunoblots of transverse tubule membrane proteins. In immunoblots of gels run under reducing conditions, the four antibodies bound specifically to a polypeptide of M_r 140,000 which has been designated α_2 by others (FIG. 2A). Under nonreducing conditions no specific staining was seen in the region of M_r 140,000, but larger proteins were detected. All four antibodies recognized a protein of M_r 165,000-170,000 and, with the exception of antibody α-DHP-R 11, the antibodies also recognized larger complexes of M_r 310,000-330,000

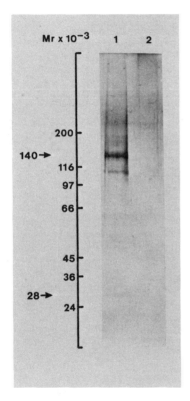

FIGURE 1. Sodium dodecyl sulfate polyacrylamide gel electrophoresis of purified 1,4-dihydropyridine receptor. Electrophoresis was carried out in 4-12% linear polyacrylamide gradient gels under reducing conditions and gels were silver stained. Lanes: 1, N-acetyl-D-glucosamine elution medium; 2, N-acetyl-D-glucosamine eluate from wheat germ agglutinin-Ultrogel column. M_r markers were: myosin (200,000), β-galactosidase (116,000), phosphorylase b (97,000), bovine serum albumin (66,000), ovalbumin (45,000), pepsin (35,000), carbonic anhydrase (29,000), and trypsinogen (24,000).

(FIG. 2B). These results were interpreted to indicate that under native conditions the α_2 subunit is disulphide-linked to a small polypeptide of approximately M_r 30,000.[25] This interpretation is supported both by immunological[27,28] and biochemical[13,14] evidence from other laboratories. In addition, the detection of larger proteins of M_r 310,000-330,000 under nonreducing conditions indicated a further complexity of the channel structure. Although our results were insufficient to distinguish between dimerization of the M_r-170,000 complex or the presence of additional polypeptide components, subsequent purification studies have shown the latter to be the case.[13,14]

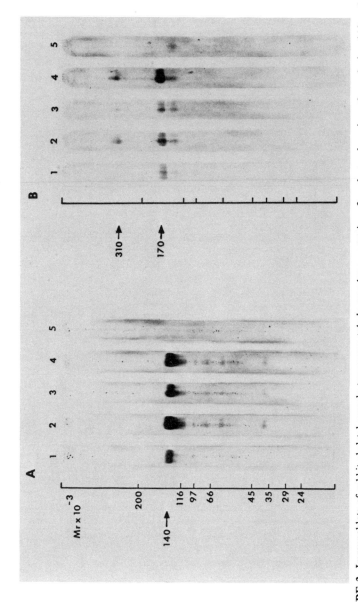

FIGURE 2. Immunoblots of rabbit skeletal muscle transverse tubule membrane proteins after electrophoresis under reducing (**A**) and nonreducing conditions (**B**). Antibody binding to the electrophoretically resolved membrane proteins was assayed by incubating the nitrocellulose paper with diluted ascitic fluids followed by a peroxidase-linked antibody detection system. Lanes: 1, α-DHP-R 11; 2, α-DHP-R 13; 3, α-DHP-R 14; 4, α-DHP-R 15; 5, anti-gyrase B control ascitic fluid (Harrison, T. M., unpublished). M_r markers as in FIGURE 1. (Reprinted from Norman *et al.*[25] by permission of *FEBS Letters*.)

IMMUNOLOGICAL CHARACTERIZATION OF THE α_2 CALCIUM-CHANNEL SUBUNIT

Using the immunoblot technique, one of the antibodies, α-DHP-R 13, has been applied to the characterization of the α_2 subunit.[25] Immunocross-reactivity has been demonstrated between the α_2 components of skeletal muscle microsomal membranes prepared from all species tested, including rabbit, rat, mouse, and frog (FIG. 3). In

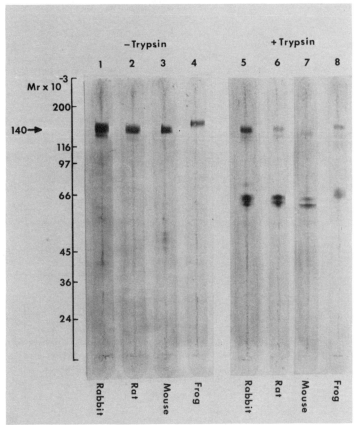

FIGURE 3. Immunoblots of CHAPS-solubilized microsomal membrane proteins from skeletal muscle of different species. α-DHP-R 13 staining on immunoblots of microsomal membrane proteins after incubation (30 min at 37°C) without enzyme (lanes 1-4) and with trypsin, 2 μg/ml (lanes 5-7) or 1 μg/ml (lane 8). Lanes 1 and 5 rabbit (160 μg protein), lanes 2 and 6 rat (100 μg protein), lanes 3 and 7 mouse (150 μg protein) and lanes 4 and 8 frog (180 μg protein). M_r markers as in FIGURE 1. (Reprinted from Norman *et al.*[29] by permission from *Biochemical Society Transactions.*)

each case, when gels were run under reducing conditions, specific staining of an α_2 component of M_r 140,000-145,000 was seen (FIG. 3, left-hand panel). Antibody staining of gels run under nonreducing conditions revealed specific staining of larger protein bands of M_r 170,000-180,000 in all cases and, with the exception of the frog membranes, additional complexes of M_r 310,000.[25] The similarity in reducing/nonreducing gel profiles of the different preparations indicates that only minor differences in the subunit molecular weights exist in each of the species investigated. Similarly, immunocross-reactivity between the α_2 subunit of rabbit skeletal muscle 1,4-dihydropyridine receptor and comparable components in rabbit heart and brain[25] and bovine chromaffin cells (unpublished results) has also been demonstrated using this antibody. Thus, although tissue-specific isotypes of the L-type calcium channel have been suggested on the basis of binding studies,[5] it is likely that the overall structures of these channel isotypes will turn out to be very similar, as will comparable channels from different species.

The structure of the α_2 component of the skeletal muscle calcium channel appears to be conserved between species not only with respect to size but also with respect to primary sequence[29] and the extent of glycosylation.[30] Similarity with respect to primary sequence has been suggested by the comparable immunoreactive band profiles obtained in immunoblots of skeletal muscle membrane proteins following limited tryptic digestion (FIG. 3, right-hand panel).[29] This suggests that the primary sequence is conserved not only with respect to antibody binding sites but also with respect to the positioning of trypsin cleavage sites.

The similarity in overall structure has also been demonstrated with respect to the extent of glycosylation of this component. A "core polypeptide" size of approximate M_r 105,000 has been deduced following essentially complete deglycosylation of the α_2 component of rabbit, rat, and mouse skeletal muscle membranes (FIG. 4). Thus, in each species tested approximately 25% of the total mass of the α_2 subunit of skeletal muscle is composed of carbohydrate residues. Deglycosylation was achieved using a chemical treatment with trifluoromethanesulfonic acid (TFMS), which hydrolyzes glycosidic bonds without significant cleavage of peptide bonds in the protein backbone.[31] These experiments were performed on membrane protein samples, because the need for prior purification was obviated by the use of antibody α-DHP-R 13, directed against the α_2 component, to identify the deglycosylated species in immunoblots. Deglycosylation was assessed by measuring the shift in apparent molecular weight of the immunoreactive species.

Repetitive treatment with TFMS did not reduce the polypeptide size below M_r 105,000, indicating that deglycosylation was apparently complete under these conditions (FIG. 4). This was confirmed by testing blots of deglycosylated protein samples for lectin binding with peroxidase-linked concanavalin A. The absence of specific staining of blots by concanavalin A-peroxidase conjugate following TFMS treatment indicated essentially complete removal of concanavalin A specific sugars from all glycosylated proteins.[30]

Interestingly, we were unable to fully deglycosylate the α_2 component using enzymatic methods even when high concentrations of glycosidase enzymes, extended incubation times and sodium dodecyl sulfate treated membrane proteins were used.[30] These results suggest that some of the carbohydrate residues are buried within the folded structure of the α_2 subunit under nonreducing conditions.

Antibody reactivity in the immunoblot technique and a combination of chemical and enzymatic treatments of membrane protein samples have contributed to the characterization of the α_2 subunit of the 1,4-dihydropyridine-sensitive calcium channel and suggest that it is structurally similar wherever it is located.

ROLE OF GLYCOSYLATED SUBUNITS IN CALCIUM-CHANNEL STRUCTURE

In view of the sequence homology of the α_1 subunit with the sodium-channel α-protein[15] and data from affinity labeling studies[16-19] that implicate a central role for

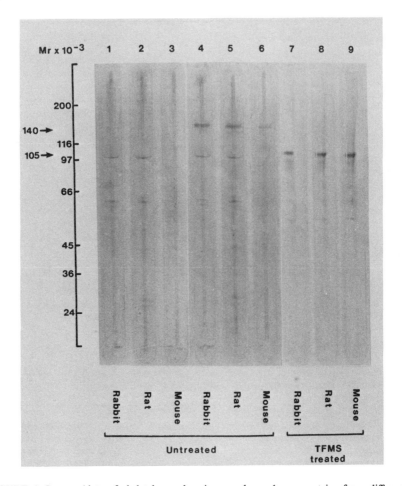

FIGURE 4. Immunoblots of skeletal muscle microsomal membrane proteins from different species following deglycosylation with trifluoromethanesulfonic acid (TFMS). TFMS treatments were as described in Burgess & Norman.[30] Anti-gyrase B control antibody (lanes 1-3) and α-DHP-R 13 (lanes 4-9) staining on immunoblots of untreated (lanes 1-6) and TFMS-treated (lanes 7-9) microsomal membrane proteins. Lanes 1, 4, and 7: rabbit (57 μg protein); lanes 2, 5, and 8: rat (100 μg protein); lanes 3, 6, and 9: mouse (48 μg protein). M_r markers as in FIGURE 1.

this polypeptide in the functional 1,4-dihydropyridine-sensitive calcium channel, questions remain concerning the degree of involvement of the glycosylated polypeptides, α_2, γ, and δ, in the channel structure. Association of the α_1 and α_2 components has been suggested both on the basis of copurification of the α_1 subunit with the lectin-bound α_2 subunit, by lentil lectin or wheat germ agglutinin affinity chromatography,[13,14] and by the coimmuneprecipitation of the two subunits from digitonin solubilized membrane extracts using anti-α_1 subunit antibodies.[14] The tightness of the association between the two subunits has, however, been questioned because they are separated in the presence of other detergents such as Triton X-100.[14] Under these conditions 1,4-dihydropyridine binding activity is lost, and hence it has been suggested, by inference, that the α_1 subunit requires the α_2 subunit to be associated for 1,4-dihydropyridine binding activity to be present.[14] We chose to investigate the involvement of the glycosylated subunits in the calcium-channel structure by observing the effect of deglycosylation of membrane proteins on the 1,4-dihydropyridine ([³H]PN 200-110) and desmethoxyverapamil ([³H]D888) binding activity of transverse tubule membranes.

Treatment of transverse tubule membranes with neuraminidase, which reduced the apparent molecular weight of the α_2 component from M_r 140,000 to M_r 130,000 on immunoblot analysis, resulted in a reduction of available 1,4-dihydropyridine and desmethoxyverapamil binding sites. After neuraminidase treatment, only 27% of the original (+)[³H]PN 200-110 binding sites and 22% of the [³H]D888 sites remained available (TABLE 1). Further deglycosylation using a mixture of neuraminidase and endoglycosidase F resulted in a broad region of antibody staining from M_r 115,000-130,000 on the blots. Only a small further decrease in specific binding sites for (+)[³H]PN 200-110 was measured under these conditions compared to neuraminidase treatment alone (TABLE 1).

Of the five putative protein subunits of the 1,4-dihydropyridine-sensitive calcium channel, three have been shown to carry significant amounts of carbohydrate, namely the large glycoprotein α_2 subunit and its disulphide-linked partner the δ subunit and the small γ subunit.[14] The sensitivity of 1,4-dihydropyridine and desmethoxyverapamil binding to enzymatic deglycosylation conditions that probably remove a substantial proportion of the terminal sialic acid and N-acetylglucosamine residues from the glycosylated subunits implicate a possible role for the heavily glycosylated α_2 subunit in specific channel, antagonist binding.[30] It should be noted that although most laboratories have reported to the contrary,[13,14] it has been suggested that the α_1 subunit bears a small amount of glycosylation.[32] Which of the subunits is responsible for the changes in antagonist binding cannot be resolved from our data. It has been proposed that binding of the 1,4-dihydropyridines to the α_1 subunit may require the presence of the glycosylated dimer of $\alpha_2\delta$.[14] At first sight, such an interpretation would be incompatible with reports of 1,4-dihydropyridine binding preparations shown to contain only M_r 140,000 and M_r 30,000 components.[8-10,22-25] However, it is known that

TABLE 1. Effect of Enzymatic Deglycosylation on Calcium-Channel Antagonist Binding to Transerve Tubule Membranes

	Calcium-Channel Antagonist Binding Activity[a]		
Ligand	Control	Neuraminidase Treated	Neuraminidase + Endoglycosidase F Treated
[³H]PN 200-110	40.3 ± 4.8 (100)	10.0 ± 2.6 (27)	9.2 ± 2.5 (22)
[³H]D888	42.2 ± 1.7 (100)	9.3 ± 2.0 (22)	n.t.

NOTE: Deglycosylation was performed as described in Burgess & Norman.[30] The apparent dissociation constants (K_D) for [³H]PN 200-110 (0.2-0.83 nM) and [³H]D888 (0.62-0.84 nM) binding were essentially unchanged following deglycosylation. Results are expressed as mean ± SEM ($n = 3$).

[a] Expressed in pmol/mg protein. Percent activity is given in parentheses; n.t. = not tested.

the α_1 subunit is much more sensitive to proteolysis by endogenous enzymes than the α_2 subunit.[14] Thus, it is possible that in these preparations the α_1 subunit is still present but in a proteolytically nicked form that retains structural integrity under nondenaturing conditions but that becomes indistinguishable after sodium dodecyl sulfate polyacrylamide gel electrophoresis.

In our experiments a proportion (approximately 25%) of the antagonist binding sites in transverse tubule membranes appeared to be relatively resistant to enzymatic deglycosylation. Because the apparent dissociation constant of the remaining sites was essentially the same as those in the original membrane sample, we conclude that these sites probably remain as a result of incomplete deglycosylation of the membrane preparation. Antagonist binding to membranes essentially completely deglycosylated using TFMS was not possible because conditions for the chemical deglycosylation are denaturing.

DEVELOPMENT OF THE α_2 CALCIUM-CHANNEL SUBUNIT

Further evidence that the α_2 polypeptide is an integral structural component of the 1,4-dihydropyridine-sensitive calcium channel has been derived from a study of its development in rat skeletal muscle.[30] Levels of the α_2 polypeptide were measured at different stages of development using the anti-α_2 subunit antibody α-DHP-R 13 to identify the polypeptide in immunoblots of skeletal muscle microsomal membranes (FIG. 5).[25] The appearance of the α_2 component was compared with the appearance of 1,4-dihydropyridine binding sites in the same samples. Levels of the α_2 subunit rose in proportion to the appearance of 1,4-dihydropyridine-binding activity from day two post partum to a plateau level at day 34. Development of the α_2 component was therefore essentially indistinguishable from the appearance of 1,4-dihydropyridine-binding activity consistent with the involvement of the large glycoprotein subunit at all stages of development of the 1,4-dihydropyridine-sensitive channel.

While our results implicate a more central role for the glycosylated subunits, and in particular the α_2 subunit, in the calcium-channel structure than has been suggested recently, further studies will be required to resolve their precise role in 1,4-dihydro-pyridine binding and indeed full calcium-channel activity. This will probably entail reconstitution experiments in which individual purified subunits are added singly or as mixtures to the central α_1 polypeptide subunit.

CONCLUSION

Antibodies against the α_2 subunit of the 1,4-dihydropyridine-sensitive calcium channel from rabbit skeletal muscle have proved useful in the characterization of this subunit. The α_2 subunit from a variety of species has been shown to be conserved with respect to antibody binding sites,[25] the positioning of trypsin cleavage sites in the primary structure,[29] and the extent of glycosylation.[30] The size of the core poly-peptide of this subunit has been determined to be M_r 105,000 indicating that this component is heavily glycosylated to approximately 25% of its total weight.[30]

The role of the α_2 component of the 1,4-dihydropyridine-sensitive calcium channel has been questioned due its apparent weak association with the α_1 subunit.[33] Immunoprecipitation experiments with anti-α_1 subunit antibodies and experiments testing the ability of lectin affinity columns to immobilize the various putative calcium-channel subunits by virtue of their association with the α_2 subunit have shown that although the five putative subunits remain associated with each other in 0.1-0.5% digitonin or 0.1% CHAPS, they are dissociated in 1% CHAPS and 0.5-1.0% Triton X-100.[14] Data suggest that the various subunits bind to the α_1 subunit with decreasing affinity in the order $\beta > \gamma > \alpha_2\delta$.[14] Such observations have led to the suggestion that the disulfide-linked glycoprotein dimer of $\alpha_2\delta$ is only weakly associated with the calcium-channel protein complex. Data from experiments to test the effect on 1,4-dihydropyridine binding activity of enzymatic deglycosylation and the demonstration that the α_2 component and 1,4-dihydropyridine binding activity codevelop in rat skeletal muscle suggest that the glycoprotein subunits, and in particular the α_2 subunit, of the calcium-channel protein complex may play a more central role in calcium-channel structure and function than has been suggested to date.

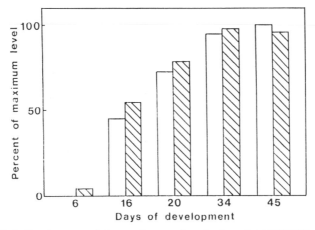

FIGURE 5. Development of the α_2 glycoprotein subunit and $(+)[^3H]PN$ 200-110 binding activity in rat skeletal muscle. Levels of α_2 subunit ($\square$) at each age were quantified on immunoblots of microsomal membrane protein (150 µg protein) by sequential incubation with α-DHP-R 13, biotinylated anti-mouse immunoglobulin G and $[^{125}I]$streptavidin as described in Burgess & Norman.[30] $(+)[^3H]$ PN 200-110 binding activity ($\boxtimes$) was determined by equilibrium binding analysis.

REFERENCES

1. HAGIWARA, S. & L. BYERLY. 1981. Annu. Rev. Neurosci. **4:** 69-125.
2. TSIEN, R. W. 1983. Annu. Rev. Physiol. **45:** 341-358.
3. MCCLESKY, E. W., A. P. FOX, D. FELDMAN & R. W. TSIEN. 1986. J. Exp. Biol. **124:** 177-190.
4. TRIGGLE, D. J. & R. A. JANIS. 1987. Annu. Rev. Pharmacol. Toxicol. **27:** 347-369.
5. GLOSSMANN, H., D. R. FERRY, A. GOLL, J. STREISSNIG & G. ZERNIG. 1985. Arzneim. Forsch. Drug Res. **35:** 1917-1935.

6. FLECKENSTEIN, A. 1977. Ann. Rev. Pharmacol. Toxicol. **17:** 149-166.
7. RIOS, E. & G. BRUM. 1987. Nature **325:** 717-720.
8. BORSOTTO, M., J. BARHANIN, R. I. NORMAN & M. LAZDUNSKI. 1984. Biochem. Biophys. Res. Commun. **122:** 1357-1366.
9. BORSOTTO, M., J. BARHANIN, M. FOSSET & M. LAZDUNSKI. 1985. J. Biol. Chem. **260:** 14255-14263.
10. CURTIS, B. M. & W. A. CATTERALL. 1984. Biochemistry **23:** 2113-2118.
11. FLOKERZI, V., H-J. OEKEN, F. HOFMANN, D. PELZER, A. CAVALIE & W. TRAUTWEIN. 1986. Nature **323:** 66-68.
12. SIEBER, M., W. NASTAINCZYK, V. ZUBOR, W. WERNET & F. HOFMANN. 1987. Eur. J. Biochem. **167:** 117-122.
13. LEUNG, A. T., T. IMAGAWA & K. P. CAMPBELL. 1987. J. Biol. Chem. **262:** 7943-7946.
14. TAKAHASHI, M., M. J. SEAGER, J. F. JONES, B. F. X. REBER & W. A. CATTERALL. 1987. Proc. Natl. Acad. Sci. USA **84:** 5478-5482.
15. TANABE, T., H. TAKESHIMA, A. MIKAMI, V. FLOCKERZI, H. TAKAHASHI, K. KANGAWA, M. KOJIMA, H. MATSUO, T. HIROSE & S. NUMA. 1987. Nature **328:** 313-318.
16. GALIZZI, J-P., M. BORSOTTO, J. BARHANIN, M. FOSSET & M. LAZDUNSKI. 1986. J. Biol. Chem. **261:** 1393-1397.
17. SHARP, A. H., T. IMAGAWA, A. T. LEUNG & K. P. CAMPBELL. 1987. J. Biol. Chem. **262:** 12309-12315.
18. STREISSNIG, J., H-G. KNAUS, M. GRABNER, K. MOOSBURGER, W. SEITZ, H. LEITZ & H. GLOSSMANN. 1987. FEBS Lett. **212:** 247-253.
19. VAGHY, P. L., J. STREISSNIG, K. MIWA, H-G. KNAUS, K. ITAGAKI, E. MCKENNA, H. GLOSSMANN & A. SCHWARTZ. 1987. J. Biol. Chem. **262:** 14337-14342.
20. IMAGAWA, T., A. T. LEUNG & K. P. CAMPBELL. 1987. J. Biol. Chem. **262:** 8333-8339.
21. NASTAINCZYK, W., A. ROHRKASTEN, M. SIEBER, C. RUDOLPH, C. SCHACHTELE, D. MARNE & F. HOFMANN. 1987. Eur. J. Biochem. **169:** 137-142.
22. NAKAYAMA, H., R. M. WITHY & M. A. RAFTERY. 1982. Proc. Natl. Acad. Sci. USA **79:** 7575-7579.
23. NAKAYAMA, N., T. L. KIRLEY, P. L. VAGHY, E. MCKENNA & A. SCHWARTZ. 1987. J. Biol. Chem. **262:** 6572-6576.
24. BARHANIN, J., T. COPPOLA, A. SCHMID, M. BORSOTTO & M. LAZDUNSKI. 1987. Eur. J. Biochem. **164:** 525-531.
25. NORMAN, R. I., A. J. BURGESS, E. ALLEN & T. M. HARRISON. 1987. FEBS Lett. **212:** 127-132.
26. MCKENNA, E. J., J. S. SMITH, J. MA, J. VILVEN, P. L. VAGHY, A. SCHWARTZ & R. CORONADO. 1987. Biophys. J. **51:** 29.
27. SCHMID, A., J. BARHANIN, T. COPPOLA, M. BORSOTTO & M. LAZDUNSKI. 1986. Biochemistry **25:** 3492-3495.
28. VANDAELE, S., M. FOSSET, J-P. GALIZZI & M. LAZDUNSKI. 1987. Biochemistry **26:** 5-9.
29. NORMAN, R. I., A. J. BURGESS, E. ALLEN & T. M. HARRISON. 1987. Biochem. Soc. Trans. **15:** 895-896.
30. BURGESS, A. J. & R. I. NORMAN. 1988. Eur. J. Biochem. **178:** 527-533.
31. EDGE, A. S. B., C. R. FALTYNEK, L. HOF, L. E. REICHER, JR. & P. WEBER. 1981. Anal. Biochem. **118:** 131-137.
32. HOSEY, M. M., J. BARHANIN, A. SCHMID, S. VANDAELE, J. PTASIENSKI, C. O'CALLAHAN, C. COOPER & M. LAZDUNSKI. 1987. Biochem. Biophys. Res. Commun. **147:** 1137-1145.
33. HOFMANN, F., W. NASTAINCZYK, A. ROHRKASTEN, T. SCHNEIDER & M. SIEBER. 1987. Trends Pharmacol. Sci. **8:** 393-398.

Antibodies against Calcium Channels in the Lambert-Eaton Myasthenic Syndrome

D. W.-WRAY,[a] B. LANG,[b] J. NEWSOM-DAVIS,[b]
AND C. PEERS[a]

[a]Department of Pharmacology
Royal Free Hospital School of Medicine
Rowland Hill Street
London NW3 2PF, England

[b]Institute for Molecular Medicine
John Radcliffe Hospital
Oxford OX3 9DU, England

INTRODUCTION

The Lambert-Eaton myasthenic syndrome (LEMS) is a presynaptic disorder of neuromuscular transmission in which patients show skeletal muscle weakness due to a reduction in the evoked release of acetylcholine (ACh).[1] The electrophysiological features seen are : (1) reduced quantal content (the number of packets of ACh released per nerve impulse) and (2) facilitation of endplate potential (epp) amplitudes at high-frequency nerve stimulation. We have previously shown that injecting mice with the immunoglobulin G (IgG) fraction of LEMS patients' plasma reproduced these features in the animals, thus providing strong evidence for an autoimmune basis for this disorder.[2,3] The site of action of LEMS IgG at the nerve terminal has been investigated. It does not have its main effect on action potentials in the nerve terminal because it still reduces ACh release even in the absence of action potentials (in solutions containing high K^+ concentration).[4] Furthermore LEMS IgG does not act intraterminally as indicated by the lack of effect of LEMS IgG in experiments using Ca^{2+}-free solutions[5] or agents known to affect intraterminal processes (such as ouabain).[6] The mechanism of action of LEMS IgG does not appear to involve late complement components, since injecting LEMS IgG into mice genetically deficient in the fifth component of complement still transferred the disorder.[7] Here we review evidence showing that the site of action of LEMS IgG is the Ca^{2+} channel at motor nerve terminals[5] and in cultured cells,[8] while the most likely mechanism of action involves cross-linking[9] of channels by IgG antibody leading to loss of channels.

METHODS

The IgG fraction of plasma from LEMS patients and from control subjects was prepared using the Rivanol-ammonium sulphate method.[10] For studies of neuromuscular transmission, mice were injected intraperitoneally (i.p.) with 60 mg/day of IgG for two days. The diaphragm muscle was then removed and intracellular recordings of miniature endplate potentials (mepps) and endplate potentials (supramaximal phrenic nerve stimulation at 0.5 Hz) were made at room temperature (mean 23°C) using glass microelectrodes as previously described.[3] Monovalent (Fab) and divalent (F(ab')$_2$) antibody fragments were prepared from the IgG fraction using papain or pepsin digestion, respectively. Isolated diaphragms were incubated *in vitro* in culture medium containing 2-4 mg/ml antibody fragment at room temperature for 24 hours before electrophysiological recordings were made. All data were stored on magnetic tape and subsequently analyzed by computer; epp quantal content being obtained by Poisson statistics.

The effects of LEMS IgG were also investigated directly on neuronal calcium-channel currents in the mouse neuroblastoma $\times$ rat glioma hybrid cell line NG 108 15. For this, undifferentiated cells were incubated with 2-4 mg/ml of LEMS or control IgG for 24-48 hours in culture medium at 37°C. The cells were then transferred to a HEPES-buffered physiological salt solution containing 10 mM Ba^{2+} as charge carrier. The medium also contained 2.5 μM tetrodotoxin and 25 mM tetraethylammonium to block current flow through Na$^+$ and K$^+$ channels, respectively. Whole-cell patch-clamp recordings were made at 24-26°C using electrodes containing an intracellular medium with Cs$^+$ replacing K$^+$ to further suppress current flowing through K$^+$ channels (composition [mM]: CsCl, 140; EGTA, 1.1; MgCl$_2$, 2; CaCl$_2$, 0.1; HEPES, 10). Cells were voltage clamped using an Axoclamp 2 amplifier (switching frequency 5-10 kHz) at holding potentials of -80 mV and -40 mV. Ca^{2+} channel currents were evoked in response to 200-msec depolarizing voltage steps applied at a frequency of 0.1 Hz. Current and voltage traces were stored on magnetic tape and later analyzed by computer. Current amplitudes were measured following linear leakage and capacitance subtraction.

Values shown are means $\pm$ SE mean, and the two-tailed Student's t-test was used to test for statistical significance.

RESULTS AND DISCUSSION

Effects of LEMS IgG at the Motor Nerve Terminal

FIGURE 1 shows epp quantal content measurements made in mice injected with control or LEMS IgG, over a wide range of Ca^{2+} concentrations. For the lower Ca^{2+} concentration range, LEMS IgG caused significant reductions in quantal content as compared with controls. However, at higher Ca^{2+} concentrations. the curves for both control and LEMS IgG-treated muscles began to approach the same saturating maximum so that at these high Ca^{2+} concentrations quantal content values were similar. Such a shift to the right in the quantal content versus Ca^{2+} concentration curve by LEMS IgG is what would be expected for an antagonist to Ca^{2+} entry at the nerve terminal, such as Mg^{2+} ions.[11]

Quantal release of ACh can be evoked in the absence of action potentials by raising extracellular K$^+$ concentration. FIGURE 2 shows a plot of ACh release (mepp frequency) at fixed high K$^+$ concentration (15.9 mM) for a range of Ca^{2+} concentrations measured in control and LEMS IgG-treated muscles. For controls, mepp frequency increased with increasing Ca^{2+} concentration up to a plateau level. For LEMS IgG treated muscles a similar pattern was seen, but at each Ca^{2+} concentration studied, mepp frequency was significantly reduced as compared with controls, so that LEMS IgG simply caused a downward shift of the curve from control values. Thus the

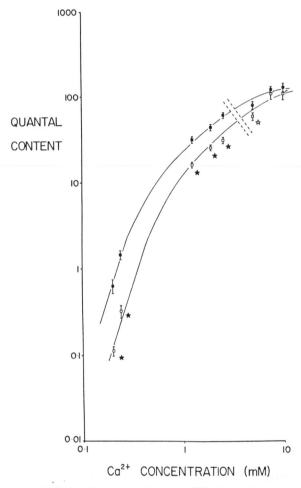

FIGURE 1. Effect of LEMS IgG on quantal content. This figure shows quantal content for mice treated with control IgG (■) and with IgG from an LEMS patient (□). Each point is the mean ± S.E. mean of sixteen to eighteen endplates from three mice. ★ $p < 0.001$, ☆ $p < 0.05$: significant reductions by LEMS IgG. For further details see Lang *et al.*[5]

saturating maximum at high Ca^{2+} concentrations was reduced, although the slope (2.2) at low Ca^{2+} concentrations was not affected. Such a downward shift by LEMS IgG is similar to the effect that might be expected for an irreversible Ca^{2+} antagonist acting at the terminal.

The extent of reduction in Ca^{2+} entry by LEMS IgG can be estimated from the shifts in the release versus Ca^{2+} concentration curves. For nerve-evoked release (FIG. 1) the shift to the right was by a Ca^{2+} concentration ratio of 1.6, while for K^+-evoked release (FIG. 2) in the low Ca^{2+} concentration range it was 1.4. Thus, the Ca^{2+} concentration had to be increased by 1.4-1.6-fold in the presence of LEMS IgG

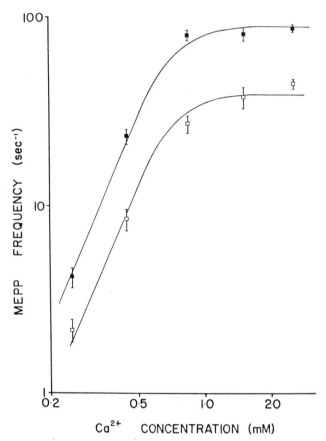

FIGURE 2. Effect of LEMS IgG on mepp frequency at high K^+ concentration. This figure[5] shows mepp frequency for mice treated with control IgG (■) and with IgG from two LEMS patients (□). Each point is the mean ± S.E. mean of eighteen to thirty-three endplates from three to six mice. Mepp frequency was reduced ($p < 0.002$) by LEMS IgG at each Ca^{2+} concentration studied.

to produce an effect similar to controls. Assuming Ca^{2+} entry is proportional to Ca^{2+} concentration,[12] LEMS IgG therefore appears to reduce Ca^{2+} entry to 63-71% (i.e., 1/1.6-1/1.4) of control values.

It is perhaps perplexing that for nerve-evoked release (FIG. 1) there was a rightward shift by LEMS IgG, while for K^+-evoked release (FIG. 2) there was a downward shift. It is important to recall that an irreversible antagonist can produce a rightward shift in the presence of "spare" Ca^{2+} channels and a downward shift otherwise. In contrast to K^+-evoked release, for nerve-evoked release the instantaneous rate of Ca^{2+} entry is high leading to saturation of intraterminal release processes at high Ca^{2+} concentrations, and hence the appearance of "spare" Ca^{2+} channels.[13] Thus the data is consistent with an action of LEMS IgG as an irreversible antagonist at Ca^{2+} channels.

This irreversible antagonism could in principle come about by blocking Ca^{2+} channels or by their loss (downregulation). However, the effect of LEMS IgG in blocking Ca^{2+} entry does not occur rapidly; muscles exposed to the antibody for up to two hours show quantal content values similar to controls.[7] It is unlikely, therefore, that LEMS IgG causes its effect by direct pharmacological block of Ca^{2+} channels. More likely is a longer term effect, such as antibody-induced channel loss. One possible way by which this may occur is by the antibody cross-linking adjacent Ca^{2+} channels, thus triggering their degradation. Such a possibility was investigated by measuring quantal content in muscles treated with whole LEMS antibody and with divalent $(F(ab')_2)$ or monovalent (Fab) LEMS antibody fragments. These were all applied *in vitro* as fragments are rapidly excreted from the body. Such *in vitro* application for 24 hours of whole LEMS IgG was as effective as i.p. injections in reducing quantal content in mouse muscles. Thus it caused a significant ($p < 0.001$) reduction in quantal content from 145 ± 11 ($n = 22$ endplates) in controls to 84 ± 7 ($n = 21$). *In vitro* application of divalent LEMS fragment $(F(ab')_2)$ also caused significant reductions ($p < 0.001$) in quantal content as compared with controls (from 145 ± 13, $n = 17$ to 59 ± 5, $n = 14$). Monovalent LEMS antibody fragment (Fab), however, did not significantly affect quantal content (controls 142 ± 12, $n = 21$; test 127 ± 26, $n = 10$). The divalent structure of LEMS IgG therefore appears to be essential in order to cause reductions in quantal content. This suggests that LEMS IgG cross-links adjacent Ca^{2+} channels to trigger their downregulation. These findings are supported by (i) immuno-electron microscopic localization of LEMS IgG binding at active zones,[14] and (ii) by electron microscope freeze-fracture studies of nerve terminals where active zone particles (the putative Ca^{2+} channels[15]) move closer together and are reduced in number by i.p. injections of mice with LEMS IgG.[14] Furthermore, a similar effect on active-zone particles is also seen following 24 hours *in vitro* application to mouse diaphragms of either whole LEMS IgG or divalent LEMS antibody fragment, but not following monovalent LEMS antibody fragment treatment.[16] Thus, the effect of LEMS IgG at the nerve terminal is to cause cross-linking of adjacent Ca^{2+} channels, leading to their degradation and loss. Our data on reduction of Ca^{2+} entry (see above) suggests that the extent of this loss in the number of Ca^{2+} channels is by 29-37%, similar to the observed loss in active-zone particles.[17]

Effects of LEMS IgG on Neuronal Ca^{2+}-Channel Currents

To look more directly at the effect of LEMS IgG on Ca^{2+} channels, whole cell patch clamp recordings were made from NG 108 15 cells after incubation with control or LEMS IgG. When cells were voltage-clamped at a holding potential of -40 mV, depolarizing step potentials evoked sustained inward currents with characteristics similar to those expected for L-type channels.[18] FIGURE 3 shows the current-voltage relationship for these currents measured in control and LEMS IgG-treated cells. Over a wide range of test potentials, LEMS IgG caused significant reductions in current amplitudes as compared with controls, indicating that the antibody acts to reduce current flow through L-type Ca^{2+} channels. This was supported by the finding that, in the presence of 5 μM nitrendipine, which blocks L-type Ca^{2+} channels, there was no significant difference in sustained current amplitudes between control and LEMS IgG-treated cells.[19] Indeed, in the presence of nitrendipine, sustained currents were outward at high test potentials, presumably due to the incomplete block of other ionic channels under the conditions used. Although there appears to be a shift in the reversal potential by LEMS IgG in the absence of nitrendipine (FIG. 3), the presence of additional outward currents as well as inward Ca^{2+}-channel currents distorts the

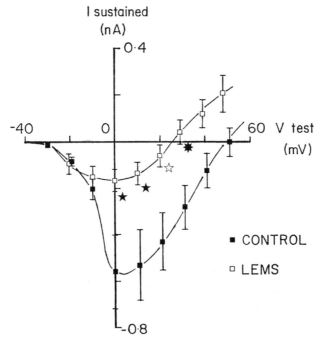

FIGURE 3. Effect of LEMS IgG on sustained inward currents (I sustained) at different test membrane potentials (V test). All currents were evoked from a holding potential of -40 mV. Values plotted[19] are mean ± S.E. mean from eighteen cells for control IgG (■) and from seventeen cells for LEMS IgG (□). ★ $p < 0.001$, ☆ $p < 0.005$, ✳ $p < 0.002$: significant reductions by LEMS IgG.

curves at large positive test potentials, and it is likely that there is no real shift in reversal potential by LEMS IgG.

Cells were also voltage clamped at a holding potential of -80 mV, and small depolarizing steps applied from this holding potential evoked only transient inward currents, presumably due to the activation of T-type channels.[18] These were unaffected by LEMS IgG; for instance, for voltage steps to -30 mV, the mean transient inward current was 0.63 ± 0.12 nA ($n = 25$ cells) for LEMS IgG, which was not significantly different from controls (0.65 ± 0.12 nA, $n = 25$). For larger step depolarizations from a holding potential of -80 mV, sustained components of the current were also seen as well as the transient component. These sustained currents (presumably due to L-type Ca^{2+} channels) were again reduced in amplitude in LEMS-treated cells.[19] Thus, in conclusion, all our data on these cells is consistent with an action of LEMS IgG to selectively block current flow through L-type channels while leaving T-type channels unaffected.

CONCLUDING REMARKS

In summary, we have shown that LEMS IgG acts on the presynaptic nerve terminal Ca^{2+} channels at the neuromuscular junction to reduce evoked quantal release of ACh. The antibody appears to act by cross-linking adjacent Ca^{2+} channels, which in turn leads to their downregulation. Also, we have shown that LEMS IgG selectively reduces current flow through L-type Ca^{2+} channels in the neuronal cell line NG 108 15, without affecting T-type channels.

The type of Ca^{2+} channel that is concerned with transmitter release at the mammalian motor nerve terminal has not yet been clearly identified. Data at present suggest that they are not L-type channels; for example organic Ca^{2+}-channel antagonists and agonists have little or no effect on nerve-evoked or high-K^+-evoked quantal release of ACh at the neuromuscular junction.[20] However, the cross-reactivity of LEMS antibody with L-type Ca^{2+} channels in NG 108 15 cells and with Ca^{2+} channels at mammalian motor nerve terminals suggests that antigenic similarities exist between these two types of channel.

LEMS IgG not only appears to cross-react with Ca^{2+} channels at motor nerve terminals and NG 108 15 cells, but also with Ca^{2+} channels in other neuronal type tissues. So for instance, LEMS IgG also cross-reacts with Ca^{2+} channels in chromaffin cells (L-type channels)[21] and in pituitary cells.[22] Furthermore, approximately 70% of LEMS patients have associated small-cell carcinoma of the lung, and it appears that in these patients the primary antigenic determinants are the Ca^{2+} channels present in these small-cell carcinoma cells that are derived from the neuroectoderm.[23] Moreover, many LEMS patients also have autonomic dysfunction (e.g., dry mouth, impotence)[24] possibly due to an action on Ca^{2+} channels at parasympathetic postganglionic nerve endings. On the other hand, L-type Ca^{2+} channels in mouse ventricular muscle appear to be unaffected by LEMS IgG,[25] as are Ca^{2+} channels in insect skeletal muscle fibers.[26] Thus, LEMS antibody appears to bind to a channel determinant that is specific for neuronal-type tissues.

The preferential action of LEMS antibody on neuronal tissues indicates that a common determinant on the Ca^{2+}-channel protein is shared between certain neuronal channel subtypes, but may be missing from other subtypes and from channels in nonneuronal tissues. Thus LEMS antibody may be useful in the further characterization of channel types.

REFERENCES

1. LAMBERT, E. H. & D. ELMQVIST. 1971. Quantal components of endplate potentials in the myasthenic syndrome. Ann. N.Y. Acad. Sci. **183**: 183-199.
2. LANG, B., J. NEWSOM-DAVIS, D. WRAY, A. VINCENT & N. MURRAY. 1981. Autoimmune aetiology for myasthenic (Eaton-Lambert) syndrome. Lancet **ii**: 224-226.

3. LANG, B., J. NEWSOM-DAVIS, C. PRIOR & D. WRAY. 1983. Antibodies to nerve terminals: An electrophysiological study of a human myasthenic syndrome transferred to mouse. J. Physiol. **344:** 335-345.

4. LANG, B., J. NEWSOM-DAVIS, C. PEERS & D. WRAY. 1985. Mechanism of action of human autoantibodies interfering with acetylcholine release in the mouse. J. Physiol. **365:** 79P.

5. LANG, B., J. NEWSOM-DAVIS, C. PEERS, C. PRIOR & D. W.-WRAY. 1987. The effect of myasthenic syndrome antibody on presynaptic calcium channels in the mouse. J. Physiol. **390:** 257-270.

6. LANDE, S., B. LANG, J. NEWSOM-DAVIS & D. WRAY. 1985. Site of action of Lambert-Eaton myasthenic syndrome antibodies at mouse motor nerve terminals. J. Physiol. **371:** 61P.

7. PRIOR, C., B. LANG, D. WRAY & J. NEWSOM-DAVIS. 1985. Action of Lambert-Eaton myasthenic syndrome IgG at mouse motor nerve terminals. Ann. Neurol. **17:** 587-592.

8. LANG, B., J. NEWSOM-DAVIS, C. PEERS & D. W.-WRAY. 1987. Selective action of Lambert-Eaton myasthenic syndrome antibodies on Ca^{2+} channels in the neuroblastoma $\times$ glioma hybrid cell line NG 108 15. J. Physiol. **394:** 43P.

9. LANG, B., J. NEWSOM-DAVIS, C. PEERS & D. W.-WRAY. 1987. The action of myasthenic syndrome antibody fragments on transmitter release in the mouse. J. Physiol. **390:** 173P.

10. HOREJSI, J. & R. SMETANA. 1956. The isolation of gamma globulins from blood serum by Rivanol. Acta Med. Scand. **155:** 65-70.

11. JENKINSON, D. H. 1957. The nature of the antagonism between calcium and magnesium ions at the neuromuscular junction. J. Physiol. **138:** 434-444.

12. SILINSKY, E. M. 1985. The biophysical pharmacology of calcium-dependent acetylcholine secretion. Pharmacol. Rev. **37:** 81-131.

13. SILINSKY, E. M. 1981. On the calcium receptor that mediates depolarization-secretion coupling at cholinergic motor nerve terminals. Br. J. Pharmacol. **73:** 413-429.

14. ENGEL, A. G., T. FUKUOKA, B. LANG, J. NEWSOM-DAVIS, A. VINCENT & D. WRAY. 1987. Lambert-Eaton myasthenic syndrome IgG: Early morphologic effects and immunolocalization at the motor endplate. Ann. N.Y. Acad. Sci. **505:** 333-345.

15. PUMPLIN, D. W., T. S. REESE & R. LLINAS. 1981. Are the presynaptic membrane particles the calcium channels? Proc. Natl. Acad. Sci. USA **78:** 7210-7213.

16. A. NAGEL, T. FUKUOKA, H. FUKUNAGA, M. OSAME, B. LANG, J. NEWSOM-DAVIS, A. VINCENT, D. W.-WRAY & C. PEERS. 1989. Motor nerve terminal calcium channels in Lambert-Eaton myasthenic syndrome: Morphologic evidence for depletion and that the depletion is mediated by autoantibodies. Ann. N.Y. Acad. Sci. This volume.

17. FUKUNAGA, H., A. G. ENGEL, B. LANG, J. NEWSOM-DAVIS & A. VINCENT. 1983. Passive transfer of Lambert-Eaton myasthenic syndrome with IgG from man to mouse depletes the presynaptic membrane active zones. Proc. Natl. Acad. Sci. USA **80:** 7636-7640.

18. NOWYCKY, M. C., A. P. FOX & R. W. TSIEN. 1985. Three types of neuronal calcium channel with different calcium agonist sensitivity. Nature **316:** 440-443.

19. LANG, B., J. NEWSOM-DAVIS, C. PEERS & D. W.-WRAY. 1989. Selective action of myasthenic syndrome antibodies on calcium channels in the rodent neuroblastoma $\times$ glioma cell line NG 108 15. In preparation.

20. BURGES, J. & D. W.-WRAY. 1989. Effect of the calcium channel agonist CGP 28392 on transmitter release at mouse neuromuscular junctions. Ann. N.Y. Acad. Sci. This volume.

21. KIM, Y. I. & E. NEHER. 1988. IgG from patients with Lambert-Eaton syndrome blocks voltage-dependent calcium channels. Science **439:** 405-408.

22. LOGIN, I. S., Y. I. KIM, A. M. JUDD, B. L. SPANGELO & R. M. MACLEOD. 1987. Immunoglobulins of Lambert-Eaton myasthenic syndrome inhibit rat pituitary hormone release. Ann. Neurol. **22:** 610-614.

23. ROBERTS, A., S. PERERA, B. LANG, A. VINCENT & J. NEWSOM-DAVIS. 1985. Paraneoplastic myasthenic syndrome IgG inhibits $^{45}Ca^{2+}$ flux in a human small cell carcinoma line. Nature **317:** 737-739.

24. RUBENSTEIN, A. E., S. H. HOROWITZ & A. N. BENDER. 1979. Cholinergic dysautonomia and Eaton-Lambert syndrome. Neurology **29:** 720-723.

25. LANG, B., J. NEWSOM-DAVIS & D. W.-WRAY. 1988. The effect of Lambert-Eaton myasthenic syndrome antibody on slow action potentials in mouse cardiac ventricle. Proc. R. Soc. B **235:** 103-110.

26. PEARSON, H. A., J. NEWSOM-DAVIS, G. LEES & D. W.-WRAY. 1989. Lack of action of Lambert-Eaton myasthenic syndrome antibody on calcium channels in insect muscle. Ann. N.Y. Acad. Sci. This volume.

Motor Nerve Terminal Calcium Channels in Lambert-Eaton Myasthenic Syndrome

Morphologic Evidence for Depletion and That the Depletion Is Mediated by Autoantibodies[a]

ANDREW G. ENGEL,[b] ALEXANDRE NAGEL,[b]
TADAHIRO FUKUOKA,[b,c] HIDETOSHI
FUKUNAGA,[b,c] MITSUHIRO OSAME,[b,c]
BETHAN LANG,[d] JOHN NEWSOM-DAVIS,[d]
ANGELA VINCENT,[d] DENNIS W.-WRAY,[e] AND
CHRISTOPHER PEERS[e]

[b] *Department of Neurology and Muscle Research Laboratory*
Mayo Clinic
Rochester, Minnesota 55905

[d] *Institute of Molecular Medicine*
University of Oxford
John Radcliffe Hospital
Oxford OX3, 9DU, England

[e] *Department of Pharmacology*
Royal Free Hospital School of Medicine
London, NW3 2PF, England

The Lambert-Eaton myasthenic syndrome (LEMS) is caused by a loss of voltage-sensitive calcium channels (VSCCs) from the motor nerve terminal (MNT). The disease is mediated by auto antibodies that bind to VSCCs and reduce their number by antigenic modulation. This paper summarizes the morphologic evidence for the loss of VSCCs from the MNT and the autoimmune origin of LEMS. To set these data in perspective, we begin with a brief overview of our current understanding of the syndrome.

[a] This work was supported by Grant NS 6277 from the National Institutes of Health, a Research Center Grant from the Muscular Dystrophy Association, and by the British Medical Research Council.

[c] Present address: 3rd Department of Internal Medicine, Kagoshima University School of Medicine, Kagoshima 890, Japan.

The characteristic symptoms are weakness and abnormal fatigue of the trunk muscles and mostly of the proximal limb muscles. Cranial muscles tend to be spared. Strength facilitates at the start of contraction. Autonomic disturbances, such as decreased salivation, lacrimation and sweating, orthostatism, impotence, and abnormal pupillary light reflexes, occur in about 75% of the cases. Two-thirds of LEMS cases are associated with neoplasm, and more than 80% of the neoplasms are small-cell carcinomas of the lung (reviewed by Engel[1] and O'Neill *et al.*[2]).

The electrophysiologic basis of LEMS is a decreased release of acetylcholine quanta from the MNT by nerve impulse.[3,4] The decreased quantal release is not due to reduced synthesis or storage of acetylcholine, because the acetylcholine content and choline acetyltransferase activity of LEMS muscle are normal.[5] Repetitive stimulation or raising the external Ca^{2+} concentration, which increase the probability of quantal release, improve the transmission defect.[4]

Indirect evidence for an autoimmune origin of LEMS includes a favorable response to immunosuppressants and plasma exchange, an association with other autoimmune disorders, HLA-B8 and DRw3 antigens, and organ-specific autoantibodies (reviewed by O'Neill *et al.*[2]). The passive transfer to mice with IgG of the electrophysiologic[6§8] and morphologic[9-12] features of the disease provides direct evidence for an autoimmune cause of LEMS.

FREEZE-FRACTURE ELECTRON MICROSCOPY STUDIES OF THE MOTOR NERVE TERMINAL

Paucity of Active Zone Particles on the MNT in Human LEMS[13]

The initial morphologic studies set out to test the hypothesis that the reduced quantal release from the MNT by nerve impulse in LEMS was caused by a depletion or disorganization of the active zone particles (AZPs) of the MNT. The hypothesis was based on the fact that opening of VSCCs increases the probability of quantal release[14-17] and the assumption that the AZPs comprise the VSCCs. Four reasons for assuming that the VSCCs and AZPs were identical structures are: (1) Synaptic vesicle exocytosis captured by quick freezing occurs near the AZPs.[18] (2) AZPs tend to specify sites of exocytosis even when dispersed by exposure to a low-Ca^{2+} medium.[19] (3) At the squid giant synapse, the latency between the onset of the calcium current and transmitter release is only 0.2 msec, and during this time Ca^{2+} ions could diffuse only 150 nm from their site of entry into the MNT.[20] (4) At the same synapse, the maximum presynaptic calcium current appears to be related to the number of AZPs.[21]

Visualization of the AZPs requires freeze-fracture electron microscopy. In the freeze-fractured MNT the AZPs are readily recognized as large particles, 10-12 nm in diameter, packed in double parallel arrays. Each array consists of two parallel rows, and each row contains a number of AZPs (FIG. 1). Freeze-fracture analysis, however, presents a number of problems: the MNTs, which occupy only a small fraction of a muscle sample, are hard to find in freeze-fracture replicas; the AZPs are best observed in face-on views of the protoplasmic (P) leaflet of the presynaptic membrane, but only one in four fractures traversing MNTs yields this view. Further, the density (number per unit area) of AZPs cannot be determined in a conventional electron micrograph because the projection of a curving membrane surface on a two-dimen-

FIGURE 1. Freeze-fracture electron micrograph of control presynaptic membrane P-face viewed face on. Ten active zones can be identified (arrows). Each zone contains two arrays; each array consists of two parallel rows of large particles. The active zones vary in length and not all rows in a given zone have the same number of particles. Asterisk indicates primary synaptic cleft. Magnification: ×100,000.

sional micrograph reduces the membrane area. These problems were resolved by fracturing many muscle segments enriched in MNTs, stereo imaging all MNTs found in the replicas, and estimating the presynaptic membrane area by a computer-assisted stereometric method.[13,22]

The freeze-fracture studies revealed several striking alterations in the LEMS MNTs: The density of the active zones and AZPs was markedly reduced; the number of particles per active zone was also reduced; and a proportion of the remaining AZPs were aggregated into clusters (FIGS. 2A, 2B, and 3). The AZPs were selectively affected, for no change occurred in the density or size distribution of those membrane particles that were not associated with either active zones or clusters.[13]

From these results and the assumption that the AZPs represent VSCCs, the physiologic defect in LEMS could now be attributed to the loss of VSCCs from the MNT.[13] This study was published in 1982 from the Muscle Research Laboratory at the Mayo Clinic. By this time, Lang and coworkers at the Royal Free Hospital have transferred the physiologic defect of LEMS to mice with IgG.[6] Therefore, one could infer that the AZPs (i.e., the VSCCs) (1) are targets of the LEMS autoantibodies, (2) aggregate because they are cross-linked by IgG, and (3) are depleted by the mechanism of antigenic modulation.[13] Subsequent joint studies by the two laboratories obtained adequate evidence for each of these postulates.

The AZPs Are Targets of LEMS Autoantibodies[9]

That the membrane lesions in LEMS are mediated by IgG was established by freeze-fracture studies on diaphragm muscles of mice treated with 10 mg control or LEMS IgG per day for 27-75 days. The LEMS IgG was from three patients. IgG from patients one and two transferred the electrophysiologic defect to mice, but IgG from patient three did not. The morphologic study, which was done blindly, demonstrated clustering and depletion of AZPs only in those mice that received IgG from patients one and two.

The AZPs Are Sufficiently Close Together to Be Cross-Linked by the Two Antigen-Binding Sites of IgG[10,11]

The spacing of AZPs in normal mouse diaphragm MNTs was determined by freeze-fracture analysis. To obviate the foreshortening of distances in two-dimensional images of three-dimensional membranes, a computer program was written for measuring interparticle distances in stereo-pair electron micrographs using the x, y, and z coordinates of each point. This study established that the least distance between AZPs in an outer row (7.3 nm), inner row (4.6 nm), or in two adjacent rows (9.6 nm) of an array is less than the 14 nm distance between the two antigen-binding sites of IgG (FIGS. 4A and 5A). Therefore, the AZPs of a given array are sufficiently close together to be cross-linked by IgG. However, the least distance between particles in two adjacent arrays of an active zone (27.8 nm) cannot be spanned by IgG as long as the rectilinear and parallel orientation of the inner rows of the arrays is maintained.

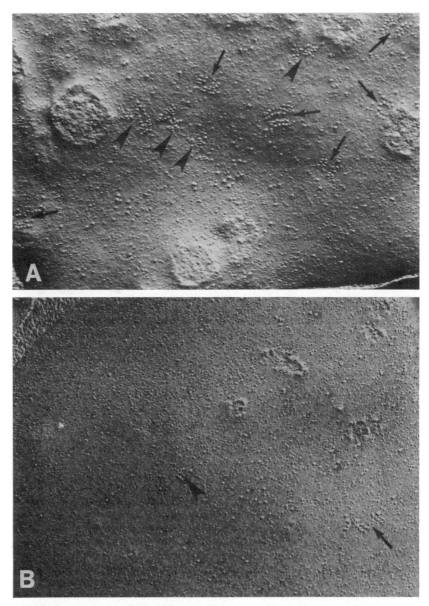

FIGURE 2. LEMS presynaptic membrane P-faces. In (**A**), membrane leaflet shows six recognizable active zones (arrows) and four clusters of particles (arrowheads). In (**B**), entire membrane contains only one short active zone and a single cluster of large particles. Compare with FIGURE 1. Magnification: (A) ×59,800; (B) ×61,100. (From Fukunaga *et al.*[13] Reprinted by permission.)

Aggregation of AZPs Is an Early Effect of LEMS IgG[10,11]

In mice treated with 120-180 mg pathogenic LEMS IgG over two days, the initial alteration in the active zone was a decrease in the distance between particles in a given row and between adjacent rows of an array; the distance between the two arrays of the active zone remained unaltered (FIGS. 4B and 5B). In more affected active zones, the parallel orientation of the rows was disturbed (FIGS. 4B and 5C) and the arrays became clusters (FIGS. 4B and 5D). After two days of treatment there was also a significant (31%) decrease in the density of large membrane particles found in active zones and clusters. The endplate potential quantal content in the same diaphragm muscles was reduced by 45-58% below the control mean.

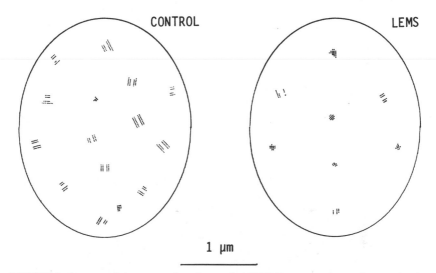

FIGURE 3. Stereometric reconstruction of control and LEMS presynaptic membranes, showing numbers, sizes, and configurations of active zones and numbers and sizes of particle clusters. The analysis is based on 83 control and 93 LEMS samples. A 5-μm^2 region is imaged in each case. Control membrane contains 13 active zones and two clusters. LEMS membrane contains three active zones and five clusters. The active zones are smaller and the clusters are larger in LEMS than in control membrane. (From Fukunaga *et al.*[13] Reprinted by permission.)

Divalency of LEMS IgG Is an Essential Requirement for the Aggregation and Depletion of AZPs[12]

All the above morphologic data were consistent with modulation of AZPs crosslinked by LEMS IgG. If this were the case, then only divalent LEMS IgG and F(ab')$_2$ should aggregate and deplete the AZPs and monovalent LEMS Fab should be without effect. To test this hypothesis, mouse diaphragms were exposed to control and LEMS IgG and IgG fragments in organ culture for 24 hours and then studied by quantitative

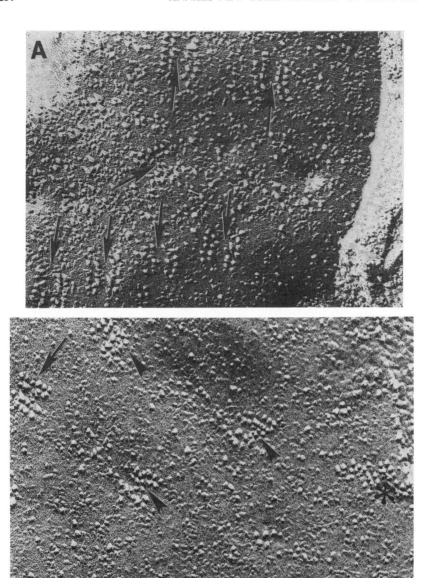

FIGURE 4. Presynaptic membrane P faces from diaphragm muscles of mice treated for two days with 120 mg of control IgG (**A**) and pathogenic LEMS IgG (**B**). Numerous normal active zones are present in (**A**) (arrows). (**B**) contains one normal-appearing active zone (arrow) and several abnormal active zones (arrowheads). In the abnormal zones the particles in individual rows are aggregated, but the two arrays remain distinct. A particle cluster, in which the space between the arrays is obliterated, is also imaged (asterisk). Magnification of (A) and (B) ×117,000. (From Engel et al.[10] Reprinted by permission.)

freeze-fracture electron microscopy. Divalent LEMS IgG and F(ab')₂ aggregated and depleted the active-zone particles, whereas monovalent Fab was without effect. The findings reconfirmed that LEMS IgG binds to the AZPs and gave direct evidence for modulation of the AZPs by LEMS IgG. The findings were also in harmony with parallel electrophysiologic studies of the effects of LEMS IgG fragments on transmitter release in the same diaphragm muscles.[23]

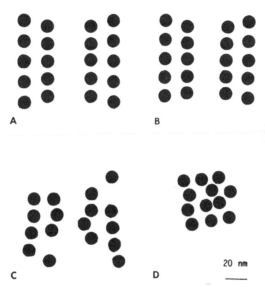

FIGURE 5. Stereometric reconstruction of the spacing and arrangement of particles in normal active zones of control mice (**A**) and in mice treated with LEMS IgG (B, C, and D). (**B**) shows reconstruction of active zones judged to be normal by simple inspection. Exact measurements, however, reveal slight reduction of particle spacing within the two arrays. (**C**) is a reconstruction of active zones within more pronounced particle aggregation. (**D**) represents a cluster. Calibration mark at the lower right indicates 20 nm. (From Engel et al.[10] Reprinted by permission.)

IMMUNOELECTRON MICROSCOPY LOCALIZATION OF LEMS IgG AT THE MOTOR ENDPLATE[10,24]

The conclusion drawn from the freeze-fracture studies was that LEMS IgG binds to the AZPs and reduces their density by antigenic modulation. This was further substantiated by the immunolocalization of LEMS IgG to the active zones. The mouse passive transfer model was used. Special problems in this study were the paucity of AZPs (about $50/\mu m^2$ normally and still lower in LEMS) and diffusion artifacts in the immunoperoxidase method. These difficulties were obviated by the use of (1) sensitive avidin-biotin detection systems, (2) both peroxidase and ferritin labels (FIGS. 6 and 7), and (3) quantitative immunoelectron microscopy and endplate morphometry. Mice treated with LEMS IgG, control IgG, and no IgG were compared. In all mice

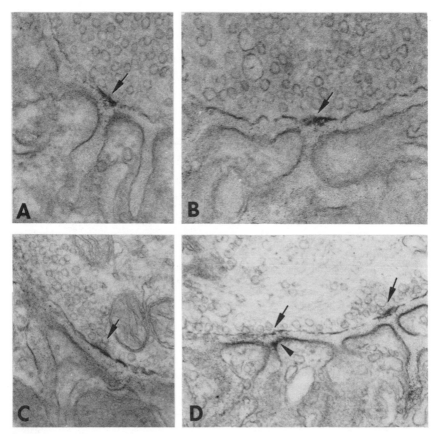

FIGURE 6. Immunoperoxidase localization of human IgG in mice treated with single intravenous (**A, B and C**) or multiple intraperitoneal (**D**) injections of LEMS IgG. Arrows indicate immunostained active zones. Postsynaptic staining vis-a-vis positive active zones (arrowhead in (**D**)) is attributed to a diffusion artifact. Magnification: (A) ×50,300; (B) ×58,200; (C) ×44,000; (D) ×43,800. (From Fukuoka *et al.*[24] Reprinted by permission.)

nonspecific background staining was found in the basal lamina covering the muscle fibers and Schwann cells. When a single dose of 10 mg IgG was injected intravenously, IgG samples from 12 patients did (FIGS. 6A-6C) and from seven patients did not produce significant immunostaining of the mouse active zones. Higher doses of IgG injected intraperitoneally (20 mg three times a day for two days, or 10 mg a day for 15 days) from each of four patients caused significant immunostaining of mouse active zones (FIGS. 6D and 7): (1) the mean density (No./μm presynaptic membrane length) of positive active zones was 0.91 in the immunoferritin and 0.72 in the immunoperoxidase study (control values: 0.12 and 0.02); (2) 43% of the ferritin particles in the primary synaptic cleft were concentrated at the active zones, and the rest were scattered randomly (control value: 5.3%) (FIG. 8). These findings clearly established that LEMS IgG binds to presynaptic membrane active zones.[10,24]

THE AZPs AND VSCCs OF THE MNT ARE IDENTICAL STRUCTURES

Recent electrophysiological studies also suggest that LEMS IgG causes a loss of VSCCs from the MNTs. When mice are treated with LEMS IgG, spontaneous quantal release from MNTs depolarized by high K^+ concentrations is significantly reduced over a range of Ca^{2+} concentrations.[25] Further, LEMS IgG has now been shown to reduce functional VSCCs in small-cell carcinoma cells,[26] anterior pituitary cells,[27] adrenal chromaffin cells,[28] and neuroblastoma-glioma hybrid cells.[29] These observations, earlier studies suggesting that AZPs comprise the VSCCs of the MNT,[18–21] and the freeze-fracture data that indicate that LEMS IgG depletes the AZPs [9–13] are sufficient evidence to assume that the AZPs and the VSCCs of the MNT are identical

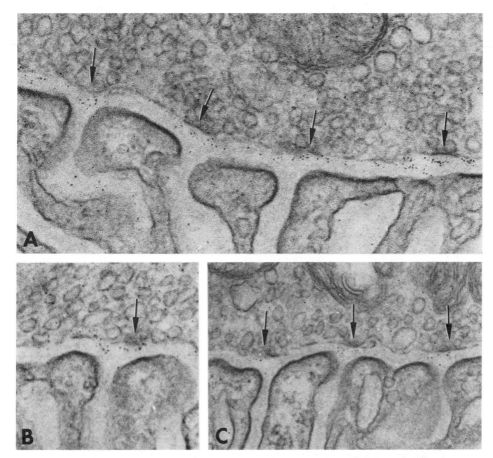

FIGURE 7. Immunoferritin localization of human IgG in mice treated with multiple intraperitoneal doses of LEMS IgG. The ferritin particles in the primary clefts are concentrated near the active zones (arrows). Magnification: (A) ×81,900; (B) ×81,100; (C) ×60,200. (From Fukuoka *et al.*[24] Reprinted by permission.)

structures. In the neuroblastoma-glioma hybrid cells, LEMS IgG has a selective effect on L-type but not T-type Ca^{2+} channels.[29] This, however, does not mean that the VSCCs of the MNT are L-type channels, as LEMS IgG might recognize homologous and antigenically similar domains in different types of Ca^{2+} channels.[29]

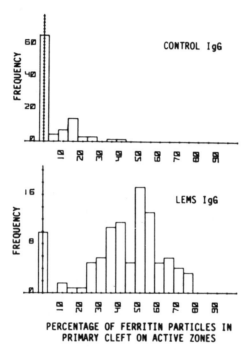

FIGURE 8. Frequency distribution of the percentage of ferritin particles in the primary clefts associated with active zones in mice treated with multiple doses of control IgG (70 endplate regions) and with LEMS IgG (122 endplate regions). Each endplate region contains one primary synaptic cleft.

REFERENCES

1. ENGEL, A. G. 1986. Myasthenic syndromes. *In* Myology. A. G. Engel & B. Q. Banker, Eds.: 1955-1990. McGraw-Hill. New York.

2. O'NEILL, J. H., N. M. F. MURRAY & J. NEWSOM-DAVIS. 1988. The Lambert-Eaton myasthenic syndrome. A review of 50 cases. Brain **111:** 577-596.

3. ELMQVIST, D. & E. H. LAMBERT. 1968. Detailed analysis of neuromuscular transmission in a patient with the myasthenic syndrome sometimes associated with bronchogenic carcinoma. Mayo Clin. Proc. **43:** 689-713.

4. LAMBERT, E. H. & D. ELMQVIST. 1971. Quantal components of end-plate potentials in the myasthenic syndrome. Ann. N.Y. Acad. Sci. **183:** 183-199.

5. MOLENAAR, P. C., J. NEWSOM-DAVIS, R. L. POLAK & A. VINCENT. 1982. Eaton-Lambert syndrome: Acetylcholine and choline acetyltransferase in skeletal muscle. Neurology **32:** 1061-1065.

6. LANG, B., J. NEWSOM-DAVIS, D. W.-WRAY, A. VINCENT & N. MURRAY. 1981. Autoimmune aetiology for myasthenic (Lambert-Eaton) syndrome. Lancet **2:** 224-226.
7. LANG, B., J. NEWSOM-DAVIS, C. PRIOR & D. W.-WRAY. 1983. Antibodies to motor nerve terminals: An electrophysiological study of a human myasthenic syndrome transferred to mouse. J. Physiol. (London) **344:** 335-345.
8. KIM, Y. I. 1985. Passive transfer of the Lambert-Eaton myasthenic syndrome: Neuromuscular transmission in mice injected with plasma. Muscle Nerve **8:** 162-172.
9. FUKUNAGA, N., A. G. ENGEL, B. LANG, J. NEWSOM-DAVIS & A. VINCENT. 1983. Passive transfer of Lambert-Eaton myasthenic syndrome IgG from man to mouse depletes the presynaptic membrane active zones. Proc. Natl. Acad. Sci. USA **80:** 7636-7640.
10. ENGEL, A. G., T. FUKUOKA, B. LANG, J. NEWSOM-DAVIS, A. VINCENT & D. W.-WRAY. 1987. Lambert-Eaton myasthenic syndrome IgG: Early morphological effects and immunolocalization at the motor endplate. Ann. N.Y. Acad. Sci. **505:** 333-345.
11. FUKUOKA, T., A. G. ENGEL, B. LANG, J. NEWSOM-DAVIS, C. PRIOR & D. W.-WRAY. 1987. Lambert-Eaton myasthenic syndrome: I. Early morphologic effects of IgG on the presynaptic membrane active zones. Ann. Neurol. **22:** 193-199.
12. NAGEL, A., A. G. ENGEL, B. LANG, J. NEWSOM-DAVIS & T. FUKUOKA. 1988. Lambert-Eaton syndrome IgG depletes presynaptic membrane active zone particles by antigenic modulation. Ann. Neurol. **24:** 552-558.
13. FUKUNAGA, H., A. G. ENGEL, M. OSAME & E. H. LAMBERT. 1982. Paucity and disorganization of presynaptic membrane active zones in the Lambert-Eaton myasthenic syndrome. Muscle Nerve **5:** 686-697.
14. KATZ, B. & R. MILEDI. 1967. The timing of calcium action during neuromuscular transmission. J. Physiol. (London) **189:** 535-544.
15. KATZ, B. & R. MILEDI. 1967. The release of acetylcholine from nerve endings by graded electrical pulses. Proc. Roy. Soc. B (London) **167:** 23-28.
16. LLINAS, R. & C. NICKELSON. 1975. Calcium in depolarization secretion coupling: An aequorin study in squid giant synapse. Proc. Natl. Acad. Sci. USA **72:** 187-190.
17. CHARLTON, M. P., S. J. SMITH & R. ZUCKER. 1982. Role of presynaptic calcium ions and channels in synaptic facilitation and depression at the squid giant synapse. J. Physiol. (London) **323:** 173-193.
18. HEUSER, J. E., T. S. REESE, M. J. DENNIS, Y. JAN, L. YAN & L. EVANS. 1979. Synaptic vesicle exocytosis captured by quick freezing and correlated with quantal transmitter release. J. Cell Biol. **81:** 275-300.
19. CECCARELLI, B., F. GROHAVAZ & W. P. HURLBUT. 1979. Freeze-fracture studies of frog neuromuscular junctions during intense release of neurotransmitter. I. Effects of black widow spider venom and Ca^{2+}-free solutions on the structure of the active zones. J. Cell Biol. **81:** 163-177.
20. LLINAS, R., I. Z. STEINBERG & K. WALTON. 1976. Presynaptic calcium currents and their relation to synaptic transmission: Voltage clamp study in squid giant synapse and theoretical model for the calcium gate. Proc. Natl. Acad. Sci. USA **73:** 2918-2922.
21. PUMPLIN, D. W., T. S. REESE & R. LLINAS. 1981. Are the presynaptic membrane particles calcium channels? Proc. Natl. Acad. Sci. USA **78:** 7210-7213.
22. ENGEL, A. G., H. FUKUNAGA & M. OSAME. 1982. Stereometric estimation of the area of the freeze-fractured membrane. Muscle Nerve **5:** 682-685.
23. LANG, B., J. NEWSOM-DAVIS, C. PEERS & D. W.-WRAY. 1987. The action of myasthenic syndrome antibody fragments on transmitter release in the mouse. J. Physiol. (London) **390:** 173P.
24. FUKUOKA, T., A. G. ENGEL, B. LANG, J. NEWSOM-DAVIS & A. VINCENT. 1987. Lambert-Eaton myasthenic syndrome: II. Immunoelectron microscopy localization of IgG at the mouse motor end-plate. Ann. Neurol. **22:** 220-211.
25. LANG, B., J. NEWSOM-DAVIS, C. PEERS, C. PRIOR & D. W.-WRAY. 1987. The effect of myasthenic syndrome antibody on presynaptic calcium channels in mouse. J. Physiol. (London) **390:** 257-270.
26. ROBERTS, A., S. PERERA, B. LANG, A. VINCENT & J. NEWSOM-DAVIS. 1985. Paraneoplastic myasthenic syndrome IgG inhibits $^{45}Ca^{2+}$ flux in a human small cell carcinoma line. Nature **317:** 737-738.

27. LOGIN, I. S., Y. I. KIM, A. M. JUDD, B. L. SPANGELO & R. M. MACLEOD. 1987. Immunoglobulins of Lambert-Eaton myasthenic syndrome inhibit rat pituitary hormone release. Ann. Neurol. **22:** 610-614.
28. KIM, Y. I. & E. NEHER. 1988. IgG from patients with Lambert-Eaton syndrome blocks voltage-dependent calcium channels. Science **239:** 405-408.
29. LANG, B., J. NEWSOM-DAVIS, C. PEERS & D. W.-WRAY. 1987. Selective action of Lambert-Eaton myasthenic syndrome antibodies on Ca^{2+} channels in the neuroblastoma $\times$ glioma hybrid cell line NG108 15. J. Physiol. (London) **394:** 43P.

Lack of Action of Lambert-Eaton Myasthenic Syndrome Antibody on Calcium Channels in Insect Muscle

H. A. PEARSON,[a] J. NEWSOM-DAVIS,[b] G. LEES,[c] AND
D. W. -WRAY[a]

[a] Department of Pharmacology
Royal Free Hospital School of Medicine
Rowland Hill Street
London NW3 2PF, England

[b] Department of Clinical Neurology
Radcliffe Infirmary
Oxford, England

[c] Department of Pesticide Dynamics
Wellcome Research Laboratories
Berkhamsted, Hertfordshire, England

The Lambert-Eaton myasthenic syndrome (LEMS) is an autoimmune disorder in man causing muscle weakness.[1] Immunoglobulin G (IgG) obtained from LEMS patients has been shown to act at Ca^{2+} channels in the motor nerve terminals of a number of mammalian preparations,[2] but has so far been untested in any invertebrate systems.

We have therefore investigated the effect of LEMS IgG on Ca^{2+} channels in skeletal muscle fibers of larvae of the moth *Plutella xylostella*. For this, we tested its effect on action potentials which, in common with those of many other insects, are based on voltage-dependent Ca^{2+} channels.[3,4]

METHODS

Intracellular microelectrode recordings were made from the same identified muscle fiber in the second abdominal segment of *Plutella xylostella* larvae in their fourth or fifth instar. Muscles were incubated in control or LEMS IgG at room temperature for 24 hours before recording. Recordings were made at 21-23°C in solutions of the following composition (mM): NaCl, 12; KCl, 30; MgCl$_2$, 18; CaCl$_2$, 10; NaHCO$_3$, 0.06; and NaH$_2$PO$_4$, 0.06. Hypertonic sucrose (800 mM) was used to prevent muscle contraction and tetraethylammonium (25 mM) was present to minimize K$^+$ currents. Action potentials were stimulated at 0.1 Hz by passing square-wave depolarizing current pulses (10-20 nA, 40-msec duration) via a single microelectrode that was also used to record action potentials via an Axoclamp-2 amplifier in discontinuous current clamp mode (4-6 kHz). Voltage records were stored on magnetic tape and analyzed by computer (digitization rate 5 kHz). The Student's t-test was used for statistical comparisons.

RESULTS

Evidence for action potentials being based on an inward movement of Ca^{2+} was provided by the following experiments.[3] First, action potentials were completely blocked by the presence of cadmium ions (1 mM), which are known to block Ca^{2+} channels. Second, action potentials were also blocked by complete removal of Ca^{2+} from the perfusing medium. Furthermore, reducing the Ca^{2+} concentration from 10.0 mM to 5.0 mM produced a significant reduction of 50% ($p < 0.05$) in the maximum rate of rise of the action potential (FIG. 1) together with a small reduction in amplitude, which, however, did not attain statistical significance.

The effect of LEMS IgG on action potentials is shown in FIGURE 2. The antibody had no significant effect on amplitude, duration, or maximum rate of rise of action potentials.

DISCUSSION

LEMS IgG produces a reduction in Ca^{2+} current by about 50% in mammalian preparations.[2] If a similar effect had occurred for the insect muscle Ca^{2+} channels, we would have expected to see a 50% reduction in maximum rate of rise of action potentials. Indeed, such a reduction was seen in control muscle when the Ca^{2+} concentration was reduced by 50%. However for LEMS IgG-treated muscles, there was no effect at all on action potentials. This result suggests that the Ca^{2+} channels involved in the action potential in these fibers are antigenically different from those mammalian Ca^{2+} channels where LEMS IgG is known to act (motor nerve terminals,[5] L-type channels in NG108,[6] and chromaffin cells[7]).

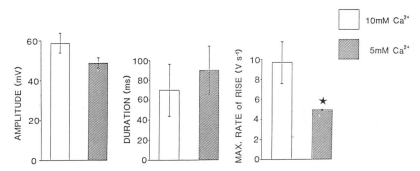

FIGURE 1. Effect of Ca^{2+} concentration on action potentials. Values represent means ± standard error for amplitude, duration, and maximum rate of rise of action potentials in eight muscle fibers, each exposed to 10 mM and 5 mM Ca^{2+} (* significant reduction, $p < 0.05$).

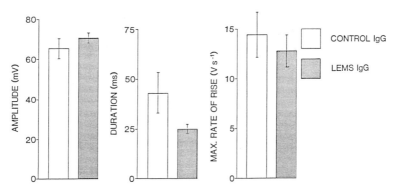

FIGURE 2. Effect of LEMS IgG on action potentials. Values represent means ± standard error for amplitude, duration, and maximum rate of rise of action potentials in 10 muscle fibers incubated with control IgG and 10 muscle fibers incubated with LEMS IgG (Ca^{2+} concentration 10 mM.)

REFERENCES

1. LAMBERT, E. H. & D. ELMQVIST. 1971. Quantal components of end-plate potentials in the myasthenic syndrome. Ann. N.Y. Acad. Sci. **183:** 183-199.
2. W.-WRAY, D., B. LANG, J. NEWSOM-DAVIS & C. PEERS. 1989. Antibodies against calcium channels in the Lambert-Eaton myasthenic syndrome. Ann. N.Y. Acad. Sci. This volume.
3. PEARSON, H. A., G. LEES & D. W.-WRAY. 1989. Characterisation of calcium channels in muscle fibres in larvae of the moth *Plutella xylostella*. In preparation.
4. ASHCROFT, F. M. 1981. Calcium action potentials in the skeletal muscle fibres of the stick insect *Carausius morosus*. J. Exp. Biol. **93:** 257-267.
5. LANG, B., J. NEWSOM-DAVIS, C. PEERS, C. PRIOR & D. W.-WRAY. 1987. The effect of myasthenic syndrome antibody on presynaptic calcium channels in the mouse. J. Physiol. **390:** 257-270.
6. LANG, B., J. NEWSOM-DAVIS, C. PEERS & D. W.-WRAY. 1987. Selective action of Lambert-Eaton myasthenic syndrome antibodies on Ca^{2+} channels in the neuroblastoma $\times$ glioma hybrid cell line NG108 15. J. Physiol. **394:** 43P.
7. KIM, Y. I. & E. Neher. 1988. IgG from patients with Lambert-Eaton myasthenic syndrome blocks voltage-dependent calcium channels. Science **439:** 405-408.

K^+-Stimulated Ca^{2+} Influx in Cell Lines Derived from Small Cell Lung Cancer and Neuronal Tumors

B. LANG, K. LEYS, A. VINCENT, AND
J. NEWSOM-DAVIS

Neurosciences Group
Institute of Molecular Medicine
University of Oxford
John Radcliffe Hospital
Oxford OX3 9DU, United Kingdom

The Lambert-Eaton myasthenic syndrome (LEMS) is an autoimmune disorder in which neurotransmitter release from motor nerve terminals is reduced. Autoantibodies are directed against the presynaptic voltage-gated calcium channels (VGCC), resulting in a decrease in the influx of Ca^{2+}, and subsequent reduction in the quantal and nonquantal release of acetylcholine.

The autoimmune cause of the disease has been established by passive transfer to mice,[1] using patients' IgG. Electrophysiological and morphological studies based on this model have indicated that the antibodies cause a reduction in the number of functional VGCC,[2-4] probably by cross-linking and accelerated internalization.[5,6]

In 60% of patients with LEMS there is an associated small-cell lung cancer (SCLC),[7,8] which is thought to be neuroectodermal in origin. We have shown[9] that a human SCLC line exhibits K^+-stimulated Ca^{2+} flux that can be blocked by VGCC antagonists nitrendipine and methoxyverapamil (D600), indicating the presence of functional VGCC.

We have now tested 25 cell lines for stimulated Ca^{2+} flux (eight SCLC, five non-SC lung carcinomas, and seven other human cell lines, three of neuronal origin and five rodent-derived lines). K^+-stimulated $^{45}Ca^{2+}$ flux was detected in all eight SCLC lines and five others, all of neural origin but in none of the non-SC lung carcinoma lines. In each of these cell lines Ca^{2+} flux was inhibited by pooled LEMS IgG but not by control IgG.

Four lines were tested extensively with IgG prepared from individual LEMS patients. In all lines the results show a highly significant reduction in the K^+-stimulated flux compared with those grown in control IgG. There did not appear to be any difference between cells grown in SCLC-LEMS or non-SCLC LEMS (see FIG. 1).

Individual LEMS IgG preparations produced a similar inhibition of the Ca^{2+} influx into each of the SCLC lines tested, but showed considerable variation in their reaction with neuronal cell lines, indicating either a heterogenous population of autoantibodies or a difference in antigenic determinants on VGCCs in these lines.

These apparent differences between VGCCs in SCLC and neuronal lines was further demonstrated by the pharmacological effectiveness of a VGCC antagonist. Nitrendipine

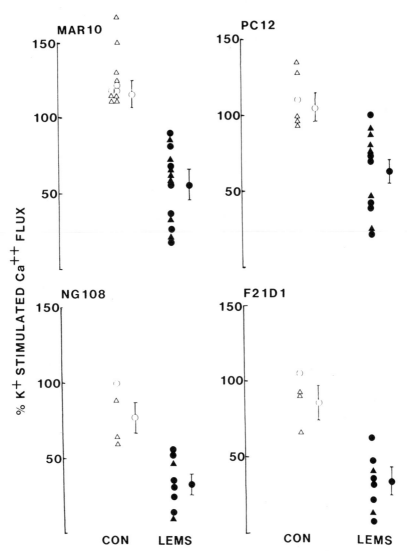

FIGURE 1. The effect of IgG (LEMS and Control) on K⁺-stimulated Ca²⁺ flux in cell lines. Cells were incubated for three to seven days (depending on the doubling time of the individual cell lines) with IgG prepared from healthy subjects (○), patients with other neurological disorders (△), and from patients with SCLC-LEMS (▲) or non-SCLC LEMS (●), at a final concentration of 4 mg per ml culture supernatant. Cells were then harvested and assayed as previously described.[9] Each value is expressed as percent K⁺-stimulated Ca²⁺ flux of cells grown concurrently in media alone. The mean value ± SE is shown for each group.

Cell lines: MAR10 (human SCLC), PC12 (rat pheochromocytoma), F21D1 (mouse neuroblastoma—human dorsal root ganglia), NG108 (mouse neuroblastoma—rat glioma).

had an IC_{50} of $2 \times 10^{-8} M$ in rodent line PC12, while in human SCLC line MAR it was $1 \times 10^{-6} M$.

We conclude that functional VGCCs appear to be a marker for cells of neuroectodermal-neuroendocrine origin and may provide a means of discriminating SCLC from non-SC lung carcinomas. LEMS IgG inhibits K^+-stimulated Ca^{2+} flux into all cells expressing functional VGCCs.

REFERENCES

1. LANG, B., J. NEWSOM-DAVIS, D. WRAY, A. VINCENT & N. MURRAY. 1981. Lancet ii: 224-226.
2. LANG, B., J. NEWSOM-DAVIS, C. PRIOR & D. WRAY. 1983. J. Physiol. (London) 344: 335-345.
3. LANG, B., J. NEWSOM-DAVIS, C. PEERS, C. PRIOR & D. W.-WRAY. 1987. J. Physiol. (London) 390: 257-270.
4. FUKUOKA, T., A. ENGEL, B. LANG, J. NEWSOM-DAVIS, C. PRIOR & D. W.-WRAY. 1987. Ann. Neurol. 22: 193-199.
5. NAGEL, A., A. G. ENGEL, B. LANG, J. NEWSOM-DAVIS & T. FUKUOKA. 1988. Ann. Neurol. In press.
6. LANG, B., J. NEWSOM-DAVIS, C. PEERS & D. W.-WRAY. 1987. J. Physiol. (London) 390: 173p.
7. LAMBERT, E. H. & D. ELMQVIST. 1971. Ann. N. Y. Acad. Sci. 1971 183: 183-191.
8. O'NEILL, J. H., N. M. F. MURRAY & J. NEWSOM-DAVIS. 1988. Brain 111: 577-596.
9. ROBERTS, A., S. PERERA, B. LANG, A. VINCENT & J. NEWSOM-DAVIS. 1985. Nature 317: 737-739.

Effect of the Calcium-Channel Agonist CGP 28392 on Transmitter Release at Mouse Neuromuscular Junctions

J. BURGES AND D. W.-WRAY

Department of Pharmacology
Royal Free Hospital School of Medicine
London NW3 2PF, England

Transmitter release at the neuromuscular junction is dependent upon the entry of Ca^{2+} into the nerve terminal via voltage-dependent Ca^{2+} channels. At least three different types of neuronal Ca^{2+} channel have been classified,[1] but those directly involved in the release of ACh at the mouse neuromuscular junction have not yet been clearly characterized. Organic Ca^{2+}-channel antagonists (which act on L-type channels) are without effect on transmitter release at the skeletal neuromuscular junction.[2,3] However, there are known to be Ca^{2+}-antagonist-sensitive Ca^{2+} channels present at motor nerve terminals.[4] Although not normally involved in transmitter release, L-type channels may play a role in release in the presence of Ca^{2+}-channel agonists. To test this, we have investigated the effect of the Ca^{2+}-channel agonist CGP 28392 on acetylcholine release at nerve terminals of mouse diaphragm muscles, comparing the effect with the antagonist nitrendipine.

METHODS

Diaphragm muscles with the phrenic nerve attached were removed from mice following cervical dislocation. Intracellular microelectrode recordings were made as previously described[5] in well-oxygenated Krebs solution (22-24°C) of the following composition (mM): NaCl, 118; KCl, 4.7; MgSO$_4$, 1.2; KH$_2$PO$_4$, 1.2; NaHCO$_3$, 25.0; CaCl$_2$, 2.52; glucose, 11.1 (pH 7.1-7.4). For endplate potential (epp) recordings, CaCl$_2$ concentration was reduced to 0.24 mM, and the nerve was stimulated supramaximally at 1.0 Hz. For miniature endplate potential (mepp) recordings, KCl concentration was increased to 11.8 mM while the NaCl concentration was reduced to 110.9 mM. Continuous recordings were made from single endplates during the application of dihydropyridines. All recordings were stored on magnetic tape and analyzed subsequently by computer (digitization rate 10 kHz). For epps, quantal content was calculated for each fiber from the average of the variance, failures, and direct methods.[6] Test and controls were compared by Student's t-test (level for insignificance: $p > 0.05$).

RESULTS

CGP 28392 (2 μM) had no significant effect on epp quantal content (FIG. 1a), nor did 2 μM nitrendipine (FIG. 1b). At a high K^+ concentration (13 mM), the agonist gave a small increase in mepp frequency (FIG. 2a), but this was not statistically significant. Nitrendipine did not significantly affect mepp frequency in high-K^+ solutions (FIG. 2b). Thus neither the agonist nor antagonist had marked presynaptic effects. Finally, neither drug significantly affected mepp amplitudes (control 1.1 $\pm$ 0.1 mV, n = 19 endplates; CGP 28392 1.4 $\pm$ 0.2mV, n = 13; nitrendipine 0.9 $\pm$ 0.2mV, n = 6), indicating lack of postsynaptic action at this concentration.

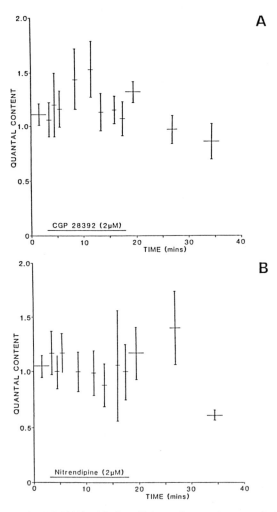

FIGURE 1. Effect of CGP 28392 and nitrendipine and quantal content in low-Ca^{2+} concentration solutions. Values represent means $\pm$ standard error from six muscles exposed to (A) 2 μM CGP 28392 and from six muscles exposed to (B) 2 μM nitrendipine.

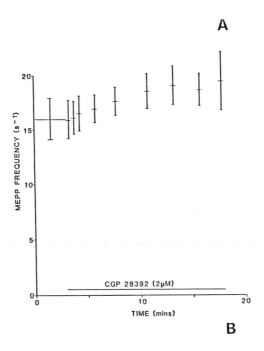

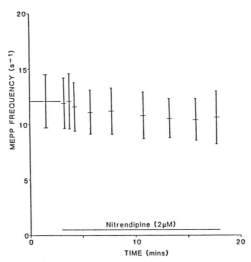

FIGURE 2. Effect of CGP 28392 and nitrendipine on mepp frequency in high-K^+ concentration solutions. Values represent means $\pm$ standard error from six muscles exposed to (**A**) 2 μM CGP 28392 and from six muscles exposed to (**B**) 2 μM nitrendipine.

CONCLUSION

These results show that L-type Ca^{2+} channels play no clear role in transmitter release at the neuromuscular junction even after enhancement of their possible involvement by the use of the Ca^{2+}-agonist CGP 28392. Although L-type channels may be present at motor nerve terminals (see above), they clearly do not have a functional role in quantal release. One possibility[7] could be that L-type channels are located some distance away from the active-zone Ca^{2+} channels and thus do not normally contribute to evoked release. Possibly for much higher K^+ concentrations than could be used in these experiments, there may be an effect of the agonist because so much Ca^{2+} might then enter the terminal via L-type channels to spill over to the release sites.[7] Indeed in our experiments there was a hint of an increase in mepp frequency by CGP 28392 in 13 mM K^+ solution.

The voltage-dependent Ca^{2+} channels located at the active zones of the mouse motor nerve, while not L-type, are not N-type either because they are not blocked by ω-conotoxin.[8] It is likely that they do not fit into an already-defined channel type.

REFERENCES

1. NOWYCKY, M. C., A. P. FOX & R. W. TSIEN. 1985. Three types of neuronal calcium channel with different calcium agonist selectivity. Nature 316: 440-443.
2. GOTGILF, M. & L. G. MAGAZANIK. 1977. Action of calcium channel blocking agents (verapamil, D-600 and manganese ions) on transmitter release from motor nerve endings of frog muscle. J. Neurophysiol. 9: 415-421.
3. NACHSEN, D. A. & M. P. BLAUSTEIN. 1979. The effects of some organic 'calcium-antagonists' on calcium influx in presynaptic nerve terminals. J. Mol. Pharmacol. 16: 579-586.
4. PENNER, R. & F. DREYER. 1986. Two different presynaptic calcium currents in mouse motor neve terminals. Pflugers Arch. 406: 190-197.
5. WRAY, D. 1981. Prolonged exposure to acetylcholine: Noise analysis and channel inactivation in cat tenuissimus muscle. J. Physiol. 310: 37-56.
6. CASTILLO, J. DEL & B. KATZ. 1954. Quantal components of the endplate potential. J. Physiol. 124: 560-573.
7. MILLER, R. J. 1987. Multiple calcium channels and neuronal function. Science 235: 46-52.
8. ANDERSON, A. J. & A. L. HARVEY. 1987. ω-Conotoxin does not block the verapamil-sensitive calcium channels at mouse nerve terminals. Neurosci. Lett. 82: 177-180.

Synaptosomal Ca^{2+} Channels Are Blocked by Pimozide and Flunarizine with Higher Affinity Than the Na^+/Ca^{2+} Exchanger[a]

C. A. M CARVALHO AND D. L. SANTOS

Center for Cell Biology
Department of Zoology
University of Coimbra
3049 Coimbra Codex, Portugal

Synaptosomes isolated from the central nervous system have been widely used as models in studying the regulation of Ca^{2+} in nerve cells and have been useful in studying the influx of Ca^{2+} through Ca^{2+} channels due to depolarization and also to Na^+/Ca^{2+} exchange, during alterations of the Na^+ gradient. However, there is no specific inhibitor of these mechanisms, and it has not been possible to distinguish between them when both take place simultaneously, such as when depolarization by K^+ is utilized, in which the Na^+ gradient is altered simultaneously with depolarization.[1-4] The organic Ca^{2+}-channel blockers, selective for Ca^{2+} influx through Ca^{2+} channels in smooth and cardiac muscle,[8,14] do not block similar fluxes in synaptosomes at concentrations that saturate the binding sites of the Ca^{2+}-channel proteins,[5-10] although there is a report to the contrary.[11]

In our work, we have explored experimental conditions that permit us to distinguish the effect of Ca^{2+}-channel blockers on Ca^{2+} entry through Ca^{2+} channels from the effect of Ca^{2+} entry through the Na^+/Ca^{2+} exchange mechanism, and it became evident that both mechanisms are equally sensitive to verapamil (FIG. 1), *d-cis*-diltiazem and nifedipine.[10] We now find that two other Ca^{2+}-channel antagonists, pimozide and flunarizine, which are much more lipophilic than the classical Ca^{2+}-channel blockers, are able to partially block Ca^{2+} channels at concentrations that have small effect on Na^+/Ca^{2+} exchange, and this effect is observed at concentrations about 100-fold lower than those that are necessary for the classical Ca^{2+} blockers to affect Ca^{2+} movements in synaptosomes.[5-10]

RESULTS AND DISCUSSION

In FIGURE 2 we depict the results of experiments performed with synaptosomes isolated in sucrose media (synaptosomes with low Na^+ content) and in synaptosomes

[a]This work was supported by the Instituto Nacional de Investigação Científica, the Junta Nacional de Investigação Cientifica e Tecnológica, and the Calouste Gulbenkian Foundation, Portugal.

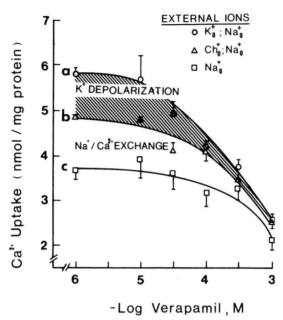

FIGURE 1. Verapamil inhibits both Ca^{2+} uptake induced by K^+ depolarization or Na^+/Ca^{2+} exchange. Synaptosomes isolated by the method of Hajós,[16] with some modifications,[17] were incubated in Na^+-rich medium as described previously.[10] The $^{45}Ca^{2+}$ uptake was initiated by transferring preincubated synaptosomes to each of the different media (in mM): (a) 60 KCl, 73 NaCl; (b) 60 choline chloride, 73 NaCl; (c) 128 NaCl, 5 KCl, all containing, additionally, 10 mM glucose, 10 mM HEPES-Tris, pH 7.4, 1 mM $^{45}CaCl_2$ (2.5 μCi/μmol) and increasing concentrations of verapamil. The reactions were performed as described previously.[10] In these synaptosomes (Na^+ rich), depolarization by 60 mM K^+, which is accompanied by an equivalent reduction of the external Na^+ (curve **a**), induces Ca^{2+} entry by K^+ depolarization (shaded area) and by Na^+/Ca^{2+} exchange (nonshaded area). However, when Na^+ is reduced to the same extent by addition of 60 mM choline rather than K^+, there is no depolarization, but the same Na^+ gradient is created and Ca^{2+} enters only by Na^+/Ca^{2+} exchange (curve **b**). Thus, the Ca^{2+} uptake due to Na^+/Ca^{2+} exchange is the difference between curves **b** and **c**, whereas Ca^{2+} uptake due to K^+ depolarization is the difference between curves **a** and **b**. Adapted from our previous results.[10]

preincubated in Na^+-rich medium (133 mM NaCl) for 10 minutes at 30°C, before the $^{45}Ca^{2+}$ influx was measured. As reported previously, the Na^+-rich synaptosomes, when depolarized in K^+ medium containing 60 mM K^+ plus 73 mM Na^+, take up Ca^{2+} due to K^+ depolarization and to Na^+/Ca^{2+} exchange,[9,10] as shown in FIGURE 1, whereas in the low Na^+ synaptosomes, the Ca^{2+} uptake occurs mainly through the Ca^{2+} channels.[9,10,12]

In FIGURE 2, we show that pimozide and flunarizine preferentially inhibit the Ca^{2+} influx in the low-Na^+ synaptosomes, which, as suggested above, occurs mainly through the Ca^{2+} channels. Thus, at 2×10^{-7} M, pimozide, which causes 50% of the maximal inhibition, produces only a relatively small effect on the fraction of Na^+/Ca^{2+} exchange that is inhibitable by the drug (FIG. 2A). Flunarizine has a similar preferential inhibitory effect on Ca^{2+} influx through Ca^{2+} channels, as compared to the Ca^{2+} influx through Na^+/Ca^{2+} exchange (FIG. 2B). The concentrations of pimozide and flunarizine that produce these effects are about 100-fold lower than those reported previously for other Ca^{2+}-channel blockers. It is observed, however, that pimozide and flunarizine, even at 3×10^{-5} M, only inhibit 50% of the total Ca^{2+} uptake (FIG. 2), which may be due to the fact that Ca^{2+} influx is occurring partially through Ca^{2+} channels that are insensitive to Ca^{2+} channel blockers, for example, N channels, which have been detected in nerve terminals.[15]

We have also found that pimozide and flunarizine have partition coefficients (K_p) in native synaptic plasma membranes (SPM) of 6.5×10^3 and 19×10^3, respectively (TABLE 1). As can also be seen in TABLE 1, these values are much higher than the K_p values found for other Ca^{2+}-channel blockers[13] (nitrendipine and ($-$)-desmethoxyverapamil), which suggests that high lipid solubility may influence the Ca^{2+}-antagonistic properties of the drugs. It is of interest that the K_p values of all the Ca^{2+} antagonists studied are considerably higher in native membranes than the K_p values obtained for the same drugs in liposomes prepared with the extracted membrane lipids[13] (TABLE 1). The higher values for K_p in SPM, as compared to liposomes, are not readily explained and raise questions as to the significance of the K_p values measured in native membranes, because, apparently, the Ca^{2+}-channel blockers in the membrane

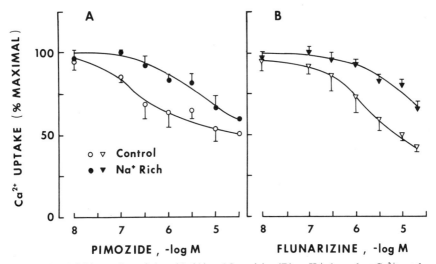

FIGURE 2. Inhibitory effect of pimozide (**A**) and flunarizine (**B**) on K^+-dependent Ca^{2+} uptake by rat brain synaptosomes. The K^+-dependent Ca^{2+} uptake was determined as the difference in uptake obtained in basal medium (in mM): 128 NaCl, 5 KCl; and in K^+-depolarizing medium (in mM): 60 KCl, 73 NaCl, both containing, additionally, 1 MgCl$_2$, 10 glucose, 10 HEPES-Tris, pH 7.4, and 1 mM ^{45}CaCl$_2$, and increasing concentrations of pimozide or flunarizine. Two types of synaptosomes were tested: low-Na$^+$ synaptosomes, which, after isolation, were maintained in ion-free sucrose medium (open symbols), and Na$^+$-rich synaptomsomes, which were preincubated in Na$^+$-rich medium, as described previously[10] (closed symbols). As reported previously, the Na$^+$-rich synaptosomes, when depolarized in K^+ medium, take up Ca^{2+} due to K^+ depolarization and to Na$^+$/Ca^{2+} exchange,[9,10] as shown in FIGURE 1, whereas in the low Na$^+$ synaptosomes, the Ca^{2+} uptake occurs mainly through the Ca^{2+} channels.[9,12] ^{45}Ca^{2+} uptake is expressed as a percentage of maximal K^+-stimulated Ca^{2+} uptake obtained in the absence of the drugs. The data are means $\pm$ SD of four experiments performed in triplicate.

are not exclusively partitioning into the lipid phase. Thus, the membrane proteins may either bind the drugs themselves, or they may create extra space for drug partitioning at the protein-lipid interfaces.

In conclusion, pimozide and flunarizine are more selective for inhibiting Ca^{2+} influx through Ca^{2+} channels than by Na^+/Ca^{2+} exchange in synaptosomes than are the other conventional Ca^{2+}-channel antagonists, and we suggest that this is related to the relatively higher lipophilicity of pimozide and flunarizine in the membranes.

TABLE 1. Apparent Partition Coefficients (K_p) of Ca^{2+}-Channel Blockers in Native Synaptic Plasma Membranes (SPM) and in SPM Liposomes

Ca^{2+} Antagonist	Native SPM	SPM Liposomes
Nitrendipine	464 ± 75 (4)	158 ± 10 (4)
(−)-Desmethoxyverapamil	361 ± 40 (4)	144 ± 14 (4)
Pimozide	6.5×10^3 (2)	1.3×10^3 (2)
Flunarizine	19×10^3 (2)	6×10^3 (2)

NOTE: The apparent partition coefficient (K_p) of the various Ca^+-channel antagonists were determined either in SPM or in liposome suspensions made from the extracted membrane lipids, by equilibrating the suspensions with the ^{3}H-labeled Ca^{2+} antagonists at a concentration of $5 \times 10^{-6}M$, for one hour, at 25°C. Samples of the suspension were filtered through Whatman GF/B filters under vacuum, and the radioactivity associated with the membranes after washing and in the incubation medium was counted by liquid scintillation spectrometry. The K_p values were calculated from the ratio of the concentration of drug in the membrane phase (lipid compartment) and in the aqueous phase. The numbers in parentheses represent the number of experiments performed in triplicate.

ACKNOWLEDGMENTS

We thank Dr. J. Leysen from Janssen Pharmaceutica for the supply of pimozide and flunarizine.

REFERENCES

1. BLAUSTEIN, M. P. 1975. J. Physiol. (London) **247:** 617-655.
2. NACHSHEN, D. A. & M. P. BLAUSTEIN. 1980. J. Gen. Physiol. **76:** 709-728.
3. COUTINHO, O. P., C. A. M. CARVALHO & A. P. CARVALHO. 1984. Brain Res. **290:** 261-271.
4. CARVALHO, A. P., O. P. COUTINHO, V. M. C. MADEIRA & C. A. M. CARVALHO. 1984. *In* Biomembranes: Dynamics and Biology. R. M. Burton & F. C. Guerra, Eds.: 327-364. Plenum Press. New York.
5. NACHSHEN, D. A. & M. P. BLAUSTEIN. 1979. Mol. Pharmacol. **16:** 579-586.
6. DANIELL, L. C., E. M. BARR & S. W. LESLIE. 1983. J. Neurochem. **41:** 1455-1459.
7. RAMPE, D., R. A. JANIS & D. J. TRIGGLE. 1984. J. Neurochem. **43:** 1688-1692.
8. MILLER, R. & S. B. FREEDMAN. 1984. Life Sci. **34:** 1205-1221.
9. CARVALHO, C. A. M., O. P. COUTINHO & A. P. CARVALHO. 1986. J. Neurochem. **47:** 1774-1784.

10. CARVALHO, C. A. M., S. V. SANTOS & A. P. CARVALHO. 1986. Eur. J. Pharmacol. 131: 1-12.
11. TURNER, T. Y. & S. M. GOLDIN. 1985. J. Neurosci. 5: 842-849.
12. CARVALHO, A. P., M. S. SANTOS, A. O. HENRIQUES, P. TAVARES & C. A. M. CARVALHO. 1987. In Cellular and Molecular Basis of Synaptic Transmission. H. Zimmerman, Ed. NATO ASI series H21: 263-284. Plenum Press. New York.
13. CARVALHO, C. A. M., C. R. OLIVEIRA, M. C. P. LIMA & A. P. CARVALHO. 1987. J. Neurochem. 48(Suppl.): 576.
14. BOLGER, G. T., P. GENGO, R. KLOCKOWSKI, E. LUCHOWSKI, H. SIEGEL, R. A. JANIS, A. M. TRIGGLE & D. J. TRIGGLE. 1983. J. Pharmacol. Exp. Ther. 225: 291-309.
15. MILLER, R. J. 1987. Science 225: 46-52.
16. HAJOS, F. 1975. Brain Res. 93: 485-489.
17. CARVALHO, C. A. M. & A. P. CARVALHO. 1979. J. Neurochem. 41: 670-676.

Voltage-Dependent Cooperative Interactions between Ca-Channel Blocking Drugs in Intact Cardiac Cells

H. PORZIG AND C. BECKER

Department of Pharmacology
University of Bern
CH-3010 Bern, Switzerland

Derivatives of 1,4-dihydropyridines, phenylalkylamines, and benzothiazepines have been shown to block voltage-sensitive L-type Ca channels by interacting with at least three distinct but allosterically coupled binding sites (for review see Godfraind *et al.*[1]). The strength of the blocking activity of these different compounds depends strongly on the membrane potential, apparently because they have a high affinity to the inactivated state and rather low affinities to the resting and open states of the Ca channel.[2-4] In the experiments reported here, we have tested whether the known allosteric interactions between the binding sites for different types of Ca-channel blockers are also voltage-dependent. We determined equilibrium binding of the dihydropyridine (+)-isradipine (PN 200-110) in myocardial cells from newborn rats at two different membrane potentials (0 and -40 mV) in the presence and absence of *d-cis*-diltiazem and ($\pm$) verapamil. In polarized cells, both diltiazem (5 μM) and verapamil (10 μM) had similarly marked positive cooperative effects on [^{3}H](+)-isradipine binding (FIG. 1). In both cases the effect was entirely due to an increase

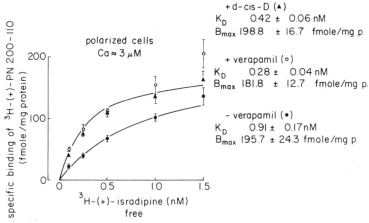

FIGURE 1. Effect of *d-cis*-diltiazem (d-cis-D) and of ($\pm$) verapamil on the specific binding of (+)-isradipine in polarized cardiac cells (resting potential about -40 mV). The experimental medium was nominally Ca-free (measured contamination ~ 3 μM Ca). The curves give computerized nonlinear least-square regressions to the experimental points obtained in the absence of blockers or in the presence of ($\pm$) verapamil. The curve fitting the data for d-cis-D has been ommitted for the sake of clarity. Calculated dissociation constants (K_D) and maximal binding capacities (B_{max}) are given for each condition on the right-hand side.

TABLE 1. Effect of d-cis-Diltiazem and of ($\pm$) Verapamil on the Dissociation Constant (K_D) and the Maximal Binding Capacity (B_{max}) for [^{3}H]($+$)-Isradipine (PN 200-110) in Intact Cardiac Cells

Condition	K_D (nM)	B_{max} (fmol/mg protein)	Number of Experiments
Polarized cells, 1.25 mM Ca^{2+}			
control	1.86 ± 0.55	323.3 ± 72.6	3
+ d-cis-diltiazem	0.47 ± 0.05	310.2 ± 19.8	3
Polarized cells, ~3 μM Ca^{2+}			
control	0.97 ± 0.14	215.4 ± 24.1	11
+ d-cis-diltiazem	0.41 ± 0.05	196.7 ± 14.4	3
+ ($\pm$) verapamil	0.28 ± 0.04	181.8 ± 12.7	3
Depolarized cells, 1.25 mM Ca^{2+}			
control	0.070 ± 0.004	197.5 ± 9.0	5
+ d-cis-diltiazem	0.069 ± 0.001	178.2 ± 0.51	3
Depolarized cells ~3 μM Ca^{2+}			
control	0.065 ± 0.006	146.7 ± 9.2	9
+ d-cis-diltiazem	0.071	184.6	1
Fragmented membranes 1.25 mM Ca^{2+}			
control	0.308	229.6	2
+ d-cis-diltiazem	0.299	235.0	2

in binding affinity. The maximal binding capacity remained unchanged. In depolarized cells verapamil had a strong negative cooperative effect on isradipine binding that was mainly caused by a decrease in binding affinity. Under the same conditions, diltiazem had no statistically significant effect on isradipine binding parameters (TABLE 1). An increase in the extracellular Ca concentrations from 3 μM to 1.25 mM enhanced the maximal binding capacity for isradipine but had no effect on the allosteric interactions between different channel ligands (TABLE 1). These results suggest that positive cooperative interactions between Ca-channel blockers can be observed under conditions where the channel conformation can fluctuate between closed, open, and inactivated states (resting potential -40 mV). They reflect the capacity of these compounds to enhance the equilibrium frequency of channels in the inactivated state. Negative cooperativity that is observed under conditions where all channels are inactivated (resting potential 0 mV) probably reflects the transformation of the high-affinity inactivated channel into a drug-induced, low-affinity conformation. The affinity for isradipine in membranes is significantly lower than in depolarized cells (TABLE 1) suggesting that the channel state is not identical in the two preparations, and hence may be subject to regulation by intracellular factors.

REFERENCES

1. GODFRAIND, T., R. MILLER & M. WIBO. 1986. Pharmacol. Rev. **38:** 321.
2. BEAN, B. P. 1984. Proc. Natl. Acad. Sci. USA **81:** 6388.
3. SANGUINETTI, M. C. & R. S. KASS. 1984. Circ. Res. **55:** 336.
4. KOKUBUN, S., B. PROD'HOM, C. BECKER, H. PORZIG & H. REUTER. 1986. Molec. Pharmacol. **30:** 571.

Modulation of Dihydropyridine Receptor Sites in Calcium Channels of Vascular Smooth Muscle

T. GODFRAIND, N. MOREL, AND M. WIBO

Catholic University of Louvain
Laboratory of General Pharmacodynamics and Pharmacology
1200 Brussels, Belgium

The inhibition of KCl-evoked contraction of vascular smooth muscle by some dihydropyridines (nisoldipine, nimodipine, and (+)PN200-110, but not nifedipine) develops gradually despite extensive preincubation with these drugs in physiological medium.[1-3]

This time-dependent pattern of inhibition of KCl-evoked contractions can be reproduced in successive contractions, separated by a resting period of 30 minutes in physiological solution in the presence of a drug. This suggests that depolarization induces a conformation of calcium channels with increased affinity for dihydropyridines. This hypothesis has been examined in isolated rat aortas and rat mesenteric arteries by studying how membrane potential influences the specific binding of (+)PN200-110 to intact arteries.

In mesenteric artery, dose-effect curves for inhibition by (+)PN200-110 measured two and 30 minutes after initiation of depolarization were characterized by IC_{50} values of 270 ± 65 pM and 33 ± 3 pM, respectively. (−)PN200-110 was less potent than (+)PN200-110, and its inhibitory potency was also time-dependent, IC_{50} values measured at two and 30 minutes in high-K^+ solution being equal to 16 ± 4 nM and 5 ± 3 nM, respectively. Thus, the effect of PN200-110 was stereoselective, the (+) enantiomer being markedly more potent than the (−) enantiomer. Similar observations were made in rat aorta, where it was found in addition that the kinetics of the development of inhibition of KCl-evoked contraction by (+)PN200-110 was similar to the kinetics of binding of (+)[^{3}H]PN200-110 to membranes isolated from the same tissue.

(+)[^{3}H]PN200-110 binding was determined in aortas and in mesenteric arteries bathed in normal and high-K^+ solutions. Nonspecific binding was not different in polarized and in depolarized arteries. On the contrary, as shown in FIGURE 1, specific binding was markedly enhanced in arteries depolarized for 30 minutes. Experimental estimates of specific binding were well fitted by one hyperbolic curve indicating a one-to-one binding of (+)PN200-110 to a single class of sites. Computer least-squares fitting of the data revealed that membrane depolarization significantly lowered apparent K_D ($p < 0.01$) while maximum binding capacity appeared not to be significantly affected. K_D value obtained in depolarized arteries (44 ± 2 pM) was similar to that found in membrane preparations and was close to the IC_{50} measured after 30 minutes of depolarization. When membrane potential of rat aorta was clamped at different levels by varying KCl concentration of the physiological solution, the apparent dissociation constant (K_{APP}) of (+)[^{3}H]PN200-110 varied as indicated in FIGURE 2.

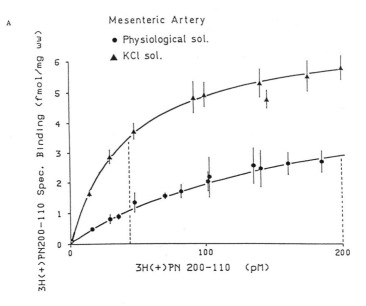

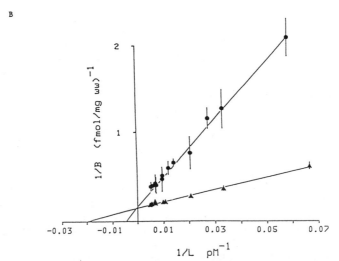

FIGURE 1. Binding of $(+)$[³H]PN200-110 to isolated mesenteric arteries. (**A**) Saturation study of the specific binding of $(+)$[³H]PN200-110 in mesenteric arteries bathed in physiological solution ($\bullet$) and in KCl-depolarizing solution ($\blacktriangle$). Each point represents measurements from 6 to 18 rings $\pm$ SEM. Dotted lines indicate K_D values as calculated with a curve-fitting program for nonlinear models (physiological solution: B_{max}: 5.8 $\pm$ 0.7 fmol/mg w.wt-K_D 200 $\pm$ 20 pM; KCl solution: B_{max}: 7.0 $\pm$ 0.1 pmol/mg w.wt-K_D 44 $\pm$ 2 pM). (**B**) Double reciprocal plot of the same data. (Modified from Morel & Godfraind.[2])

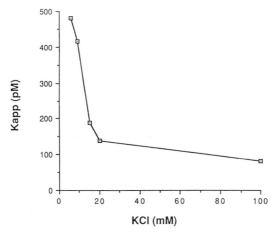

FIGURE 2. Influence of the KCl concentration in the medium on the specific binding of [³H]PN200-110 (100 p*M*) in isolated aorta. K_{APP} (apparent dissociation constant) was calculated from the relation $K_{APP} = L \dfrac{[B - b]}{b}$ with L = ligand concentration (100 p*M*); B = maximum binding capacity (4 fmol/mg); and b = specific binding of [³H]PN200-110 determined at each KCl concentration.

Our results can be interpreted by assuming the existence of two conformations of the receptor site in intact tissue, the distribution between the two conformations depending on the membrane potential. If the proportion (L) of low-affinity (K_L) sites decreases when the membrane potential is made less negative, and if the proportion (1 − L) of high-affinity (K_H) sites increases correspondingly, the apparent dissociation constant is given by:

$$K_{APP} = \frac{1}{L/K_L + [1 - L]/K_H}$$

In conclusion, the binding of PN200-110 to intact tissue, like its pharmacological effect, is stereoselective and voltage-dependent, and the time dependence of the inhibitory action in KCl-depolarized arteries is related to the time-dependence of the binding to the high-affinity conformation of the receptor. It is likely that the low- and high-affinity conformations correspond to the resting and inactivated (or open) states of the L-channels, respectively.

REFERENCES

1. GODFRAIND, T., R. C. MILLER & M. WIBO. 1986. Pharmacol. Rev. **38:** 321-416.
2. MOREL, N. & T. GODFRAIND. 1987. J. Pharmacol. Exp. Ther. **243:** 711-715.
3. WIBO, M., L. DEROTH & T. GODFRAIND. 1988. Circ. Res. **62:** 91-96.

The N-Type Ca Channel in Frog Sympathetic Neurons and Its Role in α-Adrenergic Modulation of Transmitter Release

SATHAPANA KONGSAMUT,[a] DIANE LIPSCOMBE,[a,b]
RICHARD W. TSIEN[a,b,c]

[a] Department of Cellular and Molecular Physiology
Yale University School of Medicine
New Haven, Connecticut 06510

[b] Department of Molecular and Cellular Physiology
Beckman Center
Stanford University Medical Center
Stanford, California 94305

Free cytosolic calcium ions regulate a variety of different cellular functions. In neurons, for example, the generation of an intracellular calcium transient triggers such diverse events as neurotransmitter release, short- and long-term changes in cell excitability, altered Ca-dependent enzyme activity, and altered gene expression. Given the diverse nature of these Ca-dependent responses, it is significant that several types of voltage-gated Ca channels have been described over the last few years.[1-13] With the general acceptance of the existence of multiple Ca-channel types, attention has turned toward basic questions about the physiological significance of diverse Ca channels and their control by neurotransmitters and hormones.[7,10,13]

This paper focuses on a particular class of voltage-gated Ca channels, referred to as N-type Ca channels.[13-15] In many neuronal preparations, N-type and L-type Ca channels appear to be distinct entities that contribute to high voltage activated Ca current.[8,9,14-18] Here we review information about their properties in frog sympathetic neurons, a particularly favorable preparation for studying N-type channels as distinct from L-type Ca channels.[19] We describe properties of neurotransmitter release from sympathetic neurons and the relative importance of N- and L-type Ca channels in regulating this release. Finally, we describe the selective down-modulation of N-type Ca channels by α-adrenergic agents and its possible role in inhibitory effects of norepinephrine (NE) on its own release.[20]

[c] Address for correspondence: R. W. Tsien, Department of Molecular and Cellular Physiology, Beckman Center, Stanford University Medical Center, Palo Alto, California 94305.

DISTINCTIONS BETWEEN N- AND L-TYPE CA CHANNELS

In frog sympathetic neurons, N- and L-type Ca channels are most easily identified and studied in single-channel recordings (FIGS. 1 and 2). The virtual lack of T-type Ca-channel activity in these cells,[19] as in rat sympathetic neurons,[21,22] makes the identification and separation of N and L-type Ca channels simpler than in other preparations that express T-type as well as N and L-type channels (e.g., chick sensory neurons[15]). N and L-type Ca channels differ in several respects, including single-channel conductance, voltage-dependence of activation and inactivation, and sensitivity to dihydropyridines (DHPs, see Tsien *et al.*[13] for review). In sympathetic neurons, L-type Ca channels have a unitary conductance of 25-28 pS (110 mM Ba, 20°C), are sensitive to modulation by DHPs, and are relatively resistant to inactivation with changes in holding potential. N-type Ca channels have a unitary conductance of 15-16 pS (110 mM Ba, 20°C), are insensitive to DHPs, and readily inactivate with depolarized holding potentials.

FIGURES 1 and 2 highlight the striking difference in the voltage-dependence of inactivation of N- and L-type Ca channels. Recordings from a cell-attached patch containing only L-type Ca channels are shown in FIGURE 1. The opening probability of L-type Ca channels evoked by test pulses to +20 mV is not affected by changing the holding potential from −100 mV to −40 mV. In contrast, the opening probability of N-type Ca channels decreases significantly within 10-20 seconds after the holding potential is changed from −80 mV to −40 mV. This can be seen in the two examples of N-type Ca-channel recordings shown in FIGURE 2. The recordings also illustrate that N-type Ca channels can inactivate with widely different time courses following a step depolarization. Panel A shows an example of a patch of N-type Ca channels that inactivate relatively rapidly ($\tau \sim 50$ msec) during a step depolarization to −10 mV. This may be compared with the example in panel B, which shows little or no inactivation of N-type Ca channels during a 320-msec pulse in another patch. The time course of N-channel inactivation is quite variable from patch to patch, ranging between these two extreme examples (see also Aosaki & Kasai[23]). With respect to unitary conductance and lack of sensitivity to dihyropyridines, the properties of N-type Ca channels recorded in different patches and in different regions of the neuron appear indistinguishable.[19,20]

FIGURE 3A shows an ensemble average of N-type Ca-channel current, calculated from 11 cell-attached patches recorded under identical conditions. The average current is composed of a prominently decaying component and a sustained component, reflecting the kinds of unitary activity illustrated in FIGURE 2. FIGURE 3B shows a series of whole-cell currents recorded with 2 mM external Ba for comparison with the averaged single-channel currents. The holding potentials are set at more negative values than in the single-channel recordings to allow for differences in external surface charge with 2 mM rather than 110 mM Ba. The whole-cell current is composed of different components that correspond rather well to the properties of N- and L-type channels found in single-channel recordings (FIGS. 1-3). The sustained current that remains available for activation even at a holding potential of −60 mV is dominated by current through L-type channels, which are less prone to inactivation with changes in holding potential (FIG. 1). The extra whole-cell current recruited at a holding potential of −100 mV is comprised of a prominently decaying and a more sustained current, similar in time course to the averaged currents carried by the inactivating (FIG. 2A and B) and sustained N-type Ca-channel activity (FIG. 2C and D).

The present results suggest that it is incorrect to assume that N-type Ca channels contribute only to the decaying component of whole-cell Ca-channel current, and that

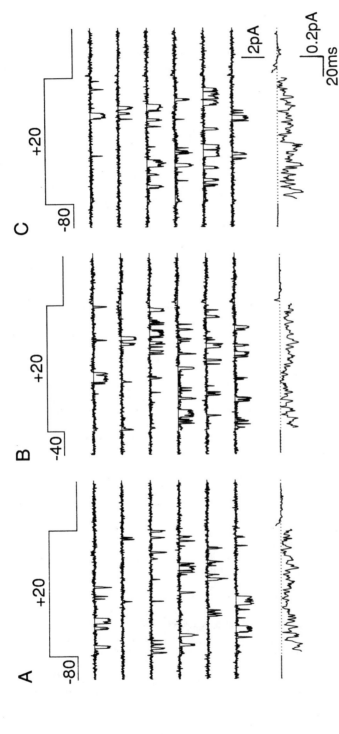

FIGURE 1. L-type Ca-channel activity is not responsive to changes in holding potential. Cell-attached patch recording containing only the L-type Ca channel, identified by its unitary conductance (25 pS between +20 mV and +40 mV), held at −80 mV (**A and C**) and −40 mV (**B**). Depolarizations to potentials more positive than +10 mV were required to activate the channel. Six sequential leak-subtracted current recordings are shown for each stimulation protocol. Average currents, plotted below the individual recordings, were calculated from more than 20 individual sweeps. The dotted line indicates zero current level. Patch pipette contained 110 mM Ba as the charge carrier and depolarizing pulses of 130-msec duration were applied every 4 seconds. No DHP was present (see Lipscombe et al.[19] for further details).

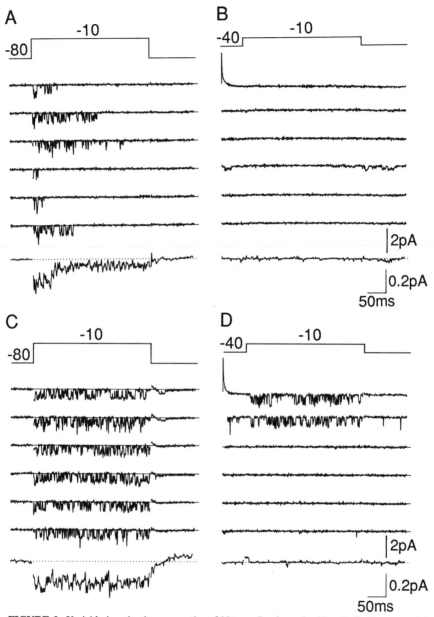

FIGURE 2. Variable inactivation properties of N-type Ca channels. (**A and C**) Six sequential sweeps showing unitary N-channel activity from two different patches evoked by a depolarization to −10 mV from a holding potential of −80 mV. (**B and D**) Six sequential sweeps from the same patches recorded immediately after a sudden change in holding potential from −80 to −40 mV (marked by upward capacity transient in first sweep). Average currents are plotted below each stimulation protocol. Average N-channel current in patch A is transient while average current in C shows little inactivation. Depolarizing pulses of 320-msec duration were applied every 10 seconds. Recording conditions as previously described in FIGURE 1 and Lipscombe *et al.*[20]

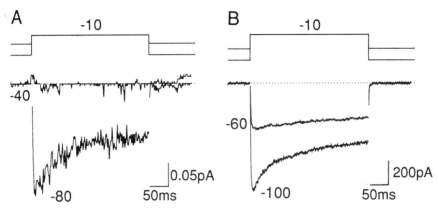

FIGURE 3. Comparison of average N-type Ca-channel activity from cell-attached patch recordings to a whole-cell current recording. (A) average of mean currents calculated from 11 cell-attached patches (including the two patches illustrated in Fig. 2) recorded under the same conditions as previously described.[19] Average currents evoked by depolarizing pulses to −10 mV from holding potentials of −80 mV and −40 mV are shown superimposed. In calculating the final average, contributions of individual patches were weighted equally. (B) Whole-cell currents recorded with 2 mM external Ba evoked by depolarizing pulses to −10 mV from holding potentials of −60 mV and −100 mV.

L-type channels underlie all of the late current. In earlier analysis of results from chick dorsal root ganglion (DRG) neurons, it was assumed as a first approximation that N-type Ca-channel current decayed rapidly and relatively completely.[16] The present data, along with single-channel analysis in rat sympathetic neurons,[22,24] PC12 cells,[24] and chicken or frog DRG neurons,[7,23] suggest that N-type Ca channels may contribute substantially to steady components of macroscopic Ca-channel current.

At present, no perfect method exists for separating N- and L-type Ca current in whole-cell recordings. Nevertheless, there is little doubt of the distinctness of these channels, given the combination of differences in single-channel conductance, kinetics, and pharmacology (see below).

STUDIES OF SYMPATHETIC TRANSMITTER RELEASE

Individual sympathetic varicosities are not readily accessible for direct study with patch-clamp methods. Thus, it has been difficult to study the role of different Ca entry pathways in the release of sympathetic transmitter and their electrophysiological basis in the same preparation. For this purpose, we took the approach of measuring radiolabeled transmitter release from cell bodies of sympathetic neurons. Evidence is available from earlier studies of Koketsu and colleagues[25,26] that cell bodies in sympathetic ganglia can release catecholamines. Miyagawa et al.[25] first suggested this possibility on the basis of electrophysiological measurements of synaptic transmission from preganglionic (cholinergic) terminals. Later, Suetake et al.[26] examined the catecholamine fluorescence intensity in cell bodies of sympathetic neurons (Fig. 4). Application of a prolonged electrical stimulation in the presence of an inhibitor of catecholamine synthesis produced a marked decrease in catecholamine content (B) relative to that found in matched ganglia that were left unstimulated (A).

FIGURE 5 illustrates our results using radiotracer methods to study the release of catecholamine from frog sympathetic ganglion cells. Most of the experiments were carried out with whole paravertebral sympathetic ganglia, although similar results have been obtained in experiments with dissociated neurons in culture (S. K. and K. R. Bley, unpublished). Sympathetic ganglia were incubated in [^{3}H]NE, washed, and stimulated with 50 mM K$^+$ to evoke secretion (see FIG. 5 legend for details). These challenges with K$^+$ allow for direct depolarization of the cells bypassing possible changes due to action potential propagation or duration. The transmitter release is steeply dependent on K$^+$ (range 30-60 mM) and extracellular Ca^{2+} (EC$_{50}$ ~0.5 mM) and is completely inhibited by cadmium ions (IC$_{50}$ ~ 10 μM). The effect of 5 μM Cd is illustrated in FIGURE 5A. In addition, ω-toxin from the marine snail *Conus geographus* is a potent inhibitor of secretion. All these observations are as expected for secretion triggered by Ca entry through voltage-gated Ca channels.[20]

In contrast to the pronounced inhibitory effects of cadmium ions and ω-conotoxin, DHP Ca-channel agonists and antagonists have little or no effect on transmitter release. FIGURE 5B and C show representative data obtained with experimental conditions chosen to favor a DHP effect. DHPs are added during a predepolarization period

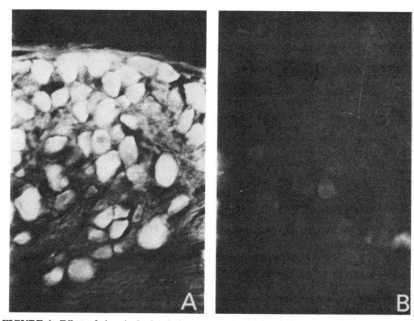

FIGURE 4. Effect of electrical stimulation on catecholamine fluorescence in the cell bodies of sympathetic neurons. Photomicrographs of formaldehyde-induced fluorescence of a pair of bull-frog paravertebral sympathetic ganglia. Ganglia were preincubated in Ringer's solution containing an inhibitor of catecholamine synthesis, α-methyl-p-tyrosine (80 μg/ml), for five hours. The specimens were exposed to formaldehyde vapor of 75% relative humidity at 80°C for three hours. (**A**) Unstimulated control ganglion. (**B**) Ganglion stimulated at 30 Hz for 60 min. (From Suetake, Kojima & Koketsu.[26] Used with permission.)

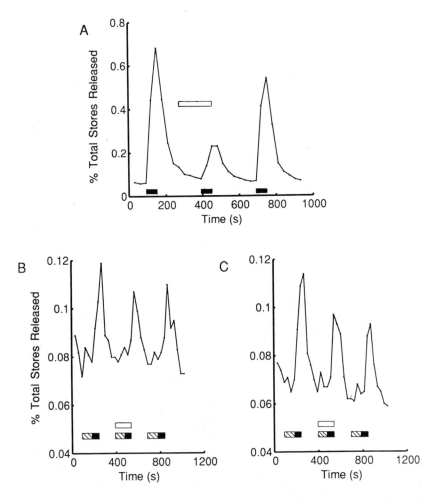

FIGURE 5. Properties of transmitter release from frog sympathetic neurons. (**A**) Effect of 5 μM Cd (open bar) on release evoked by 50 mM K$^+$ stimulation (solid bars). (**B,C**) Ganglia were superfused with low-Ca^{2+} solutions except where Ca^{2+} additions are indicated. Ganglia were predepolarized with 50 mM K$^+$ (no added Ca) for 1.5 min before evoking release (hatched bars). Release was evoked by addition of 2 mM Ca for 1 min in the continued presence of 50 mM K$^+$ (solid bars). Lack of effect of DHPs (open bars), 1 μM BAY K8644 (**B**) and 10 μM nitrendipine (**C**) on Ca^{2+}-evoked release. **Methods:** Sympathetic chains from frogs (*Rana pipiens pipiens*) were removed, incubated in a Ringer's solution (128 mM NaCl, 2 mM KCl, 10 mM HEPES, 10 mM glucose, 1 mM ascorbic acid [antioxidant], 0.1 mM pargyline [monoamine oxidase inhibitor], pH adjusted to 7.3 with NaOH) containing 10 μCi/ml [^{3}H]norepinephrine (NE) for two hours at room temperature. Excess ^{3}H was washed off by perfusing with Ringer's containing 10 μM desipramine (reuptake blocker) and 0 or 2 mM CaCl$_2$ for 35-40 minutes after which time a stable baseline release was achieved; thirty-second fractions were then collected continuously from this point. Results presented are individual representative experiments.

(without added Ca^{2+}) in order to promote DHP binding and to cause inactivation of transient Ca channels, thus favoring any release mediated by the long-lasting DHP-sensitive L-type Ca channels. Despite these measures, DHPs produce little inhibition or stimulation of release. Taken together with available electrophysiological evidence (see above), these observations indicate that release of transmitter from frog sympathetic neurons must be mediated primarily by N-type Ca channels, as has been previously reported by Hirning *et al.*[22] for rat sympathetic neurons.

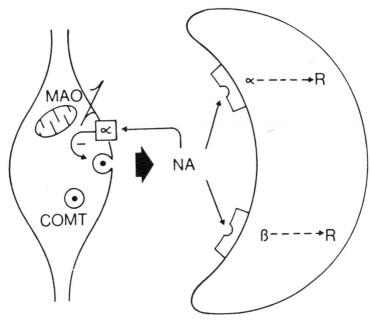

FIGURE 6. Schematic representation of a negative feedback mechanism for norepinephrine (NE) mediated by presynaptic α-adrenoceptors. Once it reaches a threshold concentration in the cleft, NE, released by nerve stimulation, activates presynaptic α-adrenoceptors leading to inhibition of further transmitter release. The presynaptic negative feedback mechanism is present both in tissues where the response (R) of the effector organ is mediated through α- or through β-adrenoceptors. MAO, monoamine oxidase; COMT, catechol-*O*-methyl transferase. (From Langer.[27] Used with permission.)

INHIBITION OF SECRETION BY NE

In the early 1970s, Langer and others proposed the idea of autoinhibition of transmitter release from sympathetic neurons.[27–33] In essence, locally released catecholamines act through α-adrenergic receptors to inhibit subsequent transmitter release through an autoinhibitory feedback loop (FIG. 6). The α-adrenergic response to released or circulating catecholamines appears to be a significant mechanism for regulating the level and spatial uniformity of transmitter output. This is the earliest and most extensively studied example of autoinhibition of release, but such a mechanism can also be seen in peripheral and central neurons containing other transmitters such as acetylcholine, ATP, dopamine, GABA, histamine, and serotonin.[29,30,32–34]

TABLE 1 and FIGURE 7 illustrate the inhibitory effect of NE on high K^+-evoked release of transmitter from frog sympathetic neurons. Experiments are performed essentially as described and illustrated in FIGURE 5A. The inhibition is reversible and dose dependent, reaching a maximal inhibition of $\sim 50\%$ at 10 to 30 μM NE (TABLE 1). The NE inhibition shows some pharmacological properties expected for an α_2 antagonist (TABLE 1; FIG. 7). Interestingly, no inhibition is seen with application of the α_2 agonist clonidine (up to 100 μM). These pharmacological properties, the lack of clonidine effect included, are characteristic of a distinct subtype of α_2 receptor previously described in sensory neurons[35] and neuroblastoma cells.[36]

Many investigators[28,29,32] have suggested that the mechanism underlying NE autoinhibition might involve altered phosphorylation of Ca channels or of intracellular or membrane-bound proteins such as synapsin I by cyclic AMP- or Ca-dependent protein kinases, as found in other systems.[37] The prevailing hypothesis may be summarized as follows:

$$\text{NE} \rightarrow \alpha_2 \text{ receptor} \rightarrow G_i \rightarrow \downarrow\text{cAMP} \rightarrow \downarrow\text{phosphorylation of Ca channels or}$$
$$\text{synapsin I or other intracellular protein} \rightarrow \downarrow\text{transmitter release}$$

TABLE 1. Effect of Norepinephrine on High K^+-Evoked Release of Transmitter from Frog Sympathetic Neurons

Treatment	[NE]	Percent Inhibition	SE	n
NE only	1	12.4	3.7	4
	3	31.9	3.8	4
	10	49.6	3.7	4
	30	50.0	5.2	7
NE + 10 μM yohimbine	10	17.2	4.3	4
	30	8.2	2.0	3
NE + 1 mM dibutyryl cyclic AMP	10	37.1	3.0	5
	30	42.8	4.8	7
NE + 300 μM H-7	30	47.5	2.4	8
NE with release stimulated by Ca addition in the presence of 10 μM ionomycin	30	12.1	5.2	4

NOTE: Experiments were performed essentially as described in the legend to FIGURE 5. Transmitter release was stimulated with three periods of depolarization by 50 mM K^+ in 2 mM Ca^{2+}: control, NE added and wash (see FIG. 5A; Lipscombe et al.[20]) with NE present for two minutes before and during the second depolarization. For each run, a fitted baseline was subtracted and the area under each peak was calculated (A_{con}, A_{NE}, and A_{wash}). Percent inhibition was calculated according to the equation: % inhibition $= 100 * [1 - (A_{NE}/(A_{con} + A_{wash})/2)]$. For ionomycin-induced release, basal release was obtained by washing with Ringer containing 2 mM EGTA and 100 nM free calcium; 10 μM ionomycin was added to the perfusion Ringer 10 minutes before collecting fractions; release was evoked by addition of 1 mM free calcium (buffered with 2 mM EGTA) in the continued presence of 10 μM ionomycin to bypass calcium entry through voltage-sensitive Ca channels. SE = standard error; n = number of runs.

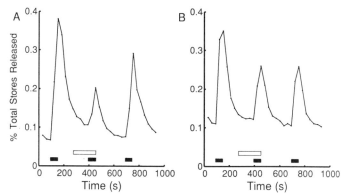

FIGURE 7. Inhibitory effect of NE on K$^+$-stimulated transmitter release. Experiments were conducted essentially as described in FIGURE 5A. Two representative examples are shown. (**A**) Inhibitory effect of 30 μM NE (open bar), present during the second K$^+$ challenge (filled bars). (**B**) Inhibitory effect of 30 μM NE (open bar) is antagonized when 10 μM yohimbine is present throughout the run.

This mechanism cannot, however, account for our findings in sympathetic neurons (TABLE 1; Lipscombe *et al.*[20]). The inhibition of NE persists in the presence of a saturating concentration (1 mM) of dibutyryl cyclic AMP, indicating that NE is not acting by reducing cyclic AMP levels (see also Johnston *et al.*[38]). Furthermore, the NE inhibition is not prevented by 300 μM H-7, a nonspecific inhibitor of several protein kinases,[39] including cyclic AMP- and calmodulin-dependent protein kinases and protein kinase C. This suggests that phosphorylation by H-7-sensitive protein kinases is not involved in the mechanism of inhibition (see also Wanke *et al.*[40]). Thus, it appears that NE can act on transmitter release without the involvement of the intracellular messengers that have previously been invoked.

To determine the locus of the inhibitory action, transmitter release was evoked with Ca in the presence of ionomycin to bypass Ca entry through voltage-gated calcium channels. Under these conditions, NE is ineffective in inhibiting transmitter release (TABLE 1 and Lipscombe *et al.*[20]), suggesting that NE does not inhibit processes subsequent to a rise in intracellular Ca. It is likely that NE acts by modulating Ca entry because NE and other catecholamines have been shown to modulate Ca currents or Ca-dependent action potentials in this (see below and Lipscombe *et al.*[20]) and other preparations of sympathetic neurons[21,41,42] in DRG neurons[21,35,43-46] and in neuroblastoma cells.[36]

NE MODULATION OF CA-CHANNEL CURRENT

In frog sympathetic neurons, NE rapidly and reversibly inhibits whole-cell Ca-channel currents. FIGURE 8A illustrates the effect of NE on whole-cell currents recorded with 10 mM Ca in the external solution. Qualitatively similar results can be obtained with either external Ca (2-10 mM) or Ba (1-2 mM) as the charge carrier. The pharmacology of the NE-mediated inhibition of Ca current is identical to the inhibitory effects on sympathetic transmitter release. The response to NE is antagonized by phentolamine and by yohimbine, but not by propranolol, a β-antagonist. As in the release studies, clonidine does not mimic the action of NE. This strong correlation between the pharmacological characteristics of the noradrenergic effect on Ca-channel currents and release is consistent with the same receptor mediating both events.[20]

Although the detailed mechanism by which α-adrenoceptor stimulation leads to inhibition of Ca channels is not clear, it is likely that a GTP-binding protein mediates at least part of the response. Thus, (1) GTP (0.3 mM) has to be present in the recording pipette in order to obtain multiple responses to NE, (2) substituting GTP with low concentrations of a nonhydrolyzable analogue, GTP-γ-S (0.05 mM), largely

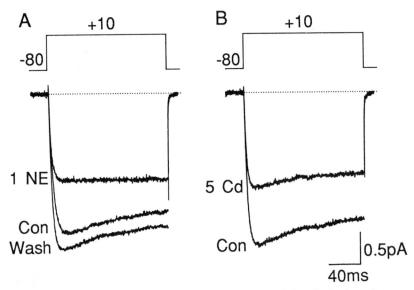

FIGURE 8. Effect of 1 μM NE (**A**) and 5 μM Cd (**B**) on whole-cell, Ca-channel currents recorded from the same cell. Currents were recorded with 10 mM external Ca as the charge carrier in solutions (internal and external) that effectively blocked other voltage-gated conductances (see Fox et al.[15] for details). Depolarizing pulses lasting 160 msec were applied every 10 sec. Leak-subtracted Ca-channel currents are shown superimposed below the appropriate voltage pulse protocols.

prevents recovery from the inhibitory effect of NE, and (3) high concentrations of GTP-γ-S (0.5 mM) mimic the inhibitory effects of NE.[20]

Selective Inhibition of N-Type Ca Channels in Whole-Cell Recordings

The inhibitory effects of NE on peak Ca-channel current are incomplete, usually not exceeding about 50% of control, even at maximally effective NE concentrations (between 10 and 100 μM NE in 2 mM external Ba, Lipscombe et al.[20]) and even in combination with intracellular GTP-γ-S. The incompleteness of the NE effect raises the question of whether one or both types of high-voltage-activated Ca channel might be partially inhibited (see also Refs. in Tsien *et al.*[13]).

FIGURE 8 shows the inhibitory effect of 1 μM NE on whole-cell Ca currents (A) and compares it to the effect of 5 μM Cd (B), a nonselective inhibitor of N- and L-type Ca channels. The inhibitory effects of these two blockers on the same cell are quite different. NE eliminates the decaying component and partially inhibits a sustained component of the control Ca current (A). The NE-sensitive current has a time course of inactivation consistent with the kinetic properties of N-type Ca channels described in FIGURE 2. In contrast, Cd inhibits the Ca current without affecting the time course of current decay compared to control, as expected for a nonselective inhibitor of both types of Ca channels. Thus, the lack of an inactivating component in the presence of NE cannot be attributed to a simple reduction in overall Ca influx and less current-dependent or Ca-dependent inactivation (*cf.* Docherty & McFadzean[36]). The most likely explanation is that NE is selectively inhibiting the N-type Ca-channel current. This idea is supported by the dependence of the NE effect on the holding potential (FIG. 9). While NE strongly reduces the current evoked from a negative holding potential of −120 mV (A), it has less effect on the sustained, largely L-type Ca current evoked from a less negative holding potential of −60 mV (B). The holding potential sensitivity of the NE-sensitive current parallels that of N-type Ca-channel currents described in FIGURES 2 and 3.

Modulation of Single N-Type Ca Channels

Unitary recordings of Ca-channel activity provide a means for testing our hypothesis that NE acts selectively on the N-type Ca channel and allows us to analyze the mechanism of inhibition in more detail. While many neurotransmitters have been shown to inhibit whole-cell, high-voltage-activated Ca currents in neurons (see Miller[10] and Tsien *et al.*[13] for references), little is known so far about how inhibition occurs at the level of single channels. One impediment has been rundown of N and L-type channels, which is particularly problematic for studies of inhibitory modulation. This may have discouraged the use of outside-out patches as applied in studies of NE modulation of low voltage-activated T-type Ca channels by Marchetti *et al.*[21]

To avoid problems of Ca channel rundown, we can take the approach of comparing large numbers of cell-attached patch recordings with and without NE in the pipette (external) solution. Because the majority of cell-attached patch recordings contain both N- and L-type Ca channels,[19] we can choose a stimulation protocol that allows them to be studied in isolation (FIG. 10). N-type Ca channels are activated selectively by stepping from a relatively negative holding potential of −80 mV to −10 mV (A). At this test potential, openings of L-type channels are extremely rare in the absence of DHP agonists. Depolarizing the holding potential from −80 mV to −40 mV largely inactivates N-type channels (B), but leaves L-type channels available for opening with stronger test depolarizations to +20 mV (C).

With these voltage protocols, we can distinguish between changes in the activity of N- and L-type Ca channels in membrane patches exposed to NE.[20] A representative example of the effect of 100 μM NE on N- and L-type Ca channels is illustrated in FIGURE 10D-F. NE inhibits the N-type Ca-channel current activated from a holding potential of −80 mV (D) while the activity of L-type Ca channels evoked by de-

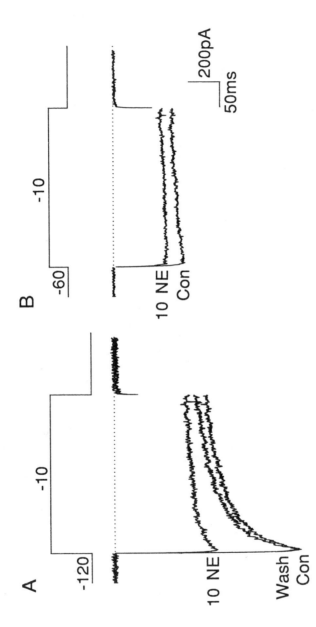

FIGURE 9. Holding potential dependence of NE-sensitive whole-cell Ca current. Whole-cell currents were recorded with 2 mM external Ba as the charge carrier. Depolarizing pulses lasting 320 msec were applied every 10 sec. The effect of 10 μM NE on currents evoked by pulses to −10 mV from holding potentials of −120 mV (**A**) and −60 mV (**B**) recorded from the same cell. Leak-subtracted records are shown superimposed below the appropriate voltage pulse protocol.

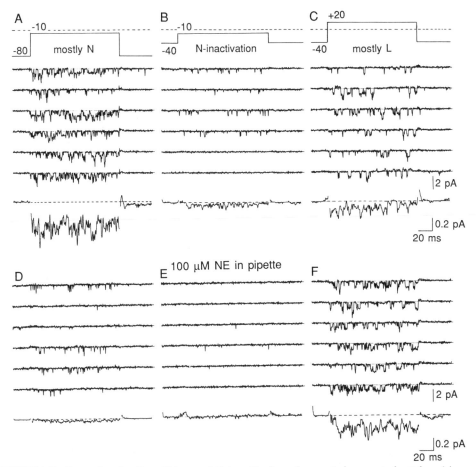

FIGURE 10. Properties of unitary N-type and L-type Ca-channel currents in a control patch and in a patch exposed to 100 μM NE present in the pipette. Recordings from two cell-attached patches with 110 mM Ba in the absence (**A**-**C**) and presence (**D**-**F**) of 100 μM NE in the patch pipette. (Exemplars of collected data from a large number of patches; see FIG. 11.) Six sequential current recordings are shown above average currents from at least 20 individual sweeps for each voltage protocol. Depolarizing pulses, 130 msec in duration, were applied every 4 sec. (**A**) N-type Ca-channel activity evoked by voltage pulses to −10 mV from a holding potential of −80 mV. (**B**) N-type Ca-channel activity mostly inactivated by changing the holding potential to −40 mV. (**C**) In the same patch recording stronger depolarizing pulses to +20 mV recruits L-type Ca-channel activity from a holding potential of −40 mV. (**D**-**F**) N-type and L-type Ca-channel activity in a cell-attached patch exposed to 100 μM NE present in the recording pipette. (**D**) Openings of N-type Ca channels are briefer and less frequent in the presence of NE compared to control (**A**). (**F**) The kinetics of L-type Ca channels are not obviously affected by the presence of NE.

polarizing pulses to $+20$ mV from a holding potential of -40 MV (F) is not significantly different from control patches (C). FIGURE 11A-D shows collected data from 33 control patches and 30 patches exposed to 10-100 μM NE.[20] Currents carried by N-type channels are strongly inhibited at 30 μM (C) and 100 μM (D) NE, though inhibition of N-type channel activity is not complete even at these high concentrations. In contrast, L-type Ca-channel activity is not significantly inhibited at any NE concentration over the range between 10-100 μM.[20]

To test whether signal transduction involves production of a readily diffusible second messenger such as cyclic AMP, cyclic GMP, or diacylglycerol, we can expose

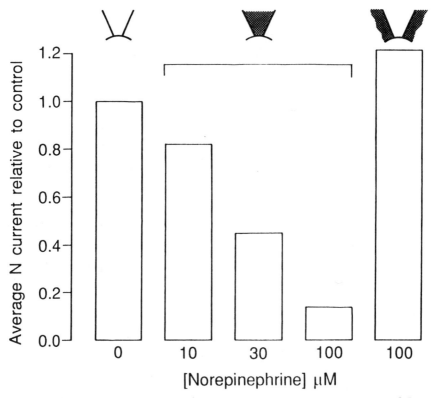

FIGURE 11. NE-mediated inhibition of N-type Ca-channel current is dose-dependent and not mediated by a readily diffusable cytosolic messenger. Averaged N-type Ca-channel currents from groups of cell-attached patches are normalized relative to currents in control patches (1.0) and plotted as bars. N-type Ca-channel activity was evoked by depolarizing pulses to -10 mV from a holding potential of -80 mV. Channel activity was identified as N-type because of its unitary conductance and sensitivity to holding potential as illustrated in FIG. 2. Illustrations at the top of the bar graph indicate control recordings with no drug (left), with 10, 30 and 100 μM NE in the recording pipette (middle), or with 100 μM NE applied to the bulk of the cell outside of the pipette (right). Mean currents from n individual patches were weighted equally to produce the average currents for each experimental condition: 0 NE inside the pipette ($n = 22$), 10 μM NE ($n = 15$), 30 μM NE ($n = 10$), and 100 μM NE ($n = 5$). The rightmost bar shows the average N-type Ca-channel activity after addition of 100 μM NE to the outside of cells ($n = 5$), normalized by activity recorded from the same patches before addition of NE.

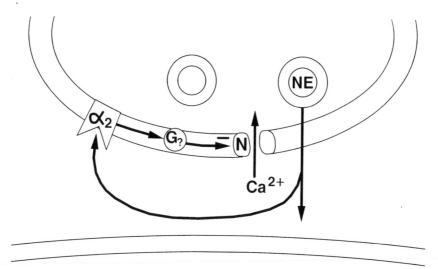

FIGURE 12. Scheme for NE-mediated autoinhibition. Depolarization causes opening of Ca channels leading to Ca entry. The Ca influx through N-type Ca channels is dominant in triggering release of vesicular sympathetic transmitter (NE) by exocytosis. Released or circulating NE can bind to an α_2 receptor to activate a G-protein leading to inhibition of N-channel activity, Ca entry and release. This scheme allows for discrete, localized regulation of Ca influx and release.

the bulk of the cell to NE and look for changes in the activity of N-type channels effectively isolated from the bathing solution by the patch pipette. In five patches, formed with pipettes of ~ 1 μm diameter, the mean current through N-type channels remains unchanged following application of 100 μM NE to the rest of the cell (FIG. 11E and F). In none of the experiments is there a detectable decrease in activity after the drug addition.[20] Thus, the inhibitory action of NE on Ca channels is not likely to be mediated by a second messenger that is free to diffuse from receptors outside the patch to channels within the patch.

Kinetic Basis of Downmodulation

NE inhibits N-type Ca-channel activity by altering its gating kinetics while having no significant effect on the unitary current amplitude (FIG. 10). The mean open time ($<t_o>$) of the N-type Ca channel decreases from 0.87 $\pm$ 0.14 msec in control to 0.38 $\pm$ 0.07 msec or 0.40 $\pm$ 0.09 msec in 30 or 100 μM NE, respectively.[20] The more than twofold abbreviation of N-type channel openings contributes substantially to the more than threefold decrease in average current ($<I>$) seen overall.

Changes in the kinetic steps leading to channel opening are also evident. The latency between the depolarizing step and the first detectable channel opening is roughly twice as large in the presence of 10-100 μM NE (21 msec) relative to the control (12 msec). Detailed kinetic interpretation of this measurement or of other closed time intervals is complicated by uncertainty about the total number of channels in individual patches. Nevertheless, the results suggest a slowing in rate constants leading to channel opening. Changes in gating kinetics on a time scale slower than the duration of the depolarizing pulses are also suggested by a significant increase in the percentage of sweeps that contain no detectable opening (8% in control, 23% in 30-100 μM NE).

DISCUSSION

In this paper we discuss information about properties of N- and L-type Ca channels, their contribution to whole-cell Ca current, and their relative importance in triggering sympathetic transmitter release. We present evidence that N-type channels are selectively inhibited by NE and describe the mechanism of inhibition at the level of single channels. Down-modulation of N-type channel activity appears to play an important role in α-adrenergic control of transmitter release.

Two Major Types of High-Voltage-Activated Ca Channels

Single-channel recordings from frog sympathetic neurons provide very clear distinctions between N-type and L-type Ca channels. As we have illustrated, these channels differ markedly in their single-channel conductance, voltage dependence, and kinetics.[13,19] Thus, it is possible with appropriate voltage protocols to study each of these channel types in isolation even if both are present in the same patch (e.g. FIG. 10). The rate of inactivation of the N-type Ca channel may vary widely during depolarizing pulses that activate the channel. Although this variability is evident in earlier published records from sympathetic neurons,[19,22] it is particularly striking in the series of cell-attached patch recordings reported here. Variability of the rate of inactivation has also been documented for cardiac L-type Ca channels by Cavalie *et al.*[47] Other groups have described rapidly decaying or slowly decaying N-type channel behavior in rat sympathetic neurons and PC12 cells[24] and chick and frog DRG neurons.[7,23] Regardless of their rate of inactivation during depolarizing pulses, N-type Ca channels differ consistently from L-type channels in their steady-state inactivation with prolonged depolarizations (FIG. 1-3).

All of this analysis suggests that it is precarious and possibly incorrect to split whole-cell current into different decaying exponential components and to assign the components to particular types of Ca channels. Rapidly or slowly inactivating N-type channel activity could also contribute to tail currents with similar deactivation kinetics following short or long depolarizing pulses.[24,48]

Selective Inhibition of N-Type Channels

Stimulation of α-receptors by NE provides a selective means of inhibiting N-type but not L-type Ca-channel activity. The sustained and decaying forms of N-type Ca-channel activity are both strongly, although not completely, inhibited. The absence of an α-adrenergic effect on L-type Ca channels was demonstrated with both whole-cell recordings (FIG. 9) and single-channel recordings from many cell-attached patches (FIGS. 10 and 11). Our results provide an explanation for the incompleteness of NE inhibition of Ca current in earlier voltage-clamp experiments.[36,41,44,45]

On the bases of whole-cell Ca-channel recordings, it seems likely that NE shares a common inhibitory mechanism of action with several neuroeffectors such as acetylcholine, LHRH, GABA, adenosine, and substance P (e.g., Refs. 10, 11, 17, 40, 49). Evidence for this hypothesis is based on (1) the similarity in the voltage and time-dependence of the transmitter-sensitive currents, (2) mutual occlusion of NE and LHRH effects in frog sympathetic neurons (K. R. Bley, unpublished observations), and (3) effects of internal GTP-γ-S, acting at low concentrations to render the inhibitory effects of the Ca current effectively irreversible[20] and at high concentrations to directly mimic neurotransmitter effects.[17,50,51]

Apart from involvement of G-proteins, the mechanism of noradrenergic modulation of Ca channels in sympathetic neurons appears to be fundamentally different from that in sensory neurons. Dunlap and colleagues[43,52,53] have demonstrated that PKC activation is required for NE-mediated downmodulation of Ca currents in chick DRG neurons. In contrast, the effect of NE on sympathetic neurons is not mimicked by agents known to stimulate protein kinase C, such as phorbol diacetate or phorbol dibutyrate (0.1-1 μM). In fact, acute application of phorbol esters consistently increases the activity of both N- and L-type Ca channels recorded in cell-attached patches by increasing their probability of opening.[54] Likewise, in earlier experiments in rat sympathetic neurons, Wanke *et al.*[40] found that ACh modulation via muscarinic receptors was not blocked by H-7 or mimicked by activators of PKC. Diversity in the signaling mechanisms linking transmitter-receptor activation and Ca-channel inhibition in different cell types have been highlighted by Wanke and Ferroni.[55] They showed, in parallel studies, that PKC activation did not mediate the effects of ACh in rat sympathetic neurons, while PKC stimulation is an important link in the signaling pathway in chick sensory neurons.[52,53]

Mechanism of NE Inhibition at the Single-Channel Level

Previously published studies of downmodulation of neuronal high-voltage-activated Ca currents by neurotransmitters have relied on whole-cell recordings. The experiments described here provide new information about the downmodulation at the single-channel level. NE produces a pronounced change in the rapid gating kinetics, which includes a marked abbreviation of N-type Ca-channel openings and an increase in the latency-to-first opening. The results clearly indicate that the mechanism of NE inhibition is not a simple elimination of channels or a mere prolongation of steps leading up to channel opening.

The signaling mechanism that links the α-receptor to the N-type Ca channel involves a GTP-binding protein. Further experiments are needed to determine whether the G-protein directly interacts with the N-type Ca channel or whether an intermediary molecule is involved. Our results indicate that the coupling between the α-adrenergic receptor and the N-type Ca channel does not involve the production of a readily diffusible second messenger (FIG. 11).

Mechanisms Controlling Transmitter Release

We have measured K^+-evoked transmitter release from cell bodies of frog sympathetic neurons. This system offers the possibility of directly measuring transmitter release and Ca currents in essentially the same preparation, an advantage over most preparations containing synaptic terminals. We have used this model system to determine which type(s) of voltage-gated Ca channel dominates transmitter release and which channel is the target of α-adrenergic mediated inhibition.

Effects of DHPs

Our findings suggest that the N-type Ca channel dominates transmitter release from frog sympathetic neurons, similar to previous findings in cultured rat superior cervical ganglion neurons.[22] [^{3}H]NE release is not affected by DHPs, agents known to act selectively on L-type channels. Insensitivity to DHPs is found even when the drug is applied during a predepolarization in the absence of external Ca; such conditions have been shown to favor DHP binding.[56,57] In contrast, Cd and ω-conotoxin, agents known to block N-type Ca channels, are effective inhibitors of K^+-evoked transmitter release.

Effect of α-Adrenergic Stimulation

Additional support for the importance of N-type Ca channels in regulating release comes from experiments studying effects of α-adrenergic stimulation with exogenous NE. NE inhibition of N-type Ca-channel activity is accompanied by a consistent and sizeable reduction of K^+-evoked transmitter release. Inhibition of Ca current and transmitter release are closely correlated in several respects. In both cases (1) the IC_{50} for NE inhibition is ~ 3 μM with 1-2 mM external Ca or Ba[20]; (2) NE block is incomplete, even at maximally effective concentrations; (3) the pharmacology is that of a subtype of α_2-like receptor that is blocked by phentolamine and yohimbine and is not stimulated by clonidine (similar to α_2-like receptors in other neuronal preparations[35,36]). Because α-stimulation fails to significantly reduce transmitter release evoked by ionomycin-induced Ca entry, we conclude that inhibition of Ca entry, particularly through N-type channels, is responsible for the reduction in K^+-evoked transmitter release.

Thus, experiments with DHPs and α-adrenergic stimulation provide complementary approaches to identifying the Ca entry system(s) that dominate transmitter release. On one hand, selective enhancement or reduction of Ca influx via L-type Ca channels by DHPs has little or no effect on transmitter release; on the other hand, selective reduction of Ca influx via N-type Ca channels by α-adrenergic agents strongly decreases release.

Because L-type channels have been shown to coexist with N-type channels on the cell body of sympathetic neurons,[19] these results reinforce the idea that L-type channels are somehow at a disadvantage in triggering catecholamine release. The reasons for this are not obvious because DHP-sensitive L-type Ca channels make a substantial contribution to the macroscopic whole-cell current and to Ca transients detected with fura-2.[22] Global measurements of intracellular calcium transients and calcium-channel currents do not necessarily predict which Ca delivery systems are important for evoking transmitter release. In this respect it is notable that caffeine induces sizeable intra-cellular Ca transients[58] but is completely ineffective in evoking sympathetic transmitter release (SK, unpublished).

We do not know yet why Ca entry through N-type Ca channels is so effective in producing transmitter release. One possibility is that the transmitter release mechanism has fast on and off kinetics and a low affinity for Ca so that it fails to respond to global Ca transients in the micromolar range. Such release mechanisms might be anchored near the mouths of N-type channels where the local concentration of Ca ions might transiently achieve much higher levels. Similar ideas have been invoked to explain the speed of transmitter release at presynaptic terminals.[59-61]

Mechanism(s) of Autoinhibition

Our results support the following mechanism for noradrenergic inhibition of K^+-evoked transmitter release:

NE $\rightarrow$ α_2-like receptor $\rightarrow$ $G_?$ $\rightarrow$ $\downarrow$N-type Ca channel activity $\rightarrow$ $\downarrow$release

However, it remains possible that the physiological effect of NE on transmitter released by action potentials involves other mechanisms. For example, one hypothesis might be as follows:

NE $\rightarrow$ α_2 receptor $\rightarrow$ G_i $\rightarrow$ $\downarrow$cAMP $\rightarrow$ $\downarrow$phosphorylation $\rightarrow$ $\uparrow$K channel activity $\rightarrow$ $\downarrow$release

This scenario is based on evidence in other neurons where K conductances are increased by stimulation of a clonidine-sensitive α_2-receptor (e.g. Williams *et al.*[62]) and inhibited by increases in cyclic AMP.[63] A mechanism based on potassium-channel modulation would modify release through changes in action potential duration. This is an electrical effect that would spread along axonal varicosities over distances of the order of the electrical space constant. In contrast, α-adrenergic modulation of N-type Ca channels might provide short-range feedback control: signaling from receptor to channel and from Ca entry to transmitter release would be highly localized.

CONCLUSIONS

Frog sympathetic neurons allow a relatively direct comparison between electro-physiological measurements of Ca-channel properties and radiotracer measurements of transmitter release. Selective α-adrenergic inhibition of N-type Ca-channel activity is seen in both whole-cell and single-channel recordings, and provides an attractive explanation for autoinhibition of sympathetic transmitter release, especially because transmitter release is thought to be triggered selectively by Ca influx through N-type Ca channels. The downmodulation of N-type Ca channels is mediated by a G-protein and is dominated by changes in rapid gating kinetics on a millisecond time scale. Overall, our experiments support the idea that N- and L-type channels are distinct entities, differing in their importance for catecholamine release and their responsiveness to neurotransmitters.

REFERENCES

1. HAGIWARA, S. & L. BYERLY. 1980. Ann. Rev. Neurosci. **4:** 69-125.
2. LLINAS, R. & Y. YAROM. 1981. J. Physiol. **315:** 569-584.
3. DEITMER, J. W. 1984. J. Physiol. **355:** 137-159.
4. CARBONE, E. & H. D. LUX. 1984. Nature **310:** 501-502.
5. CARBONE, E. & H. D. LUX. 1984. Biophys. J. **46:** 413-418.
6. ARMSTRONG, C. M. & D. R. MATTESON. 1985. Science **277:** 65-66.
7. BEAN, B. P. 1989. Ann. Rev. Physiol. **51:** 367-384.
8. DUPONT, J.-L., J.-L. BOSSU & A. FELTZ. 1986. Pfluegers Arch. **406:** 433-435.
9. KOSTYUK, P. G., Y. M. SHUBA & A. N. SAVCHENKO. 1988. Pfluegers Arch. **411:** 661-669.
10. MILLER, R. J. 1987. Science **235:** 46-52.
11. GROSS, R. A. & R. L. MACDONALD. 1987. Proc. Natl. Acad. Sci. USA **84:** 5469-5473.
12. AUGUSTINE, G. J., M. P. CHARLTON & S. J. SMITH. 1987. Ann. Rev. Neurosci. **10:** 633-693.
13. TSIEN, R. W., D. LIPSCOMBE, D. V. MADISON, K. R. BLEY & A. P. FOX. 1988. Trends Neurosci. **11:** 431-437.
14. NOWYCKY, M. C., A. P. FOX & R. W. TSIEN. 1985. Nature **316:** 440-443.
15. FOX, A. P., M. C. NOWYCKY & R. W. TSIEN. 1987. J. Physiol. **394:** 173-200.
16. FOX, A. P., M. C. NOWYCKY & R. W. TSIEN. 1987. J. Physiol. **394:** 149-172.
17. DOLPHIN, A. C. & R. H. SCOTT. 1987. J. Physiol **386:** 1-17.
18. COTTRELL, G. A. & K. A. GREEN. 1987. J. Physiol. **392:** 32P.
19. LIPSCOMBE, D., D. V. MADISON, M. POENIE, H. REUTER, R. Y. TSIEN & R. W. TSIEN. 1988. Proc. Natl. Acad. Sci. USA **85:** 2398-2402.
20. LIPSCOMBE, D., S. KONGSAMUT & R. W. TSIEN. 1989. Submitted.
21. MARCHETTI, C., E. CARBONE & H. D. LUX. 1986. Pfluegers Arch. **406:** 104-111.
22. HIRNING, L. D., A. P. FOX, E. W. MCCLESKEY, B. M. OLIVERA, S. A. THAYER, R. J. MILLER & R. W. TSIEN. 1988. Science **239:** 57-61.
23. AOSAKI, T. & H. KASAI. 1989. Pfluegers Arch. In press.
24. PLUMMER, M. R., D. E. LOGOTHETIS & P. HESS. 1989. Neuron. In press.
25. MIYAGAWA, M., S. MINOTA & K. KOKETSU. 1981. Brain Res. **224:** 305-313.
26. SUETAKE, K., K. KOJIMA, K. INANAGA & K. KOKETSU. 1981. Brain Res. **205:** 436-440.
27. LANGER, S. Z. 1977. Br. J. Pharmacol. **60:** 481-497.
28. LANGER, S. Z. 1981. Pharmacol. Rev. **32:** 337-362.
29. STARKE, K. 1987. Rev. Physiol. Biochem. Pharmacol. **107:** 73-146.
30. VIZI, E. S. 1979. Progr. Neurobiol. **12:** 181-290.
31. GILLESPIE, J. S. 1980. *In* Handbook of Experimental Pharmacology. **54(I):** 353-425.
32. MULDER, A. H., A. L. FRANKHUYZEN, J. C. STOOF, J. WEMER & A. N. M. SCHOFFEL-MEER. 1984. *In* Catecholamines: Neuropharmacology and Central Nervous System—Theoretical Aspects: 47-58. Alan Liss. New York.
33. ILLES, P. 1986. Neuroscience **17:** 909-928.
34. ALBERTS, P., T. BARTFAI & L. STJARNE. 1981. J. Physiol. **312:** 297-331.
35. CANFIELD, D. R. & K. DUNLAP. 1984. Br. J. Pharmacol. **82:** 557-563.
36. DOCHERTY, R. J. & I. MCFADZEAN. 1989. Eur. J. Neurosci. In press.
37. LLINAS, R., T. L. MCGUINNESS, C. S. LEONARD, M. SUGIMORI & P. GREENGARD. 1985. Proc. Natl. Acad. Sci. USA **72:** 187-190.
38. JOHNSTON, H., H. MAJEWSKI & I. F. MUSGRAVE. 1987. Br. J. Pharmacol. **91:** 773-781.
39. HIDAKA, H., M. INAGAI, S. KAWAMOTO & Y. SASAKI. 1984. Biochemistry: **23:** 5036-5041.

40. WANKE, E., A. FERRONI, A. MALGAROLI, A. AMBROSINI, T. POZZAN & J. MELDOLESI. 1987. Proc. Natl. Acad. Sci. USA **84**: 4313-4317.
41. GALVAN, M. & P. R. ADAMS. 1982. Brain Res. **244**: 135-144.
42. MCAFEE, D. A., B. K. HENON, J. P. HORN & P. YAROWSKY. 1981. Fed. Proc. **40**: 2246-2249.
43. DUNLAP, K. & G. D. FISCHBACH. 1981. J. Physiol. **317**: 519-535.
44. FORSCHER, P. & G. S. OXFORD. 1985. J. Gen. Physiol. **85**: 743-763.
45. FORSCHER, P., G. S. OXFORD & D. SCHULTZ. 1986. J. Physiol. **379**: 131-144.
46. HOLZ, G. G. IV, S. G. RANE & K. DUNLAP. 1986. Nature **319**: 670-672.
47. CAVALIE, A., D. PELZER & W. TRAUTWEIN. 1986. Pfluegers Arch. **406**: 241-258.
48. SWANDULLA, D. & C. M. ARMSTRONG. 1988. J. Gen. Physiol. **92**: 197-218.
49. MADISON, D. V., A. P. FOX & R. W. TSIEN. 1987. Biophys. J. **51**: 30a.
50. DOLPHIN, A. C., S. R. FORDA & R. H. SCOTT. 1986. J. Physiol. **373**: 47-61.
51. BLEY, K. R. & R. W. TSIEN. 1988. Biophys. J. **53**: 253a.
52. RANE, S. G. & K. DUNLAP. 1986. Proc. Natl. Acad. Sci. USA **83**: 184-188.
53. ANDERSON, C. S. & K. DUNLAP. 1988. Soc. Neurosci. Abstr. **14**: 644.
54. LIPSCOMBE, D., K. R. BLEY & R. W. TSIEN. 1988. Soc. Neurosci. Abstr. **14**: 153.
55. WANKE, E. & A. FERRONI. 1988. *In* Neurotransmitters and Cortical Function. M. Avoli, T. A. Reader, R. W. Dykes & P. Gloor, Eds.: 277-286. Plenum. New York.
56. BEAN, B. P. 1984. Proc. Natl. Acad. Sci. USA **81**: 6388-6392.
57. KOKUBUN, S., B. PROD'HOM, C. BECKER, H. PORZIG & H. REUTER. 1986. Mol. Pharmacol. **30**: 571-584.
58. LIPSCOMBE, D., D. V. MADISON, M. POENIE, H. REUTER, R. W. TSIEN & R. Y. TSIEN. 1988. Neuron **1**: 355-365.
59. SIMON, S. M. & R. LLINAS. 1985. Biophys. J. **48**: 485-498.
60. ZUCKER, R. L. & A. L. FOGELSON. 1986. Proc. Natl. Acad. Sci. USA **83**: 3032-3036.
61. SMITH, S. J. & G. J. AUGUSTINE. 1988. Trends Neurosci. **11**: 458-464.
62. WILLIAMS, J. T., G. HENDERSON & R. A. NORTH. 1985. Neuroscience **14**: 95-101.
63. DUNLAP, K. 1985. Pfluegers Arch. **403**: 170-174.

Multiple Types of Calcium Channels in Heart Muscle and Neurons

Modulation by Drugs and Neurotransmitters[a]

BRUCE P. BEAN

Department of Neurobiology
Harvard Medical School
Boston, Massachusetts 02115

Channels that are opened by depolarization and are selectively permeable to Ca ions are present in virtually all excitable cells. Many years ago, a long series of experiments by Hagiwara and colleagues showed that Ca channels in different cell types demonstrate a great diversity.[1] Most of these early experiments were done with various invertebrate preparations. More recently, the development of the suction pipette voltage clamp[2] and of the patch-clamp technique[3] have made it possible to characterize Ca channels in a wide range of vertebrate cells. These experiments, by many investigators, have led to the discovery that most vertebrate excitable cells possess multiple types of Ca channels (for review, see Tsien *et al.*[4] and Bean[5]). The different types of Ca channels can be distinguished by their single-channel properties, voltage-dependence, ionic selectivity, and pharmacology.

One of the most interesting features of Ca channels is that, in many cell types, the size of the current carried through Ca channels can be modulated by neurotransmitters and hormones. This modulation can be produced, in various cells, by virtually all known transmitters. Both enhancement and inhibition of Ca current have been described. Modulation by transmitters plays a vital role in many important physiological processes. For example, β-adrenergic agonists enhance the Ca current in cardiac muscle cells[6] and this modulation is directly involved in the increased rate and strength of the heartbeat that is produced by sympathetic stimulation. In the case of neurons, the best known transmitter effect is an inhibition of Ca current in sensory neurons produced by α-adrenergic agonists,[7] an effect that probably underlies a form of adrenergic presynaptic inhibition. With the discovery of multiple Ca channel types, it has become important to find out which channel types can be modulated by various transmitters.

The pharmacology of Ca channels is also of considerable interest. The most potent blockers of Ca channels are drugs of the dihydropyridine family, used clinically to treat hypertension and angina, in which role they act by blocking Ca channels in vascular smooth muscle cells. An interesting question is why the effects of dihydropyridines on vascular muscle are accompanied by so little apparent effect on the

[a] This work was supported by grants from the National Institutes of Health (HL-35034), the American Heart Association, and the Rita Allen Foundation. B.P.B. is partially supported by an Established Investigator Award from the American Heart Association.

function of heart cells, neurons, and other cells whose operation depends on Ca channels. At least part of the answer to this question is that different tissues contain different types of Ca channels with different pharmacological properties.

In this article, I summarize recent work I have done to compare the types of Ca channels in heart muscle cells and neurons, particularly with regard to their modulation by dihydropyridine drugs and neurotransmitters.

TWO TYPES OF CA CHANNELS IN HEART CELLS

Compared with neurons, categorization of the Ca channels in heart muscle cells has been relatively straightforward. Work in a number of laboratories on a variety of preparations has led to a consistent picture.[8-12] Two types of Ca channels have been described so far. The predominant type in most cells requires relatively large depolarizations to be activated (colloquially, "high threshold"), inactivates relatively slowly, and has a relatively large single-channel conductance (~ 15-25 pS with ~ 100 mM Ba). These channels are now commonly called L-type channels, following the nomenclature of the Tsien laboratory. The other type of cardiac Ca channel, called the T-type, has a lower threshold for activation, inactivates relatively rapidly, and has a smaller single-channel conductance (~ 8 pS with ~ 100 mM Ba). In general T-type channels are either missing or contribute very little current in ventricular cells and generate a larger fraction of the overall current in atrial cells or cells of the sinoatrial node.

FIGURE 1 shows the difference in the voltage dependence and kinetics of T and L components of whole-cell current in a canine atrial cell, studied with a near-physiological (for a dog) concentration of external Ca. The T channels generate a component of current that activates positive to about -50 mV and which inactivates quickly even at fairly negative potentials. This component of current can be selectively eliminated by holding the cell at a holding potential of -30 mV. In contrast, the current from L channels, which remains at a holding potential of -30 mV (and can thus be recorded in isolation) requires depolarization positive to about -30 mV to be activated, and, at least for moderate depolarizations, inactivates relatively slowly. However, at relatively large depolarizations, even L current inactivates fairly rapidly; it is by no means a noninactivating current.

One reason for believing that these two components of whole-cell current actually arise from different channel types is their different pharmacology. As shown in FIGURE 2, L channels are much more potently blocked by dihydropyridines than are T channels. In records taken at the same time from the same cell, 3 μM nitrendipine had little effect on T current, reducing it by only 15% or so, while inhibiting L current by about 70%.

The potency of dihydropyridine block of cardiac L channels depends on the holding potential of the cell,[13-15] with block being more potent at more depolarized holding potentials. This effect is shown in FIGURE 3. This experiment was done using a rabbit atrial cell, a preparation that apparently contains only L channels. Application of 300 nM nitrendipine blocked the L current elicited from a holding potential of -70 mV by about 40% but completely blocked the current elicited from -30 mV.

It is important to consider the voltage-dependence of dihydropyridine block when comparing effects on different channel types. In principle, the relative lack of effect on T channels might simply reflect the more negative holding potentials from which T currents are typically elicited. However, in cardiac atrial cells, it was found that

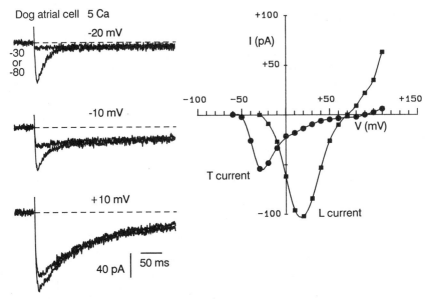

FIGURE 1. Two components of Ca current in a canine atrial cell. Currents were recorded using the whole-cell variant of the patch-clamp technique (Hamill *et al.*[3]). In the left panels, the larger currents are those elicited from −80 mV. The right panel shows the voltage dependence of L current (defined as the current elicited from −30 mV) and T current (the difference between current from −80 mV and that from −30 mV). External solution: 5 mM CaCl$_2$, 154 tetrae-thylammonium (TEA) Cl, 2 MgCl$_2$, 10 glucose, 10 HEPES, pH 7.4 with TEA OH. Internal solution: 120 CsCl, 10 EGTA, 5 MgCl$_2$, 10 HEPES, pH 7.4 with CsOH. (Redrawn from experiment in Bean.[8])

eliciting T current from more depolarized potentials, at which the T current was partially inactivated, did not significantly potentiate block of T current by nitrendipine.[8] Also, the block of L current is much stronger than of T current even with both elicited from very negative potentials.

Cardiac Ca currents are enhanced by β-adrenergic stimulation,[6] acting through cAMP-dependent protein kinase. β-Adrenergic enhancement of Ca current seems selective for L channels over T channels, as shown in FIGURE 4. In a canine atrial cell, the beta agonist isoproterenol produced a large enhancement of the L current (elicited in isolation from a holding potential of −30 mV) but had little effect on the current (predominantly T current) elicited by a step to −20 mV from a holding potential of −80 mV; the very small increase seen in this current was consistent with enhancement of a small component of L current.

MULTIPLE CHANNEL TYPES IN NEURONS: EFFECTS OF DIHYDROPYRIDINES

Neurons also possess multiple types of Ca channels.[16–29] In fact, it was in neurons that multiple types of Ca channels were first described in mammalian cells.

FIGURE 5 shows whole-cell currents in a freshly dissociated neuronal cell body from a dorsal root ganglion of an adult bullfrog. As in cardiac muscle cells, two components of Ca-channel current are evident. One component activates at relatively negative membrane potentials and inactivates fairly rapidly (with a time constant of 5-10 msec near 0 mV). This component of current disappears when the holding potential is changed to −50 mV, presumably because the channels become completely inactivated. A second component of Ca-channel current requires somewhat larger depolarizing pulses to become maximally activated, inactivates much more slowly, and is present at holding potentials of −50 or −40 mV as well as more negative holding potentials.

The low-threshold current is the component of neuronal Ca current that can be identified and characterized most unambiguously. It seems virtually identical to T current in cardiac muscle cells. In frog dorsal root ganglion cells—as in chick and rat dorsal root ganglion cells—single-channel recordings show a distinctive low-conductance channel that activates at relatively negative potentials, inactivates fairly quickly, and disappears at holding potentials of −40 mV or so. Averaging single-channel records of this low-conductance channel type yields records much like those of the transient component of current in FIGURE 5.

T channels in neurons, like those in heart muscle cells, are relatively resistant to dihydropyridine drugs. FIGURE 6 shows records from an experiment in which the dihydropyridine "agonist" BAY K 8644 was applied to a frog dorsal root ganglion neuron. The drug had no effect on the T-type Ca-channel current, which was elicited by stepping to −80 mV from a holding potential of −110 mV. However, in the same neuron, the drug dramatically enhanced the current elicited by a step from −80 mV to −40 mV. (In this cell, a holding potential of −80 mV was positive enough to completely inactivate the T component of current.)

FIGURE 2. Differential dihydropyridine sensitivity of T and L current in a canine atrial cell. T current records were obtained by subtracting currents elicited from holding potentials of −80 mV and −30 mV. External solution: 110 BaCl₂, 10 m*M* HEPES, pH 7.4 with Ba(OH)₂. Internal solution: 120 CsCl, 10 EGTA, 5 MgCl₂, 10 HEPES, pH 7.4 with CsOH. (Redrawn from experiment in Bean.[8])

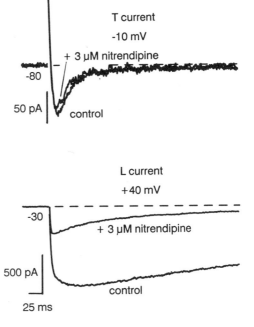

Dog atrial cell   115 Ba

T current

-10 mV

+ 3 µM nitrendipine

-80

50 pA

control

L current

+40 mV

-30

+ 3 µM nitrendipine

500 pA

control

25 ms

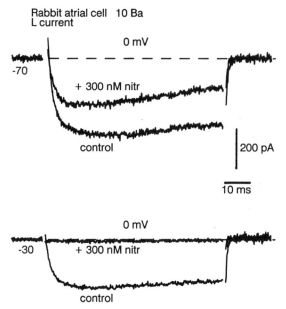

FIGURE 3. Voltage-dependent dihydropyridine block of cardiac L current in a rabbit atrial cell. External solution: 10 BaCl$_2$, 160 TEA Cl, 10 HEPES, pH 7.4 with TEA OH. Internal solution: 128 Cs glutamate, 4.5 MgCl$_2$, 9 EGTA, 9 HEPES, 4 MgATP, 14 creatine phosphate (tris salt), 1 GTP (tris salt), and 50 U/ml creatine phosphokinase, pH 7.45 with CsOH. (Unpublished experiment.)

The classification of high-threshold Ca channels in neurons has been difficult and controversial. Most groups have assumed that the high-threshold current in whole-cell records arises from a single-channel type. However, Nowycky, Fox, and Tsien[20,23,24] have reported that chick dorsal root ganglion cells contain two distinct types of high-threshold channels, which they have named "N" and "L" channels. According to their work, N and L channels can be distinguished at the single-channel level by their different conductances (N channels having a conductance smaller than L channels but larger than T channels) and by the sensitivity of L channels, but not N channels, to dihydropyridine drugs. In whole-cell records, N channels are considered to give rise to a component of high-threshold current that inactivates with a time constant of about 30-100 msec near 0 mV, slower than T channels but much faster than L channels.

According to the picture developed by Nowycky, Fox, and Tsien, dihydropyridine drugs should be useful for separating N and L components of high-threshold current: Current through L channels should be blocked by dihydropyridine antagonists and enhanced by dihydropyridine agonists, while current through N channels should not be affected. FIGURES 7-9 show effects of dihydropyridine drugs on neuronal Ca-channel currents that are not easily reconciled with this picture.

There is little doubt that most sensory neurons do contain Ca channels that are sensitive to dihydropyridine drugs. Most frog DRG neurons, like most chick DRG neurons,[30] respond to BAY K 8644 with an increase in Ca-channel current. FIGURE 6 shows a typical example, in which 300 nM BAY K 8644 produced a sevenfold increase in the size of the current elicited by a step to −40 mV. In common with heart muscle cells, the effects of BAY K 8644 on neuronal Ca-channel currents are largest on currents elicited by small depolarizations (but not on T currents). However, a major difference between effects of BAY K 8644 on cardiac cells and neurons is that peak Ca-channel current (usually occurring for depolarizations to near 0 mV) is frequently not affected in neurons (even when the current at −40 to −30 mV is greatly enhanced), while in heart cells current is usually dramatically enhanced at all potentials; current at the peak of the I-V curve is frequently increased by 2-10-fold in heart cells. FIGURE 7 shows effects of BAY K 8644 on high-threshold Ca currents in a frog DRG neuron. As in FIGURE 6, current produced by a step to −40 mV is greatly enhanced (by a factor of about eight) by BAY K 8644. However, the current elicited by a step to 0 mV is actually inhibited by BAY K 8644. Whether the different effects at −40 mV and 0 mV reflect opposite effects on two different channel types remains to be determined. Another possibility is that the effects at both potentials reflect a combination of enhancement and block (perhaps from the two optical isomers present in the BAY K 8644) and that the relative contribution of each effect somehow depends on the test potential.

The effects of the dihydropyridine blockers on neuronal Ca currents are even more different than on heart cells. Many neurons do have components of Ca current that are blocked by dihydropyridines. FIGURE 8 shows an example of a bullfrog DRG

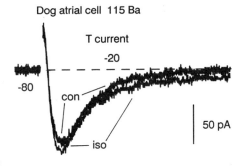

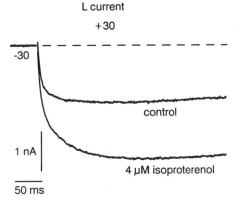

FIGURE 4. Differential effects of β-adrenergic stimulation on T and L currents in canine atrial cell. External solution: 110 BaCl$_2$, 10 mM HEPES, pH 7.4 with Ba(OH)$_2$. Internal solution: 120 CsCl, 10 EGTA, 5 MgCl$_2$, 10 HEPES, pH 7.4 with CsOH. (Redrawn from experiment in Bean.[8])

neuron. Application of a relatively low concentration of the dihydropyridine blocker nimodipine blocks about 30% of the current elicited by a step from -80 mV to -10 mV. There is a major difference from cardiac muscle cells in the voltage-dependence of dihydropyridine action; when the cell was held at -30 mV (which partially inactivated the Ca current), the effect of nimodipine was only modestly enhanced. In a cardiac muscle cell, such a change in holding potential has a much more dramatic

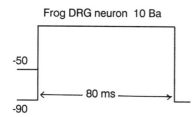

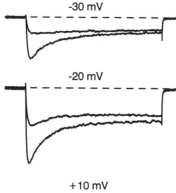

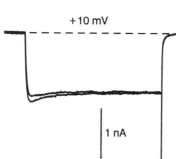

FIGURE 5. Two components of Ca current in a freshly dissociated DRG neuron from an adult bullfrog. External solution: 10 $BaCl_2$, 160 TEA Cl, 10 HEPES, pH 7.4 with TEA OH. Internal solution: 128 Cs glutamate, 4.5 $MgCl_2$, 9 EGTA, 9 HEPES, 4 MgATP, 14 creatine phosphate (tris salt), 1 GTP (tris salt), and 50 U/ml creatine phosphokinase, pH 7.4 with CsOH. (Unpublished experiment.)

effect on dihydropyridine action, with current from depolarized holding potentials being completely wiped out by similar concentrations of drug (FIG. 3).

Even more strikingly, in many neurons dihydropyridine blockers have little effect even on current that would identified as L-type current by its slow inactivation (as in FIGURE 9, where there is little decay during a six-second depolarization) and even when the drugs are applied at depolarized holding potentials.

Frog DRG neuron 3 Ba

T current

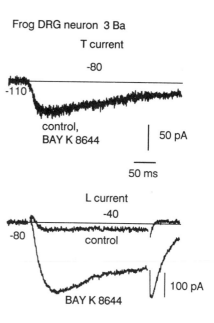

FIGURE 6. Effects of 300 nM BAY K 8644 (racemic mixture) on low- and high-threshold Ca current in bullfrog DRG neuron. External solution: 3 BaCl₂, 160 TEA Cl, 10 HEPES, pH 7.4 with TEA OH. Internal solution: 128 Cs glutamate, 4.5 MgCl₂, 9 EGTA, 9 HEPES, 4 MgATP, 14 creatine phosphate (tris salt), 1 GTP (tris salt), and 50 U/ml creatine phosphokinase, pH 7.4 with CsOH. (Unpublished experiment.)

frog DRG neuron   3 Ba

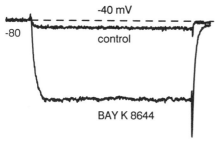

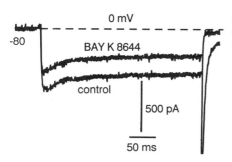

FIGURE 7. Effects of 300 nM BAY K 8644 (racemic mixture) on high-threshold current at two potentials in a bullfrog DRG neuron. External solution: 3 BaCl₂, 160 TEA Cl, 10 HEPES, pH 7.4 with TEA OH. Internal solution: 128 Cs glutamate, 4.5 MgCl₂, 9 EGTA, 9 HEPES, 4 MgATP, 14 creatine phosphate (tris salt), 1 GTP (tris salt), and 50 U/ml creatine phosphokinase, pH 7.4 with CsOH. (Unpublished experiment.)

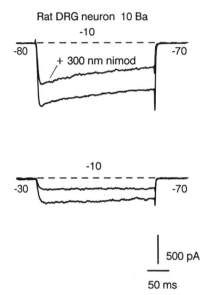

Rat DRG neuron 10 Ba

FIGURE 8. Effect of 300-nm nimodipine on Ca current elicited from two holding potentials in a rat DRG neuron. External solution: 10 BaCl$_2$, 160 TEA Cl, 10 HEPES, pH 7.4 with TEA OH. Internal solution: 115 CsCl, 5 MgATP, 5 creatine phosphate (tris salt), 0.3 GTP (tris salt), 10 BAPTA, 10 HEPES, pH 7.4 with CsOH. (Unpublished experiment.)

At least in the cases of bullfrog and rat DRG neurons, almost all cells have at least a component of current—sometimes as in FIGURE 9, most of the current—that has the voltage dependence and kinetic properties of L current, but which is not blocked by dihydropyridines. The existence of such a "long-lasting" current that is not blocked by dihydropyridines does not fit with the classification of channels proposed by Nowycky, Fox, and Tsien[20,23,24] and suggests that this classification may be incomplete.

ADRENERGIC MODULATION

Dunlap and Fischbach[7] showed that Ca current in DRG neurons can be inhibited by norepinephrine, an effect that is mediated by α-adrenergic receptors.

FIGURE 9. Effect of 3 μM nimodipine on slowly inactivating Ca current in a rat DRG neuron. External solution: 10 BaCl$_2$, 160 TEA Cl, 10 HEPES, pH 7.4 with TEA OH. Internal solution: 115 CsCl, 5 MgATP, 5 creatine phosphate (tris salt), 0.3 GTP (tris salt), 10 BAPTA, 10 HEPES, pH 7.4 with CsOH. (Unpublished experiment.)

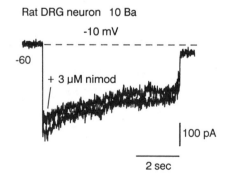

Rat DRG neuron 10 Ba

Frog DRG neuron 2 Ba

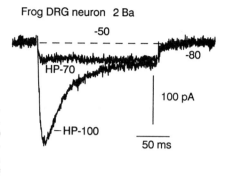

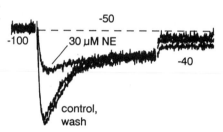

FIGURE 10. Norepinephrine inhibition of T-type current in a bullfrog DRG neuron. External solution: 2 BaCl₂, 160 TEA Cl, 10 HEPES, pH 7.4 with TEA OH. Internal solution: 128 Cs glutamate, 4.5 MgCl₂, 9 EGTA, 9 HEPES, 4 MgATP, 14 creatine phosphate (tris salt), 1 GTP (tris salt), and 50 U/ml creatine phosphokinase, pH 7.4 with CsOH. (Unpublished experiment.)

Both low- and high-threshold components of Ca current in DRG neurons can be depressed by norepinephrine. Examples are shown in FIGURES 10 and 11. FIGURE 10 shows records from a bullfrog DRG neuron in which there appeared to be selective inhibition of T-type current. This cell had a large component of T current, as shown by the presence of a large, rapidly inactivating component of current elicited by a step to -50 mV from a holding potential of -100 mV but not from -70 mV. As shown in the bottom panel, this rapidly inactivating component of current was almost abolished by the application of 30 μm norepinephrine, and the effect of norepinephrine was completely reversible. However, the current remaining at the end of the depolarization was unaffected by the application of norepinephrine, suggesting that the channel type carrying this current was unaffected, at least in this cell.

Frog DRG neuron 10 Ba

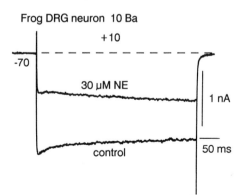

FIGURE 11. Norepinephrine inhibition of high-threshold current in a bullfrog DRG neuron. External solution: 10 BaCl₂, 160 TEA Cl, 10 HEPES, pH 7.4 with TEA OH. Internal solution: 128 Cs glutamate, 4.5 MgCl₂, 9 EGTA, 9 HEPES, 4 MgATP, 14 creatine phosphate (tris salt), 1 GTP (tris salt), and 50 U/ml creatine phosphokinase, pH 7.4 with CsOH. (Unpublished experiment.)

In other cells, there were large effects of norepinephrine on high-threshold Ca current as well. FIGURE 11 shows an example from a cell that had no T current; the current elicited by a step from -70 to -10 mV was greatly (and reversibly) inhibited by 30 μM norepinephrine.

Thus, at least two types of Ca channels in DRG neurons can be modulated by norepinephrine (see also Marchetti et al.[31]). In some neurons, it has been proposed that transmitters selectively inhibit an inactivating component of high-threshold current identified as N current.[32,33] This is clearly not the case for norepinephrine inhibition of high-threshold current in DRG neurons, since the current inhibited by NE in the cell of FIGURE 11 is primarily a relatively slowly inactivating current that would conventionally be identified as mostly L current. Whether this type of current is actually L-type current as defined by dihydropyridine sensitivity remains to be seen.

Many more experiments with blockers, transmitters, and modulating agents will be needed to identify all the various channel types that are present in neurons. It will be important to integrate pharmacological dissection of whole-cell currents with distinctions between channel types made at the single-channel level. The complete categorization of channel types in neurons is likely to take many experiments and many years. The current evidence for three channel types[20,23,24,26] may be only the beginning.

REFERENCES

1. HAGIWARA, S. 1983. Membrane Potential-Dependent Ion Channels in Cell Membrane: 5-47. Raven Press. New York.
2. KOSTYUK, P. G. & O. A. KRISHTAL. 1977. Separation of sodium and calcium currents in the somatic membrane of mollusc neurones. J. Physiol. (London) **270:** 545-568.
3. HAMILL, O. P., A. MARTY, E. NEHER, B. SAKMANN & F. J. SIGWORTH. 1981. Improved patch-clamp techniques for high-resolution current recording from cells and cell-free membrane patches. Pfluegers Arch. **391:** 85-100.
4. TSIEN, R. W., D. LIPSCOMBE, D. V. MADISON, K. R. BLEY & A. P. FOX. 1988. Multiple types of neuronal calcium channels and their selective modulation. Trends Neurosci. **11:** 431-438.
5. BEAN, B. P. 1989. Classes of calcium channels in vertebrate cells. Ann. Rev. Physiol. **51:** 367-384.
6. REUTER, H. 1967. The dependence of the slow inward current on external calcium concentration. J. Physiol. (London) **192:** 479-492.
7. DUNLAP, K. & G. D. FISCHBACH. 1981. Neurotransmitters decrease the calcium conductance activated by depolarization of embryonic chick sensory neurones. J. Physiol. (London) **317:** 519-535.
8. BEAN, B. P. 1985. Two kinds of calcium channels in canine atrial cells. Differences in kinetics, selectivity, and pharmacology. J. Gen. Physiol. **86:** 1-30.
9. NILIUS, B., P. HESS, J. B. LANSMAN & R. W. TSIEN. 1985. A novel type of cardiac calcium channel in ventricular cells. Nature **316:** 443-446.
10. BONVALLET, R. 1987. A low-threshold calcium current recorded at physiological Ca concentrations in single frog atrial cells. Pflugers Archiv. **408:** 540-542.
11. HAGIWARA, N., H. IRISAWA & M. KAMEYAMA. 1988. Contribution of two types of calcium currents to the pacemaker potentials of rabbit sino-atrial node cells. J. Physiol. **395:** 233-253.
12. MITRA, R. & M. MORAD. 1986. Two types of calcium channels in guinea-pig ventricular myocytes. Proc. Natl. Acad. Sci. USA **83:** 5340-5344.
13. SANGUINETTI, M. C. & R. S. KASS. 1984. Voltage-dependent block of calcium channel current by dihydropyridine calcium channel antagonists. Circ. Res. **55:** 336-348.
14. UEHARA, A. & J. R. HUME. 1985. Interactions of organic calcium channel antagonists with calcium channels in single frog atrial cells. J. Gen. Physiol. **85:** 621-647.

15. BEAN, B. P. 1984. Nitrendipine block of cardiac calcium channels: High affinity binding to the inactivated state. Proc. Natl. Acad. Sci. USA **81:** 6388-6392.

16. CARBONE, E. & H. D. LUX. 1984. A low voltage-activated calcium conductance in embryonic chick sensory neurones. Biophys. J. **46:** 413-418.

17. CARBONE, E. & H. D. LUX. 1984. A low voltage-activated, fully inactivating Ca channel in vertebrate sensory neurones. Nature **310:** 501-502.

18. BOSSU, J. L., A. FELTZ & J. M. THOMANN. 1985. Depolarization elicits two distinct calcium currents in vertebrate sensory neurones. Pflugers Archiv. **403:** 360-368.

19. FEDULOVA, S. A., P. G. KOSTYUK & N. S. VESELOVSKY. 1985. Two types of calcium channels in the somatic membrane of newborn rat dorsal root ganglion neurones. J. Physiol. **359:** 431-446.

20. NOWYCKY, M. C., A. P. FOX & R. W. TSIEN. 1985. Three types of neuronal calcium channel with different calcium sensitivity. Nature **316:** 440-443.

21. CARBONE, E. & H. D. LUX. 1987. Kinetics and selectivity of a low-voltage-activated calcium current in chick and rat sensory neurones. J. Physiol. **386:** 547-570.

22. CARBONE, E. & H. D. LUX. 1987. Single low-voltage-activated calcium channels in chick and rat sensory neurones. J. Physiol. **386:** 571-601.

23. FOX, A. P., M. C. NOWYCKY & R. W. TSIEN. 1987. Kinetic and pharmacological properties distinguishing three types of calcium currents in check sensory neurones. J. Physiol. **394:** 149-172.

24. FOX, A. P., M. C. NOWYCKY & R. W. TSIEN. 1987. Single-channel recordings of three types of calcium channels in chick sensory neurones. J. Physiol. **394:** 173-200.

25. HIRNING, L. D., A. P. FOX, E. W. MCCLESKEY, B. OLIVERA, S. A. THAYER, R. J. MILLER & R. W. TSIEN. 1988. Dominant role of N-type Ca^{2+} channels in evoked release of norepinephrine from sympathetic neurons. Science **239:** 57-61.

26. KOSTYUK, P. G., M. F. SHUBA & A. N. SAVCHENKO. 1987. Three types of calcium channels in the membrane of mouse sensory neurones. Pflugers Arch. **411:** 661-669.

27. MCCLESKEY, E. W., A. P. FOX, D. H. FELDMAN, L. J. CRUZ, B. M. OLIVERA, R. W. TSIEN & D. YOSHIKAMI. 1987. ω-Conotoxin: Direct and persistent blockade of specific types of calcium channels in neurons but not muscle. Proc. Natl. Acad. Sci. USA **84:** 4327-4331.

28. MILLER, R. J. 1987. Calcium channels in neurones. *In* Structure and Physiology of the Slow Inward Calcium Channel. D. J. Triggle, J. C. Venter, Eds.: 161-246. Alan R. Liss. New York.

29. MILLER, R. J. 1987. Multiple calcium channels and neuronal function. Science **235:** 46-52.

30. NOWYCKY, M. C., A. P. FOX & R. W. TSIEN. 1985. Long-opening mode of gating of neuronal calcium channels and its promotion by the dihydropyridine calcium agonist Bay K 8644. Proc. Natl. Acad. Sci. USA **82:** 2178-2182.

31. MARCHETTI, C., E. CARBONE & H. D. LUX. 1986. Effects of dopamine and noradrenaline on Ca channels of cultured sensory and sympathetic neurons of chick. Pflugers Arch. **406:** 104-111.

32. GROSS, R. A. & R. L. MACDONALD. 1987. Dynorphin A selectively reduces a large transient (N-type) calcium current of mouse dorsal root ganglion neurons in cell culture. Proc. Natl. Acad. Sci. USA **84:** 5469-5473.

33. WANKE, E., A. FERRONI, A. MALGAROLI, A. AMBROSINI, T. ROZZAN & J. MELDOLESI. 1987. Activation of a muscarinic receptor selectively inhibits a rapidly inactivated Ca^{2+} current in rat sympathetic neurons. Proc. Natl. Acad. Sci. USA **84:** 4313-4317.

Modulation of Ca Channels in Peripheral Neurons[a]

E. CARBONE

Dipartimento di Anatomia e Fisiologia Umana
Corso Raffaello 30
I-10125 Torino, Italy

H. D. LUX

Max-Planck-Institute fur Psychiatrie
Am Klopferspitz 18A
D-8033 Planegg, Federal Republic of Germany

INTRODUCTION

Modulation of voltage-dependent Ca channels is of great significance for the control of Ca-dependent neuronal activities. Neuronal Ca channels[1-6] can be modulated in many ways[7-14] but, very likely, only a few common mechanisms underly these actions (see Refs. 15-18 for recent reviews). An external receptor site is generally thought to be coupled to the functional parts of Ca channels through cyclic-nucleotide-mediated mechanisms. Occupancy of the receptor site by an agonist induces a cascade of reactions that lead either to activation, inhibition or modifications of the open-closed kinetics of the pore. Recent studies begin to reveal some details of the receptor-Ca channel interactions.[19-24]

Alternatively, Ca-channel modulation by external ligands has proved to be useful for investigating the side groups controlling channel gatings and ion permeability (see Tsien *et al.*[25] for a recent review). In this paper we will review three types of modulatory effects that apparently escape the "canonical schemes" of Ca-channel modulation. Two of them (dopamine and menthol) relate to specific modifications of the activation and inactivation gates of high-threshold (HVA; L, N) Ca channels.[11,26,27] In peripheral neurons, dopamine drastically slows the activation kinetics of HVA Ca channels, while menthol speeds up their time-dependent inactivation. The third modulatory effect concerns the ability of permeant ions (Ca^{2+} or Na^+) to influence the binding of the neurotoxic peptide ω-conotoxin (ω-CgTX) to high-threshold Ca channels.[28] ω-CgTX blocks the channels slowly and persistently when Ca^{2+} is the main current-carrying ion[29-32] but blocks them quickly and reversibly when Na^+ crosses the pore,[28] as if the toxin were able to distinguish Ca- from Na-permeable states of the same channel.

[a] This work was partially supported by NATO (Grant No. 576187) and the Consiglio Nazionale delle Ricerche (Grant No. 88.00458.04).

DOPAMINE MODULATION OF HIGH-THRESHOLD
CA CHANNELS

The effects of 10 μM dopamine (DA) on Ca currents of chick sensory neurons are illustrated in FIGURE 1. At large membrane potentials (0, +10 mV), DA slows

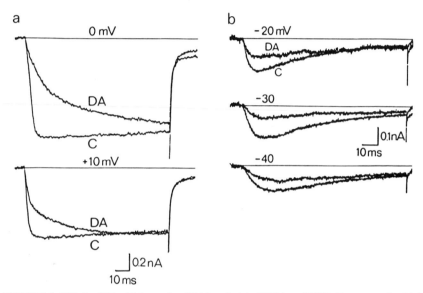

FIGURE 1. Effects of 10 μM dopamine (DA) on isolated HVA and LVA Ca currents in chick dorsal root ganglion neurons. The currents were recorded from the same neuron using the patch-clamp technique.[33] Details on cells preparation, pulse protocols, and pipette fabrication are given elsewhere.[34-36]

(a) HVA Ca currents recorded at 0 and +10 mV before (C) and during (DA) application of dopamine. (b) LVA Ca currents recorded at the potential indicated. After washing, the HVA current kinetics recovered nearly completely. The bath contained (mM): 5 CaCl$_2$, 120 choline Cl, 20 glucose, 10 Na-HEPES (pH 7.3). The pipette filling solution contained (mM): 130 CsCl, 20 TEACl, 0.25 CaCl$_2$, 5 EGTA-OH, 10 glucose, 10 Na-HEPES (pH 7.3). Holding potential, −70 mV. Temperature: 22°C (after Marchetti, Carbone & Lux[11]).

down the kinetics of HVA (L,N) Ca currents activation with little change to both the steady-state amplitude and the time course of Ca current deactivation. Below −20 mV, DA nearly halves the size of the low-voltage-activated (LVA, T) Ca currents without affecting their activation-inactivation kinetics. Interestingly, the action on HVA currents is fully reversible while that on LVA currents persists even after prolonged washing. Thus, dopamine, like other neurotransmitters,[13,14,21,22] exerts different actions on different types of neuronal Ca channels.

Several lines of evidence suggest that the action of DA on Ca currents reflects a genuine interaction of the catecholamine with Ca channels. First, DA also slows down the activation kinetics of HVA channels in sympathetic neurons (FIG. 2a) and in clonal AtT-20 pituitary cells (C. Marchetti and A. M. Brown, personal communication). Because both types of cells lack of LVA (T) channels and AtT-20 cells are known to possess only HVA L-type Ca currents,[37] these findings suggest that DA mainly delays the turning on of L-type Ca channels rather than blocking transiently activated N-type Ca channels. This conclusion is strengthened by the observation that the action of DA and other neurotransmitters[13,14] is insensitive to the holding potential. Significantly, following long conditioning pulses (10 sec) to −30 mV, which largely inactivate T and N channels, somatostatin (50 nM) and Leu-enkephalin (25 nM) also prolong the activation time course of the high-threshold Ca current in neuroblastoma-glioma NG108-15 cells.[13]

Although little is known about the possible mechanisms involved in DA-induced inhibition of Ca channels, some evidence suggests that G-proteins and/or second messengers might be implicated in this process. First, most of the identified neurotransmitter receptors are shown to be linked to Ca channels via GTP-binding proteins.[18–22,24,25] The link might be direct[20–22,24] or through second messengers like cAMP or DAG.[16–19] Second, the action of DA on HVA Ca channels is mimicked by the nonhydrolyzable G-protein activator, GTP-γ-S. In sensory and sympathetic neurons, internal application of GTP-γ-S prolongs remarkably the activation kinetics of HVA channels[20–22] independently of the holding potential (C. Marchetti, personal communication). Third, the irreversible action of DA on LVA channels in intact DRG neurons becomes fully reversible in excised membrane patches (FIG. 2b), suggesting that the loss of some internal factor might destabilize the tight binding of the catecholamine to its receptor site.

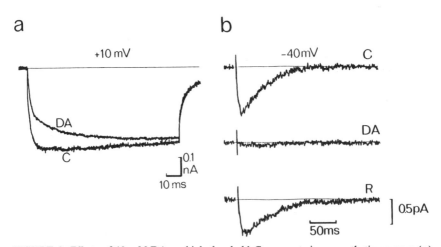

FIGURE 2. Effects of 10 μM DA on high-threshold Ca currents in sympathetic neurons (**a**) and on averaged Ca currents obtained from an outside-out patch of chick DRG (**b**). Step depolarizations were as indicated. Holding potential −70 (a) and −90 mV (b). In b the traces represent sample averages of 10 single-channel recordings, before (C), during (DA) and after (R) application of dopamine. (After Marchetti, Carbone & Lux[11]).

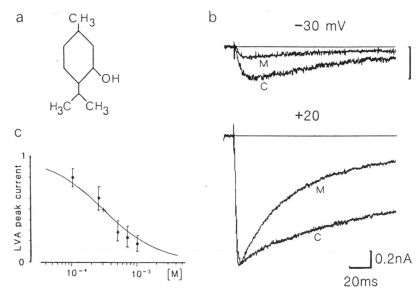

FIGURE 3. Action of menthol on Ca currents in a chick DRG cell. (**a**) Chemical structure of menthol. (**b**) LVA and HVA currents recorded before (C) and during application (M) of 0.5 mM menthol at the potentials indicated. The currents recovered completely after five minutes of washing. Holding potential, -80 mV. The bath contained (mM): 120 NaCl, 20 CaCl$_2$, 2 MgCl$_2$, 10 Na-HEPES (pH 7.3) plus 0.3 μM TTX. (**c**): Dose-response curve of normalized peak LVA current versus menthol concentration. The continuous curve was drawn assuming that menthol interacts with a receptor site following a first-order chemical reaction (after Swandulla, Carbone, Schafer & Lux[26]).

MENTHOL ACCELERATES CA-CHANNEL INACTIVATION

Like dopamine, menthol (a cyclic alcohol derived from peppermint oil) also exerts specific action on low- and high-threshold Ca channels in vertebrate neurons.[26] At 0.5 mM, menthol reduces by about 70% the size of LVA currents with little change in their time course and produces a three- to sixfold acceleration of HVA current inactivation (FIG. 3b). The action of menthol on LVA currents is dose dependent (K_D 300 μM) and fully reversible (FIG. 3c). Blockade of LVA (T) channels is neither voltage- nor time-dependent and is also not due to a negative shift of the slow steady-state current inactivation.[26]

Menthol action on HVA current inactivation has some interesting features. Menthol accelerates and enhances HVA current inactivation (FIG. 3b, bottom). The time constant of inactivation decreases from 120 to 40 msec and becomes rather insensitive to both membrane potential and Ca current amplitudes (see Fig. 5 in Swandulla *et al.*[26]). As previously suggested, it is likely that menthol facilitates the inactivation gating of HVA channels, thereby suppressing the inactivating action of intracellular Ca^{2+}. Indeed, preliminary experiments have shown that menthol-induced inactivation is strongly reduced when Na ions flow through open HVA Ca channels,[26] suggesting that either menthol action is sensitive to the permeability states (Na- and Ca-mode, see below) of HVA channels or that menthol acts on the Ca-dependent mechanisms of HVA channel inactivation.

The action of menthol seemed unrelated to the presence of the fast inactivating current observed in neurons (N-current)[5] and could not be due to a transformation of slowly inactivating (L-type) HVA channels into fast inactivating ones (N-type). This was concluded from the observation that: (i) menthol also accelerates the HVA current inactivation in cells with weakly inactivating Ca currents (FIG. 4a), (ii) the action of menthol on Ca inactivation was identical if the holding potential was changed from −80 to −50 mV (FIG. 4b), and (iii) in menthol-treated cells, HVA currents showed no sign of voltage-dependent inactivation between +10 and +40 mV.

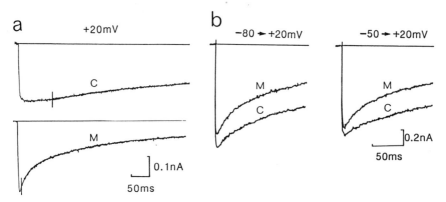

FIGURE 4. The inactivating action of menthol is unrelated to the presence of fast inactivating (N) high threshold Ca currents. (a) acceleration of the HVA current inactivation with menthol. The two traces were recorded before (C) and during application of 500 μM menthol (M). The records were fitted with a single exponential function with time constant 300 (C) and 117 msec (M). The vertical bars indicate the beginning of the fitting. Step depolarizations were from −80 to +20 mV. (b) The menthol-induced inactivation was nearly unchanged when changing the holding potential from −80 (left) to −50 mV (right) (after Swandulla, Carbone, Schafer & Lux[26]).

BLOCKADE OF CA CHANNELS BY ω-CgTX IS MODULATED BY PERMEANT IONS

As pointed out elsewhere,[29-32] the blocking action of ω-conotoxin (ω-CgTX) on Ca currents deviates considerably from that of other Ca antagonists. ω-CgTX blocks selectively and with high affinity the HVA Ca currents in a number of neurons. The action of the toxin, however, is slow and irreversible. The onset of the block occurs within several minutes with 5 μM of toxin. Prolonged washing produces almost no recovery of persistently blocked Ca channels. Due to these features, ω-CgTX is an ideal tool for isolating and biochemically purifying neuronal high-threshold Ca channels.[38]

ω-CgTX also shows another interesting property: Its blocking action is modulated by the type of permeant cation[28] (FIG. 5). Application of 5 μM ω-CgTX to the bath

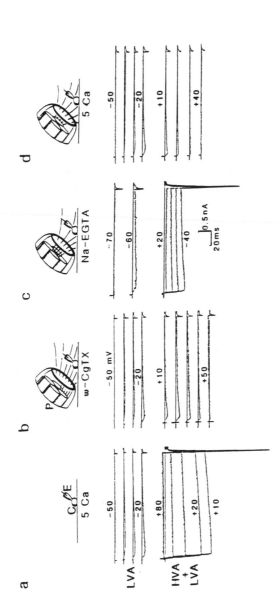

FIGURE 5. Partial relief of ω-CgTX blockade in low-Ca^{2+} solutions. Upper part: Schematic representation of the experimental conditions. Cell (C), patch-electrode (E) and four-way perfusion pipette (P).

Lower part: LVA and HVA currents recorded from a chick DRG cell in 5 mM external Ca (**a**), after application of 5 μM ω-CgTX (**b**), during washing with a Na-EGTA (pCa 7) (**c**), and on returning to 5 mM Ca^{2+} (**d**). Step depolarizations as indicated. Holding potential, -90 mV. (Modified from Carbone & Lux[28]).

inhibits HVA Ca currents in a dose-dependent manner with little change in the size and time course of LVA Ca currents (FIG. 5b). HVA channel blockade by ω-CgTX is persistent and hardly recovers after washing for several hours. However, block of HVA channels is promptly relieved when Ca^{2+} is replaced by a toxin-free, low-Ca^{2+} solution (FIG. 5c). Despite the persistent block of the toxin, HVA Na currents can be resumed at potentials positive to −40 mV. The size of these currents vary from cell to cell and seems to depend on the slow loss of Ca channels induced by cell dialysis.[39,40]

Appearance of HVA Na currents in CgTX-treated cells is unlikely to be a consequence of the unbinding of the toxin from its receptor site. On returning to normal Ca^{2+} toxin-free solution (FIG. 5d), HVA Ca currents are blocked despite continuous washing. If toxin unbinding occurs during application of toxin-free solutions, at least a small fraction of HVA Ca currents should be detected on returning to normal Ca conditions. This, however, has never been observed.

Several lines of evidence suggest that appearance of HVA Na currents is the consequence of a reversible modification of the channel-toxin complex and not the result of the turning on of some CgTX-insensitive current or due to interference of EGTA molecules with Ca channels. The former possibility is ruled out by the observation that CgTX-insensitive Na currents possess the same kinetics and pharmacology of HVA Na currents recorded in cells not treated with toxin.[28] The latter is contradicted by the following arguments. If appearance of toxin-resistant Na currents is linked to the formation of a EGTA-CgTX complex, the first-order reaction:

$$EGTA + CgTX \underset{K_{-a}}{\overset{K_a}{\rightleftharpoons}} EGTA\text{-}CgTX \tag{1}$$

predicts that Na currents would appear at a rate ($[EGTA]\, K_a + K_{-a}$), higher than that of Ca current block (K_{-a}). This, however, is at variance with the experimental observation. CgTX-resistant Na currents develop at a much lower rate than the Ca current block.

TWO SITES OF ACTION OF ω-CONOTOXIN

Conotoxin-resistant Na currents that resume during application of toxin-free Na-EGTA solutions can be quickly blocked in a dose-dependent manner by further additions of the toxin (FIG. 6). Compared to the persistent action on HVA Ca currents, the blocking of HVA Na currents occurs reversibly, on a faster time scale and independently of the saturating conditions of the Ca-current block. The onset of the Na-current block develops exponentially with time constants that decrease with increasing toxin concentration (insets in FIG. 6). The partial block of Na current reaches steady-state conditions in 30 seconds with 1 μM of toxin and nearly 10 seconds with ten times more toxin. The offset of CgTX inhibition also proceeds exponentially but independently of the degree of Na-current block (τ_{off} about 34 sec).

The ability of ω-CgTX to block Na currents through HVA Ca channels allows the pharmacological dissection of LVA Na currents over a wide range of membrane potentials (−80, +70 mV). FIGURE 7 shows families of Na currents obtained from two DRG cells at control (panel a) and in the presence of 5 μM ω-CgTX (panel b). As shown, the amplitude of the HVA Na currents is remarkably reduced by the toxin and most of the Na current flows through unaffected LVA Ca channels.

DISCUSSION AND CONCLUSIONS

The data presented here offer a broad spectrum of possible interactions between chemicals and Ca channels. The modulatory effects of dopamine, menthol, and permeant ions are only three examples of a rapidly increasing number of ligand-channel interactions. Dopamine, like other neurotransmitters, modulates neuronal Ca channels

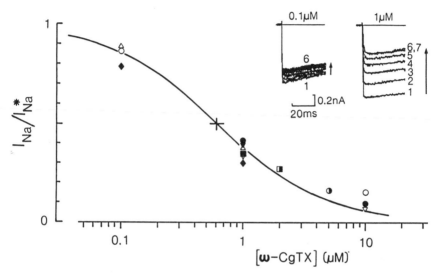

FIGURE 6. Dose-response characteristics of conotoxin-induced block of HVA Na currents in low external Ca^{2+}. The values are normalized Na currents versus the logarithm of ω-CgTX concentration. I_{Na}^* and I_{Na} are the current amplitudes before and during application of the toxin, respectively. Both values were measured at the end of a 40-msec pulse to -20 mV from -90 mV holding potential. The continuous curve was calculated for $K_D = 0.7 \mu M$, from equation:

$$I_{Na}/I_{Na}^* = K_D/(K_D + [T])$$

where K_D represents the equilibrium dissociation constant of the receptor-toxin complex and [T] is the toxin concentration. **Inset:** HVA Na currents recorded after application of 0.1 and 1 μM ω-CgTX in 1×10^{-8} M free external Ca^{2+}. The numbers indicate the sequential order of recordings. The time interval between traces was 4 and 10 sec with 1 and 0.1 μM ω-CgTX, respectively. (Modified from Carbone & Lux[28].)

by inhibiting their functioning. The action of dopamine is significant but never drastic and complete even at large doses (100 μM). This feature is shared by most neurotransmitters,[7-14,19,20,22] irrespective of their mechanism of action. There are no obvious explanations for this. The only available data of transmitter-induced inhibition are obtained from whole-cell recordings and little has been done at the single-channel level to clarify the kinetic details of the phenomenon. This work is needed in view of the significant role that Ca-channel modulation plays in neuronal functioning.

Concerning the modulatory effects of permeant ions on Ca channels, our data support the view that both channels might possess two distinct modes of ion permeation (Ca- and Na-mode).[41-47] Comparative studies of the activation-inactivation kinetics of Na and Ca currents[41-43,48] suggest that the two permeability states of Ca channels are characterized by different state conformations of the same molecule (see also Refs. 49-51) with similar activation gatings but different open-closed bursting distribution and selectivity to ions.

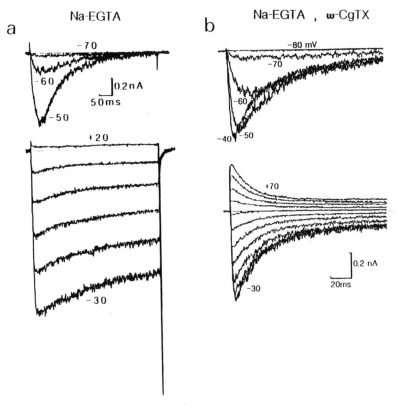

FIGURE 7. Time course of LVA and HVA Na currents recorded from two chick DRG neurons at control (**a**) and in the presence of 5 μM ω-CgTX (**b**). Membrane depolarizations as indicated, with increments of 10 mV. Holding potentials: -90 (a) and -100 mV (b). The bath contained (in mM): 140 NaCl, 4 CaCl$_2$, 5 EGTA, 10 Na-HEPES (pH 7.5) and 0.3 μMTTX.

The idea of two modes of ion permeation fits a number of experimental observations. It explains, for instance, why micromolar additions of external Ca^{2+} depresses macroscopic Na currents by increasing the probability that a Ca channel shuts (Ca-mode) rather than reducing the average time the channel opens for Na ions (Na-mode).[41,42] This is consistent with the view that Ca^{2+} specificity of Ca channels derives from the presence of an external[44,45] or internal[43] Ca-regulatory site whose occupation by Ca terminates Na-ion permeation. Consistent with this is also the observation that sudden elevations of H^+ concentration[46,47] or step decreases or extracellular Ca^{2+} concentration[52] produce a transient inward Na current flowing through Ca channels that are transformed from a voltage-gated Ca-permeable state to a proton-gated Na-permeable state. Finally, ion-channel interactions might also be involved in ion-induced changes in the response of neuronal[53] and cardiac[54] L-type Ca channels to dihydropyridine derivatives. Bay K8644 enhances the activity of high-threshold Ca channels when Ba^{2+} is the main current-carrying cation, but depresses the same type of current when Ba is replaced by Ca^{2+}.

REFERENCES

1. LLINAS, R. & Y. YAROM. 1981. Properties and distribution of ionic conductances generating electroresponsiveness of mammalian inferior olivary neurons *in vitro.* J. Physiol. **315:** 569-584.

2. CARBONE E. & H. D. LUX. 1984. A low voltage-activated, fully inactivating Ca channel in vertebrate sensory neurones. Nature **310:** 501-502.

3. FEDULOVA, S. A., P. G. KOSTYUK, & N. S. VESELOVSKY. 1985. Two types of calcium channels in the somatic membrane of newborn rat dorsal root ganglion neurones. J. Physiol. **359:** 431-446.

4. BOSSU, J. L., A. FELTZ & J. M. THOMANN. 1985. Depolarization elicits two distinct calcium currents in vertebrate sensory neurons. Pflugers Arch. **403:** 360-368.

5. NOWYCKY, M. C., A. P. FOX & R. W. TSIEN. 1985. Three types of neuronal calcium channels with different agonist sensitivity. Nature **316:** 443-446.

6. MATTESON, D. R. & C. M. ARMSTRONG. 1985. Two distinct populations of calcium channels in a clonal line of pituitary cells. Science **227:** 65-67.

7. DUNLAP, K. & G. D. FISCHBACH. 1981. Neurotransmitters decrease the calcium conductance activated by depolarization of embryonic chick sensory neurones. J. Physiol. **317:** 519-535.

8. GALVAN, M. & P. R. ADAMS. 1982. Control of calcium current in rat sympathetic neurons by norepinephrine. Brain Res. **244:** 135-144.

9. FORSCHER, P. & G. S. OXFORD. 1985. Modulation of calcium channels by norepinephrine in internally dialyzed avian sensory neurons. J. Gen. Physiol. **85:** 743-764.

10. DEISZ, R. A. & H. D. LUX. 1985. γ-Aminobutiric acid-induced depression of calcium currents of chick sensory neurons. Neurosci. Lett. **56:** 205-210.

11. MARCHETTI, C., E. CARBONE & H. D. LUX. 1986. Effects of dopamine and noradrenaline on Ca channels of cultured sensory and sympathetic neurons of chick. Pflugers Arch. Eur. J. Physiol. **406:** 104-111.

12. MACDONALD, R. L. & M. A. WERZ. 1986. Dynorphin A decreases voltage-dependent Ca conductance of mouse dorsal root ganglion neurones. J. Gen. Physiol. **377:** 237-250.

13. TSUNOO, A., Y. MITSUNOBU & T. NARAHASHI. 1986. Block of calcium channels by enkephalin and somatostatin in neuroblastoma-glioma hybrid NG108-15 cells. Proc. Natl. Acad. Sci. USA **83:** 9832-9836.

14. IKEDA, S. R., G. G. SCHOFIELD & F. F. WEIGHT. 1987. Somatostatin blocks a calcium current in acutely isolated adult rat superior cervical ganglion neurons. Neurosci. Lett. **81:** 123-128.

15. REUTER, H. 1983. Calcium channel modulation by neurotransmitters, enzymes and drugs. Nature **301:** 569-574.

16. ROSENTHAL, W. & G. SCHULTZ. 1987. Modulations of voltage dependent ion channels by extracellular signals. Trends Neurosci. **8:** 351-354.

17. HOFMANN, F., W. NASTAINCZYK, A. ROHRKASTEN, T. SCHNEIDER & M. SIEBER. 1987. Regulation of the L-type calcium channel. Trends Neurosci. **8:** 393-398.

18. DUNLAP, K., G. G. HOLZ & G. R. STANLEY. 1987. G proteins as regulators of ion channel function. Trends Neurosci. **10:** 242-244.

19. HOLZ, G. G., S. G. RANE & K. DUNLAP. 1986. GTP binding proteins mediate transmitter inhibition of voltage-dependent calcium channels. Nature **319**: 670-672.

20. LEWIS, D. L., F. F. WEIGHT & A. LUINI. 1986. A guanine nucleotide-binding protein mediates the inhibition of voltage-dependent calcium current by somatostatin in a pituitary cell line. Proc. Natl. Acad. Sci. USA **83**: 9035-9039.

21. WANKE, E., A. FERRONI, A. MALGAROLI, A. AMBROSINI, T. POZZAN & J. MELDOLESI. 1987. Activation of a muscarinic receptor selectively inhibits a rapidly inactivated Ca^{2+} current in rat sympathetic neurons. Proc. Natl. Acad. Sci. USA **84**: 4313-4317.

22. DOLPHIN, A. C. & R. H. SCOTT. 1987. Calcium channel currents and their inhibition by ($-$)-baclofen in rat sensory neurones: Modulation by guanine nucleotides. J. Physiol. **386**: 1-17.

23. HIRNING, L. D., A. P. FOX, E. W. MCCLESKEY, B. M. OLIVERA, S. A. THAYER, R. J. MILLER & R. W. TSIEN. 1988. Dominant role of N-type Ca^{2+} channels in evoked release of norepinephrine from sympathetic neurons. Science **239**: 57-61.

24. EWALD, D. A., P. C. STERNWEIS & R. J. MILLER. 1988. Guanine nucleotide-binding protein G_o-induced coupling of neuropeptide Y receptors to Ca^{2+} channels in sensory neurons. Proc. Natl. Acad. Sci. USA **85**: 3633-3637.

25. TSIEN, R. W., D. LIPSCOMBE, D. V. MADISON, K. R. BLEY & A. P. FOX. 1988. Multiple types of neuronal calcium channels and their selective modulation. Trends Neurosci. **11**: 431-438.

26. SWANDULLA, D., E. CARBONE, K. SCHAFER & H. D. LUX. 1987. Effect of menthol on two types of Ca currents in cultured sensory neurons of vertebrates Pflugers Arch. Eur. J. Physiol. **409**: 52-59.

27. SWANDULLA, D., K. SCHAFER & H. D. LUX. 1986. Ca channel current inactivation is selectively modulated by menthol. Neurosci. Lett. **68**: 23-28.

28. CARBONE, E. & H. D. LUX. 1988. ω-Conotoxin blockade distinguishes Ca from Na permeable states in neuronal calcium channels. Pflugers Arch. Eur. J. Physiol. **413**: 14-22.

29. KERR, L. M. & D. YOSHIKAMI. 1984. A venom peptide with a novel presynaptic blocking action. Nature (London) **308**: 282-284.

30. CRUZ, L. J. & B. M. OLIVERA. 1986. Calcium channel antagonists. ω-Conotoxin defines a new high-affinity site. J. Biol. Chem. **261**: 6230-6233.

31. FELDMAN, D. H., B. M. OLIVERA & D. YOSHIKAMI. 1987. Omega *Conus geographus* toxin: A peptide that blocks calcium channels. FEBS Lett. **214**: 295-300.

32. MCCLESKEY, E. W., A. P. FOX, D. H. FELDMAN, L. J. CRUZ, B. M. OLIVERA, R. W. TSIEN & D. YOSHIKAMI. 1987. ω-Conotoxin: Direct and persistent blockade of specific types of calcium channels in neurons but not in muscle. Proc. Natl. Acad. Sci. USA **84**: 4327-4331.

33. HAMILL, O. P., A. MARTY, E. NEHER, B. SAKMANN & F. J. SIGWORTH. 1981. Improved patch-clamp techniques for high-resolution current recording from cells and cell-free membrane patches. Pflugers Arch. Eur. J. Physiol. **391**: 85-100.

34. CARBONE, E. & H. D. LUX. 1984. A low voltage-activated calcium conductance in embryonic chick sensory neurons. Biophys. J. **46**: 413-418.

35. CARBONE, E. & H. D. LUX. 1986. Sodium channels in cultured chick dorsal root ganglion neurons. Eur. Biophys. J. **13**: 259-271.

36. CARBONE, E. & H. D. LUX. 1987. Kinetics and selectivity of a low voltage-activated Ca^{2+} current in chick and rat sensory neurones. J. Physiol. (London) **386**: 571-601.

37. LUINI, A., D. LEWIS, S. GUILD, G. SCHOFIELD & F. WEIGHT. 1986. Somatostatin, an inhibitor of ACTH secretion, decreases cytosolic free calcium and voltage-dependent calcium current in a pituitary cell line. J. Neurosci. **6**: 3128-3132.

38. BARHANIN, J., A. SCHMID & M. LAZDUNSKI. 1988. Properties of structure and interaction of the receptor for ω-Conotoxin, a polypeptide active on Ca^{2+} channels. Biochem. Biophys. Res. Commun. **150**: 1051-1062.

39. FENWICK, E. M., A. MARTY & E. NEHER. 1982. Sodium and calcium channels in bovine chromaffin cells. J. Physiol. (London) **331**: 599-635.

40. CAVALIÈ, A., R. OCHI, D. PELZER & W. TRAUTWEIN. 1983. Elementary currents through Ca^{2+} channels in guinea pig myocytes. Pflugers Arch. Eur. J. Physiol. **398**: 284-297.

41. LUX, H. D. & E. CARBONE. 1987. External Ca ions block Na conducting Ca channel by

promoting open to closed transitions. *In* Receptors and Ion Channnles. Y. A. Ovchinnikov & F. Hucho, Eds. Vol **1**: 149-155. Walter de Gruyter & Co. Berlin.

42. CARBONE, E. & H. D. LUX. 1987. External Ca^{2+} ions block unitary Na^+ currents through Ca^{2+} channels of cultured chick sensory neurones by favouring prolonged closures. J. Physiol. **382**: 125P.

43. LUX, H. D., E. CARBONE & H. ZUCKER. 1988. Block of Na currents through a neuronal Ca channel by external Ca and Mg ions. *In* The Calcium Channel: Structure, Function and Implications. M. Morad, W. G. Nayler, S. Kazda & M. Schramm, Eds.: 128-137. Springer Verlag. Heidelberg.

44. KOSTYUK, P. G., S. L. MIRONOV & Y. M. SHUBA. 1983. Two ion-selecting filters in the calcium channel of the somatic membrane of mollusc neurons. J. Membr. Biol. **76**: 83-93.

45. KOSTYUK, P. G. & S. L. MIRONOV. 1986. Some predictions concerning the calcium channel model with conformational states. Gen. Physiol. Biophys. **6**: 649-659.

46. KONNERTH, A., H. D. LUX & M. MORAD. 1986. Proton-induced transformation of calcium channel in chick dorsal root ganglion cells. J. Physiol. **386**: 603-633.

47. DAVIES, N. W., H. D. LUX & M. MORAD. 1988. Site and mechanism of activation of proton-induced sodium current in chick dorsal root ganglion neurones. J. Physiol. **400**: 159-187.

48. CARBONE, E. & H. L. LUX. 1988. Na currents through neuronal Ca channels: Kinetics and sensitivity to Ca-antagonists. *In* The Calcium Channel: Structure, Function and Implications. M. Morad, W. G. Nayler, S. Kazda, M. Schramm, Eds.: 115-127. Springer Verlag, Hidelberg.

49. LAUGER, P. 1985. Ionic channels with conformational substates. Biophys. J. **47**: 581-590.

50. CIANI, S. 1984. Coupling between fluxes in one-particle pores with fluctuating energy profiles. Biophys. J. **46**: 249-252.

51. PIETROBON, D., B. PROD'HOM & P. HESS. 1988. Conformational changes associated with ion permeation in L-type calcium channels. Nature **333**: 373-376.

52. HABLITZ, J. J., U. HEINEMANN & H. D. LUX. 1986. Step reductions in extracellular Ca^{2+} activate a transient inward current in chick dorsal root ganglion cells. Biophys. J. **50**: 753-757.

53. BOLL, W. & H. D. LUX. 1985. Action of organic antagonists on neuronal calcium currents. Neurosci. Lett. **56**: 335-339.

54. NILIUS, B., P. HESS, J. B. LANSMAN & R. W. TSIEN. 1985. A novel type of neuronal calcium channels with different calcium agonist sensitivity. Nature **316**: 443-446.

Calcium Channels in Vertebrate Neurons

Experiments on a Neuroblastoma Hybrid Model[a]

D. A. BROWN, R. J. DOCHERTY, AND
I. McFADZEAN

Department of Pharmacology
University College London
London, WC1N1AX, United Kingdom

INTRODUCTION

The presence of Ca currents in vertebrate neurons has been known for a long time (see Llinas & Walton[1] and Kostyuk[2] for reviews). They perform such functions as mediating transmitter release and controlling excitability and spike-firing patterns[1]; they may also be involved in growth and differentiation.[3] The types of neuronal Ca currents and their sensitivities to drugs and transmitters have been extensively reviewed by Miller.[4]

We have previously been involved in studies on Ca currents in mammalian central neurons[5-9] but have recently turned our attention to the control of Ca currents in a neuroblastoma hybrid cell line, the NG108-15 mouse neuroblastoma × rat glioma cell,[10] as a model for studying transmitter regulation of neuronal Ca currents. The reason for this—apart from convenience—is that these cells possess a variety of transmitter receptors whose biochemical effects have been studied extensively[11] so that one can attempt to relate changes in Ca-channel behavior with possible underlying biochemical responses. Further, when differentiated, NG108-15 cells show many of the excitable characteristics and ionic currents of the parent sympathetic neurons,[12] including the presence of Ca spikes[13,14] and robust Ca currents under voltage-clamp recording conditions.[13,15-19]

EXPERIMENTAL METHODS

Our technique for recording Ca currents from NG108-15 cells has been described previously.[19] In brief, cells were grown in Eagle's minimal essential medium (EMEM)

[a]This work was supported by a grant from the United Kingdom Medical Research Council. I. M. was supported by the Beit Memorial Foundation.

with 10% added fetal calf serum and differentiated by adding 10 μM PGE_1 plus 50 μM IBMX. Three to eight days later cells were voltage-clamped by the "whole-cell" variant of the patch-clamp technique using a switching amplifier cycling between voltage recording and current injection at 3-8 KHz. To isolate Ca currents, cells were bathed in a bicarbonate-buffered Krebs solution at pH 7.4 containing 2.5 mM Ca, 0.5 μM TTX to suppress Na currents, and 120 mM TEA-Cl replacing NaCl to suppress K currents. The internal (pipette) solution contained 136 mM CsCl, 3 mM $MgCl_2$, 3 mM EGTA, and 40 mM HEPES (pH 7.4). No appreciable "run-down" of Ca currents over one to two hours' recording was experienced with this solution. Electrode resistances were 2-10 MΩ and seal resistances usually about 5 GΩ. Cells with short processes were selected for study, to minimize space-clamp problems.

RESULTS AND DISCUSSION

When clamped at relatively negative potentials (-80 to -100 mV) and subjected to depolarizing commands of 0.5-1-sec duration, undifferentiated cells showed relatively small inward Ca currents. These usually inactivated completely within about 50-100 msec (FIG. 1). The activated threshold for these currents was about -60 to -50 mV, and peak current was attained at command potentials of between -30 to -20 mV. Hence, these currents kinetically resembled the rapidly inactivating current studied by Bodewei *et al.*[15,16]

Following differentiation, currents were usually larger and now consisted of two components—an early inactivating phase and a more sustained phase (FIG. 2). Comparison of peak and steady-state current-voltage curves (FIG. 2, right side) suggested that the inactivating component had a lower threshold (about -50 mV) and peaked at a more negative potential (about -10 mV) than the sustained component (threshold about -30 mV, peak about $+10$ mV). This accords with the presence of at least two kinetically different Ca currents in these cells, akin to "type I" and "type II" currents reported in these[17] and in neuroblastoma cells of the N1E-115 line by Narahashi and his colleagues[20] (see also Fishman & Spector[13]). Further tests (below) suggest, however, that the "transient" component was a composite of two currents with differing inactivation characteristics.

Inorganic Ca-Channel Blockers

Ca currents were readily blocked by low concentrations (2-10 μM) of cadmium ions. Comparison of scaled currents before and after Cd block (FIG. 3a) suggested that both inactivating and sustained components were equally affected; thus, we were unable to confirm the clear differential sensitivities of these two components to Cd reported in N1E-115 cells.[20] In contrast, the trivalent lanthanide gadolinium produced a striking alteration in the current trajectory, such that the residual current showed a very clear separation between a rapidly inactivating (tau $\sim$40 msec) and a non-inactivating component (FIG. 3b). Our interpretation of this[19] is that Gd selectively blocked a third, more slowly inactivating (tau 0.5-1 sec) component of current, revealed by subtracting the residual current in the presence of a full blocking concentration ($\leq$ 10 μM) of Gd from the initial current (FIG. 4). This component had a threshold of about -30 mV and peaked at about 0 mV.[19]

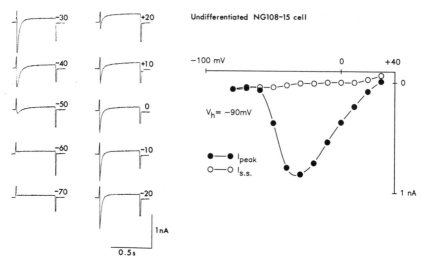

FIGURE 1. Inward Ca currents recorded from an undifferentiated NG108-15 cell (i.e., not treated with PGE_1/IBMX, see METHODS) evoked by 0.5-sec depolarizing commands from a holding pontential (V_h) of -90 mV to the command potentials indicated. The graph on the right shows peak ($\bullet$, I_{peak}) current and current at the end of the command pulse ($\bigcirc$, "steady-state current," I_{ss}) plotted against command potential. See METHODS and Docherty[19] for technical details.

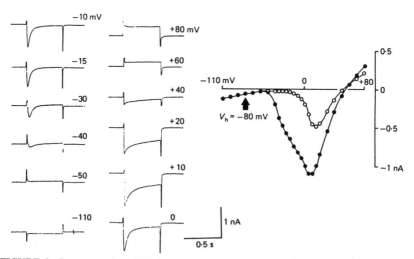

FIGURE 2. Ca currents in a differentiated NG108-15 cell (see METHODS) evoked by 0.5-sec commands from V_h -80 mV. The graph shows peak ($\bullet$) and sustained ($\bigcirc$) components of current plotted against command potential. Note the larger sustained component than in the undifferentiated cell in FIGURE 1.

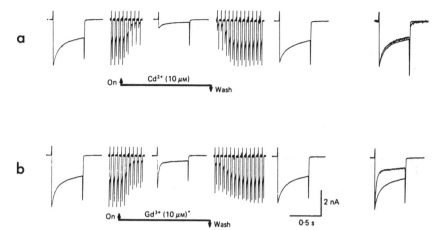

FIGURE 3. Effects of (**a**) 10 μM cadmium and (**b**) 10 μM gadolinium on Ca currents in a differentiated NG108-15 cell evoked by 0.5-sec voltage steps to $+11$ mV from -89 mV applied at 25-sec intervals. The recorder was slowed $\times 100$ for addition and washout of ion. Records on the right show superimposed currents recorded before and after adding Cd or Gd, normalized to the peak control current: Note that the form of the current was changed in Gd but not in Cd. (Adapted from FIG. 2 in Docherty.[19])

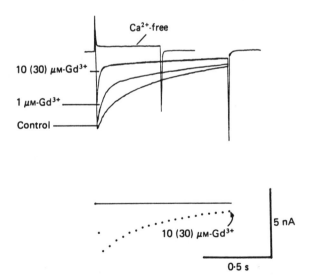

FIGURE 4. Upper records: Ca currents recorded before (control) and after adding 1, 10, and 30 μM gadolinium. The top record shows the leak current in Ca-free solution. Currents were generated by 0.5 (leak) or 1-sec steps to $+11$ mV from -89 mV. Lower trace: gadolinium-sensitive component of current revealed by subtracting the currents recorded in 10 and 30 μM Gd from the control current. This component decayed mono-exponentially with a time constant of 0.58 sec. (Adapted from FIG. 7 in Docherty.[19])

The residual rapidly inactivating component of current seen in Gd solution could be selectively inhibited by nickel (50-500 μM) (FIG. 5a), so that, after a combination of Gd and Ni only a high-threshold (-20 mV) sustained current remained. Conversely, initial application of Ni reduced the initial peak of inward current but left an inactivating component that was further inhibited by superimposed Gd (FIG. 5b). This suggests that Gd and Ni inhibit different components of current with differing inactivation rates and different activation ranges (cf. FIG. 5c).

It is clearly tempting to draw an analogy between these components of Ca current in NG108-15 cells with the three components in sensory cells designated T, N, and L by Tsien and his colleagues.[21,22] In this case, the rapidly inactivating, low-threshold, Ni-sensitive component might correspond to the T-current (see also Refs. 23-25); the Gd-sensitive, more slowly inactivating component might correspond to the N-current (or, more accurately, to the more slowly inactivating variant of this current reported in sympathetic neurons[26]); while the sustained current remaining after both Gd and Ni might correspond to the L-current. However, the amplitudes of the different components in NG108-15 cells varied considerably, and it was rarely possible to obtain a clear separation with preinactivating pulse protocols; in particular, the distinction between the contributions of "N" and "L" components to the total current was difficult to ascertain, and it was also usually difficult to select an inactivating prepulse which could permit a firm identification of the contribution of the "T" current to the initial component of the current envelope.

Dihydropyridines

Freedman et al.[27] have previously reported that the dihydropyridine nitrendipine reduces K-stimulated ^{45}Ca fluxes in differentiated NG108-15 cells. In accord with

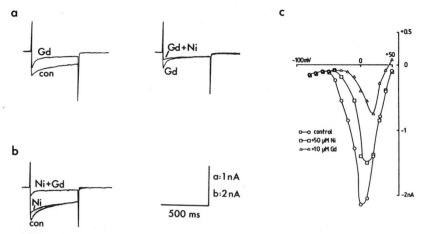

FIGURE 5. Effects of sequential additions of (a) 5 μM Gd and 5 μM Gd plus 500 μM Ni, and (b) 100 μM Ni and 100 μM Ni plus 10 μM Gd on Ca currents. Two different cells. Pulse protocols: a, holding potential -86 mV, command potential $+5$ mV; b, holding potential -90 mV, command potential $+10$ mV. Graphs in (c) show peak currents from the experiment illustrated in b recorded before (o, control) and after adding 50 μM Ni ($\square$) then 50 μM Ni plus 10 μM Gd ($\triangle$) plotted against command potential. Note that Ni removes a low-threshold component of current, and that Gd reduces the residual current with less shift in threshold.

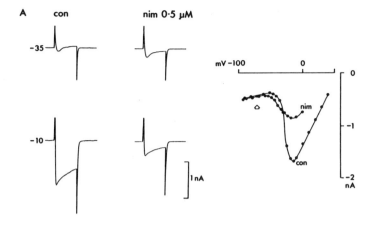

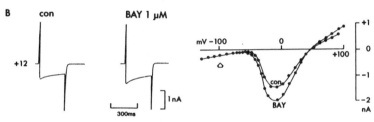

FIGURE 6. Effects of (A) 0.5 μM nimodipine (nim) and (B) 1 μM BAY K8644 (BAY) on Ca currents in two NG108-15 cells. Records on the left show currents activated by commands to A −35 and −10 mV from −70 mV, and B +12 mV from −90 mV. Graphs on the right show peak currents plotted against command potential in the absence (con) and presence (nim, BAY) of the dihydropyridine.

this, dihydropyridine Ca-channel blockers such as nimodipine reduced the peak amplitude of the Ca current in our experiments (FIG. 6A). However, the current-voltage curve was also shifted to the left so that currents evoked by just suprathreshold voltage steps were actually *increased*. The dihydropyridine Ca-agonist, BAY K8644, also tended to shift the current-voltage curve to the left, but simultaneously *increased* the peak current (FIG. 6B). Shifts in the voltage-dependence of activation by both agonist and antagonist dihydropyridines,[28,29] and specifically the ability of DHP antagonists to increase Ca current amplitudes at relatively negative command potentials,[30] have been noted previously.

Neurotransmitters: The Role of Adenylate Cyclase

NG108-15 cells possess receptors for a variety of neurotransmitters and neuro-modulators, including acetylcholine, noradrenaline, opiates, prostaglandins, adenosine, and some peptides such as bradykinin and neurotensin.[10,11,31,32] These induce several different biochemical changes such as activation or inhibition of adenylate cyclase,[10,11,33] elevation of cyclic GMP,[11] stimulation of phosphatidylinositide breakdown[31] or acceleration of Na-H exchange.[34] We have been particularly interested in the effects of transmitters capable of inhibiting adenylate cyclase: these include opiates,[10] noradrenaline[11,33] and acetylcholine (via muscarinic receptors[35]). The ability of opiates to inhibit Ca currents in these cells (via delta receptors) has been described previously.[17,18,36] We have confirmed this, and have also found an equivalent and highly reproducible effect of the α-adrenoceptor-agonist noradrenaline.[37] In the following section we describe the action of these agents and consider the relationship between Ca-current inhibition and adenylate cyclase inhibition.

Action of Noradrenaline

Noradrenaline (up to 5 μM) produced a rapid, reversible inhibition of the peak total Ca current in NG108-15 cells activated from hyperpolarized holding potentials of up to about one-third (mean maximum 32 $+/-$ 3 %, $n = 18$) with an IC$_{50}$ of 0.18 μM (FIG. 7a). Dopamine was about 60 times weaker and isoprenaline (see FIG. 7b) and 5-hydroxytryptamine had no effect at concentrations up to 10 μM. The action of noradrenaline was inhibited by phentolamine and yohimbine with pA$_2$ values of 6.78 and 6.75, respectively. Prazosin (pA$_2$ 5.40) was much less effective, suggesting a receptor resembling the alpha-2 type. However, strict classification as an alpha-2

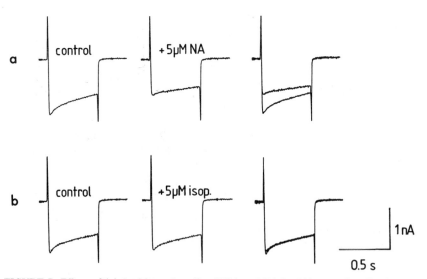

FIGURE 7. Effects of (a) 5 μM noradrenaline (NA) and (b) 5 μM isoprenaline (isop) on Ca currents evoked by 0.5-sec commands to +4 mV from a holding potential of −96 mV in the same NG108-15 cell. Records on the right show superimpositions of currents recorded before and after adding drug. The current is reduced by noradrenaline but not by isoprenaline, implying an alpha receptor.

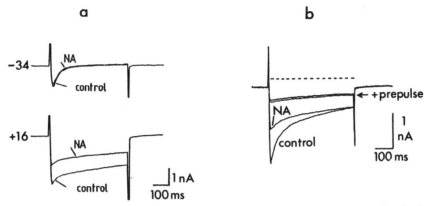

FIGURE 8. Different pulse protocols to reveal differential effects of 5 μM noradrenaline (NA) on different components of Ca current. Currents recorded in the absence and presence of noradrenaline are superimposed in each record. In (a) a cell was held at -80 mV and commanded to -34 mV (upper record) to activate the low-threshold, transient current, or to $+16$ mV to activate additional, more sustained components (see Fig. 2). In (b) the potential of another cell was stepped for 0.5 sec to 0 mV, either directly from a holding potential of -90 mV or 20 msec after a 0.5-sec prepulse to -10 mV (designated "$+$ prepulse").

receptor is precluded by the fact that the alpha-2 agonist clonidine has neither agonist nor antagonist activity at concentrations up to 10 μM.

We have not been able to firmly categorize the action of noradrenaline as exclusively directed at any one particular component of the Ca current envelope in these cells, though some of our observations might be interpreted to suggest a preferential inhibition of the "N-type" current, as suggested for amphibian sympathetic neurons[38] (which appear to lack a T-current). Thus, as shown in FIGURE 8a, the low-threshold transient component of current activated by small depolarizing steps was relatively (though not totally) resistant to block by noradrenaline. Likewise, the transient current seen in undifferentiated cells was not reduced by noradrenaline, though this may equally reflect a paucity of receptors or receptor-linked transduction molecules (see below). Also, when the transient components of current were partially inactivated by a 0.5-sec depolarizing prepulse, the residual, more slowly inactivating component (L-current?) showed a reduced amount of block (FIG. 8b). Further evidence that the sustained current is relatively resistant to noradrenaline may be adduced from FIGURE 9a, if it is assumed that BAY K8644 selectively enhances the L current,[21] since the total amount of current blocked by noradrenaline was not increased in proportion to the current enhancement. On the other hand, the same did not apply when the current was reduced by nimodipine, since, in this case, the effect of noradrenaline was reduced proportionately (FIG. 9b). Further, noradrenaline remained effective after the intermediate component of current was inhibited by Gd (FIG. 9c). If the effect of Gd is indeed to selectively inhibit the N current, then we have to consider alternative explanations to a preferential action on the N-current *per se* for the effects shown in FIGURE 7. One possibility is that there is a degree of time- and voltage-sensitivity to the action of noradrenaline itself. This has also been suggested for opiate action, on the basis that the residual current showed a remarkably slowed activation.[17] We have noted a similar effect with noradrenaline, particularly when using Ba as a charge carrier (FIG. 10); although it might be argued that this results from the selective

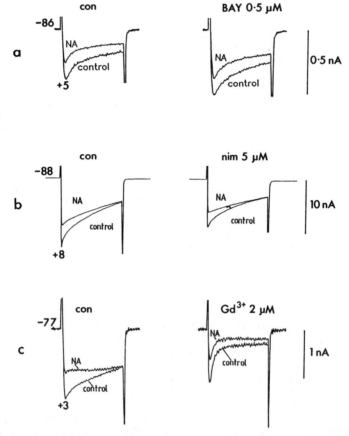

FIGURE 9. Effects of (a) 0.5 μM BAY K8644 (BAY), (b) 5 μM nimodipine (nim) and (c) 2 μM gadolinium (Gd) on the inhibition of Ca currents produced by 5 μM noradrenaline (NA). Three different cells: holding and command potentials as indicated. Each record shows superimposed currents recorded before and after adding noradrenaline; records on the left show responses in normal bathing medium, those on the right after adding the appropriate Ca-current modifier.

elimination of the N-current, the activation of the residual current was remarkably slow; moreover, in some experiments the peak current at the end of the command actually *exceeded* that obtained before adding noradrenaline, suggesting either a time- and voltage-dependent relief of block[17] or a change in current kinetics.[39]

Comparison and Interaction with Opiates

As reported previously[17,18,36] opiates also inhibit Ca currents in NG108-15 cells, probably through a delta type of receptor. In our experiments, the opiate DADLE

(D-ala^2-D-leu^4-enkephalin) produced a comparable inhibition, of equal peak magnitude, to that produced by noradrenaline (FIG. 11). This effect was also replicated by somatostatin (FIG. 11; see Tsunoo *et al.*[17]). Further, following a peak inhibition with noradrenaline, no further inhibition could be produced by a superimposed application of DADLE or somatostatin, suggesting that they blocked the same component of current. However, one difference from noradrenaline is that the responses to DADLE and somatostatin rapidly desensitized; this appeared to be a receptor-specific phenomenon because, when the responses to DADLE or somatostatin had desensitized, noradrenaline could still inhibit the current (FIG. 11).

Transduction Mechanism

GTP-Binding Protein

Transmitter-induced inhibition of adenylate cyclase is normally mediated by the activation of an inhibitory GTP-binding protein ("G-protein"), G_i.[40] Firm evidence for the role of a G-protein in mediating the inhibitory effect of opiates on I_{Ca} in NG108-15 cells has been provided by Hescheler *et al.*[36] We[41] have obtained comparable evidence for intermediation by a G-protein in mediating the inhibition of the Ca current by noradrenaline. Thus, addition of the GTP analogue, GTPγS to the patch

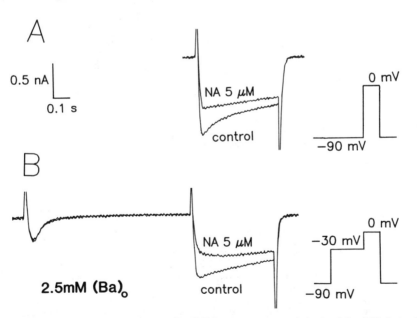

FIGURE 10. Effects of 5 μM noradrenaline (NA) on currents recorded using 2.5 mM Ba instead of Ca as charge carrier. **A** shows currents recorded during a 0.5-sec step from −90 to 0 mV; in **B** the command to 0 mV was preceded by a one-second step to −30 mV to activate, then inactivate, the low-threshold, transient current (*cf.* FIG. 2).

pipette solution prevented reversal of the inhibition by noradrenaline; reversal requires hydrolysis of the bound GTP by the GTPase activity of the alpha-subunit, but GTPγS is not hydrolyzed, so the G-protein becomes "locked" in the activated and dissociated GTP-bound form.[40] Conversely Bordetella pertussis toxin (PTX, islet-activating protein[42]) maintains G_i in the inactive state and prevents the inhibition of adenylate cyclase in NG108-15 cells by opiates, noradrenaline, and carbachol.[43] In our experiments, preincubation of cells for two to four hours with 500 ng/ml PTX totally prevented inhibition of the Ca current by subsequent addition of noradrenaline or DADLE (FIG. 12).

Role of Adenylate Cyclase Inhibition

In order to see whether the Ca-current inhibition might result from the previously reported adenylate cyclase inhibition, we tested whether the inhibition could be reversed by applying the adenylate cyclase activator forskolin (10 μM) in combination with a phosphodiesterase inhibitor, IBMX (100 μM), or by the cyclic AMP analogues, dibutyryl cyclic AMP (1 mM) or 8-bromo-cyclic AMP (100-200 μM). However,

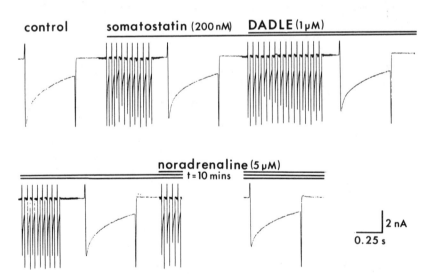

FIGURE 11. Ca current inhibition by successive superimpositions of 200 nM somatostatin, 1 μM DADLE (D-Ala²-D-leu⁴-enkephalin) and 5 μM noradrenaline. Addition of somatostatin transiently reduced the current but this recovered in the continued presence of somatostatin. Superimposition of DADLE again transiently reduced the current, by about the same amount, but this again recovered. Subsequent addition of noradrenaline (lower trace) in the presence of both somatostatin and DADLE, was still capable of reducing the current; the effect of noradrenaline persisted for 10 minutes. Holding potential, −90 mV; command potential, 0 mV; 0.5-sec commands at 25-sec intervals. Note: The concentrations of agonist used in this experiment were all supramaximal; increasing the concentration of the desensitizing agonist did not restore Ca-current block.

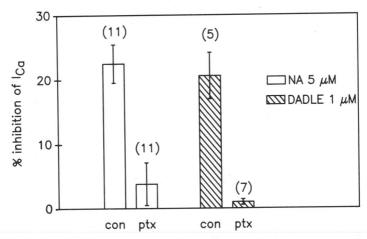

FIGURE 12. Inhibition of I_{Ca} by noradrenaline (NA) or DADLE is prevented by 2-4-hr pretreatment with 500 ng/ml pertussis toxin (ptx). Histograms show mean percentage inhibition of peak current evoked by 0.5-sec commands to about 0 mV from about −90 mV cells without (con) and with (ptx) pertussis treatment. Open blocks, 5 μM noradrenaline; hatched blocks, 1 μM DADLE. Bars show SEM (number of cells in brackets).

none of these compounds consistently modified the control currents or reversed or prevented the inhibition of I_{Ca} by noradrenaline. Thus, and in contrast to previous observations on these cells,[17] the Ca current in our sample of NG108-15 cells did not appear to be affected by phosphorylation by protein kinase A: possibly the channels were already phosphorylated during the differentiation procedure and not dephosphorylated thereafter. The currents were also insensitive to protein kinase C, since they were unmodified by phorbol 12,13-dibutyrate (0.1-1 μM; nine cells) and responses to noradrenaline were not reduced by staurosporin (0.4 μM; $n = 3$), an inhibitor of protein kinase C.[44] Hence, unlike chick sensory ganglion cells,[45] inhibition of the Ca current in NG108-15 cells by noradrenaline does not appear to be mediated through activation of protein kinase C.

CONCLUSION

Although there is a superficial parallel between the ability of noradrenaline to reduce the Ca current in NG108-15 cells and to inhibit adenylate cyclase in that both involve the coupling of an alpha receptor to a PTX-sensitive G-protein, our results suggest that these are not causally related. One possibility, therefore, is that inhibition of the Ca current might involve a more direct coupling of the G-protein to the Ca channels without the intermediation of an enzymatically generated second messenger, as suggested for the inhibitory action of noradrenaline on chick sensory neurons.[46] If so, the question then arises whether the two effects of noradrenaline are mediated by the same or different G-proteins. NG108-15 cells contain at least two PTX-sensitive

G-proteins, G_i and G_o.[47] Adenylate cyclase is perferentially inhibited by G_i[48] and the stimulation of GTPase activity in NG108-15 cells by opiates is inhibited by antibodies to the COOH-terminal region of G_i but not by antibodies of G_o.[49] However, Hescheler et al.[36] have shown that the inhibitory action of opiates on I_{Ca} in PTX-treated NG108-15 cells is preferentially reconstituted by a G-protein more closely resembling G_o. We have not yet identified the G-protein responsible for inhibition of I_{Ca} by noradrenaline, but the mutual occlusion between opiates and noradrenaline suggest that they may interact through a common G-protein. This is unlikely to be the same G-protein as that responsible for adenylate cyclase inhibition because we have been unable to replicate the effects of opiates or noradrenaline on I_{Ca} with muscarinic agonists, which are equally capable of inhibiting adenylate cyclase.[35] If G_o is indeed the common transducer for opiate, noradrenaline, and (presumably) somatostatin inhibition of I_{Ca}, then it seems rather unlikely that the molecular forms of the opiate and noradrenaline receptor responsible for this effect are the same as those responsible for G_i-mediated inhibition of adenylate cyclase: further pharmacological tests, together with tests using specific antibodies to different G-proteins,[49] may assist in resolving these options.

ACKNOWLEDGMENTS

We thank Dr. R. J. Miller for the supply of NG108-15 cells.

REFERENCES

1. LLINAS, R. & K. WALTON. 1970. Voltage-dependent calcium conductances in neurones. In The Cell Surface and Neuronal Function. C. W. Cotman, G. Poste & G. L. Nicolson, Eds.: 87-118. Elsevier. Amsterdam.
2. KOSTYUK, P. G. 1981. Calcium channels in the neuronal membrane. Biochem. Biophys. Acta 650: 128-150.
3. KATER, S. B., M. P. MATTSON, C. COHAN & J. CONNOR. 1988. Calcium regulation of the neuronal growth cone. Trends Neurosci. 11: 315-321.
4. MILLER, R. J. 1987. Calcium channels in neurones. In Structure and Physiology of the Slow Inward Calcium Channel. D. J. Triggle, Ed.: 161-246. A. R. Liss. New York.
5. BROWN, D. A. & W. H. GRIFFITH. 1983. Persistent slow inward current in voltage-clamped hippocampal neurones of the guinea-pig. J. Physiol. 337: 303-320.
6. BROWN, D. A., A. CONSTANTI, R. J. DOCHERTY, M. GALVAN, B. GAHWILER & J. V. HALLIWELL. 1984. Pharmacology of calcium currents in mammalian central neurones. Proc. IUPHAR 9th Int. Congr. Pharmacol. 2: 343-348. Macmillan. London.
7. DOCHERTY, R. J. & D. A. BROWN. 1986. Interaction of 1,4-dihydropyridines with somatic Ca currents in hippocampal CA$_1$ neurones of the guinea pig in vitro. Neurosci. Lett. 70: 110-115.
8. GAHWILER, B. H. & D. A. BROWN. 1987. Effects of dihydropyridines on calcium currents in CA3 pyramidal cells in slice cultures of rat hippocampus. Neuroscience 20: 731-738.
9. GAHWILER, B. H. & D. A. BROWN. 1987. Muscarine affects calcium currents in rat hippocampal pyramidal cells in vitro. Neurosci. Lett. 76: 301-306.
10. KLEE, W. R. & M. NIRENBERG. 1973. A neuroblastoma × glioma hybrid cell line with morphine receptors. Proc. Natl. Acad. Sci. USA 71: 3474-3477.

11. HAMPRECHT, B. 1977. Structural, electrophysiological and pharmacological properties of neuroblastoma-glioma cell hybrids in cell culture. Int. Rev. Cytol. **49:** 99-170.

12. BROWN, D. A. & H. HIGASHIDA. 1988. Voltage- and calcium-activated potassium currents in mouse neuroblastoma × rat glioma hybrid cells. J. Physiol. **397:** 149-165.

13. FISHMAN, M. C. & I. SPECTOR. 1981. Potassium current suppression by quinidine reveals additional calcium currents in neuroblastoma cells. Proc. Natl. Acad. Sci. USA **78:** 5245-5249.

14. FURUYA, K., S. FURUYA, & YAMAGISHI. 1983. Developmental time courses of Na and Ca spikes in neuroblastoma × glioma hybrid cells. Dev. Brain Res. **11:** 229-234.

15. BODEWEI, R., S. HERING, B. SCHUBERT & A. WOLLENBERGER. 1985. Sodium and calcium currents in neuroblastoma × glioma hybrid cells before and after morphological differentiation by dibutyryl cyclic AMP. Gen. Physiol. Biophys. **4:** 113-127.

16. HERING, S., R. BODEWEI, B. SCHUBERT, K. RHODE & A. WOLLENBERGER. 1985. A kinetic analysis of the inward calcium current in 108CC15 neuroblastoma × glioma hybrid cells. Gen. Physiol. Biophys. **4:** 129-141.

17. TSUNOO, A., M. TOSHII & T. NARAHASHI. 1986. Block of calcium channels by enkephalin and somatostatin in neuroblastoma × glioma hybrid NG108-15 cells. Proc. Natl. Acad. Sci. USA **83:** 9832-9836.

18. SHIMAHARA, T. & C. ICARD-LIEPKALNS. 1987. Activation of enkephalin receptors reduces calcium conductance in neuroblastoma cells. Brain Res. **415:** 357-361.

19. DOCHERTY, R. J. 1988. Gadolinium selectively blocks a component of calcium current in rodent neuroblastoma × glioma hybrid (NG108-15) cells. J. Physiol. **398:** 33-47.

20. NARAHASHI, T., A. TSUNOO & M. YOSHII. 1987. Characterization of two types of calcium channels in mouse neuroblastoma cells. J. Physiol. **383:** 231-249.

21. NOWYCKY, M. C., A. P. FOX & R. W. TSIEN. 1985. Three types of neuronal calcium channels with different calcium agonist sensitivity. Nature **316:** 440-443.

22. FOX, A. P., M. C. NOWYCKY & R. W. TSIEN. 1987. Kinetic and pharmacological properties distinguishing three types of calcium current in chick sensory neurones. J. Physiol. **394:** 149-172.

23. CARBONE, E. & H. D. LUX. 1984. A low voltage-activated, fully inactivating Ca channel in vertebrate sensory neurones. Nature **310:** 501-502.

24. BOSSU, J. L., A. FELTZ & J. M. THOMANN. 1985. Depolarization elicits two distinct calcium currents in vertebrate sensory neurones. Pflug. Arch. **403:** 360-368.

25. FEDULOVA, S. A., P. G. KOSTYUK & N. S. VESELOVSKY. 1985. Two types of calcium channels in the somatic membrane of new-born rat dorsal root ganglion neurones. J. Physiol. **359:** 431-446.

26. HIRNING, L. D., A. P. FOX., E. W. MCLESKEY, B. M. OLIVERA, S. A. THAYER, R. J. MILLER & R. W. TSIEN. 1988. Dominant role in N-type Ca^{2+} channels in evoked release of norepinephrine from sympathetic neurones. Science **239:** 57-61.

27. FREEDMAN, S. B., G. DAWSON, M. L. VILLEREAL & R. J. MILLER. 1984. Identification and characterization of voltage-sensitive calcium channels in neuronal clonal cell lines. J. Neurosci. **4:** 1453-1467.

28. SANGUINETTI, M. C. & R. S. KASS. 1984. Regulation of cardiac calcium channel current and contractile activity by the dihydropyridine Bay K 8644 is voltage-dependent. J. Mol. Cell. Cardiol. **16:** 667-670.

29. HESS, P., J. B. LANSMAN & R. W. TSIEN. 1984. Different modes of Ca channel gating behaviour favoured by dihydropyridine Ca agonists and antagonists. Nature **311:** 538-544.

30. BROWN, A. M., D. L. KUNZE & A. YATANI. 1986. Dual effects of dihydropyridines on whole cell and unitary calcium currents in single ventricular cells of guinea-pig. J. Physiol. **379:** 495-514.

31. YANO, K., H. HIGASHIDA, R. INOUE & Y. NOZAWA. 1984. Bradykinin-induced rapid breakdown of phosphatidylinositol 4,5-bisphosphate in neuroblastoma × glioma hybrid NG108-15 cells. J. Biol. Chem. **259:** 10201-10207.

32. NAKAGAWA, Y., H. HIGASHIDA & N. MIKI. 1984. A single class of neurotensin receptors with high affinity in neuroblastoma × glioma NG108-15 hybrid cells that mediate facilitation of synaptic transmission. J. Neurosci. **4:** 1653-1664.

33. SABOL, S. L., & M. NIRENBERG, 1979. Regulation of adenylate cyclase of neuroblastoma

          × glioma hybrid cells by α-adrenergic receptors. Proc. Natl. Acad. Sci. USA **254:**
          1913-1920.
34. ISOM, L. L., E. J. CRAGOE & L. L. LIMBIRD. 1987. Multiple receptors linked to inhibition
     of adenylate cyclase accelerate N^+/H^+ exchange in neuroblastoma × glioma cells via
     a mechanism other than decreased cyclic AMP accumulation. Proc. Natl. Acad. Sci.
     USA **262:** 17504-17509.
35. NATHANSON, N. M., W. L. KLEIN & M. NIRENBERG. 1978. Regulation of adenylate
     cyclase activity mediated by muscarinic acetylcholine receptors. Proc. Natl. Acad. Sci.
     USA **75:** 1788-1791.
36. HESCHELER, J., W. ROSENTHAL, W. TRAUTWEIN & G. SCHULTZ. The GTP-binding
     protein, G_o, regulates neuronal calcium channels. Nature **325:** 445-447.
37. DOCHERTY, R. J. & I. McFADZEAN. 1987. Noradrenaline decreases voltage-dependent
     calcium current in neuroblastoma × glioma hybrid (NG108-15) cells. J. Physiol. **390:**
     81P.
38. LIPSCOMBE, D. & R. W. TSIEN, 1987. Noradrenaline inhibits N-type Ca channels in frog
     sympathetic neurones. J. Physiol. **390:** 84P.
39. BEAN, B. P., 1987. Norepinephrine changes the voltage-dependence of calcium channels
     in sensory neurons. Biophys. J. **51:** 30a.
40. GILMAN, A. G. 1987. G-proteins: Transducers of receptor-generated signals. Ann. Rev.
     Biochem. **56:** 615-649.
41. DOCHERTY, R. J. & I. McFADZEAN. 1988. On the mechanism of α-adrenergic inhibition
     of voltage-sensitive calcium currents in NG108-15 neuroblastoma × glioma hybrid cells.
     Neurosci. Lett. Suppl. **32:** S22.
42. UI, M. 1984. Islet-activating protein, pertussis toxin: A probe for functions of the inhibitory
     guanine nucleotide regulatory component of adenylate cyclase. Trends Pharmacol. Sci.
     5: 277-279.
43. KUROSE, H., T. KATADA, T. AMANO & M. UI. 1983. Specific uncoupling by islet-activating
     protein, pertussis toxin, of negative signal transduction via α-adrenergic, cholinergic, and
     opiate receptors in neuroblastoma × glioma hybrid cells. J. Biol. Chem. **258:** 4870-4875.
44. TAMAOKI, T., H. NOMOTA, T. TAKAHASHI, Y. KATO., M. MORIMOTO & F. TOMITA.
     1986. Staurosporin, a potent inhibitor of phospholipid/calcium-dependent protein kinase.
     Biochem. Biophys. Res. Commun. **135:** 397-402.
45. RANE, S. G. & K. DUNLAP. 1986. Kinase C activator 1,2-oleoylacetylglycerol attenuates
     voltage-dependent calcium current in sensory neurons. Proc. Natl. Acad. Sci. USA **83:**
     184-188.
46. FORSCHER, P., G. S. OXFORD & D. SCHULZ. 1986. Noradrenaline modulates calcium
     channels in avian dorsal root ganglion cells through tight receptor-channel coupling. J.
     Physiol. **379:** 131-144.
47. MILLIGAN, G., P. GIERSHIK., A. M. SPIEGEL & W. A. KLEE. 1986. The GTP-binding
     regulatory protein of neuroblastoma × glioma, NG108-15, and glioma, C6, cells. FEBS
     Lett. **195:** 225-230.
48. ROOF, D. J., M. L. APPLEBURY & P. C. STERNWEISS. 1985. Relationships within the
     family of GTP-binding proteins isolated from bovine central nervous system. J. Biol.
     Chem. **260:** 16242-16249.
49. McKENZIE, F. R., E. C. H. KELLY, C. G. UNSON, A. M. SPIEGEL & G. MILLIGAN. 1988.
     Antibodies which recognize the C-terminus of the inhibitory guanine-nucleotide binding
     protein (G_i) demonstrate that opioid peptides and foetal calf serum stimulate the high-
     affinity GTPase activity of two separate pertussis-toxin substrates. Biochem. J. **249:**
     653-659.

Direct G-Protein Regulation of Ca^{2+} Channels

A. M. BROWN,[a] A. YATANI,[a] Y. IMOTO,[a] J. CODINA,[b] R. MATTERA,[b] AND L. BIRNBAUMER[b]

[a] *Department of Physiology and Molecular Biophysics*
[b] *Department of Cell Biology*
Baylor College of Medicine
Houston, Texas 77030

INTRODUCTION

When we first identified specific guanine nucleotide binding, or G proteins, that could directly gate muscarinic cholinergic atrial K$^+$ channels,[1,2] we wondered how widespread the phenomenon was. K$^+$ channels in other tissues were prime candidates if they, like muscarinic atrial K$^+$ channels, could also be activated by neurotransmitters or hormones in a cyclic adenosine 3′,5′-cyclic monophosphate (cAMP)-independent manner. Our expectations were confirmed when we found that the same exogenous G protein, G$_k$, that was effective in heart mimicked the effects of somatostatin and acetylcholine on specific K$^+$ channels in GH3 cells, a clonal anterior pituitary cell line.[3,4] The GH3 K$^+$ channels had a larger conductance than the atrial K$^+$ channels indicating that direct gating by G proteins was not restricted to one type of K$^+$ channel and this encouraged us to broaden the scope of our enquiry by asking whether a completely different category of channels might be involved. We turned to Ca^{2+} channels for the following reasons: (1) guanosine triphosphate (GTP)-altered dihydropyridine (DHP) binding in cardiac sarcolemmal and skeletal muscle t-tubule vesicles[5,6]; (2) G proteins were implicated in the coupling of a variety of neurotransmitters to Ca^{2+} channels[7,8]; and (3) G proteins were shown to couple opioid receptors to Ca^{2+} channels.[9] The situation for DHP-sensitive Ca^{2+} channels is more complicated because they are also modulated by second messengers (FIG. 1), and this mechanism has to be excluded when the presence of direct pathways is being determined. Heart and skeletal muscle were satisfactory tissues because at least three types of pertussis toxin (PTX) substrate[10–13] and two types of cholera (CTX) substrate[13] have been reported for cardiac sarcolemma, and two types of PTX and CTX substrates have been found in skeletal muscle t-tubules.[13] We found that G proteins could directly gate DHP-sensitive Ca^{2+} channels in heart[14,15] and skeletal muscle[16] and we proved, using recombinant DNA techniques, that a single G protein, G$_s$, so named because it is the stimulatory regulator of adenylyl cyclase, could also gate Ca^{2+} channels.[17–19]

METHODS

G proteins were purified from membranes of human erythrocytes and bovine brain using column chromatography[20–22] and the α subunit of interest, α_s or α_i-3, was

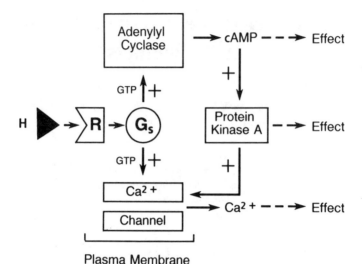

FIGURE 1. Proposed mechanism of G_s effects on Ca^{2+} channels. H, hormone (or neurotransmitter); R, receptor; +, stimulatory effect. There are two pathways, membrane delimited or direct and second messenger or cytoplasmic.

separated from the holomeric $\alpha\beta\gamma$ protein using an additional chromatography step. The purified G proteins were preactivated with GTPγS and the free GTPγS was removed by dialysis. The G proteins were added to static baths at dilutions of 100 to 10,000 and, under these conditions, the buffers used to maintain the G proteins were by themselves ineffective. Specific α subunits were expressed from recombinant αcDNAs following the scheme of FIGURE 2. The α-subunit cDNAs of short human α_s were engineered into pT7 vectors,[23–26] and the recombinant plasmids were isolated and transfected into *E. coli* K38 cells. The open reading frame encoded downstream from the T7 promoter was expressed after infection of the cells with the phage mGP1-2. The cells were lysed and the recombinant α subunits were activated and partially purified by chromatography.[18,19] With this system, 5-8% of bacterial total protein was composed of the α-subunit polypeptide.

The assays of single-channel activity were designed to exclude cytoplasmic second messenger involvement. G proteins were added to channels that had been incorporated into planar lipid bilayers[15,16] or channels in excised inside-out membrane patches.[14]

RESULTS AND DISCUSSION

Our first experiments were done on inside-out membrane patches excised from adult guinea pig ventricle.[14] Single Ca^{2+}-channel currents ceased within about 60 seconds of excision, but a large burst of activity could be produced when GTPγS at 100 μM was in the bath. This effect was Mg^{2+}-dependent and independent of the cAMP pathway and had two requirements: (1) Isoproterenol (Iso) had to be present

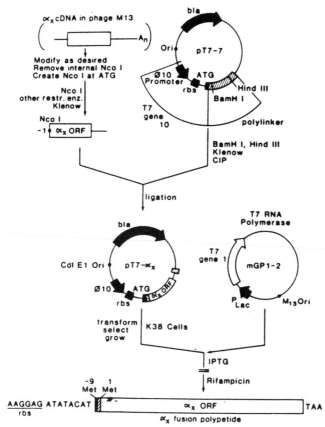

FIGURE 2. Strategy for bacterial expression of cloned proteins as applied to α subunits of G proteins (Tabor & Richardson[24] and Tabor *et al.*[25]). cDNAs were inserted into the *Bam*HI site of pT7-7 plasmid and recombinant PT7-αx plasmids with cDNAs inserted in either the antisense (3′ to 5′, open arrows) or sense (5′ to 3′, filled arrows) direction were isolated, transfected into *E. coli* K38 cells and the open reading framed encoded downstream of the T7 promoter expressed after infection of the cells with phage mGP1-2. Recombinant α_s was the Asp[71] short version. The cells were lysed and the recombinant α subunits were activated and purified partially by chromatography over DEAE-Sephacel (0 to 400 m*M* NaCl gradient) and activated by GTPγS. (From Brown *et al.*[18] Reproduced by permission of *Cold Spring Harbor Symposia on Quantitative Biology.*)

in the patch pipette; and (2) either Iso or the DHP agonist BAY K 8644 had to be present in the bath. The requirement for Iso in the pipette suggested that a G_s protein was involved. We then found that preactivated G_s, G_s^*, or its preactivated α subunit, α_s^*, at picomolar concentrations could simulate the effect of GTPγS. In these cases, Iso was not required, but it was essential that activity had not ceased before excision; otherwise the G proteins or GTPγS were ineffective. Unlike the results reported in GH_3 cells, we were unable to rescue activity[27] once it had stopped. The patch-clamp experiments were unsatisfactory because the Ca^{2+} currents were so unstable. To circumvent this difficulty, we used Ca^{2+} channels from cardiac sarcolemmal (SL) vesicles after incorporation into planar lipid bilayers. For reasons that are not understood, stable recordings of single Ca^{2+}-channel currents are possible for long periods although it is necessary to add BAY K 8644 to obtain measurable activity. The behavior of incorporated cardiac SL Ca^{2+} channels has been reported elsewhere in this volume and in Rosenberg et al.[28] The properties we observed were similar (FIG. 3) to those already described. Under our solution conditions (cis or extracellular isoBa^{2+}; trans 0 Ba^{2+}) the single-channel conductance was ~20 pS, the gating was strongly voltage-dependent, and the open times were exponentially distributed. With time there was a decrease in activity but G_s^* (FIG. 4) or α_s^* at picomolar concentrations increased activity greatly. These effects were completely independent of substrate and were due to an increase in the frequency of opening; neither channel conductance nor channel open time were changed. The effects were asymmetrical; the G proteins were effective only from the trans or intracellular side. Preactivated G_k, G_k^*, was completely ineffective even at nanomolar concentrations. The results summarized in TABLE 1 show that GTPγS, G_s^*, and α_s^* activate DHP-sensitive cardiac Ca^{2+} channels directly and equivalently. Of course, activated G_s can activate these same channels indirectly via the cAMP cascade,[28,29] and we will deal with the relative effects of these two pathways later.

While the cardiac studies illustrated the direct G_s effect, the requirement for BAY K 8644 left open the possibility that direct effects could only be elicited in the presence

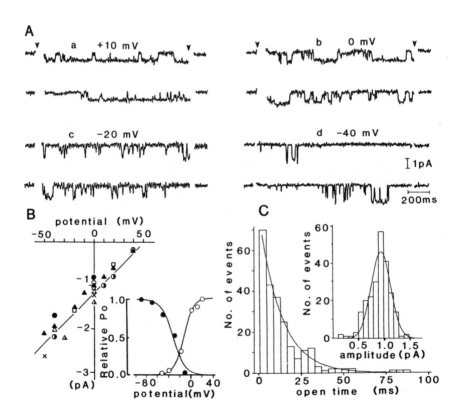

of DHP agonist. To circumvent this problem, we turned to another preparation, namely Ca^{2+} channels from skeletal muscle T-tubule vesicles which, like the cardiac SL Ca^{2+} channel, were incorporated into planar lipid bilayers. It is possible in this preparation to obtain stable activity even in the absence of BAY K 8644.[30] We began our studies in the presence of BAY K 8644 but found that with time activity actually increased, presumably the result of increased incorporation. To increase background stability, we washed the *cis* chamber thoroughly after incorporation and as a result were able to record stable activity for periods of 30 minutes (FIG. 5). Note that the channels are inside-out in these vesicles so that the *cis* chamber to which the vesicles were always added was now intracellular using the standard current recording convention. Three conductance levels were present in our solutions (*cis*, iso-Ba^{2+}; *trans* 0 Ba^{2+}), a majority conductance of 10 pS, and two others of 5 and 12 pS at about 0 mV (FIG. 6); all three were DHP-sensitive. The majority conductance value also agrees with values reported for the DHP-sensitive slow Ca^{2+} channel present in clonal BC3H1 cells and tissue-cultured satellite cells of skeletal muscle (C2 cells), both of which have DHP-sensitive skeletal muscle Ca^{2+}-current phenotypes.[31–33] Activity was continuous at potentials of 0 mV making depolarizing steps that were required for cardiac Ca^{2+} channels unnecessary. Under these conditions $G_s^{\cdot}$ and $\alpha_s^{\cdot}$ activated the DHP-sensitive Ca^{2+} channels at picomolar concentrations (FIG. 7). As for cardiac cells, to be effective the G proteins had to be added to the intracellular chamber. $G_k^{\cdot}$ was ineffective even at nanomolar concentrations and had no effect on $G_s^{\cdot}$-activated channels. Endogenous G protein was co-incorporated with the Ca^{2+} channels because

---

FIGURE 3. Single cardiac Ca^{2+}-channel currents recorded from cardiac sarcolemmal vesicles incorporated into planar lipid bilayers. *Cis* solution (mM); NaCl, 40; BaCl$_2$, 100; MgCl$_2$, 2; HEPES, 10 (pH 7.4 with NaOH). *Trans* solution (mM); NaCl, 40; MgCl$_2$, 2; HEPES, 10 (pH 7.4 with NaOH). Vesicles are right side out, *cis* solution is extracellular and Ba^{2+} currents are inward.

(A) Currents produced by depolarizing steps to $+10$ mV (a), 0 mV (b), -20 mV (c), and -40 mV (d) from a holding potential of -70 mV. Duration pulses of 1.5 seconds were applied every 5 seconds. Leakage and capacitive currents were subtracted. Arrows indicate the onset and the end of the pulses. Traces are low-pass filtered at 100 Hz. Averaged opening probabilities, P_os for -40, -20, 0, and $+10$ mV were 0.03, 0.29, 0.49, and 0.51, respectively.

(B) Current-voltage relationship of single Ca^{2+}-channel currents. Each symbol represents a different experiment and the values were averaged from at least 10 openings. Straight line is a linear least-squares fit ($\gamma^2 = 0.953$ with single-channel conductance of 24.9 pS, between -50 and $+40$ mV). Inset figure shows voltage dependence of Ca^{2+}-channel activation ($\bigcirc$) and inactivation ($\bullet$) with relative P_o normalized to P_{max} on the ordinate. P_os were determined by comparing averaged P_os obtained from the current responses to 10 pulses (1500-msec duration) at each test potential, and the value was normalized to maximum P_o. For the inactivation curve, the test potential was $+10$ mV, and prepulse duration was 3.5 seconds. For the activation curve the holding potential was -70 mV. Solid curve was drawn according to:

$$\text{Relative } P_o = \frac{P_o}{P_{max}} = [1 + \exp\{(V_m - V_h)\}]^{-1}$$

V_h = midpotential, k = slope factor, and P_{max} is the maximum opening probability observed. For activation curve, $V_h = -12.7$ mV and k = -8.7 mV. For inactivation curve, $V_h = -32.5$ mV and k = 10.8 mV.

(C) Open time and amplitude histograms of single-channel currents at 0 mV (holding potential, -50 mV). Open time τ was 12.7 $\pm$ 0.77 mx (SD). The mean and SD of the amplitudes were 0.95 pm 0.2 pA. (From Imoto *et al.*[15] Reproduced by permission of the *American Journal of Physiology.*)

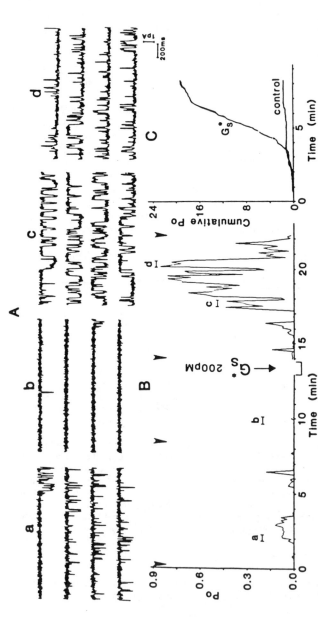

FIGURE 4. G_s^* activates single cardiac Ca^{2+} channels incorporated into a planar lipid bilayer. In the presence of BAY K 8644 and with 100 mM Ba^{2+} as the charge carrier, the conductance between -50 and $+20$ mV was ~20 pS. The channel opened in bursts, and the probability of opening was voltage-dependent. The mean open times were fit to an exponential distribution ($\tau\sim12\pm3$ msec, $n=9$, at test potentials between 0 and $+20$ mV). The values are consistent with those obtained for BAY K 8644-stimulated channels in cell-attached patches. Current traces produced by depolarizing clamp steps to 0 mV from a holding potential of -40 mV are shown in (A) before (1 and 2) and after (3 and 4) addition of G_s^*. Pulses were applied every 30 seconds for 20 seconds. Leakage and capacitive currents were subtracted. Traces 1 to 4 were taken at the times indicated in (B) and are two-second segments from the 20-second pulses. (**B**) The entire experimental record. P_o is proportion of open time per record and is equivalent to average current per record. Note the decrease in activity with time in control in (B) before addition of G_s^* (100 pM) to the *trans* chamber. The ordinate is given as P_o because N was 1 in this experiment. (C) Cumulative P_os obtained between the arrows in (B). Cum P_o (G_s^*)/cum P_o (control) is 11. (From Yatani *et al.*[14] Reprinted by permission from *Science*.)

TABLE 1. Effects of GTPγS, G_s^*, α_s^*, and α_k^* on Mean Open Time and Opening Probability of Ca^{2+} Channel

Procedure	Mean Open Time (msec)		P_o		P_o(after)/P_o(before)	n
	Before	After	Before	After		
Control[+]	16.6 ±5.8	12.4 ±4.6	0.20 ±0.14	0.02 ±0.02	0.10 ±0.10	10
GTPγS (8-100 μM)	10.2 ±5.1	27.7 ±9.7	0.16 ±0.05	0.54 ±0.15[a]	3.73 ±1.77	5
G_s^* (10-200 pM)	14.2 ±10.1	20.6 ±8.0	0.08 ±0.05	0.41 ±0.24[a]	6.11 ±3.27	5
α_s^* (10-100 pM)	19.0 ±7.2	39.5 ±21.7	0.13 ±0.09	0.41 ±0.13[b]	4.14 ±1.83	4
α_k^* (100-400 pM)	15.0 ±8.3	10.2 ±4.7	0.20 ±0.18	0.02 ±0.02	0.11 ±0.09	5

NOTE: Mean open time and opening probability (P_o) were measured before and after addition of the agent for periods of 2-4 minutes each. Depolarizing pulses had durations between 1.5 to 5 seconds and were to 0 or +10 mV. The holding potentials were between −40 and −70 mV. All data were from single-channel experiments. Ratio of P_os before and after addition of agent was calculated in each experiment and the averaged values are shown in the column of P_o(after)/P_o(before). Control[+] compares the values from initial and late periods (3 min each) of recordings of at least six minutes' duration.

[a] $p < 0.05$.
[b] $p < 0.005$ after correction for the background decrease in opening frequency. The magnitude of this decrease is shown in the first and last rows of this table.

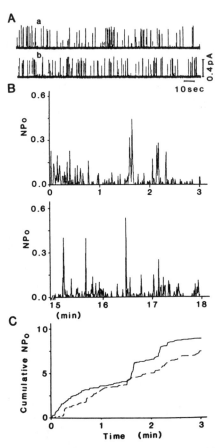

FIGURE 5. Representative single Ca^{2+}-channel currents and resulting average activities expressed as NP_os and cumulative NP_os during control and mock test periods. *Cis* and *trans* solutions as in FIGURE 3 legend. However, the t-tubule vesicles are inside-out, *cis* solution is intracellular, and Ba^{2+} currents are outward.

Panel A: Single-channel openings and closings as seen at between 0 and 3 minutes (upper tracing) and 17 and 20 minutes (lower tracing) at +20 mV.

Panel B: Diaries of cyclical channel opening and closing events (NP_os) as seen in consecutive 0.5-second periods for the first three minutes after *cis* chamber washout (upper) and 15 minutes later (lower).

Panel C: Accumulation of NP_o values (cumulative NP_os) obtained throughout the early (continuous tracing) and late (dashed tracing) periods of activity. Note that the slopes of cumulative NP_os as a function of elapsed time are cumulative NP_o/min and constitute the measure of average Ca^{2+}-channel activity under control and experimental conditions. Cumulative NP_o/min values for early and late activities were 3.0 and 2.5, respectively. BAY K 8644 (3 μM) was present throughout. (From Yatani *et al.*[16] Reproduced by permission from *Journal of Biological Chemistry.*)

GTPγS at 100 μM produced the same effects as G_s^*. In 1-10% of the experiments the β-adrenoreceptor was also co-incorporated and the combination of *trans* Iso and *cis* GTP activated the channels. Additionally, it was possible to activate G_s with CTX, G_s^{CTX}, in the presence of GTP which activated the Ca^{2+} channels. The results are summarized in FIGURE 8. The effects of G_s^* were additive with the effects of cAMP as well as with the effects of BAY K 8644 (FIG. 6 in Yatani *et al.*[16]). These experiments

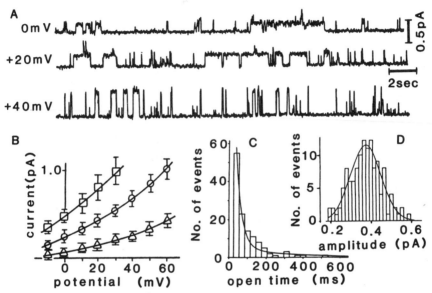

FIGURE 6. Analysis of properties of skeletal muscle t-tubule Ca^{2+}-channel activities incorporated into planar lipid bilayers in the presence of 3 μM BAY K 8644.

Panel A: Representative single-channel records obtained at holding potentials of 0, +20, and +40 mV.

Panel B: Current-voltage relationships for the three types of single-channel currents observed in lipid bilayers. Between −10 and +10 mV the slope conductances were 4.6 pS, 10 pS, and 13 pS, of which the middle value corresponds to most frequently observed, majority channel.

Panel C: Open-time frequency histogram of the majority channel at +20 mV fitted by the sum of two first-order decay functions with τ of 27 and 243 msec in the proportion of 73 to 27%, respectively.

Panel D: Amplitude frequency histogram of the 10 pS channel at +20 mV fitted by a single Gaussian distribution function with a maximum at 0.37 pA. (From Yatani *et al.*[16] Reproduced by permission from *Journal of Biological Chemistry.*)

were done in the presence of BAY K 8644 and did not establish if G_s^* was effective in the absence of other activators; however, it was possible to record limited activity in the absence of BAY K 8644, and in these cases G_s^* and α_s^* had their greatest effects relative to control (FIG. 7). Thus, there are three distinct sites that were modulated in these experiments, a DHP site, a protein kinase A site, and a G_s site.

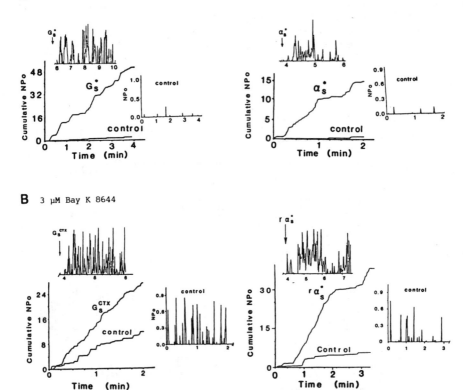

FIGURE 7. Representative experiments showing direct regulation of the skeletal muscle t-tubule dihydropyridine-sensitive Ca^{2+} channel purified human erythrocyte G_s. Experiments were carried out both in the absence (**A**) and in the presence (**B**) of the dihydropyridine agonist BAY K 8644 at 3 μM in the *cis* chamber. Shown are the NP_o diaries and the cumulative NP_o curves before and after addition of GTPγS-activated G_s, G_s^* GTPγS-activated α_s, α_s^* (**A**) and CTX-treated G_s (in the presence of GTP) and GTPγS-activated recombinant α_s of the short type (**B**). NP_o is the probability of opening for the N channels in the patch. The holding potentials were either 0 or +20 mV, and all experiments were done with $BaCl_2$ in the *cis* chamber. Average stimulation of activity was between 10- and 20-fold in the absence and two- to threefold in the presence of 3 μM BAY K 8644. The Ca^{2+} channels incorporated into bilayers can be stimulated either by exogenously activated G_s, or by activation of an endogenous t-tubule G protein, presumably also G_s, co-incorporated with the Ca^{2+} channel. In other experiments it was established that to be effective GTPγS had to be added to the *cis* chamber, confirming the sidedness of the t-tubule membranes.

We have shown that the G protein G_s activates DHP-sensitive Ca^{2+} channels independently of second messengers. Neither this nor our studies on K^+ channels prove direct G protein gating of ionic channels; rather a membrane-delimited, non-cytoplasmic pathway has been established. This necessary first step also preceded the demonstration of direct effects of G proteins on adenylyl cyclase and cyclic guanosine 3',5'-cyclic monophosphate (cGMP) phosphodiesterase.[34] Further support for direct gating comes from a recent report[35] of a direct inhibitor effect of bovine brain G_o on solubilized DHP receptors from skeletal muscle.

While the direct G_s effect has been revealed by our experiments, the better known effect of G_s on Ca^{2+} channels is via the cAMP cascade. We evaluated the relative effects of these two pathways in whole cell experiments on adult guinea pig ventricular myocytes. Supramaximal concentration of 8 bromo-cAMP and Forskolin produced three- to fourfold increases in Ca^{2+} currents, confirming the large effects of the second-messenger pathway. Iso at $10^{-5}M$ had larger effects but the difference was small. In other experiments it was possible to block the effects of cAMP and Forskolin completely by perfusion with a synthetic peptide PKA inhibitor. In this situation Iso also produced a small but definite increase in Ca^{2+} current. Finally, Ca^{2+} currents always run down with time in these cells, and we found that neither cAMP nor Forskolin were effective when only about 10-20% of the current remained. Under these circumstances Iso was effective and the effect relative to second messengers was now very large. These results confirm a direct effect of G_s on Ca^{2+} channels, and it is possible that G_s is especially important for basal Ca^{2+} currents that do not appear to be under the influence of the cAMP cascade.[29]

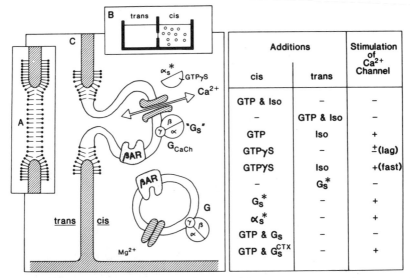

	Additions		Stimulation of Ca²⁺ Channel
	cis	trans	
	GTP & Iso	–	–
	–	GTP & Iso	–
	GTP	Iso	+
	GTPγS	–	±(lag)
	GTPγS	Iso	+(fast)
	–	G_s^*	–
	G_s^*	–	+
	α_s^*	–	+
	GTP & G_s	–	–
	GTP & G_s^{CTX}	–	+

FIGURE 8. Experiments that showed stimulation of skeletal muscle t-tubule Ca^{2+} channels on activation of coincorporated t-tubule G protein or addition of activated G_s. t-Tubule vesicles are inside-out, and cardiac sarcolemmal vesicles are right-side-out and both retain these orientations after incorporation. Vesicle membranes are asymmetrical with respect to orientation of receptors, G proteins, and ion channels. Vesicles are added to the *cis* chamber, which is extensively washed after initial incorporation of Ca^{2+} channels to prevent further incorporation during test periods. For t-tubules the *cis* side corresponds to intracellular side. The table summarizes the results. To satisfy the asymmetries, nucleotides and G proteins were added to *cis* chamber, and Isoproterenol (Iso) was effective only when added to the *trans* chamber. G_s^*, G_s activated with GTPγS and Mg^{2+}; α_s^*, resolved α_s-GTPγS complex; G_s^{CTXH}, G_s ADP-ribosylated with cholera toxin (CTX), and NAD in presence of ADP-ribosylation factor, G_{CaCh}, a G protein (possibly G_s) which activates the Ca^{2+} channel directly; βAR: β-adrenoreceptor. Addition of G proteins, 100-200 pM; of nucleotides, 100 μM. BAY K 8644 (1 μM) was present throughout. (Reprinted by permission from *American Journal of Physiology.* 1988. **23**(3): H401-H410.)

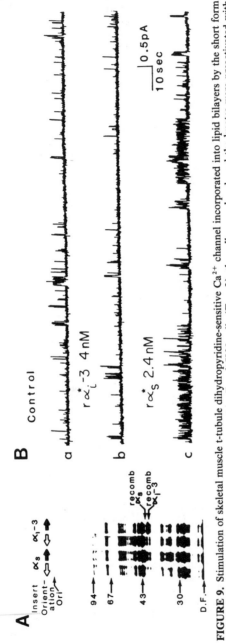

FIGURE 9. Stimulation of skeletal muscle t-tubule dihydropyridine-sensitive Ca^{2+} channel incorporated into lipid bilayers by the short form of recombinant α_s, α_s-Polypeptides were expressed in *E. coli* K38 cells (FIG. 2), the cells were lysed, and the lysates were preactivated with a combination of NaF, AlCl$_3$, MgCl$_2$, and GDP or MgCl$_2$ and GTPγS. The treated lysates were centrifuged and diluted aliquots of the supernatants were analyzed by SDS in 10% polyacrylamide gel slabs. The slabs were stained with Coomassie blue and photographed (A). Note the appearance of insert-specific and orientation-dependent α-subunit bands; these constituted between 5 and 8% of the total cell protein. α_s-1: short human α_s of 380 amino acids with Asp in position 71 and Ser in position 72 followed by Glu; α_i-3: PTX substrate of 354 amino acids cloned from a human liver cDNA library; α_o: PTX substrate of 354 amino acids identical to rat brain and bovine retina α_o cloned a rat heart cDNA library. (B) Recombinant α_s (Asp^{71} version) and rα_i-3 prepared in their GTPγS-activated form and partially purified by DEAE-Sephacel chromatography. To avoid Ba^{2+} effects on the recombinant polypeptides, the Ba^{2+} gradient was reversed from *cis* to *trans* positive to *cis* to *trans* negative and the current-recording connection was also reversed. Ba^{2+} currents are outward. Single-channel Ba^{2+} currents before (control) and after addition first of GTPγS-activated recombinant α_i^{-3} (B) and then of GTPγS-activated recombinant α_s. Holding potential $+20$ mV; Bay K 8644 at 3 μM was present in bath chambers. The single-channel currents are from the majority 10 pS channel.

In addition to showing a direct effect of G proteins on yet another type of ionic channel, these experiments have shown that the direct effects may be of two types: obligatory for the K^+ channels and modulatory as for the Ca^{2+} channels. In the latter case, the G proteins by themselves cannot activate the channels; a depolarized membrane is the necessary condition. Another finding of importance to the physiology of G proteins is that G_s had two effectors, adenylyl cyclase and the Ca^{2+} channel. One possibility, however, was that two different G proteins mediated these effects. Thus, the α subunit of G_s occurs in four forms[34] that arise from alternate splicing of a single gene. Two differ by 15 amino acids and two by one serine residue at the 3' end of the insertion/deletion. Human red blood cell (rbc) G_s is of the short form but the two varieties cannot be separated from each other. To settle the issue, we expressed the short serine deletion α subunit in *E. coli* and tested the effects of the recombinant α polypeptide on both effectors. We found that a single type of $r\alpha_s$ preactivated with either GTPγS or ALF^{2-} activated Ca^{2+} channels (FIG. 9) and adenylyl cyclase (not shown) at nanomolar concentrations. These were 10-100 times the concentrations required of native proteins and a similar result was reported for a different $r\alpha_s$ tested on AC.[36] We are uncertain as to the reason for the lower potency, but this does not detract from the important conclusion that a single α_s has two effectors.

CONCLUSIONS

G proteins can directly activate two different K^+ channels and, as this report has shown, two different types Ca^{2+} channels. Direct G protein gating of ionic channels is likely to be widespread. Furthermore, it is clear that a single G protein may have more than one effector. The significance of direct gating probably lies in the fact that it is more direct than second messengers and in some cases may be the sole coupling mechanism.

REFERENCES

1. YATANI, A., J. CODINA, A. M. BROWN & L. BIRNBAUMER. 1987. Science **235**: 207-211.
2. CODINA, J., A. YATANI, D. GRENET, A. M. BROWN & L. BIRNBAUMER. 1987. Science **236**: 442-445.
3. YATANI, A., J. CODINA, R. D. SEKURA, L. BIRNBAUMER & A. M. BROWN. 1987. Mol. Endocrinol. **1**(4): 283-289.
4. CODINA, J., G. GRENET, A. YATANI, L. BIRNBAUMER & A. M. BROWN. 1987. FEBS Lett. **216**(1): 104-106.
5. GALIZZI, J.-P., M. FOSSETT & M. LAZDUNSKI. 1984. Eur. J. Biochem. **144**: 211-215.
6. TRIGGLE, D. J., A. SKATTEBOL, D. RAMPE, A. JOCLYN & P. GENGO. 1986. *In* New Insights Into Cell and Membrane Transport Processes. G. Poste & S. T. Crooke, Eds.: 125.
7. DUNLAP, K., G. G. HOLZ IV & S. G. RANE. 1987. TINS **10**: 241-246.
8. SCOTT, R. H. & A. C. DOLPHIN. 1987. J. Physiol. **386**: 1-17.
9. HESCHELER, J., W. ROSENTHAL, W. TRAUTWEIN & G. SCHULTZ. 1987. Nature **325**: 445-447.
10. HAZEKI, O. & M. UI. 1981. Eur. J. Pharmacol. **56**: 179-180.
11. KUROSE, H. & M. UI. 1983. J. Cyclic Nucl. Protein Phosphor. Res. **9**: 305-318.

12. HALVORSEN, S. W. & N. M. NATHANSON. 1984. Biochemistry **23:** 5813-5321.
13. SCHERER, N., M.-J. TORO, M. L. ENTMAN & L. BIRNBAUMER. 1987. Arch. Biochem. Biophys. **259:** 431-440.
14. YATANI, A., J. CODINA, Y. IMOTO, J. P. REEVES, L. BIRNBAUMER & A. M. BROWN. 1987. Science **238:** 1288-1292.
15. IMOTO, Y., A. YATANI, J. P. REEVES, J. CODINA, L. BIRNBAUMER & A. M. BROWN. 1988. Am. J. Physiol. **255:** H722-H728.
16. YATANI, A., Y. IMOTO, J. CODINA, S. L. HAMILTON, A. M. BROWN & L. BIRNBAUMER. 1988. J. Biol. Chem. **263(20):** 9887-9895.
17. BIRNBAUMER, L., J. CODINA, R. MATTERA, A. YATANI, R. GRAF, J. OLATE, J. SANFORD & A. M. BROWN. 1988. Receptor-effector coupling by G proteins, purification of human erythrocyte G_i-2 and G_i-3 and analysis of effector regulation using recombinant α subunits synthesized in *E. coli.* Cold Spring Harbor Symp. Quant. Biol. **53:** In press.
18. BROWN, A. M., A. YATANI, Y. IMOTO, G. KIRSCH, H. HAMM, J. CODINA, R. MATTERA & L. BIRNBAUMER. 1988. Direct coupling of G proteins to ionic channels. Cold Spring Harbor Symp. Quant. Biol. **53:** In press.
19. MATTERA, R., M. P. GRAZIANO, A. YATANI, Z. ZHOU, R. GRAF, J. CODINA, L. BIRNBAUMER & A. M. BROWN. 1988. Science **243:** 804-807.
20. CODINA, J., J. D. HILDEBRANDT, R. D. SEKURA, M. BIRNBAUMER, J. BRYAN, C. R. MANCLARK, R. IYENGAR & L. BIRNBAUMER. 1984a. J. Biol. Chem. **259:** 5871-5586.
21. CODINA, J., W. ROSENTHAL, J. D. HILDEBRANDT, R. D. SEKURA & L. BIRNBAUMER. 1984. J. Receptor Res. **4:** 411-442.
22. CODINA, J., J. D. HILDEBRANDT, R. IYENGAR, L. BIRNBAUMER, R. D. SEKURA & C. R. MANCLARK. 1983. Proc. Natl. Acad. Sci. USA **80:** 4276-4280.
23. TABOR, S. 1987. Dissertation of the bacteriophage T7 DNA replication by the overproduction of its essential genetic components. Ph.D. thesis, Harvard University, Boston, MA.
24. TABOR, S. & C. C. RICHARDSON. 1985. Proc. Natl. Acad. Sci. USA **82:** 1074-1078.
25. TABOR, S., H. E. HUBER & C. C. RICHARDSON. 1987. J. Biol. Chem. **262:** 16212-16223.
26. MATTERA, R., A. YATANI, Z. ZHOU, R. GRAF, J. CODINA, L. BIRNBAUMER & A. M. BROWN. 1989. J. Biol. Chem. **264(1):** 465-471.
27. ARMSTRONG, D. & R. ECKERT. 1987. Proc. Natl. Acad. Sci. USA **84:** 2518-2522.
28. ROSENBERG, R. L., P. HESS, J. P. REEVES, H. SMILOWITZ & R. W. TSIEN. 1986. Science **231:** 1564-1566.
29. KAMEYAMA, M., J. HESCHELER, F. HOFMANN & W. TRAUTWEIN. 1986. Pfluegers Arch. **407:** 123-128.
30. AFFOLTER, H. & R. CORONADO. 1985. Biophys. J. **48:** 341-347.
31. CAFFREY, J. M., A. M. BROWN & M. D. SCHNEIDER. 1987. Science **236:** 570-573.
32. CAFFREY, J. M., M. D. SCHNEIDER & A. M. BROWN. 1989. J. Neurosci. In press.
33. CAFFREY, J. M., A. M. BROWN. & M. D. SCHNEIDER. 1988. Biophys. J. **53:** 18a.
34. GILMAN, A. G. 1987. Ann. Rev. Biochem. **56:** 615-650.
35. HORNE, W. A., M. ABDEL-GHANG, E. RASHEN, G. A. WEILAND, R. E. OSWALD & R. A. CERRIONE. 1988. Proc. Natl. Acad. Sci. USA **85:** 3718-3722.
36. GRAZIANO, M. P., P. J. CASEY & A. GILMAN. 1987. J. Biol. Chem. **262:** 11375-11381.

Modulation of Ca^{2+}-Channel Currents in Sensory Neurons by Pertussis Toxin-Sensitive G-Proteins^a

ANNETTE C. DOLPHIN AND RODERICK H. SCOTT

Department of Pharmacology
St. George's Hospital Medical School
London SW17 ORE, United Kingdom

Pertussis toxin-sensitive guanine nucleotide binding proteins (G-proteins) are involved in coupling a variety of neurotransmitter receptors with Ca^{2+} channels in cultured rat dorsal root ganglion neurons. Activation of G-proteins by intracellular application of GTP-γ-S (guanosine 5′-0-3-thiotriphosphate; 20-500 μM) resulted in inhibition of the transient component of the whole cell I_{Ba}, leaving a residual noninactivating Ca^{2+}-channel current with several characteristics of an L-type current.[1-3] We thus examined the effect on this current of several classes of Ca-channel ligand, including 1,4-dihydropyridines, phenylalkylamines, and benzothiazepines.[4] In control cells, nifedipine (5 μM), D$_{600}$ (10 μM), and diltiazem (30 μM), inhibited Ca^{2+}-channel currents, although nifedipine and D$_{600}$ induced initial transient increases in a proportion of cells. In contrast, in the presence of internal GTP-γ-S all the ligands potentiated the residual sustained Ca^{2+}-channel current by 1.5- to fourfold (FIG. 1). This effect was voltage-dependent,[5] agonist responses in the presence of GTP-γ-S being less marked at depolarized holding potentials.[4] The rate of activation of the maximum I_{Ba} was slowed by GTP-γ-S, and this was subsequently increased by application of either D$_{600}$ or nifedipine. D$_{600}$ decreased τ_{act} from 9.5 ± 2.3 msec to 4.1 ± 0.7 msec (mean ± SEM, $n = 6$). Internal GDP-β-S had no effect on the response to the Ca^{2+}-channel ligands, but preincubation of the cells with pertussis toxin prevented both the transient agonist effect to D$_{600}$ in control cells, and the more sustained agonist response in GTP-γ-S-containing cells. The agonist response to (±)-BAY K 8644 was greater in GTP-γ-S-containing than in control cells, and was also prevented by pertussis toxin, being converted into an antagonist response.[6] Associated with its agonist response in the presence of GTP-γ-S, nifedipine, like BAY K 8644, induced a hyperpolarizing shift of 5-10 mV in the current-voltage relationship and a 20 mV hyperpolarizing shift in the steady-state inactivation curve for I_{Ba}. In contrast, agonist responses to D$_{600}$ were

^{a}This work was supported by the Medical Research Council.

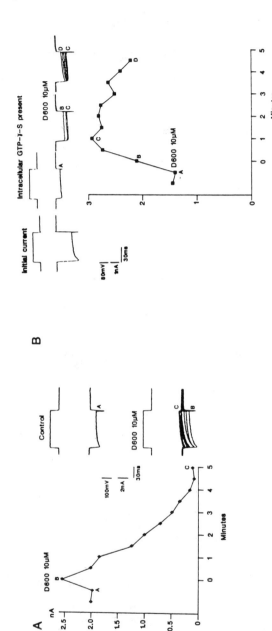

FIGURE 1. Activation of a G-protein by GTP-γ-S promotes an agonist response to D_600. DRGs from two-day-old rats were dissociated and maintained in culture for up to six weeks.[1] Cells were voltage clamped using the whole-cell recording technique. The recording medium contained (in mM): NaCl 130, KCl 3, MgCl_2 0.6, NaHCO_3 1.0, Hepes 10, glucose 4, BaCl_2 2.5, tetraethylammonium bromide 25, and tetrodotoxin (TTX) 0.0025. The pH was 7.4 and the osmolarity 320 mosM. Patch pipettes (2–5MΩ) were filled with a solution containing (in mM): Cs acetate 140, CaCl_2 0.1, EGTA 1.1, MgCl_2 2, ATP 2 and Hepes 10. The pH was 7.2 and osmolarity 310 mosM. (**A**) The effect of D_600 on Ca^{2+}-channel currents in a control cell. The graph shows the time course of action of D_600 (10 μM) on the maximum I_{Ba} recorded every 30 sec from a control cell. The arrow indicates the time (0 min) at which D_600 application began. (**A**) is the maximum I_{Ba} recorded before application of D_600. (**B**) is the first current activated after the start of pressure ejection of D_600. (**B**) The effect of D_600 on Ca^{2+}-channel currents in a GTP-γ-S-containing cell. The initial current was recorded within 30 sec of establishing whole-cell recording, and the cell was then allowed to equilibrate with 500 μM GTP-γ-S in the patch pipette for five minutes. The graph illustrates the time course of action of D_600 on the amplitude of the maximum I_{Ba} recorded every 30 sec, with the arrow indicating the time at which D_600 application began. (**A**) shows the maximum Ca^{2+}-channel current markedly attenuated by internal GTP-γ-S. (**B**) and (**C**) show the initial D_600-induced increase in I_{Ba}. (**C**) and (**D**) show potentiation relative to (**A**) lasting for at least five minutes in the continued presence of D_600, although the I_{Ba} diminished between one and five minutes. All currents are shown following subtraction of scaled linear leakage and capacitance currents. The holding potential was −80 mV.

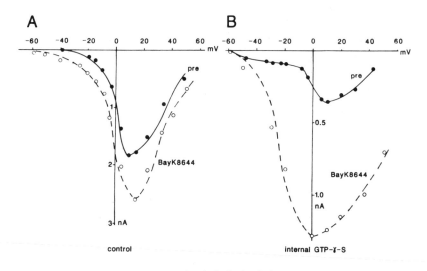

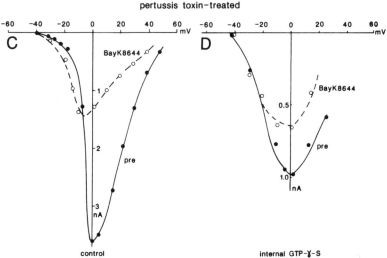

FIGURE 2. The agonist effect of BAY K 8644 on neuronal Ca^{2+}-channel currents is promoted by G-protein activation and blocked by pertussis toxin.

Current-voltage relationships are shown for I_{Ba} before and during application of BAY K 8644. Racemic BAY K 8644 was dissolved in 70% ethanol (10 mM stock) and diluted in recording medium immediately before use. It was applied at 5 μM by low-pressure ejection (less than 1 psi) from a wide (10 μm) tipped pipette positioned about 50 μm from the cell. BAY K 8644 was applied under conditions of very low illumination.

A shows data from a control cell in the absence ($\bullet$) and presence of BAY K 8644 ($\bigcirc$).

B illustrates the effect of BAY K 8644 on a cell recorded in the presence of internal GTP-γ-S (500 μM in the patch pipette). The maximum I_{Ba} was increased by 26 $\pm$ 8% for control cells and 107 $\pm$ 37% ($n = 10$) for GTP-γ-S containing cells. **C** and **D** show the inhibitory action of BAY K 8644, in the absence and presence, respectively, of intracellular GTP-γ-S, on cells that were pretreated with pertussis toxin for 2.5-5 hr (500 ng ml^{-1} at 37°C). Under these conditions BAY K 8644 reduced the maximum I_{Ba} by 47 $\pm$ 10% ($n = 6$) in control cells and 58 $\pm$ 7% ($n = 5$) in GTP-γ-S containing cells.

less sustained, and not clearly associated with such voltage shifts. The promotion of agonist responses to Ca-channel ligands by G-protein activation may result from changes in the gating kinetics of L-type channels.

REFERENCES

1. DOLPHIN, A. C. & R. H. SCOTT. 1987. Calcium channel currents and their inhibition by
   (−)-baclofen in rat sensory neurones: Modulation by guanine nucleotides. J. Physiol. **386:**
   1-17.
2. DOLPHIN, A. C., J. F. WOOTTON, R. H. SCOTT & D. R. TRENTHAM. 1988. Photoactivation
   of intracellular guanosine triphosphate analogues reduces the amplitude and slows the
   kinetics of voltage-activated calcium channel currents in sensory neurones. Pflügers Arch.
   411: 628-636.
3. FOX, A. P., M. C. NOWYCKY & R. W. TSIEN. 1987. Kinetics and pharmacological properties
   distinguish three types of calcium currents in chick sensory neurones. J. Physiol. **394:**
   149-172.
4. SCOTT, R. H. & A. C. DOLPHIN. 1987. Activation of a G-protein promotes agonist responses
   to calcium channel ligands. Nature **330:** 760-762.
5. BROWN, A. M., D. L. KUNZE & A. YATINI. 1986. Dual effects of dihydropyridines on
   whole cell and unitary calcium currents in single ventricular cells of guinea-pig. J. Physiol.
   (London) **379:** 495-514.
6. SCOTT, R. H. & A. C. DOLPHIN. 1988. The agonist effect of Bay K 8644 on neuronal
   calcium channel currents is promoted by G-protein activation. Neurosci. Lett. **89:** 170-175.

Phosphorylation and the Identification of a Protein Kinase Activity Associated with the Dihydropyridine Receptor Isolated from Rabbit Heart and Skeletal Muscle[a]

BRIAN J. MURPHY AND BALWANT S. TUANA[b]

Department of Pharmacology
University of Toronto
Toronto, Canada M5S 1A8

Electrophysiological studies have shown that the voltage-regulated Ca^{2+} channel can be modulated by protein phosphorylation.[1] Studies on the isolated dihydropyridine (DHP) receptor from skeletal muscle have indicated that the 170- (α_1) and 52-kDa (β) subunits can be phosphorylated by added cAMP-dependent protein kinase, and this phosphorylation modulates the activity of the reconstituted Ca^{2+} channel.[2–4] We have been interested in the question of whether there are endogenous protein kinases that can phosphorylate the putative subunits of the dihydropyridine receptor. In this study we investigated whether the isolated DHP receptor contains endogenous protein kinase activity that can phosphorylate the subunits of the receptor.

Receptor-enriched membranes were isolated from rabbit skeletal and cardiac muscle. The membranes were labeled with [³H]PN200-110, solubilized with CHAPS, and the receptors purified on wheat germ agglutinin columns followed by sucrose density gradient centrifugation.[5–6] The dihydropyridine receptor fraction isolated from skeletal muscle indicated a polypeptide composition of 170, 150, 108, 56, 50, and 32 kDa in nonreducing SDS-PAGE. The 170-kDa polypeptide shifted to an apparent M_r of 140 kDa in reducing conditions.[5] The cardiac [³H]PN200-110 binding sites indicated a polypeptide composition of 170, 150, 130, 100, and 50 kDa.[6] The 150-kDa polypeptide was specifically photoaffinity labeled with [³H]azidopine in both the cardiac and skeletal forms of the receptor.[7] The 150-kDa polypeptide in our preparation migrates as a 170-kDa protein under different polyacrylamide concentrations and is thus equivalent to the α_1 (170-kDa) subunit.[8–10] FIGURE 1 shows the phosphorylation of the isolated dihydropyridine receptor. Both types of receptor fractions contain a protein

[a]BST was supported by a scholarship from the Heart & Stroke Foundation of Ontario. BJM was supported by a predoctoral studentship from the Canadian Heart Foundation. This research was funded by the Heart and Stroke Foundation of Ontario and the Medical Research Council.

[b]Address for correspondence: Dr. B. S. Tuana, Department of Pharmacology, University of Ottawa, 451 Smyth, Ottawa, Canada.

391

kinase that phosphorylates the 150- and 50-kDa polypeptides (endogenous lanes). Cyclic AMP, Ca^{2+}, or calmodulin were without effect on the endogenous protein kinase. The pH optima for the kinase activity in both preparations was found to be 7.0. Addition of 0.5 μg of the catalytic subunit of cAMP protein kinase further stimulates the ^{32}P incorporation into the 150- and 50-kDa subunits in both types of preparations. Thus the endogenous kinase as well as the cAMP-dependent kinase phosphorylate the same receptor subunits in the two types of muscle.

One of the features of protein kinases is that they contain an ATP-binding domain. In order to identify the polypeptide exhibiting protein kinase activity, the receptor fractions were covalently labeled with [^{32}P]azido-ATP by irradiation with UV light. FIGURE 2 shows the covalent incorporation of [^{32}P]azido ATP into a protein band of 50 kDa in both receptor fractions. The labeling was inhibited by the presence of excess unlabeled ATP during incubation with 50 nM [^{32}P]azido-ATP. No labeling was observed in the absence of UV illumination. These results suggest that a 50-kDa polypeptide in the receptor fraction contains the ATP binding sites and this is a good candidate for the protein kinase found in these fractions. The 50-kDa polypeptide is not glycosylated, therefore its presence in the isolated receptor fraction can only be accounted for by its interaction with the receptor complex because this fraction was purified on lectin columns. Whether the 50-kDa phosphoprotein and the ATP-binding protein are equivalent to the β subunit of the Ca^{2+} channel remains to be established.

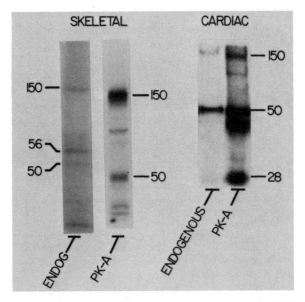

FIGURE 1. Phosphorylation of the isolated dihydropyridine receptors. The DHP receptors were isolated from rabbit heart and skeletal muscle as described previously.[5,6] Phosphorylation was carried out in 0.1 ml of 50 mM-Tris-HCl buffer, pH 7.4, 10 mM MgCl$_2$, 2.5 mM EGTA, and 5 μg isolated receptor. The reaction was initiated by the addition of 10 μM [^{32}P]ATP, incubated for two minutes at 25°C and terminated by the addition of SDS-PAGE loading buffer. cAMP-dependent phosphorylation was carried out in the presence of 0.25-1.0 μg of the catalytic subunit of cAMP-dependent protein kinase. Gels were stained, dried, and subjected to autoradiography using Kodak X-omat AR film.

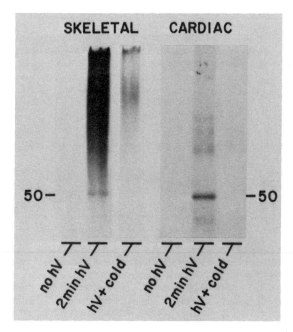

FIGURE 2. Affinity labeling of the dihydropyridine receptors with Azido-ATP. Assay conditions were similar to these described for the phosphorylation reactions in FIGURE 1 except 10 nM [^{32}P]azido-ATP replaced ATP. Samples were incubated at room temperature for five minutes in the presence or absence of excess unlabeled ATP, then placed on ice and irradiated for two minutes at a distance of 3 cm using a Mineralight UV lamp (Model UVS-58). The reactions were terminated with SDS loading buffer, and the samples were subjected to SDS-PAGE and autoradiography.

Isolated triads from skeletal muscle have recently been shown to contain a protein kinase that can phosphorylate the 170- and 52-kDa subunits of the receptor.[3] The function of the intrinsic protein kinase remains to be elucidated. Because the subunit structure of the dihydropyridine receptor from both types of muscle appears to be similar and both show similar phosphoprotein patterns, it is possible that the intrinsic protein kinase exerts a common regulatory role on cardiac and skeletal muscle Ca^{2+} channels.

REFERENCES

1. REUTER, H. 1983. Nature **301:** 569-574.
2. CATTERALL, W. A., M. J. SEAGER & M. TAKAHASHI. 1988. J. Biol. Chem. **263:** 3535-3538.
3. IMAGAWA, T., A. T. LEUNG & K. P. CAMPBELL. 1987. J. Biol Chem. **262:** 8333-8339.
4. FLOCKERZI, V., H. J. OEKEN, F. HOFMANN, D. PELZER, A. CAVALIE & W. TRAUTWEIN. 1986. Nature **323:** 66-68.
5. TUANA, B. S., B. J. MURPHY & Q. YI. 1988. Mol. Cell. Biochem. **80: 133-146.**
6. TUANA, B. S., B. J. MURPHY & Q. YI. 1987. Mol. Cell. Biochem. **76:** 173-184.

7. MURPHY, B. J. & B. S. TUANA. 1988. J. Mol. Cell. Cardiol. **20:** 7.
8. FERRY, D. R., A. GOLL & H. GLOSSMAN. 1987. **243:** 127-135.
9. LEUNG, A. T., T. IMAGAWA & K. P. CAMPBELL. 1987. J. Biol. Chem. **262:** 7943-7946.
10. TAKAHASHI, M., M. SEAGER, J. JONES, B. REBER & W. CATTERALL. 1987. Proc. Natl. Acad. Sci. U.S.A. **84:** 5478-5482.

Identification of the cAMP-Specific Phosphorylation Site of the Purified Skeletal Muscle Receptor for Calcium-Channel Blockers

A. RÖHRKASTEN,[a,b] W. NASTAINCZYK,[a]
H. E. MEYER,[c] M. SIEBER,[a] T. SCHNEIDER,[a]
ST. REGULLA,[a] AND F. HOFMANN[a]

[a] Institut für Physiologische Chemie
Medizinische Fakultät Universität des Saarlandes
D-6650 Homburg/Saar, Federal Republic of Germany

[c] Institut für Physiologische Chemie
Abt. Biochemie Supramolekularer Systeme
Ruhr Universität Bochum
D-4630 Bochum, Federal Republic of Germany

Cyclic AMP-dependent protein kinase increases the opening probability of voltage-dependent L-type calcium channels in a number of tissues (see Hofmann et al.[1] for review). The receptor for calcium-channel blockers (CaCB) purified from rabbit skeletal muscle that contains three polypeptides of apparent M_r of 165, 55, and 32 kDa[2] is phosphorylated by cAMP-dependent protein kinase.[3] Reconstitution of the CaCB receptor suggested that cAMP-dependent phosphorylation increases the opening probability of the calcium channel.[4] Cyclic AMP-dependent protein kinase preferentially phosphorylated the 165-kDa subunit. Approximately 1 mol and 2 mol phosphate per mol 165-kDa protein are incorporated within 10 minutes and two hours, respectively, at physiological concentrations of cAMP-dependent protein kinase.[5] Tryptic digestion of the 165-kDa protein yields one major phosphopeptide after five minutes and up to three phosphopeptides after two hours on reverse-phase HPLC (FIG. 1). The presence of multiple phosphorylation sites was further supported by phosphoamino acid analysis. Hydrolysis of peptide I yielded only phosphoserine, whereas peptide II contained phosphoserine and some phosphothreonine. The rapidly phosphorylated amino acid of tryptic peptide I was identified by sequential Edman degradation of the corresponding phosphopeptides I, L1, and G1.[6] These sequences were compared with the recently published amino acid sequence of the 165-kDa protein.[7] The sequences of the isolated phosphopeptides corresponded to the amino-acid residues 685-688, 686-693, and 681-697 (TABLE 1). In each case the radioactive phosphate was located on Ser-687. This major phosphorylation site of the 165-kDa CaCB-receptor protein is localized on the putative cytoplasmic domain between the transmembrane regions

[b] Present address: Beecham-Wülfing GmbH, Klinische Forschung, Stresemannallee 6, D-4040 Neuss, F.R.G.

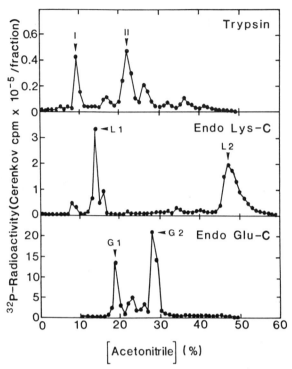

FIGURE 1. Phosphopeptide pattern of the phosphorylated 165-kDa subunit of the CaCB-receptor. The purified CaCB receptor was phosphorylated with cAMP-dependent protein kinase for two hours at 37°C. ^{32}P-labeled subunits were separated by SDS-PAGE[5] or by gel exclusion chromatography.[2] A portion of the labeled 165-kDa protein was digested either with TPCK-treated trypsin or endoproteinase Lys-C or endoproteinase Glu-C. The digests were fractionated by reverse-phase HPLC. Fractions of 1 ml were collected, and the radioactivity was determined. Fractions (I, II, L1, L2, G1, and G2 as indicated by the arrows) were collected for further analysis (see text for details).

TABLE 1. Amino Acid Sequence of ^{32}P-Labeled Phosphopeptides

Peptide	Digestion Enzyme	Fragment	Length
I	Trypsin	[685]KMS (P) R [688]	4aa[a]
L1	Endo Lys-C	[686]MS (P) RGLPDK [693]	8aa[a]
G1	Endo Glu-C	[681]RKRRKMS (P) RGLPDKTEEE [697]	17aa[a]
II	Trypsin	[615]TNS (P) LPPVMANQR [1626]	12aa[b]
L2	Endo Lys-C	[1551]DTVQIQAGLRTIEEEAAP........	≤166aa[b]
G2	Endo Glu-C	[1614]RTNS (P) LPPVMANQRPLQFAE [1632]	19aa[a]

NOTE: The amino-acid sequence of each phosphopeptide was determined either by sequential Edman degradation or predicted from the recently published deduced amino acid sequence of the CaCB-receptor.[7]

[a] Determined.

[b] Predicted.

II and III. The phosphorylation of Ser-687 may be essential for activation of the calcium channel.[4]

Tryptic phosphopeptide analysis indicated that the 165-kDa subunit contains a second slowly phosphorylated site. A peptide corresponding to this site was detected both in the digest with endoproteinase Lys-C and Glu-C. The sequence of peptide G2 is located in the large cytoplasmic domain of the 165-kDa protein near the carboxy-terminus and corresponds to residues 1614-1632 containing phosphorylated Ser-1617 (TABLE 1). The phosphorylation of this sequence may not be related to the *in vivo* function of the calcium channel because it occurred only after prolonged incubation.

REFERENCES

1. HOFMANN, F, W. NASTAINCZYK, A. RÖHRKASTEN, T. SCHNEIDER & M. SIEBER. 1987. TIPS **8**: 393-398.
2. SIEBER, M., W. NASTAINCZYK, V. ZUBOR, W. WERNET & F. HOFMANN. 1987. Eur. J. Biochem. **167**: 117-122.
3. CURTIS, B. M. & W. A. CATTERALL. 1985. Proc. Natl. Acad. Sci. USA **82**: 4255-4263.
4. FLOCKERZI, V., H-J. OEKEN, F. HOFMANN, D. PELZER, A. CAVALIÉ & W. TRAUTWEIN. 1986. Nature **323**: 66-68.
5. NASTAINCZYK, W., A. RÖHRKASTEN, M. SIEBER, C. RUDOLPH, C. SCHÄCHTELE, D. MARME & F. HOFMANN. 1987. Eur. J. Biochem. **169**: 137-142.
6. RÖHRKASTEN, A., H. E. MEYER, W. NASTAINCZYK, M. SIEBER & F. HOFMANN. 1988. J. Biol. Chem. **263**: 15325-15329.
7. TANABE, T., H. TAKESHIMA, A. MIKAMI, V. FLOCKERZI, H. TAKAHASHI, K. KANGAWA, M. KOJIMA, H. MATSUO, T. HIROSE & S. NUMA. 1987. Nature **328**: 313-318.

L and N Ca^{2+} Channels Coupled to Muscarinic Receptors in Rat Sensory Neurons

E. WANKE, A. SARDINI, AND A. FERRONI

Department of General Physiology and Biochemistry
University of Milan
Milano, Italy

Various neurotransmitters[1-6] are known to modulate different types of Ca^{2+} channels present in vertebrate neurons.[9,10] We found, in rat dorsal root ganglion neurons (DRG), that the long-lasting and the transient, high-voltage activated components of the Ca^{2+} currents that resemble L- and N-type channels can be reversibly inhibited in different cells through activation of muscarinic receptors.

The two typical responses found after acetylcholine (ACh) application are shown in FIGURE 1. In the upper panels the current (CON curve) elicited by a command pulse to +10 mV is rapidly inhibited to that labeled ACh. The difference between control and ACh-response currents is shown with the label CON-ACh. The significant difference between the two responses (type-A and type-B neuron in the left and right panels, respectively) can be roughly summarized by saying that the Ca^{2+}-current components (transient versus steady-state) exchange their roles of modulated or unmodulated currents during drug application. The response to ACh of type-A neuron (mostly inhibition of a transient component) is similar to that shown, (1) by Wanke et al.[4] in sympathetic (SCG) neurons to ACh itself, (2) by Gross et al.[6] in DRG neurons to Dynorphin A, and (3) by Lipscombe et al.[5] in bullfrog SCG neurons to NA. On the other hand, the response of type-B neurons (inhibition of a steady component) is similar to that shown (1) by Forscher et al.,[1] Holz et al.,[2] and Marchetti et al.[3] to NA and by Wanke et al.[8] to ACh in chick DRG neurons and (2) by MacDonald et al.[9] in mouse DRG to Dynorphin A.

It is known[12] that in chick DRG the activation threshold of N channels is smaller than the threshold of L channels. If the response of A- and B-neurons corresponds to the inhibition of N- and L-channels, selective activations are expected (with ACh) in the current-voltage I-V relationships. We used the linear ramp command voltage to analyze in a fast and qualitative way the I-V plots, but the same results can also be obtained with more classical protocols. As shown in FIGURE 1 (lower panels), the I-V curve with drug (ACh) shifts right (L-channels are mostly unchanged) in the A-neuron, and is not shifted but only depressed at right (no effect on N-channels) in the B-neuron. Moreover the inhibited component of the I-V relation (CON-ACh) is right-shifted in the B-neuron (L-channels are inhibited) and peaked at left in the A-neuron (N-channels inhibited). The physiological relevance of this data can be reasonably explained in terms of the different populations of sensitive neurons present in spinal ganglia.[13-15] The substance-P-releasing neurons[16] (mostly nociceptive) are of type-B while type-A neurons will probably be those responsible for the miotactic reflex.

We do not know at present if different receptors or alternative second-messenger couplings are responsible for these responses. As far as we know this is the (see Marchetti *et al.*[3]) second evidence that the same neurotransmitter modulates different Ca^{2+} channels in cells of the same ganglion.

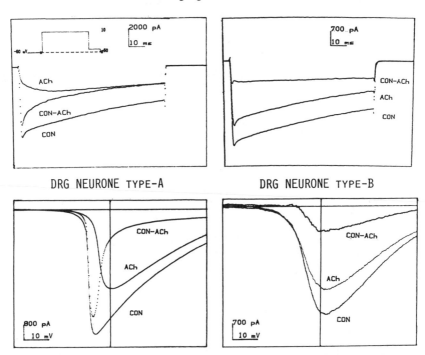

FIGURE 1. Action of acetylcholine on the Ca^{2+} currents in different sensitive rat DRG neurons. **Upper panels:** effect of 50 μM ACh on the whole-cell Ca^{2+} currents (test pulse to 10 mV). Three curves are shown: control (CON), during ACh application (ACh), and the difference (the inhibited current, CON-ACh), which is mainly transient (with a minor long-lasting component) in the neuron of type A (left panel), or constant in the neuron of type B (right panel). **Lower panels:** I-V current-voltage curves, before (CON) and during ACh application on DRG neurons of type A (left panel) and B (right panel). The curve labeled CON-ACh corresponds to the inhibited current. Data were obtained from experiments where linear-ramp command voltages were used. Methods: DRG neurons were dissected from two-day-old rats and used after culturing as described in Wanke *et al.*[4] Experiments done with the whole-cell patch-clamp technique[11] were always completely reversible at 36°C. The responses were the following: 30% type A, 30% type B, 40% no response; in B-neurons, responses were almost always associated with a transient increase of a leak current. The holding potential was −60 mV in order to inactivate the eventual channels of the T type. The ramp command voltages lasted 250 msec. Intracellular chelator BAPTA was used instead of EGTA; 50 μM Cd^{2+} was used for the subtraction of the Cd-insensitive currents; the Axon Intruments pClamp hard-software package was used for experiments and analysis.

REFERENCES

1. FORSCHER, P., *et al.* 1985. J. Gen. Physiol. **85:** 743.

2. HOLZ, G. G., *et al.* 1986. Nature **319:** 670.
3. MARCHETTI, C., *et al.* 1986. Pflügers Archiv. **406:** 104.
4. WANKE, E., *et al.* 1987. Proc. Natl. Acad. Sci. USA **84:** 4313.
5. LIPSCOMBE, D., *et al.* 1987. J. Physiol. **384:** 97P.
6. GROSS, R. A., *et al.* 1987. Proc. Natl. Acad. Sci. USA **84:** 5469.
7. WANKE, E., *et al.* 1988. Neurotransmitters and Cortical Function. M. Avoli *et al.*, Eds.: 277. Plenum. New York.
8. MACDONALD, R. L., *et al.* 1986. J. Physiol. **377:** 237.
9. CARBONE, E., *et al.* 1984. Nature **310:** 501.
10. NOWYCKY, M. C., *et al.* 1985. Nature **316:** 440.
11. HAMILL, O. P., *et al.* 1981. Pflügers Archiv. **391:** 85.
12. FOX, A. P., *et al.* 1987. J. Physiol. **394:** 149.
13. FITZGERALD, M. 1985. J. Physiol. **364:** 1.
14. HARPER, A. A., *et al.* 1985. J. Physiol. **359:** 47.
15. FULTON, B. 1986. J. Physiol. **377:** 80P.
16. RANE, G. G., *et al.* 1987. Pflügers Archiv. **409:** 361.

The Sarcoplasmic Reticulum Calcium Channel of Skeletal Muscle

Modulation by Cytoplasmic Calcium[a]

PHILIP M. BEST AND WAI-MENG KWOK

Department of Physiology and Biophysics
University of Illinois
Urbana, Illinois 61801

Several studies have suggested the existence of a Ca^{2+}-dependent inactivation of sarcoplasmic reticulum (SR) Ca^{2+} channels in striated muscles. Results from investigations on intact skeletal muscle fibers by Baylor *et al.*[1] and Simon *et al.*[2] indicated the possibility of a Ca^{2+}-dependent inactivation of the Ca^{2+}-release channels. Fabiato's work[3] on skinned Purkinje cells suggested an SR Ca^{2+} channel with time- and Ca^{2+}-dependent activation and inactivation.

In the current study, the role of myofilament space Ca^{2+} in modulating the SR Ca^{2+} channel of skeletal muscle of the frog was investigated. Ca^{2+} release rates were optically determined with antipyrylazo III (0.4 mM) by simultaneous absorption measurements at 720 and 790 nm in single fibers that had been mechanically skinned (sarcolemma removed) and preincubated in solutions of varying Ca^{2+} concentrations. Solutions contained 2 mM MgATP, 15 mM creatine phosphate, 1 mM Mg, 100 mM monovalent cations, and approximately 35 mM MOPS buffer (pH = 7.0 at 10°C) at an ionic strength of 0.15 M. A standard procedure was used to deplete and load the SR with Ca^{2+} before Ca^+ release was stimulated by 5.0 (7.5) mM caffeine. The test release rate was compared to the average rate of the two bracketing controls. Solutions with pCa = 8.4 were used as controls. The test solutions contained varying Ca^{2+} concentrations ranging from pCa = 6.1 to 7.4.

In the range of pCa = 6.1 to 7.4, SR Ca^{2+} release rates varied in a biphasic manner. Maximum inactivation occurred at pCa = 6.9 where the release rate was 76 ± 3% (mean ± SEM) of control. Release rates at pCa = 6.1 and 7.4 were 85 ± 4 and 98 ± 3, respectively. The effect of a decreased driving force on Ca^{2+} ion was investigated to determine whether it could account for the change in release rates seen. At pCa = 6.1, the highest concentration tested, the measured effects of a change in driving force could account for at most 8% of the observed inactivation. Similar controls ruled out any significant effect of the increased exposure to myofilament space Ca^{2+} on SR loading. The effects of varying myofilament space concentrations of Ba^{2+} and Sr^{2+} on Ca^{2+} release rates were also investigated. In the range where Ca^{2+}-dependent inactivation of Ca^{2+} release occurred, both Ba^{2+} and Sr^{2+} had no inactivating effect on the Ca^{2+} release rates. However, there was a noticeable activation of Ca^{2+} release rate at pSr = 7.0.

[a] This work was supported by the National Institutes of Health (AR32062) and the M.D.A.

These results suggest that the conductance of the SR Ca^{2+} channel is modulated by Ca^{2+} at physiological concentrations. The specificity for Ca^{2+} in producing an inactivation is similar to results reported for a number of plasmalemmal Ca^{2+} channels.

REFERENCES

1. BAYLOR, S. M., W. K. CHANDLER & M. W. MARSHALL. 1983. J. Physiol. **344:** 625-666.
2. SIMON, B. J., M. F. SCHNEIDER & G. SZUCS. 1985. J. Gen. Physiol. **86:** 36a.
3. FABIATO, A. 1985. J. Gen. Physiol. **85:** 247-289.

Calcium Currents in Insulin-Secreting β-Cells[a]

I. FINDLAY,[b] F. M. ASHCROFT,[c] R. P. KELLY,[c]
P. RORSMAN,[d] O. H. PETERSEN,[e] AND G. TRUBE[f]

[b] Laboratoire de Physiologie Comparee
Universite de Paris IX
Orsay, France

[c] University Laboratory of Physiology
Oxford University
Oxford, United Kingdom

[d] Department of Medical Physics
University of Goteborg
Göteborg, Sweden

[e] Department of Physiology
University of Liverpool
Liverpool, United Kingdom

[f] Max Planck Institut fur Physiologische Chemie
Göttingen, Federal Republic of Germany

INTRODUCTION

In vivo, insulin release is controlled by fluctuations in the concentration of circulating nutrients (in particular, glucose) and is modulated by various hormones and transmitters. Glucose homeostasis involves rapid integration of all these signals by the β-cell. It is believed that this is achieved primarily by changes in β-cell electrical activity.[1-2]

Glucose influences β-cell electrical activity on at least two levels. First, by causing the closure of ATP-sensitive K channels, it produces membrane depolarization[3]; if the glucose concentration exceeds about 7 mM, this depolarization is sufficient to trigger electrical activity. Second, glucose influences the intensity of electrical activity. In the β-cell, electrical activity follows a characteristic pattern consisting of oscillations (bursts or slow waves) between a depolarized plateau potential, on which Ca-dependent action potentials are superimposed, and a more hyperpolarized silent phase. As the glucose concentration is raised, the duration of the slow waves increases and the interval between them decreases until finally the plateau potential is sustained and action potentials fire continuously.[1-2] It has also been suggested that glucose might exert a third level of modulation through a direct effect on the Ca current that underlies the action potential.[4]

[a] FMA was supported by British Diabetic Association and the Wellcome Trust. FMA is a Royal Society 1983 University Research Fellow.

The application of the patch-clamp technique[5] to isolated cells has enabled the individual ionic currents that underlie β-cell electrical activity to be studied (for review see Petersen & Findlay[2]). Whole-cell recordings have shown that voltage-dependent Ca and K currents contribute to the action potentials that arise from the plateau level. The ionic mechanisms underlying the plateau remain unresolved, although there is some evidence that Ca currents may be involved. In this paper, we review our present knowledge of the Ca currents of the β-cell.

WHOLE-CELL STUDIES

Calcium currents have been recorded in neonatal,[6-7] adult rat β-cells,[8] in mouse β-cells,[9-11] in human β-cells,[12] and in the clonal insulin-secreting cell lines RINm5F[13-14] and HIT T15.[7] It is worth pointing out that whereas membrane potential recordings have been almost exclusively carried out on mouse β-cells, these patch-clamp studies have used a wide variety of β-cell types. Furthermore it is not clear whether isolated β-cells and cell lines retain the same pattern of electrical activity as in the intact islet. Thus, although action currents can be recorded from cell-attached patches, a regular bursting pattern is often absent.

Inward currents recorded in the presence of outward K currents are comparatively small, as they are partially masked by the outward currents. Therefore the intracellular solution used in all studies of β-cell Ca currents has contained an impermeant K^+ substitute, such as Cs^+ or N-methyl glucamine (NMG), to reduce outward K currents. In addition, MgATP has usually been usually included both to prevent activation of ATP-sensitive K channels[3] and to facilitate maintenance of Ca-channel activity,[15] and the intracellular Ca concentration has been buffered in the nanomolar range to prevent Ca-dependent inactivation of Ca currents.[10]

Inward currents carried by divalent cations in β-cells are activated by voltage pulses to potentials less negative than -50 mV. Peak Ca currents occur at around 0 mV, and no reversal of these currents is seen with potentials less negative than 50 mV in cells with little leakage conductance. Several groups have reported that the maximum amplitude of the inward current is very variable, even when normalized to the cell capacitance.[8-10] The amplitude of the inward currents varies with the external Ca concentration. The current (I)-voltage (V) relation is also shifted to more negative potentials in low-Ca solutions, an effect that may be due to a reduction in screening of membrane surface charges by divalent cations. In physiological Ca concentrations, about 2.6 mM Ca^{2+}, the I-V relation is shifted by about -10 mV with respect to that recorded in 10 mM Ca.

Inactivation

All studies agree that β-cell Ca currents show some degree of inactivation, although the extent of this inactivation is variable. There is good evidence that inactivation is Ca dependent.[10] This may be summarized as follows. First, inactivation is most marked at potentials at which Ca currents are largest.[9-10] Second, two pulse experiments show that the voltage-dependence of inactivation is U-shaped. Inactivation is absent at potentials at which there is no Ca current, is maximal at potentials at which Ca influx is greatest, and decreases at more positive potentials as the Ca-equilibrium potential is approached. In a careful study, Plant[10] has shown that the extent of inactivation of mouse β-cell Ca currents is dependent only on the amount of Ca entry during the first pulse. Third, little inactivation is found in either single-pulse[10] or two-pulse experiments[12] when Ba^{2+} carries inward current. Finally, with Na^+ as the charge carrier, inward currents showed no inactivation during 150-msec pulses at any potential. The lack of inactivation seen with Na^+ as a charge carrier suggests that voltage-dependent inactivation is not significant during pulses shorter than 150 msec.[10]

Pharmacology

Like other Ca channels, β-cell Ca currents are inhibited by the transition metals Co^{2+}, Cd^{2+}, and Mn^{2+} [8-10]: Cd^{2+} is effective at micromolar concentrations. They are also sensitive to the dihydropyridines (DHP), being inhibited by the Ca-channel antagonists nitrendipine (5 μM[10]) and nifedipine (5 μM[9]) and increased by the Ca-channel agonists BAY K 8644 (1 μM[11]) and CGP 28392 (5 μM[10]). The effect of the Ca-channel agonists was greatest at more negative membrane potentials. This is in part due to a negative shift in the voltage dependence of Ca-current activation in the presence of the drug and in part because the drug is less effective at positive potentials (where it may even act as an inhibitor[16]).

Tetrodotoxin is without effect on β-cell Ca currents at micromolar concentrations.

Selectivity

Barium is 1.5 to 2 times more permeant than Ca^{2+} through β-cell Ca channels.[8,10,11] The I-V relation is also shifted to more negative membrane potentials in Ba solutions, probably because Ba^{2+} ions are less effective at screening or binding to membrane surface charge. In the absence of divalent cations, large inward currents carried by Na^+ can be recorded. The kinetics and pharmacology of these currents suggest that they flow through the Ca channel—they are sensitive to dihydropyridines and not blocked by tetrodotoxin.[10] The ability of Na^+ to permeate Ca channels when Ca^{2+} is reduced to micromolar levels has been demonstrated in a number of other tissues.[17-18]

SINGLE-CHANNEL STUDIES

Few studies of single Ca-channel currents in β-cells have yet appeared.[11,19-21] Because Ca-channel currents are too small to be recorded under physiological conditions, these studies have used a high concentration of divalent cation (110 mM) to increase the current amplitude and Ba^{2+} as a charge carrier, as it is about twice as permeant as Ca^{2+} through β-cell Ca channels. With the exception of one short report,[19] only a single type of Ca channel was seen that had a single-channel conductance of between 20 and 30 pS.

The properties of this 25-pS channel correspond to those of the whole-cell Ca currents. Like the whole-cell current, single-channel currents were inhibited by micromolar Cd^{2+},[11,19] by nifedipine,[19] verapamil,[19] and D600.[21] The DHP Ca-channel agonist BAY K 8644 prolonged single-channel openings, thus accounting for its ability to increase the whole-cell Ca current. Channel openings showed little inactivation during the course of a 200-msec pulse, consistent with the lack of inactivation of whole-cell Ba currents.[11] In outside-out patches, channel activity declined with time, whereas it was stable in cell-attached patches, suggesting that soluble cytoplasmic constituents may be required to maintain channel activity.[11] A similar decline in Ca-current amplitude (rundown or washout) is also found in whole-cell recordings.[7-10]

The single Ca-channel conductance of mouse β-cells was 24 pS in 110 mM Ba^{2+}. Because whole-cell inward currents increase approximately 10-fold when 110 mM Ba^{2+} is substituted for 2.6 mM Ca^{2+}, the single-channel conductance in physiological Ca concentrations (2.6 mM) is likely to be around 2 pS.[11] This value is in reasonable agreement with that of 5 pS estimated from voltage noise measurements.[22]

Combining single-channel and whole-cell data gave a lower estimate of 500 Ca channels per β-cell and, taking the β-cell diameter as 13 μm, a density of about 1 channel/μm^2.[11] This density is comparable to that found in chromaffin cells (3–5 μm^2).[23]

IS MORE THAN ONE TYPE OF CA CHANNEL PRESENT IN THE β-CELL?

Three main types of Ca channel have been described in vertebrates, of which two are known to be present in non-neuronal tissues, the T-type and the L-type. The properties of these channels have been reviewed in detail elsewhere.[24]

There is general agreement that β-cells possess an L-type Ca channel, as evidenced by whole-cell Ca-currents that (1) are long-lasting, (2) show Ca-dependent inactivation, (3) are sensitive to micromolar cadmium, (4) are dihydropyridine-sensitive, and (5) show rundown. Single-channel recordings showing a 25-pS channel (110 mM Ba^{2+}) with similar properties are also consistent with this classification.

The presence of an additional type of Ca channel remains controversial. The T-type, or transient channels, are characterized by activation at more negative membrane potentials and by fast, voltage-dependent inactivation.[24] Their presence has usually been determined by the use of different holding potentials. Holding potentials of around −90 mV allow activation of both T and L-type channels, whereas only L-type channels can be activated from −40 mV. In the β-cell, no change in the amplitude or time course of the Ca current is found when the holding potential is changed between −50 mV and −100 mV.[10]

Another indication of two types of Ca channel, with different activation properties, is the presence of a shoulder on the Ca current-voltage relation.[24] In general, this is not seen in β-cells. Satin and Cook,[7] however, do describe such a shoulder on the I-V relation recorded from both neonatal rat β-cells and HIT T15 cells. They further showed that 20 μM Cd^{2+} had little effect on the amplitude of currents at −40 mV or −30 mV but greatly reduced the amplitude of currents at more positive potentials. Because T-type Ca channels are known to be less sensitive to Cd^{2+}, they interpreted this as evidence for a low-threshold Ca channel. Their data, however, is also consistent with a single type of Ca-channel, because single-channel recordings show that Cd^{2+}

blocks L-type Ca channels less effectively at more negative membrane potentials.[25] It appears that Cd^{2+} is slightly permeant through L-type channels.

L-type Ca channels are selectively affected by the dihydropyridines. At first sight, this would suggest that a good method for distinguishing the presence of another type of Ca channel would be to look at the current that remains after inhibition of the L-type channels with a DHP Ca-channel antagonist. Unfortunately there are at least two problems with this. First, dihydropyridines may not completely block the L-type current, even at high concentration.[24] Second, the rate of inactivation of the L-type channels may be increased by DHPs.[16,26] Despite these caveats, the β-cell Ca currents remaining after DHP inhibition have a similar time course and voltage dependence to those found in the absence of the Ca-channel blocker.[10]

Hiriart and Matteson[8] have suggested that two types of Ca channel exist in adult rat β-cells, on the basis of differences in the tail currents. They found that two exponentials are required to fit the Ca tail currents following short (10-msec) pulses. These two components differed in their activation and inactivation properties, in their Ca/Ba selectivity and in their sensitivity to washout. The slow tail current component was activated at more negative membrane potentials, was inactivated rapidly by the voltage pulse, insensitive to washout and did not discriminate between Ba^{2+} and Ca^{2+}, properties resembling those of T-type channels. They concluded that the β-cell has two types of Ca channel that they termed SD (slow-deactivating, T-type) and FD (fast-deactivating, L-type). By contrast, Rorsman and Trube[9] found only a single population of Ca channels in mouse β-cells. In their experiments, tail currents usually followed a single exponential with a time constant similar to that of the FD current. Furthermore, the tail currents following pulses of different duration (0.6 to 20 msec) had the same time course so that it is unlikely that, as Hiriart and Matteson suggest,[8] the slow component had inactivated by the end of their pulses. It is possible that the differences reported by these two groups is a species difference (rat versus mouse) or related to differences in the composition of the intracellular solution (Cs-glutamate[8]; NMG-Cl[9]).

The question is whether the two components of tail current described by Hiriart and Matteson[8] are due to two distinct populations of Ca channels or whether they result from a single population with two open states (as L-type channels are known to possess[26]). At present, it seems that all the results can be explained by either hypothesis. It would be interesting to know whether the slow tail currents are insensitive to dihydropyridines or low concentrations of Cd^{2+}, as might be expected if they belong to the T-type family.

Probably the clearest way to demonstrate that β-cells possess two types of Ca channel is by single-channel recording. One brief report of a Ca channel with properties resembling those of the T-type has appeared: In cell-attached patches on RINm5F cells, this channel had a conductance of 8 pS (110 mM Ba^{2+}), was inactivated by holding potentials more positive than -40 mV, and was insensitive to nifedipine.[19] On the other hand, only L-type Ca channels were observed in cell-attached[11] and outside-out patches[21] on normal, adult mouse β-cells. Single Ca-channel currents have not yet been recorded from rat β-cells.

Finally, we point out that alpha$_2$ cells, which comprise 10-15% of islet cells, contain two types of Ca channel.[27] It is therefore important to ensure that one is recording from a β-cell when working with isolated islet cells, particularly when the response is confined to only a few cells. This problem does not arise with cell lines, but these cells have the disadvantage that their channels may not be identical with those of normal β-cells. Several of the studies described in this review have used β-cells identified by their ability to respond to glucose with insulin secretion[8] or electrical activity.[10]

MODULATION OF Ca CHANNELS BY SECRETOGOGUES

It has been suggested that glucose metabolism increases Ca influx by a direct effect on the Ca channel.[4] Two studies have addressed this question by recording single Ca-channel activity from cell-attached membrane patches, a configuration in which the cell metabolism remains intact.[20-21] Both of the studies used Ba^{2+} as a charge carrier to increase the current amplitude (see above). Glyceraldehyde stimulates insulin release from RIN m5F cells. When added to the bath solution, this secretogogue (10 mM) increased the frequency of single Ca-channel openings and prolonged their duration.[20] A similar increase in Ca-channel open time was produced by 20 mM glucose in mouse β-cells, which is illustrated elsewhere in this volume.[21] Further analysis of the effect of glyceraldehyde on Ca-channel open probability revealed that the activation curve was shifted to more negative potentials. This led to a large increase in Ca-channel activity over the voltage range −30 mV to −10 mV.[20]

PHYSIOLOGICAL ROLE OF Ca CHANNELS

The properties of the β-cell L-type Ca current suggest that this current underlies the action potentials that arise from the plateau level. Its size is consistent with the upstroke velocity of the action potential.

Several observations support the idea that a Ca current also underlies the plateau potential. Microelectrode studies have shown that a voltage-dependent process is involved in the generation and maintenance of the slow waves.[28] The ability of Ca-channel blockers to inhibit the slow waves[29] and of DHP derivatives that prolong Ca-channel opening to increase slow wave duration[30] suggest that a voltage-dependent Ca-current might be involved. This is supported by the increase in slow-wave amplitude found when external Ca is raised.[29] The possibility that a low-threshold (T-type) Ca current might be responsible for initiating the plateau has been widely suggested.[2] As we discuss above, the existence of such a channel in β-cells is not fully resolved. The SD channels described by Hiriart and Matteson[8] would be candidates, although they are unlikely to contribute to the maintenance of the plateau because of their rapid inactivation properties. Moreover, the sensitivity of the plateau to dihydropyridines points to the involvement of an L-type channel in maintaining the plateau.

Finally, the ability of DHP agonists to increase insulin release[30] and of DHP antagonists to reduce secretion[31] is consistent with the Ca influx for insulin release flowing through L-type Ca channels.

REFERENCES

1. HENQUIN, J. C. & H. P. MEISSNER. 1984. Experientia **40:** 1043-1052.
2. PETERSEN, O. H. & I. FINDLAY. 1987. Phys. Rev. **67:** 1054-1116.
3. ASHCROFT, F. M. 1988. Ann. Rev. Neurosci. **11:** 97-118.
4. HENQUIN, J. C. 1988. Diabetologia.
5. HAMILL, O. P., A. MARTY, E. NEHER, B. SAKMANN & F. SIGWORTH. 1981. Pflugers Archiv. **391:** 85-100.

6. SATIN, L. S. & D. L. COOK. 1985. Pflugers Arch. **404:** 385-387.
7. SATIN, L. S. & D. L. COOK. 1988. Pflugers Arch. **411:** 401-409.
8. HIRIART, M. & D. R. MATTESON. 1988. J. Gen. Physiol. **91:** 617-639.
9. RORSMAN, P. & G. TRUBE. 1986. J. Physiol. **374:** 531-550.
10. PLANT, T. D. 1988. J. Physiol. **404:** 731-747.
11. RORSMAN, P., F. M. ASHCROFT & G. TRUBE. 1988. Pflugers Arch. **412:** 597-603.
12. KELLY, R. P. & F. M. ASHCROFT. Unpublished observations.
13. FINDLAY, I. & M. J. DUNNE. 1985. FEBS Lett. **189:** 281-285.
14. RORSMAN, P., P. ARKHAMMAR & P. O. BERGRREN. 1986. Am. J. Physiol. **251:** C912-919.
15. ARMSTRONG, D. & R. ECKERT. 1987. Proc. Natl. Acad. Sci. USA **84:** 2518-2522.
16. SANGUINETTI, M. C., D. S. KRAFTE & R. S. KASS. 1986. J. Gen. Physiol. **88:** 369-392.
17. KOSTYUK, P. G. & O. A. KRISHTAL. 1977. J. Physiol. **270:** 569-580.
18. ALMERS, W., E. W. MCCLESKEY & P. T. PALADE. 1984. J. Physiol. **353:** 565-583.
19. VELASCO, J. M. 1987. J. Physiol. **398:** 15P.
20. VELASCO, J. M., J. U. H. PETERSEN & O. H. PETERSEN. 1988. FEBS Lett **231:** 366-370.
21. ASHCROFT, F. M., P. RORSMAN & G. TRUBE. 1989. Ann. N.Y. Acad. Sci. This volume.
22. ATWATER, I., C. M. DAWSON, G. T. EDDLESTONE & E. ROJAS. 1981. J. Physiol. **314:** 195-212.
23. FENWICK, E. M., A. MARTY & E. NEHER. 1982. J. Physiol. **331:** 599-635.
24. MCCLESKEY, E. W., A. P. FOX, D. FELDMAN & R. W. TSIEN. 1986. J. Exp. Biol. **124:** 177-190.
25. LANSMAN, J. B., P. HESS & R. W. TSIEN. 1986. J. Gen. Physiol. **88:** 321-347.
26. HESS, P., J. B. LANSMAN & R. W. TSIEN. 1984. Nature **311:** 538-544.
27. RORSMAN, P. 1988. J. Gen. Physiol. **91:** 243-254.
28. COOK, D. L., W. E. CRILL & D. PORTE. 1980. Nature **286:** 404-406.
29. MEISSNER, H. P., M. PREISSLER & J. C. HENQUIN. 1980. *In* Diabetes. 1979. W. K. Waldhaus, Ed.: 166-171 ICS 500 Excerpta Medica. Amsterdam.
30. HENQUIN, J. C., W. SCHMEER, M. NENQUIN & H. P. MEISSNER. 1985. Biochem. Biophys. Res. Commun. **131:** 980.
31. MALAISSE, W. J. & A. C. BOSCHERO. 1977. Horm. Res. **8:** 203-209.

Single Calcium Channel Activity in Mouse Pancreatic β-Cells

F. M. ASHCROFT,[a] P. RORSMAN,[b] AND G. TRUBE[c]

[a] University Laboratory of Physiology
Oxford University
Oxford, United Kingdom

[b] Department of Medical Physics
Göteborg University
Göteborg, Sweden

[c] Max Planck Institute für Biophysikalische Chemie
Göttingen, Federal Republic of Germany

Glucose-induced insulin secretion from pancreatic β-cells is known to involve an influx of calcium through voltage-dependent Ca channels. Recently we have recorded single-channel currents flowing through β-cell Ca channels using the outside-out configuration of the patch-clamp technique and Ba^{2+} as the charge carrier.[1] Only one type of Ca channel was observed. This had a single-channel slope conductance of 24 pS in 110 mM Ba^{2+}, was sensitive to dihydropyridines, and was blocked by micromolar concentrations of Cd^{2+}. These properties are consistent with the view that this Ca channel is of the L-type (for terminology see Nowycky et al.[2]). The outside-out patch configuration involves the replacement of the cytosol with an artificial solution and is therefore unsuitable for investigations of the metabolic regulation of the Ca channel. For these studies we have used cell-attached patches, because Ca channels are then exposed to their normal internal environment. We present here preliminary cell-attached recordings of single Ca channels in mouse pancreatic β-cells, using previously published methods.[1]

FIGURE 1 shows single-channel currents recorded from a cell-attached patch on a β-cell immersed in normal external solution containing (in mM): 138 NaCl, 5.6 KCl, 1.2 $MgCl_2$, 2.6 $CaCl_2$, and 10 HEPES-NaOH (pH 7.4). The pipette was filled with (in mM) 100 $BaCl_2$, 10 TEACl, 10 HEPES-Ba(OH)$_2$ (pH 7.4). In the absence of glucose, most channel openings were of brief duration and long openings were observed only rarely. One minute after the addition of 20 mM glucose to the bath solution, channel activity was greatly increased and long openings predominated. The associated mean current is consequently considerably larger in the presence than in the absence of glucose. The channel openings have the same amplitude in the presence and absence of the sugar, excluding the possibility that the effect is due to β-cell depolarization. In fact, the single-channel current (I)-voltage (V) relations recorded under both conditions superimpose. After three minutes in high glucose, the β-cell fired action potentials, which frequently elicited Ca-channel openings. These preliminary findings suggest that glucose might modulate Ca-channel activity in the β-cell. A related phenomenon has been described in an insulin-secreting cell line.[3]

The single-channel slope conductance was 20 ± 2 pS ($n = 6$) when measured in

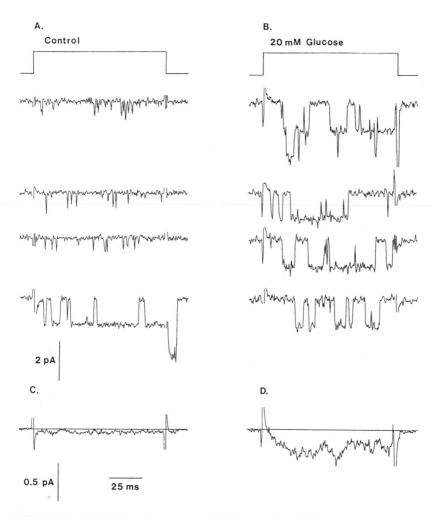

FIGURE 1. Single Ca-channel currents recorded from a cell-attached patch on a mouse β-cell immersed in normal external solution. Pulses were applied from a pipette holding potential of 0 mV to a pipette potential of −70 mV (assuming a β-cell resting potential of −70 mV, this corresponds to a membrane potential of 0 mV) at a frequency of 0.5 Hz. Records were filtered at 0.5 Hz. (**A**) Ba currents recorded in the absence of glucose. (**B**) Ba currents recorded one minute after increasing the bath glucose concentration to 20 m*M*. (**C,D**) Associated mean currents for the data in A,B obtained by averaging approximately 50 individual sweeps.

normal extracellular solution. I-V relations were also recorded in a depolarizing, high K^+ solution, which allows better control of membrane potential (in mM: 125 KCl, 30 KOH, 1 MgCl$_2$, 2 CaCl$_2$, 10 EGTA and 5 HEPES-KOH, pH 7.15). In this solution, the single-channel conductance was 25 ± 2 pS ($n = 4$) and the I-V relation was shifted by 70 mV with respect to that recorded in normal external solution, consistent with the usual β-cell membrane potential of around -70mV. Ca-channel activity was markedly increased by 5 μM BAY K 8644 and inhibited by 25 μM D600.

Our preliminary studies suggest that the use of cell-attached patches will be valuable in future studies concerning the regulation of Ca-channels in β-cells, and corroborate the view that, at least in mouse β-cells, L-type Ca channels contribute to most of the Ca current involved in the process of insulin release.

REFERENCES

1. RORSMAN, P., F. M. ASHCROFT & G. TRUBE. 1988. Pflugers Arch. **412:** 597.
2. NOWYCKY, M. C., A. P. FOX & R. W. TSIEN. 1985. Nature **316:** 400.
3. VELASCO, J. M., J. U. H. PETERSEN & O. H. P. PETERSEN. 1988. FEBS Lett. **231:** 366.

Functions for Calcium Channels in Pituitary Cells[a]

S. A. DeRIEMER

Department of Biological Science
Columbia University
New York, New York 10027

The cells of the anterior pituitary secrete hormones under the control of intrinsic and extrinsic factors. The existence of multiple calcium channels in endocrine cells (TABLE 1) raises the question of what roles specific calcium-channel types have in secretory cells, and what has driven the evolution of multiple calcium-channel subtypes.

Patch-clamp recordings were made from identified anterior pituitary cells[1] and secretion was monitored in a perifused cell column or in 35-mm tissue culture dishes. Hormones were assayed by radioimmunoassay using materials supplied by NIDDK.

L-type currents (measured from a holding potential of -50 mV) were similar in the two cell types studied despite considerable cell-to-cell variability. In one series of experiments using normal male rats and 2 mM calcium, the peak L current values were 8.0 ± 4.0 pA/pF (SD, $n = 7$) for lactotrophs and 7.4 ± 4.4 ($n = 5$) for somatotrophs. T-type currents (measured as the difference between the values obtained from holding potentials of -90 vs. -50 mV) were larger in lactotrophs, 8.1 ± 6.0 pA/pF, than in somatotrophs, 3.3 ± 1.8, in the same series of cells.

The involvement of T and L currents in regulating secretion was examined in lactotrophs using dihydropyridines to modulate L channel function and ethanol to inhibit T channels.[2] The relative efficacy of 0.001 to 0.1 mM EtOH on secretion in cells held at different resting potentials by changing $[K^+]_o$ was compared with the current-voltage dependence of the T channel measured electrophysiologically. Inhibition of prolactin secretion was maximal at 5 mM K_o, which corresponds to approximately -40 mV and the maximal T-type calcium current. These results along with data showing effects of BAY K8644 on basal prolactin secretion[3] suggest that when a cell has both channels, they are both used, although the relative importance may vary with the state of the cell. The first phase (0-5 min) of the secretory response to thyrotropin releasing hormone (TRH) was unaffected by BAY K8644 or EtOH treatment, consistent with it being due to calcium release from intracellular stores.[4] In the second phase (5-30 min) however, an effect of both drugs could be observed. This data, and results from other cells listed in TABLE 1 suggest that responses to stimulants may involve no voltage-sensitive calcium channels, and/or both T and L, and/or novel calcium channels.[17]

[a] Supported by the National Science Foundation Grant # DCB-86-15840 and March of Dimes Grant #5-611.

TABLE 1. Voltage-Gated Calcium Channels in Endocrine Cells

Cell Type	Channel Types	Ref.
Mast cells	—	5
Somatotroph	L	1 (cf. 6)
Gonadotroph	L	7
Adrenal chromaffin	L	8
Pancreatic β	L	9 (cf. 10)
AtT20	L	11
Lactotroph	L & T	1,6,12,13
Corticotroph	L & T	7
Melanotrophs	L & T	14
Pancreatic α_2	L & T	15
GH$_3$/GH$_4$Cl	L & T	16
Aplysia bag cells	L^a & TPA-induced	17

^{a} Not dihydropyridine sensitive.

REFERENCES

1. DeRiemer, S. A. & B. Sakmann. 1986. Exp. Brain Res. Ser. **14:** 139-154.
2. Llinas, R. & Y. Yarom. 1986. Soc. Neurosci. Abstr. **12(1):** 174.
3. Enyeart, J. J., S. -S. Sheu & P. M. Hinkle. 1987. J. Biol. Chem. **262(7):** 3154-3159.
4. Gershengorn, M. C. 1986. Ann. Rev. Physiol. **48:** 515-526.
5. Penner, R., G. Matthews & E. Neher. 1988. Nature **334:** 499-504.
6. Lewis. D. L., M. B. Goodman, P. A. St. John & J. L. Barker. 1988. Endocrinology **123:** 611-621.
7. Marchetti, C., G. V. Childs & A. M. Brown. 1987. Am. J. Physiol. **252 (3 pt 1):** P E340-346.
8. Fenwick, E. M., A. Marty & E. Neher. 1982. J. Physiol. **331:** 599-635.
9. Rorsman, P. & G. Trube. 1986. J. Physiol. **374:** 531-550.
10. Hiriart, M. & D. R. Matteson. 1988. J. Gen. Physiol. **91(5):** 617-639.
11. Nowycky, M. C. 1987. Soc. Neurosci. Abstr. **13(2):** 793.
12. Lingle, C. J., S. Sombati & M. E. Freeman. 1986. J. Neurosci. **6(10):** 2995-3005.
13. Cobbett, P., C. D. Ingram & W. T. Mason. 1987. Neuroscience **23:** 661-677.
14. Cota, G. 1986. J. Gen. Physiol. **88:** 83-105.
15. Rorsman, P. 1988. J. Gen. Physiol. **91(2):** 243-254.
16. Armstrong, C. M. & D. R. Matteson. 1985. Science **227:** 65-67.
17. Strong, J. A., A. P. Fox, R. W. Tsien & L. K. Kaczmarek. 1987. Nature **325(6106):** 714-717.

Receptor-Stimulated Calcium Mobilization and Calcium Influx Pathways in Pituitary Cells[a]

PATRICE MOLLARD,[b,c] BENOÎT P. WINIGER,[d]
PIERRE VACHER,[b] BERNARD DUFY,[b] AND
WERNER SCHLEGEL[d]

[b] *Laboratoire de Neurophysiologie, UA CNRS 1200, and*
[c] *UFR de Sciences Pharmaceutiques*
Université de Bordeaux II
Bordeaux 33076 Cedex, France

[d] *Fondation pour Recherches Médicales*
Département de Médecine
Université de Genève
CH-1211 Genève 4, Switzerland

The importance of calcium ions (Ca^{2+}) in the control of the secretory activity of endocrine cells is well recognized. It is also known that the regulation of cytosolic free Ca^{2+} concentration, $[Ca^{2+}]_i$, may involve entry of Ca^{2+} ions through voltage-activated Ca^{2+} membrane channels as well as mobilization of intracellularly sequestered Ca^{2+} pools. The GH3 pituitary cells, a rat clonal pituitary cell line that releases prolactin, are known to spontaneously display Ca^{2+} action potentials.[1] We have recently demonstrated that action potentials are effective in eliciting $[Ca^{2+}]_i$ transients of a magnitude sufficient to trigger $[Ca^{2+}]_i$-dependent hormone release.[2]

In this work, we have studied the regulation of cytosolic free Ca^{2+} concentration, $[Ca^{2+}]_i$, by the releasing factor thyrotropin-releasing hormone (TRH) at the single-cell level using tight-seal whole-cell recording techniques associated with microfluorimetric monitoring of $[Ca^{2+}]_i$. The Ca^{2+} probe fura-2 was used in these experiments.

GH3/B6 (GH3 subclone) cells were cultured on glass cover slides for three to five days under standard conditions. Electrophysiological recordings were obtained at room temperature using the whole-cell recording configuration of the patch-clamp technique, as described previously[3]; whole-cell patch pipettes (resistance 3–10 megohms) contained the following solution (in mM) K^+ gluconate (140), $MgCl_2$ (1), EGTA (1.1), HEPES (5), pH 7.1, and fura-2 (30 μM). Buffering of $[Ca^{2+}]_i$ obtained under these conditions is somewhat variable, such that the resting $[Ca^{2+}]_i$ values observed after equilibration of the cells with the pipette were found to vary within the range of 0.01 to 0.2 μM. Excitation light alternated rapidly (50 Hz) between λ_1

[a] This work was supported by CNRS (UA 1200), INSERM (87 4003), Fonds National Suisse (3.824-0.86), European Science Foundation, and Foundation Carlos and Elsie de Reuter.

415

= 350 nm and λ_2 = 380 nm (SPEX fluorimeter, Glen Creston, London). Fluorescence was monitored by photon counting through a circular diaphragm slightly larger than the cells and an interference filter (λ = 500 nm). The ratio R = F_{350}/F_{380} is representative for $[Ca^{2+}]_i$, which can be calculated from calibration constants derived in intact cells as described previously.

Local application of TRH (100 nM) to a cell current clamped at resting level caused a characteristic pattern of changes in membrane excitability closely correlated with changes in $[Ca^{2+}]_i$ (FIG. 1). In a first phase, TRH triggered a large $[Ca^{2+}]_i$

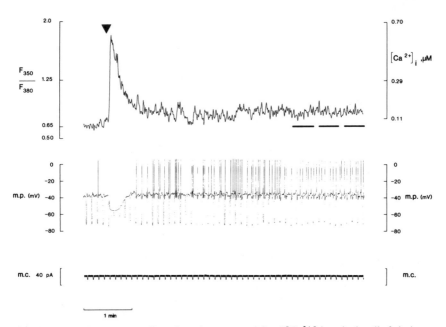

FIGURE 1 Simultaneous recording of membrane potential and $[Ca^{2+}]_i$ in a single cell of pituitary line GH3/B6. This combined experiment was performed as described previously.[2] The ratio R = F350/F380 is representative for $[Ca^{2+}]_i$, which can be calculated from calibration constants derived in intact cells.[2] Shown are the ratio of fura-2 fluorescence calibrated for $[Ca^{2+}]_i$ (top trace, left and right scales, respectively), the membrane potential (m.p., middle trace) obtained under current clamp conditions, and the monitoring of current (m.c., bottom trace) that was periodically applied to measure membrane conductance. TRH (100 nM, triangle) was briefly applied to the cell recorded from a micropipette placed close to the cell membrane (10 μm). The dashed line represents the initial $[Ca^{2+}]_i$.

transient monitored by the fura-2 signal; concomitantly there was a hyperpolarization associated with a decrease in membrane input resistance. Pharmacological studies performed under voltage clamp revealed that this hyperpolarization involved an outward K^+ current that was activated by a TRH-induced release of Ca^{2+} from internal stores. After a short lag time, TRH then caused a series of $[Ca^{2+}]_i$ oscillations correlated with low-amplitude (2-3 mV) fluctuations of the membrane potential and enhanced firing of action potentials. As has been shown previously,[2] the latter may cause the $[Ca^{2+}]_i$ oscillations that represent Ca^{2+} entry through voltage-dependent Ca^{2+} channels.

In summary, the combination of patch-clamp techniques with the monitoring of $[Ca^{2+}]_i$ at subsecond time resolution offers a novel tool for studying directly the link between $[Ca^{2+}]_i$ and plasma membrane ion channel activities in intact cells. This combined approach enabled us to clearly ascribe the TRH-evoked $[Ca^{2+}]_i$ oscillations seen in single unpatched GH3/B6 cells[4] to a dual mechanism: Ca^{2+} mobilization and Ca^{2+} influx. We are currently investigating the biochemical pathways involved using internal perfusion[5] of putative second messengers.

REFERENCES

1. DUFY, B., J. D. VINCENT, H. FLEURY, P. DU PASQUIER, D. GOURDJI & A. TIXIER-VIDAL. 1979. Science **204:** 509-511.
2. SCHLEGEL, W., B. P. WINIGER, P. MOLLARD, P. VACHER, F. WUARIN, G. R. ZAHND, C. B. WOLLHEIM & B. DUFY. 1987. Nature **329:** 719-721.
3. HAMILL, O. P., A. MARTY, E. NEHER, B. SAKMANN & F. J. SIGWORTH. 1981. Pflugers Arch. **391:** 85-100.
4. WINIGER, B. P. & W. SCHLEGEL. 1988. Biochem. J. **855:** 161-167.
5. KIMURA, J., A. NOMA & H. IRISAWA. 1986. Nature **319:** 596-597.

Appearance and Function of Voltage-Dependent Ca^{2+} Channels during Pre- and Postnatal Development of Cardiac and Skeletal Muscles[a]

J. F. RENAUD, M. FOSSET, T. KAZAZOGLOU,
M. LAZDUNSKI, AND A. SCHMID

FONDAX-Groupe de Recherche SERVIER
92800 Puteaux, France
and
Centre de Biochimie du CNRS
Parc Valrose, 06034 Nice cedex, France

INTRODUCTION

The most prominent feature of many excitable cells is their ability to respond to voltage applied across their plasma membranes. Usually, the response consists of a change in membrane permeability to specific cations due to the opening or closing of ionic channels.

It has been shown that the ionic properties of most of the cell types studied are often different at the early stage of development than in the adult.[1] Furthermore, during the course of development, the ionic response changes in reaction to a change in the cationic channels involved.[2,7] These observations have raised several questions: (1) Are the changes observed common to all excitable cell types? (2) Are these changes developmentally regulated, and if so what mechanism controls them? (3) What is the role of these ionic channels in the differentiation process and in maturation of the cells?

This review tries to provide partial answers to these questions using the example of the voltage-sensitive, slow Ca^{2+} channel, which is sensitive to dihydropyridines,[8] and those throughout the development of cardiac and skeletal muscles. These two cell types were chosen for the study of the differentiation process, because many pharmacological and biochemical properties were already known for adult cells and tissues.[9,12] This has made it easier to investigate the time course and appearance of structures that are responsible for cell excitability and to detail changes in the biochemical and functional properties of the slow, voltage-dependent Ca^{2+} channels from the earliest to the latest stage of development.

[a] This work was supported by the Centre National de la Recherche Scientifique, the Association Française de Lutte contre les myopathies, and the Foundation pour la Recherche Médicale.

DEVELOPMENT OF SKELETAL MUSCLE CELLS AND APPEARANCE OF DIHYDROPYRIDINE-SENSITIVE CA^{2+} CHANNELS

Myogenic cells first appear as mononucleated myoblasts that originate from a pool of mesenchymal cells that also have the potential to differentiate into chondrocytes and fibroblasts.[13] Myoblasts fuse into multinucleated myotubes, which then gives rise to mature muscle fibers.[5,6] A few myoblasts do not undergo therminal differentiation and coexist with muscle fibers in the adults.[14] These so called "satellite cells" still have the potential to differentiate into muscle fibers and also can contribute to the regeneration of normal and diseased muscle. Isolated satellite cells retain the ability to divide and to fuse *in vitro* into multinucleated myotubes.[14] Appearance of voltage-sensitive Ca^{2+} channels during *in vitro* and *in vivo* myogenesis of skeletal muscle cells has been followed using both [^{3}H]nitrendipine binding and ^{45}Ca^{2+} flux measurement.

In Vitro *Development of Chick Skeletal Muscle*

Embryonic skeletal muscle differentiates spontaneously in cell culture from mononucleated myogenic cells (myoblasts) to multinucleated cross-striated muscle fibers (myotubes), displaying electrical and contractile activity.[5,6] The nitrendipine receptor is absent in the earliest, undifferentiated myoblasts. [^{3}H]Nitrendipine binding sites appear after 20-hour of culture. The number of [^{3}H]nitrendipine binding sites then increases rapidly to reach a plateau value of 150 fmol/mg of protein in 55-hour-old myotubes.

The time course of appearance of [^{3}H]nitrendipine binding sites closely follows the onset of fusion. The K_D value for the nitrendipine receptor complex remains at 0.4 ± 0.1 nM during all stages of cell culture. This value is similar to that measured from inhibition of dihydropyridine-sensitive ^{45}Ca^{2+} flux with myotubes in culture. The nitrendipine-sensitive component of Ca^{2+} flux is undetectable in 20-hour-old myoblasts in culture, whereas it is present in 72-hour-old myotubes differentiated in culture.[15] Inhibition of fusion of myoblasts into myotubes by adding EGTA or the protein synthesis inhibitor cycloheximide prevents the appearance of [^{3}H]nitrendipine binding sites.

The chronology of appearance of the nitrendipine-sensitive Ca^{2+} channel is at first sight similar to that found for the voltage-dependent Na$^+$ channel, which is also absent in myoblasts and present in myotubes. However, EGTA, which removes Ca^{2+}, blocks fusion of myoblasts into myotubes and prevents the appearance of the nitrendipine-sensitive Ca^{2+} channel, whereas it does not prevent the development of the Na$^+$ channel[16] or of other ion-transport systems such as the Na$^+$, K$^+$-ATPase.[17] Cycloheximide, which is a protein synthesis inhibitor, also inhibits fusion and prevents the appearance of the nitrendipine-sensitive Ca^{2+} channel. These results taken together indicate that the appearance of the nitrendipine receptor is linked both to the fusion process and to protein synthesis control.

The reason for this difference in behavior may be due to the particular localization of nitrendipine-sensitive Ca^{2+} channels. Both electrophysiological techniques and the titration of [^{3}H]nitrendipine receptor[15,20] have indicated that these channels are essentially, if not uniquely, present in T-tubules. *In vitro* investigations with cells in culture have shown that T-tubules are absent in mononucleated myoblasts, they only appear after conversion of myoblasts into myotubes.[21] In myoblasts where fusion is blocked, there appears to be a correlation between inhibition of T-tubule formation and the inhibited synthesis of nitrendipine receptors.

In Vivo *Development of Skeletal Muscle Cells*

The *in vivo* appearance of the chick nitrendipine-sensitive Ca^{2+} channel shows the existence of two different phases.[15] A first phase starts near day 10 of *in ovo* development. The level of Ca^{2+} channels then remains stable between day 11 and day 17 *in ovo*. At this stage, the maximum level of nitrendipine receptors (124 ± 10 fmol/mg of protein) and the K_D value (0.4 nM) are very similar to the level attained in the course of the *in vitro* development. This first phase of development corresponds to the *in vivo* fusion of myoblasts into myotubes. The second phase of development starts a little before hatching, and the change in a number of [^{3}H]nitrendipine receptors is particularly fast between hatching and seven days postnatal, when another plateau phase is reached corresponding to 900 fmol/mg of protein. This change in number of nitrendipine receptors is accompanied by a change in the equilibrium dissociation constant of the nitrendipine receptor complex. K_D values increase by factors of 4 to 10 when passing from the first phase to the second phase of development. The appearance of [^{3}H]nitrendipine binding sites during postnatal development of rat skeletal muscle has also been studied.[22] As for chick muscle, there is a continuous increase in the number of [^{3}H]nitrendipine binding sites after birth. However, a plateau level is reached after about six days for chick muscle, while it is reached after about 20 days for rat muscle. These differences in time course of development are probably related to the fact that very soon after birth the chick must feed, drink and walk, whereas the rat is born helpless and develops a mature behavioral pattern after only two to three weeks of postnatal life.

Regulation of Ca^{2+}-Channel Properties by Innervation

Surgical denervation was carried out in order to know whether innervation is responsible for the second phase of development of nitrendipine-sensitive Ca^{2+} channels. Denervation has a biphasic effect on nitrendipine receptors of rat and chick muscle.[15,23] During the first phase of denervation, the number of nitrendipine receptors increase twofold after 15 days, then the number of [^{3}H]nitrendipine binding sites declines. These changes in the number of sites occur without any change in affinity of nitrendipine for its receptor. Denervation produces different effects on voltage-dependent Na^+ channels. It decreases the level of Na^+ channels with a high tetrodotoxin affinity[15,23] and renders Na^+ channels resistent to tetrodotoxin in mammalian muscle.[24] The initial increase in the number of [^{3}H]nitrendipine receptors following denervation may be due to an over-production of T-tubule membranes,[25] where Ca^{2+} channels seem to be localized.

Regulation of Ca²⁺-Channel Activity and Ca²⁺-Channel Number by Isoproterenol and cAMP

The β-adrenergic system seems to play an important role in the regulation of the level of the dihydropyridine receptor sites during chick skeletal muscle development after hatching. *In vivo* treatments of seven-day-old chicks with reserpine, which inhibits norepinephrine and epinephrine synthesis, or with alprenolol, which is a potent β-adrenergic antagonist, induces a decrease in the number of nitrendipine receptors.[26] This decrease in receptor number is accompanied by a four- to fivefold decrease in the affinity of nitrendipine for its receptor. These effects on the nitrendipine receptor were prevented by simultaneous injection of isoproterenol. Therefore, skeletal muscle nitrendipine receptors in reserpine and alprenolol-treated animals are more similar to "fetal" nitrendipine receptors than to the postnatal nitrendipine receptors. These results clearly favor the conclusion that the physiological stimulation of β-adrenergic receptors regulates both the number of specific nitrendipine receptors and the affinity of the receptor for dihydropyridine. This hypothesis is confirmed by short-term and long-term stimulation of β-adrenergic receptors *in vitro*.

The effects of short-term stimulation of β-adrenergic receptors and elevations in intracellular cyclic AMP on nitrendipine-sensitive voltage-dependent Ca²⁺ channels of skeletal muscle cells *in vitro* has been studied using both the ⁴⁵Ca²⁺ flux technique and [³H]nitrendipine binding experiments.[26] Isoproterenol increased the nitrendipine-sensitive ⁴⁵Ca²⁺ influx under depolarizing conditions. The effects of isoproterenol were additive to those of depolarization and were antagonized by alprenolol. Half-maximal inhibition of ⁴⁵Ca²⁺ influx induced both by depolarization and by isoproterenol occurred at a nitrendipine concentration of 1 nM. Treatments that resulted in an increased level of intracellular cyclic AMP, such as treatment with 1-methyl 3-isobutylxanthine, theophylline, dibutyryl cyclic AMP or 8-bromocyclic AMP, also resulted in an activation of the nitrendipine-sensitive Ca²⁺ channel.

In cardiac cells,[27] phosphorylation of the Ca²⁺ channel in skeletal muscle *via* a cAMP-dependent kinase increases the probability of finding the Ca²⁺ channel in the open form.

In contrast, long-term treatment of myotubes in culture with isoproterenol and other compounds that increased intracellular cyclic AMP led to a very large increase in the number of nitrendipine receptors. This increase was accompanied by a 4-10-fold decrease in the affinity of the receptors for nitrendipine. Alprenolol inhibited the long-term effects of isoproterenol.

DEVELOPMENT OF CARDIAC MUSCLE CELLS AND APPEARANCE OF DIHYDROPYRIDINE-SENSITIVE CA²⁺ CHANNELS

Cardiac muscles, in contrast to skeletal muscles, are composed of single cells. Cardiac muscle cells differentiate early during development and the first beats are observed at two days and 9-10 days' development for chick embryo and mouse or rat embryo, respectively.

In Vitro *and* in Vivo *Development of Chick Cardiac Muscle*

There are two main stages of differentiation of electrical excitability of cardiac cells during development *in ovo* and *in vitro*. Myocardial cells in young (three days) heart have slowly rising action potentials (10-15 V/sec), which are believed to be generated by a Na^+ current through slow Na^+ channels that are insensitive to tetrodotoxin and Mn^{2+} ions.[2,3] At this stage of development, functional, slow Ca^{2+} channels seem to be absent.[28] Conversely, adult-like cells (11-days *in ovo*) have an action potential with a fast rate of rise (100-150 V/sec), which is generated mainly by Na^+ influx through fast Na^+ channels that are blocked by tetrodotoxin; slow Ca^{2+} channels are present and are responsible for the plateau phase of action potential. These channels are insensitive to tetrodotoxin, and they are blocked by verapamil, D_{600}, and inorganic Ca^{2+} blockers.[8,28] The transition between the early and the late stage of development occurs between five days and seven days.[28] Differentiation in this way is also observed when cardiac cells are cultured as aggregates.[3] The expression of the slow Ca^{2+} channel in adult-like cells (11-day aggregates) was observed after blockade of the fast Na^+ channel by tetrodotoxin and after addition of isoproterenol, which is known to increase the size of the calcium current.[27] Under these conditions nitrendipine (50 nM) completely abolished both the slowly rising action potential, induced by isoproterenol, and contraction. Addition of nitrendipine at a concentration about 100 times higher than K_D (100 nM) to the three-day-old aggregates (representing the earliest stage of differentiation) modified neither the slow action potential nor the contraction. These results indicate that the slow Ca^{2+} channel identified as a nitrendipine receptor is either absent at the youngest embryonic stage studied or present in a nonfunctional form, whereas it has been observed that at this stage of development tetrodotoxin-insensitive slow action potentials can be blocked by D_{600} at concentrations ranging between 0.1 and 1 μM.[28] We have found that unlike nitrendipine, D_{600}, or verapamil acts by depolarization of the membrane of three-day-old heart.

The appearance of the nitrendipine-sensitive Ca^{2+} channel has been studied in more detail using [^3H]nitrendipine-binding experiments at different stages of embryonic and postnatal development of the heart from chick and rat.[22,29] Results from chicks have shown that [^3H]nitrendipine receptors are already present at the youngest embryonic stage studied. If nitrendipine receptors do correspond to Ca^{2+} channel structures as previously shown, then these channel structures exist at the early stage of development but do not seem to be physiologically functional.[29] Tetrodotoxin receptors and muscarinic receptors have also been shown to correspond to Na^+ channels and muscarinic receptors that are not pharmacological functional.[13,30] These Na^+ channels, which are nonelectrically functional, have been called silent Na^+ channels.[3]

The number of nitrendipine binding sites gradually increases from 40 fmol/mg of protein to 100 fmol/mg of protein between day three and day 14 of development and stays nearly stable until day 18. During this period, two events happen: the number of nitrendipine receptors representing the Ca^{2+} channel structure increases, and silent Ca^{2+} channels convert into functional Ca^{2+} channels. This event is detectable by electrophysiology and blocked by nitrendipine. This is due to the synthesis or the assembly of a new component of the Ca^{2+}-channel structure, distinct from the nitrendipine-binding component that would be essential for the expression of Ca^{2+}-channel activity inhibited by nitrendipine.

After day 18 *in ovo*, the number of sites increases again rapidly from 100 fmol/mg of protein to 210 fmol/mg of protein at day four after hatching, when a new plateau value is reached. K_D values remain essentially invariant at 0.50 ± 0.07 nM

at all stages of development. Increases in [^{3}H]nitrendipine binding sites during embryonic development *in ovo* follows the ontogenic appearance of the voltage-dependent Na$^+$ channels measured with a [^{3}H]en-TTX derivative.[3] The ratio of maximum binding capacities for [^{3}H]nitrendipine and [^{3}H]en-TTX used as Ca^{2+} channel and Na$^+$ channel markers, respectively, is between 1.2 and 1.4 throughout development *in ovo*. After birth, the number of [^{3}H]nitrendipine binding sites continues to increase, whereas [^{3}H]en-TTX binding sites remain at a lower level. Under these conditions the ratio of maximum binding capacities for [^{3}H]nitrendipine and [^{3}H]en-TTX is between 2.2 and 2.5. The second increase in [^{3}H]nitrendipine binding sites coincides with the hatching process. It has previously been reported that the amount of protein in the chick embryo heart cell does not seem to increase with age and that ventricular growth during the embryonic and early neonatal period is largely the result of an increase in cell number without an appreciable change in the protein-to-DNA ratio of the tissue. The increase in receptor number appears to be directly linked to an increase in the number of receptors per cell.

Regulation of Ca^{2+}-Channel Properties by Innervation

In order to know whether the autonomic nervous system and/or the capacity of neurotransmitter release are important in the sequence of expression of [^{3}H]nitrendipine receptors in the heart plasma membrane, chronic injections of 6-hydroxydopamine and reserpine were performed in seven-day-postnatal chicks at a concentration of 100 mg/kg. At this concentration it has been shown that young (two-week-old) chick heart fluorescent terminal adrenergic nerves disappeared nearly completely within 24 hours and that terminal adrenergic fibers first appeared around seven days after treatment. Complete recovery of adrenergic nerves was observed between 20 days and 30 days after treatment. Under these conditions 6-hydroxydopamine in high doses causes the loss of both nerve endings and the disappearance of norephinephrine storage vesicles in nerve terminals,[31,32] whereas reserpine does not destroy adrenergic transmission but depletes norephinephrine storage.[33] Treatment with 6-hydroxydopamine leads to a 35% increase in [^{3}H]nitrendipine binding as compared to the control level and in parallel with a significant loss (30%) of high-affinity [^{3}H]TTX binding sites that have been shown to correspond to the functional form of the voltage-dependent Na$^+$ channels in chick heart. Under these new conditions the ratio of maximum binding capacities for [^{3}H]nitrendipine and [^{3}H]en-TTX is near 4.3. Treatment with 6-hydroxydopamine also results in a very significant increase in the level of muscarinic cholinergic receptors. Intramuscular and intraperitoneal injections gave the same result. Treatment with reserpine was studied to investigate whether the change in the level of nitrendipine receptors was due to the loss of sympathetic transmission. Reserpine does not provoke any significant variation for either maximum binding capacities or K_D values for the three ligands assayed. These results taken together favor the conclusion that the effects of 6-hydroxydopamine treatment are presumably due to the loss of innervation rather than to the disappearance of a functional norepinephrine release system. These results also suggest that the second increase in [^{3}H]nitrendipine-binding sites that correspond to the hatching period is probably independent of the onset of the adrenergic neurotransmission that becomes functional on the 21st day of embryonic life, which corresponds to the day of hatching.[34] Furthermore, results obtained with embryonic heart cells cultured either as monolayers or as aggregates (that are nerve free) are in agreement with the idea that there exists a regulation of the level of nitrendipine-sensitive Ca^{2+} channels by innervation, since the number of [^3H]nitrendipine binding sites are 29% and 45% higher in heart cells cultured as aggregates and monolayers, respectively, than in the cells of intact hearts taken at the same age of embryonic development.

The nature of the neuronal factors that regulate levels of tetrodotoxin-sensitive Na^+ channels, nitrendipine-sensitive Ca^{2+} channels, and muscarinic receptor in cardiac cells remains to be determined.

CONCLUSION

Comparison of the time course of appearance of Ca^{2+} channels that are involved in membrane electrogenesis and cell function in skeletal muscle and cardiac muscle indicate differences and similarities.

The main differences are the following: (1) Ca^{2+} channels are functional as soon as they appear during skeletal muscle development, whereas in cardiac muscle they first appear in nonfunctional form. (2) Ca^{2+} channels in skeletal muscle are upregulated by β-adrenergic agonists and cAMP. In cardiac muscle the precise mechanisms that modulate the expression of Ca^{2+} channels during ontogenesis remain to be determined. (3) Ca^{2+} channels in skeletal muscle appear to be independent of the innervation process, whereas in cardiac muscle, their number seems to be controlled by neuronal factors. The main similarities are that Ca^{2+} and Na^+ channels are regulated in an opposite way by innervation in both cell types and that Ca^{2+} channel function in both cardiac and skeletal muscle is dependent on the phosphorylation process.

ACKNOWLEDGMENTS

Thanks are due to Dr. Horstmann, Dr. Müller, Dr. Traber, and Dr. Zellerhoff from Bayer AG (Federal Republic of Germany) and Dr. Perin from Bayer Pharma (France) for the gift of nitrendipine, Mrs. A. M. Lamy from the Biosedra (France) and Knoll AG (Federal Republic of Germany) laboratories for the gift of verapamil and D_{600}.

The authors are indebted to Kate Chadwick for preparing and typing the manuscript.

REFERENCES

1. SPITZER, N.C. 1979. Ann. Rev. Neurosci. **2**: 363-397.
2. SPERELAKIS, N. 1981. *In* Cardiac Toxicology. T. Balazs, Ed. Vol. **1**: 39-108. CRC Press. Boca Raton, FL.
3. RENAUD, J. F., G. ROMEY, A. LOMBET & M. LAZDUNSKI. 1981. Proc. Natl. Acad. Sci. USA **78**: 5348-5352.
4. HAGIWARA, S. 1983. Membrane Potential-Dependent Ion Channels in Cell Membrane Phylogenetic and Developmental Approaches. Vol. **3**: 118. Raven Press. New York.

5. KANO, M. 1975. J. Cell Physiol. **86:** 503-510.
6. SPECTOR, I. & J. M. PRIVES. 1977. Proc. Natl. Acad. Sci. USA **74:** 5166-5170.
7. KANO, M., K. WAKUTA & R. SATOH. 1987. Dev. Brain Res. **32:** 233-240.
8. FLECKENSTEIN, A. 1977. Ann. Rev. Pharmacol. Toxicol. **17:** 149-166.
9. HAGIWARA, S. & L. BYERLY. 1981. Ann. Rev. Neurosci. **4:** 69-125.
10. LAZDUNSKI, M. & J. F. RENAUD. 1982. Ann. Rev. Physiol. **44:** 463-473.
11. FOSSET, M. & M. LAZDUNSKI. 1988. *In* Structure and Physiology of Slow Inward Calcium Channels. J. Craig Venter & D. Triggle, Eds.: 141-159. Alan R. Liss. New York.
12. TRAUTWEIN, W. 1973. Physiol. Rev. **53:** 793-835.
13. HOLTZER, H. 1978. *In* Cell Lineages, Stem Cells, and the Quantal Cell Cycle Concept. B. I. Lord, C. S. Potten & R. J. Cole, Eds.: 1-27. Cambridge University Press. London.
14. LINKHART, T. A., C. H. CLEGG & S. D. HAUSCHKA. 1981. Dev. Biol. **86:** 19-30.
15. SCHMID, A., J. F. RENAUD, M. FOSSET, J. P. MEAUX & M. LAZDUNSKI. 1984. J. Biol. Chem. **259:** 11366-11372.
16. FRELIN, C., A. LOMBET, P. VIGNE, G. ROMEY & M. LAZDUNSKI. 1981. J. Biol. Chem. **256:** 12355-12361.
17. VIGNE, P., C. FRELIN & M. LAZDUNSKI. 1982. J. Biol. Chem. **257:** 5380-5384.
18. POTREAU, D. & G. RAYMOND. 1980. J. Physiol. **307:** 9-22.
19. ALMERS, W., R. FINK & P. T. PALADE. 1981. J. Physiol. **312:** 177-207.
20. FOSSET, M., E. JAIMOVICH, E. DELPONT & M. LAZDUNSKI. 1983. J. Biol. Chem. **258:** 6086-6092.
21. EZERMAN, B. E. & H. ISHIKAWA. 1967. J. Cell. Biol. **35:** 405-420.
22. KAZAZOGLOU, T., A. SCHMID, J. F. RENAUD & M. LAZDUNSKI. 1983. FEBS Lett. **164:** 75-79.
23. SCHMID, A., T. KAZAZOGLOU, J. F. RENAUD & M. LAZDUNSKI. 1984. FEBS Lett. **172:** 114-118.
24. REDFERN, P. & S. THESLEFF. 1971. Acta Physiol. Scand. **82:** 70-78.
25. PELLGRINO, C. & C. FRANZINI. 1963. J. Cell Biol. **17:** 327-332.
26. SCHMID, A., J. F. RENAUD & M. LAZDUNSKI. 1985. J. Biol. Chem. **260:** 13041-13046.
27. REUTER, H. 1983. Nature **301:** 569-574.
28. SPERELAKIS, N. 1980. *In* The Slow Inward Current and Cardiac Arrhythmics. D. P. Zipes, J. C. Bailey & V. Elharrar, Eds.: 221-262.
29. RENAUD, J. F., T. KAZAZOGLOU, A. SCHMID, G. ROMEY & M. LAZDUNSKI. 1984. Eur. J. Biochem. **139:** 673-681.
30. RENAUD, J. F., J. BARHANIN, D. CAVEY, M. FOSSET & M. LAZDUNSKI. 1980. Dev. Biol. **78:** 184-200.
31. KOSTZEWA, R. M. & D. M. JACOBOWITZ. 1974. Pharmacol. Rev. **26:** 199-288.
32. BENNETT, T. & T. MALMFORS, J. L. S. COBB. 1973. Z. Zellforsch. Mikroskop. Anat. **142:** 103-130.
33. MEISHERI, K. D., T. E. TENNER & J. H. MCNEILL. 1979. Life Sci. **24:** 473-480.
34. PAPPANO, A. J. 1977. Pharmacol. Rev. **29:** 3-33.

Calcium Channels in Bacteria

Purification and Characterization

TAKU MATSUSHITA,[a,b] HAJIME HIRATA,[c] AND
IWAO KUSAKA[a,d]

[a] Institute of Applied Microbiology
University of Tokyo
Tokyo 113, Japan

[c] Department of Biochemistry
Jichi Medical School
Tochigi 329-04, Japan

Many studies on Ca^{2+} channels in eukaryotic cells have been pursued, but the idea that Ca^{2+} channels could be functioning in prokaryotic cells such as bacteria has until now never been considered. These most primitive organisms, which presumably lack a nervous system, are, after all, an unlikely organism in which to study Ca channels. In this short paper, we will describe the Ca^{2+} channel in bacterial cells for the first time.

METHODS

The right-side-out membrane vesicles and membrane proteins were prepared from *Bacillus subtilis* W23 as described in a previous paper.[1,2] ^{45}Ca uptake by membrane vesicles or reconstituted proteoliposomes was driven by a membrane potential, inside-negative, due to K^+ efflux via valinomycin. Proteins responsible for the Ca^{2+} uptake were highly purified by chromatography on Fractogel TSK DEAE and Fractogel TSK Butyl, and were also reconstituted into planar phospholipid bilayers formed by the method of Montal and Mueller.[3] Electrophysiological analysis was performed by using a List (West Germany) patch-clamp amplifier. Data were recorded on a beta video cassette recorder (Sony) after digitization with a pulse-code modulator (Sony PCM-501-ES).

[b] Present address: Department of Chemical Engineering, Faculty of Engineering, Kyushu University, Higashiku, Fukuoka, 812, Japan.

[d] Address for Correspondence: I. Kusaka, Institute of Applied Microbiology, University of Tokyo, Tokyo 113, Japan.

RESULTS AND DISCUSSION

Ca^{2+} was accumulated by right-side-out membrane vesicles of *Bacillus subtilis* when an inside-negative membrane potential was imposed. The voltage-dependent Ca^{2+} uptake was inhibited by Ca^{2+} channel blockers such as nitrendipine, verapamil, $LaCl_3$, and ω-conotoxin as shown in TABLE 1. Proteins responsible for the Ca^{2+} uptake were highly purified by chromatography and contained several polypeptides of 32, 35, and 42 kDa. The voltage-dependent Ca^{2+} uptake by reconstituted proteoliposomes was also inhibited by Ca^{2+}-channel blockers and activated by Ca^{2+}-channel agonist BAY K 8644 (data not shown). When the highly purified proteins responsible

TABLE 1. Effect of Ca^{2+}-Channel Blockers on the Voltage-Dependent Ca^{2+} Uptake System (Ca^{2+} Channel) in *Bacillus subtilis*

Ca^{2+}-Channel Blocker	Concentration (μM)	Activity (nmol/min/mg)	Percent
No addition	—	1.6	100
ω-Conotoxin	5	0.98	61
ω-Conotoxin	10	0.35	22
Nitrendipine	1.0	0.61	38
Nitrendipine	10	0.37	23
$LaCl_3$	100	0.32	20
$LaCl_3$	250	0	0
Verapamil	25	1.58	99
Verapamil	50	0.77	48

NOTE: Ca^{2+} uptake was driven by a K^+ diffusion potential in right-side-out membrane vesicles of *Bacillus subtilis*. K^+-loaded vesicles were suspended in 25 mM Tris/H_2SO_4 (pH 7.2) containing choline chloride as a osmoregulater and several drugs. Uptake was initiated by the addition of valinomycin.

for the voltage-dependent Ca^{2+} uptake were reconstituted into planar phospholipid bilayers, a typical single-channel current fluctuation was observed (FIG. 1 A). The conductance of this Ca^{2+}-channel was calculated from I-V curves to be about 25 pS for Ba^{2+} as a charge carrier, and the mean open time was 20 msec. The Ca^{2+} (Ba^{2+}) current through the Ca^{2+}-channel was inhibited by Ca^{2+}-channel blockers such as ω-conotoxin (FIG. 1 B) and $LaCl_3$, and enhanced by BAY K 8644 (data not shown).

Furthermore, Ca^{2+}-channel blockers also inhibited chemotactic behavior of *Bacillus subtilis* toward L-alanine without any effect on either cell growth or motility.[4]

These results suggest that Ca^{2+} enters the cell through a voltage-dependent Ca^{2+} channel as in the excitable membranes of eukaryotes, and they also suggest that internal Ca^{2+} plays an important role in the sensory system of bacterial chemotaxis.

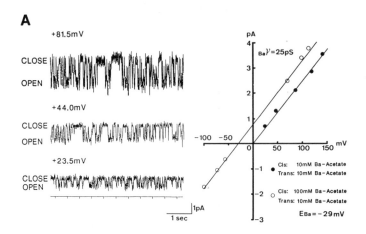

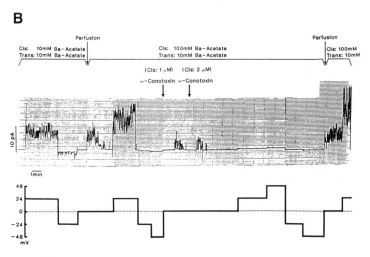

FIGURE 1. (A) Current-voltage relationships of the Ca^{2+} (Ba^{2+}) channel. The highly purified Ca^{2+} channel from *Bacillus subtilis* was reconstituted into a planar phospholipid (asolectin) bilayer by the direct insertion method using β-octylglucoside as a detergent. The left panel shows the single-channel fluctuations with symmetrical barium acetate. The right panel shows the I-V relationships of single-channel current under symmetrical ($\bullet$), or asymmetrical ($\circ$) barium acetate concentrations on the two sides of the bilayer. The single-channel current reversed at the calculated Ba^{2+} reversal potentials, $E_{Ba} = -29$ mV, indicating that the channel was Ba^{2+} (Ca^{2+}) selective.

(B) Inhibition of Ca^{2+} (Ba^{2+}) current by ω-conotoxin. The Ca^{2+} (Ba^{2+}) currents through the Ca^{2+} channel were inhibited by a Ca^{2+} channel blocker, ω-conotoxin. In this experiment, the planar phospholipid bilayer contained multiple Ca^{2+} channels (about ten molecules). The Ca^{2+} (Ba^{2+}) current was recovered by removal of ω-conotoxin performed by a perfusion of the *cis* chamber. The bold lines in the lower section show the holding potential in this experiment.

REFERENCES

1. KUSAKA, I. & T. MATSUSHITA. 1987. J. Gen. Microbiol. **133:** 1337-1342.
2. MATSUSHITA, T., T. UEDA & I. KUSAKA. 1986. Eur. J. Biochem. **156:** 95-100.
3. MONTAL, M. & P. MUELLER. 1972. Proc. Natl. Acad. Sci. USA **69:** 3561-3566.
4. MATSUSHITA, T., H. HIRATA & I. KUSAKA. 1988. FEBS Lett. **236:** 437-440.

Mitogenic Receptor-Regulated Ca Channel in a Tumor Cell Line

YVES CHAPRON,[a] CLAUDE COCHET,[b] SERGE
CROUZY,[a] THIERRY JULLIEN,[c] MICHELLE
KERAMIDAS,[b] AND JEAN VERDETTI[c]

[a] *Laboratoire de Biophysique Moleculaire et Cellulaire*
UA 520 du CNRS, DRF/LBIO, C.E.N.G., 85X
38041 Grenoble cedex, France

[b] *Laboratoire de Biochimie des Regulations Cellulaires Endocrines*
INSERM U244, DRF/LBIO, C.E.N.G., 85X
38041 Grenoble cedex, France

[c] *Laboratoire de Physiologie Cellulaire, UA 632 du CNRS*
Universite Joseph Fourier, BP 68
38402 Saint Martin-d'Heres, France

Extracellular calcium ions are required for the proliferation of a number of cell lines.[1,2] Furthermore, it has been demonstrated that cytosolic free calcium increases in a variety of stimulated transformed and untransformed cells.[3-4] In some cases, the increase in $[Ca]_i$ has been correlated with phosphoinositide breakdown.[5-6] In view of these observations, a study was undertaken to test the putative role of calcium channels and IP3 messenger in carcinomal cells stimulated by a mitogenic agent. We have applied the gigaohm-seal patch-clamp method to carcinomal A 431 cells stimulated by the epidermal growth factor (EGF).[7]

MATERIALS AND METHODS

The A 431 cells were grown in Dubelcco's modified Eagle's medium supplemented with 10% fetal calf serum. Single-channel currents were recorded from cell-attached and excised inside-out patches of membrane using the patch-clamp technique. In the case of cell-attached experiments, the cells were immersed in a Tyrode's solution containing NaCl, 140 mM; KCl, 5mM; MgCl$_2$, 1mM; NaH$_2$PO$_4$, 0.5 mM; CaCl$_2$, 1 mM; glucose, 11 mM; taurine, 10 mM; Hepes, 10 mM, pH 7.4. For inside-out patch studies, the cells were immersed in an intracellular-like solution containing KCl, 150 mM; CaCl$_2$, 0.55 mM; MgCl$_2$, 2 mM; EGTA, 1.1mM; Hepes, 10 mM (pH 7.4); (Ca-free, 10-7 M). For these two situations, the patch pipette was filled with BaCl$_2$, 50 mM; saccharose, 150 mM; Hepes, 10 mM (pH 7.4).

RESULTS

A unitary conductance of 8pS was observed upon single-channel recording. When EGF was added to the bathing solution, bursting multichannel activity occurred (FIG. 1a); FIGURE 1b shows that activation of the channels by EGF is dose dependent. These channels were identified as calcium channels (inhibition by lanthanum, E rev positive). Because the seal of the pipette on the membrane is very tight, the existence of an intracellular messenger between the EGF receptor and the channel was suspected.

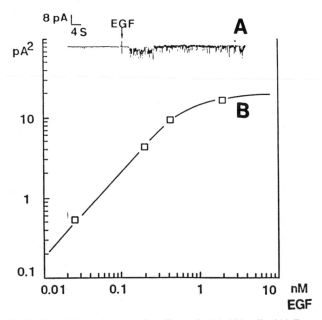

FIGURE 1. Single-channel inward current in cell-attached A 431 cells. (A) Example of EGF (20 nM) channel activation. The membrane is at the resting potential. (B) Effect of dose of EGF on the energy of the inward current.

To determine precisely the putative role of IP3, we have exposed the cytoplasmic side of the membrane to an intracellular solution containing IP3. FIGURE 2a shows that IP3 (2.10-6 M) induces a large inward current. Binomial analysis showed nonindependent behavior, therefore we did the kinetic analysis by means of noise analysis.[7] In this case the values of the single-channel conductance and the corner frequency of the spectral density curves (FIG. 2b) were similar to the values determined for EGF-stimulated cells. In both cases we conclude that kinetic constants are probably not affected, but the number of active Ca channels is increased.

DISCUSSION

We have identified calcium channels that are activated by EGF via IP3. This observation agrees with a report that has shown that EGF induces an increase in IP3 in A 431 cells.[8] The function of these channels remains obscure, but it could be suggested that Ca-channel activation by growth factors may be a step in the mitogenesis induction.

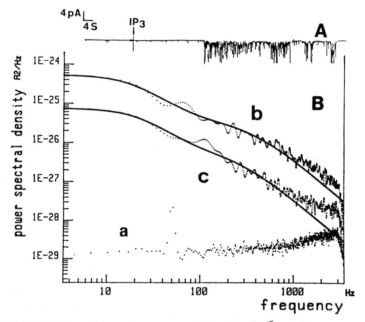

FIGURE 2. Single-channel inward current recorded in inside-out patches excised from A 431 cells. (A) Example of IP3 (2×10^{-7}) channel activation. The membrane was at zero potential. (B) Spectral density curves obtained before any addition (a) and after EGF (20 nM) (b) or IP3 (2×10^{-7}) (c).

REFERENCES

1. HENNINGS, H., D. MICHAEL, C. CHENG, P. STEINERT, K. HOLBROOK & S. H. YUSPA. 1980. Cell **19:** 144-146.
2. METCHALFE, J. C., J. P. MOORE, G. A. SMITH & T. R. HESKETH. 1986. Br. Med. Bull. **42(4):** 405-412.
3. MOOLENAAR, W. H., R. J. AERTS, L. G. J. TERTAKEN & S. W. deLAAT. 1986. J. Biol. Chem. **261:** 279-284.
4. GONZALEZ, F. A., D. J. GROSS, L. A. HEPPEL & W. W. J. WEBB. 1988. Cell Physiol. **135:** 269-276.

5. KUNO, M., J. GORONZY, C. M. WEYAND & P. GARDNER. 1986. Nature **323:** 269-273.
6. KUNO, M. & P. GARDNER. 1987. Nature **326:** 301-304.
7. CHAPRON, Y., *et al.* 1989. Biochem. Biophys. Res. Commun. **158(2):** 527-533.
8. TILLY, B. C., P. A. VAN PARIDON, I. VERLAAN, S. W. DELAAT & W. H. MOOLENAAR. 1988. Biochem. J. **252:** 857-863.

Calcium Agonists and Antagonists Modulate Ischemic Injury via Vascular and Myocardial Mechanisms

F. Th. M. VAN AMSTERDAM, M. HAAS, M. S. VAN
AMSTERDAM-MAGNONI, N. C. PUNT,
AND J. ZAAGSMA

Department of Pharmacology and Therapeutics
University of Groningen
9713 AW Groningen, the Netherlands

Global ischemia has far-reaching consequences for the beating heart. With cessation of the coronary flow, an immediate lack in O_2 and glucose occurs, leading to a reduction of CP and ATP. Furthermore, a deterioration of intracellular Ca^{2+}-resequestration as well as an increase of extracellular potassium may induce increased intracellular Ca^{2+} levels that activate Ca^{2+}-ATPases and accelerate the reduction of ATP levels. In the rat heart, ATP depletion results after about 10 minutes in a highly strengthened interaction between actin and myosin, causing an elevation of the end diastolic pressure.[1-3]

We have investigated the concentration-dependent effects of the Ca antagonists nifedipine (NIF), (+)- and (−)-gallopamil (GAL) and the agonist (−)-BAY K 8644 (BAY) on the end diastolic left ventricular pressure (EDP) in the Langendorff-perfused rat heart subjected to global ischemia. Without drugs the EDP increased from zero to a maximum of 6.18 ± 0.12 kPa after 15.1 ± 1.2 minutes of ischemia, with a concomitant cessation of contractility within one to two minutes. After 30 minutes of ischemia the reperfusion was started, resulting in a quick rise of EDP, followed by a slow decrease to control level. Recovery of EDP with 90% (tR90) takes 90 ± 12 minutes. Contractility usually returned within 10 minutes. NIF, (−)- and (+)-GAL at negative inotropic concentrations (pEC_{50} values: 6.7, 7.4, and 5.6, respectively) were found to (a) delay the time at which EDP was maximal (tEDPm), (b) decrease EDPm itself, and (c) shorten the time of recovery during reperfusion (tR90, FIG. 1). On the other hand, NIF and (+)-GAL at low, vasodilating concentrations (pEC_{50} values 8.4 and 6.7, respectively), were selectively active in reducing the time of recovery at reperfusion, almost without affecting tEDPm and EDPm. The data suggest a close relation between (1) the negative inotropic (energy saving) activity of the drugs and the protection against the rise in EDP during ischemia and (2) the vasodilating activity and the reduction of tR90 during reperfusion. This was confirmed by the observation that GAL showed a high stereoselectivity factor (sf) toward tEDPm and EDPm, but not toward tR90, which parallels a high sf for the inotropic and a low sf for the vasodilating activity of the drug in normoxic hearts.[4] BAY showed opposite activities in that the rise of EDP was accelerated and enhanced, and recovery was retarded at the optimal calcium agonistic concentration of 100 nmol/1. At lower and higher concentrations a complex behavior was observed, which may be explained by considering the balance between the effects of vasoconstriction and positive inotropic activity. In FIGURE 2 it is shown that 10 nmol/1 BAY quickly induces a strong

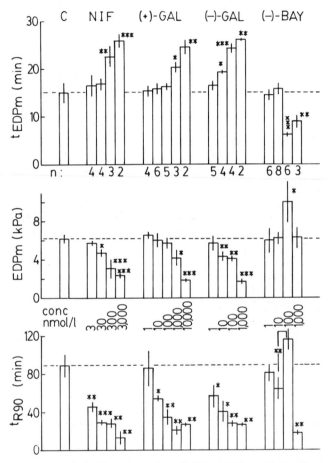

FIGURE 1. Influence of the calcium antagonists nifedipine, (+)- and (−)-gallopamil and the calcium agonist (−)-BAY K 8644 on the ischemic contracture in the rat heart. Drug concentrations and number of experiments (*n*) are indicated in the figure. Control values ± SEM amount to 15.1 ± 1.2 minutes for t_{EDPm} (*n* = 11), 6.18 ± 0.12 kPa for EDPm (*n* = 11) and 89.8 ± 12.0 minutes for t_{R90} (*n* = 7). Significance levels of differences with respect to the control values were tested with the unpaired Student's *t*-test: $p < 0.05$: * $p < 0.01$: ** and $p < 0.001$: ***.

vasoconstriction. The developing negative inotropic activity (probably a result of flow-limited supply of nutrients and O_2) is slowly inverted to a small positive inotropic effect. Complexively, energy expenditure was not increased and the ischemic contracture was not aggravated. At the highest concentration, however, the fast vasoconstriction was overcome, probably via metabolic autoregulation, due to the strong positive inotropic effect of the drug, inducing a strong ischemic contracture, but also protection at reperfusion. A direct vasodilating activity of calcium antagonists and metabolic autoregulation (e.g., release of adenosine) induced by the positive inotropic effect of the agonist may both be a reason for the improved recovery of EDP.

In conclusion, the opposite and (stereo)selective ischemia-modulating activity of these drugs may be explained by their different selectivity toward coronary vascular and myocardial tissue, as well as indirect metabolic factors.

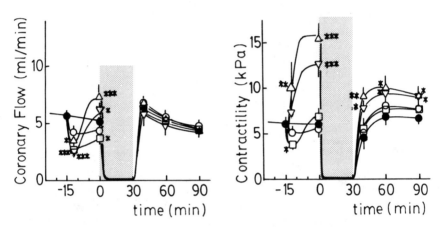

FIGURE 2. Concentration-dependent effects of (−)-BAY K 8644 on coronary flow and contractility (left ventricular max systolic minus end diastolic pressure) before and after 30 minutes of global ischemia (shaded area). The concentrations used are 1 (O), 10 (□), 100 (△), 1000 nmol/1 (▽); Control flow: (●). Significance levels of differences with respect to control values are as indicated in FIGURE 1.

REFERENCES

1. VARY, T. C., D. K. REIBEL & J. R. NEELY. 1981. Ann. Rev. Physiol. **43:** 419-430.
2. NAYLER, W. G., S. PANAGIOTOPOULOS, J. S. ELZ & W. J. STURROCK. 1987. Am. J. Cardiol. **59:** 75B-83B.
3. VAN AMSTERDAM, F. TH. M., N. C. PUNT, M. HAAS & J. ZAAGSMA. 1988. Submitted.
4. VAN AMSTERDAM, F. TH. M. & J. ZAAGSMA. 1988. Naunyn-Schmiedeberg's. Arch. Pharmacol. **337:** 213-219.

Intracellular Ca Measured with Indo-1 in Single Smooth Muscle Cells from Guinea Pig Ureter Held under Voltage Clamp

C. D. BENHAM AND P. I. AARONSON

Department of Pharmacology
Smith Kline & French Research Ltd.
The Frythe, Welwyn, Herts, and
St. George's Hospital Medical School
London SW 17 ORE, United Kingdom

The action potential of the guinea pig ureter differs from that of other smooth muscles inasmuch as it displays a plateau phase resembling that observed in cardiac ventricular cells. Because the plateau phase is attenuated by procedures that reduce the transmembrane Na gradient,[1] it is possible that the plateau is mediated by an electrogenic Na/Ca exchange, especially as Aickin *et al.* have recently demonstrated the presence of Na/Ca exchange in this tissue.[2]

We have further characterized the processes underlying the action potential in this tissue using the whole-cell patch-clamp technique and microspectrofluorimetry to measure Na/Ca exchange and voltage-gated Ca channel-mediated currents. For microspectrofluorimetry the Ca-sensitive dye Indo-1 (100 μM) was introduced into the cells through the patch pipette. Intracellular [Na] was set at 11 mM, and 2 mM ATP was added to the pipette solution. The intracellular free-Ca concentration (Ca_i) was calculated according to the ratio method.[3]

For measurement of Na/Ca exchange, nifedipine (10 μM) was added to the solution to abolish the voltage-gated Ca current. Under these conditions depolarizations of greater than five seconds evoked a slow rise in Ca_i and an associated outward current that became progressively larger and faster with increasing depolarization even beyond the reversal potential for voltage-gated Ca currents (FIG. 1), consistent with Na/Ca exchange operating in reverse mode. For very long depolarizations, the rise in Ca_i often reached a plateau, as did the outward current. Reduction of Na_o elicited an elevation of Ca_i at the resting membrane potential and potentiated the increase in Ca_i in response to depolarization.

In the absence of nifedipine, short (100–500 msec) depolarizations evoked a transient rise in Ca_i with the same bell-shaped voltage dependence as voltage-gated Ca currents. In the absence of Na_o, the recovery phase of these Ca transients was prolonged, suggesting that Na/Ca exchange is partly responsible for extruding Ca after such a Ca load. Electrogenic Ca extrusion following Ca entry during the upstroke of the action potential would result in an inward current that might explain the Na-dependent plateau phase of the action potential.

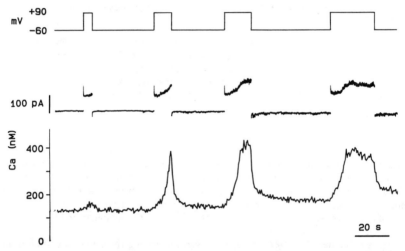

FIGURE 1. Na/Ca exchange-mediated membrane current (middle trace) and increases in Ca_i (bottom trace) during depolarization of guinea pig ureter cells. Cells were held at -60 mV and stepped for 5-25 seconds to $+90$ mV (upper trace) in K-free physiological saline solution in which NaCl was reduced to 67 mM by partial substitution with LiCl. Nifedipine (10 μM) was present to block voltage-gated Ca currents. Stepping to $+90$ mV resulted at first in slow rises in Ca_i and outward current (5-sec step); the rate of rise of both parameters then increased over 5-10 sec (10-sec step). With longer (15, 25 sec) steps, both parameters became nearly constant. These data suggest that increases in Ca_i cause an increase in the rate of Ca influx via "reverse-mode" Na/Ca exchange, leading to an acceleration of exchange activity. As the Ca gradient becomes increasingly favorable for Ca extrusion, "forward-mode" exchange increases, resulting in a new steady state.

The voltage-gated Ca current was studied at a holding potential of −60 mV. The current showed little inactivation for up to two seconds at potentials negative to −10 mV; some inactivation occurred on stronger depolarization (FIG. 2).

The presence of Na/Ca exchange in these cells, as well as shortening of the action potential duration by Na-gradient reduction,[1] suggest that it is Na/Ca exchange, rather than the presence of a sustained potential-dependent Ca current, that underlies the action potential plateau. In this case, the plateau would represent the removal of Ca from the cell via electrogenic Na/Ca exchange, rather than Ca influx through voltage-gated channels.

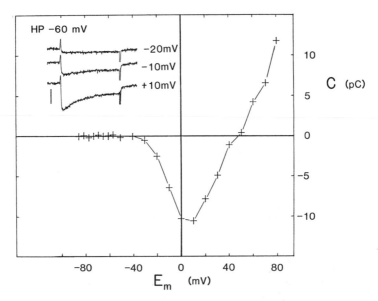

FIGURE 2. Voltage-gated Ca currents in control physiological saline solution, measured in the absence of intracellular NaCl in order to minimize Na/Ca exchange. The cell was held at −60 mV and stepped to more positive or negative potentials for 500 msec. The ordinate represents the total charge crossing the membrane over 500 msec. Comparison of this value with typical increases in free Ca_i obtained under similar conditions indicated that less than 0.1% of the Ca entering the cells contributed to the rise in Ca_i. Representative currents elicited at −20, −10, and +10 mV are shown; the vertical line represents 20 pA.

REFERENCES

1. SHUBA, M. F. 1977. J. Physiol. **264:** 837-851.
2. AICKIN, C. C., A. F. BRADING & T. V. BURDYGA. 1984. J. Physiol. **347:** 411-430.
3. GRYNKIEWICZ, G., M. POENIE & R. Y. TSIEN. 1985. J. Biol. Chem. **260:** 3440-3450.

A Comparison of the Actions of BAY K-8644 and Angiotensin II on Extracellular Calcium-Dependent Contractile Responses Mediated by Postjunctional α_2-Adrenoceptors

C. J. DALY, W. R. DUNN, J. C. McGRATH, AND
V. G. WILSON

Autonomic Physiology Unit
Institute of Physiology
University of Glasgow
Glasgow G12 8QQ Scotland

Pressor responses mediated by stimulation of postjunctional α_2-adrenoceptors in the pithed rat are inhibited by both calcium antagonists (e.g. nifedipine) and angiotensin-converting enzyme inhibitors (e.g. teprotide).[1] Because calcium antagonists do not affect resting blood pressure in the pithed rat, unlike angiotensin-converting enzyme inhibitors, these two groups of agents appear to produce inhibition by different mechanisms. To investigate this further, we have examined the effects of agents that potentiate the relevant systems, angiotensin II (AII) and the calcium-channel facilitator BAY K-8644, against α_2-adrenoceptor-mediated contractile responses in the rabbit isolated saphenous vein (RSV).

METHODS

Three- to four-millimeter segments of the RSV were prepared for isometric tension recording as previously described.[2] Three experiments were performed. (1) Following completion of a control cumulative concentration-response curve (CRC) to noradrenaline (NA), preparations were exposed to either saline, or 2.5 μM rauwolscine five minutes before the addition of 0.3 μM phenoxybenzamine (RP-treated). After 30 minutes the preparations were washed seven times with physiological salt solution (PSS) over 45 minutes and the NA CRC repeated. The NA CRC was repeated again 40 minutes later in the presence of either 0.1 μM prazosin, 2.5 μM rauwolscine, 30 nM AII, or 30 nM BAY K-8644. (2) Maximal contractile responses to NA in normal and RP-treated preparations were elicited 10 minutes after exposure to Ca^{2+}-buffered PSS containing 0.25 mM Ca^{2+} and 0.5 mM EGTA. (3) Ca^{2+} CRCs were constructed following exposure to nominally Ca^{2+}-free PSS with 65 mM KCl (exchanged with

NaCl on an equimolar basis). This was repeated in the presence of either 30 nM AII or 30 nM BAY K-8644. The negative logarithm of the molar concentration of Ca^{2+} required to cause 50% of the maximum contraction (pD_2) was determined and differences between mean values ($p < 0.05$; paired Student t-test) considered statistically significant.

RESULTS

In normal preparations, 0.1 μM prazosin produced a 10-fold parallel displacement of the NA CRC and 2.5 μM rauwolscine produced a nonparallel displacement (FIG. 1a). This suggests the presence of both α_1- and α_2-adrenoceptor subtypes. Following

a

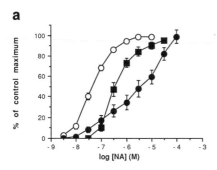

FIGURE 1. The effects of prazosin and rauwolscine on noradrenaline-induced contractions in the rabbit isolated saphenous vein before (**a**) and after (**b**) isolation of the postjunctional α_2-adrenoceptor subtype. (**a**) Responses to noradrenaline in untreated preparations in the absence (○) and in the presence of either 0.1 μM prazosin (■) or 2.5 μM rauwolscine (●). (**b**) Responses to noradrenaline before (○) and after exposure to 0.3 μM phenoxybenzamine and 2.5 μM rauwolscine followed by seven washes over 45 minutes (□). The post phenoxybenzamine-rauwolscine responses to noradrenaline were then repeated in the presence of 0.1 μM prazosin (■) and 2.5 μM rauwolscine (●). All responses have been expressed as a percentage of the maximum response to NA in the initial control concentration-response curve and are shown as the mean of six to eight observations ± the standard error of the mean (vertical bars).

b

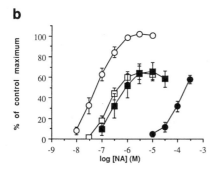

RP treatment, the maximum responses to NA were reduced by approximately 40%, 0.1 μM prazosin produced less than a twofold displacement of the NA CRC and 2.5 μM rauwolscine caused a 100-fold parallel displacement (FIG. 1b)—thus isolation of postjunctional α_2-adrenoceptors was achieved. In Ca^{2+}-buffered PSS, contractile responses to NA in normal and RP-treated preparations were transient, returning to baseline within 10 minutes and peak responses were 27.5 ± 2.0% ($n = 8$) and 7.7 ± 1.9% ($n = 8$), respectively, of the responses in PSS. The transient response to NA in normal preparations was prazosin-sensitive, rauwolscine-resistant (α_1-mediated). In view of the small size of the transient response to NA in RP-treated preparations, no attempt was made to determine the subtype involved. Following selective isolation of α_2-adrenoceptors by RP-treatment, responses to NA were potentiated by both 30 nM AII and 30 nM BAY K-8644 (two- to threefold leftward displacement of the CRC and an increase in the maximum response; see FIGS. 2a and 2b). In nominally Ca^{2+}-free, high-K^+ PSS, a small contraction was evident in most

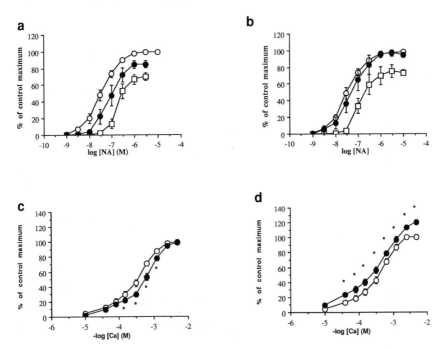

FIGURE 2. The effects of angiotensin II and BAY K-8644 on α_2-adrenoceptor-mediated contractions and Ca^{2+}-induced contractions in the rabbit isolated saphenous vein. (**a and b**) Responses to NA following isolation of the α_2-adrenoceptor-mediated component with phenoxybenzamine/rauwolscine ($\square$) were repeated in the presence of (**a**) 30 nM angiotensin II ($\bullet$) and (**b**) 30 nM BAY K-8644 ($\bullet$). All responses have been expressed as a percentage of the maximum responses to NA before the isolation procedure ($\bigcirc$) and are shown as the mean of six observations ± the standard error of the mean (vertical bars). (**c and d**) Ca^{2+}-induced contractions (following exposure to Ca^{2+}-free, high-K^+ PSS) before ($\bigcirc$) and after exposure to (**c**) 30 nM angiotensin II ($\bullet$) and (**d**) 30 nM BAY K-8644 ($\bullet$). All responses have been expressed as a percentage of the maximum response to Ca^{2+} in the initial concentration-response curve and are shown as the mean of eight to nine observations ± the standard error of the mean. *-denotes a significant difference ($p < 0.05$) between Ca^{2+} CRCs in the presence and absence of either 30 nM angiotensin II or 30 nM BAY K-8644.

preparations (between 5-10% of the maximum responses to the readdition of 2.5 mM Ca^{2+}), and consecutive Ca^{2+} CRCs were superimposable (pD$_2$ for Ca^{2+}—3.49 ± 0.07 and 3.48 ± 0.08, $n = 10$; for 1st and 2nd curves, respectively). 30 nM AII caused a significant rightward displacement of the Ca^{2+} CRC (pD$_2$—3.49 ± 0.06 and 3.29 ± 0.07, $n = 9$; $p < 0.005$. See also FIG. 2c) without altering the maximum response. In marked contrast, 30 nM BAY K-8644 failed to alter the sensitivity of the depolarized preparation to Ca^{2+} (pD_2—3.47 ± 0.06 and 3.46 ± 0.06, $n = 8$) but significantly increased the contractile response to all concentrations of Ca^{2+} (FIG. 2d).

CONCLUSIONS

In conclusion, α_2-adrenoceptor-mediated contractions in the RSV are dependent upon the influx of Ca^{2+} through an ion channel. Although both BAY K-8644 and AII potentiate these responses, this appears to be the result of different mechanisms; the action of AII, unlike that of BAY K-8644, does not involve an increase in the function of voltage-operated Ca^{2+} channels. These observations are consistent with the known differences in the action of calcium antagonists and angiotensin-converting inhibitors *in vivo*.[1] However, the basis of the small inhibitory effect of AII on voltage-operated Ca^{2+} channels is not presently known.

REFERENCES

1. O'BRIEN *et al.* 1985. Clin Sci. **68**(Suppl 10): 99s-104s.
2. DALY *et al.* 1987. Br. J. Pharmacol. **92**: 775P.

Modulation of the α_1-Adrenoceptor Response in Taenia Caeci of the Guinea Pig

A. DEN HERTOG, S. A. NELEMANS, J. VAN DEN
AKKER, AND A. MOLLEMAN

Department of Pharmacology and Clinical Pharmacology
University of Groningen
9713 BZ Groningen, the Netherlands

Activation of α_1-adrenoceptors or P_2-purinoceptors in smooth muscle cells of guinea pig taenia caeci causes mobilization of calcium, probably from a plasma membrane-bound store and subsequent replenishment of this store with calcium from the extracellular space.[1] Calcium mobilization is linked with the opening of potassium channels and possibly also coupled with activation of calcium-dependent calcium channels. This study was carried out in the absence of external calcium to exclude a contribution of calcium-induced calcium currents. The process of calcium mobilization and replenishment is represented by membrane depolarization if the potassium channels are blocked by apamin. Activation of potassium channels is accompanied by potassium efflux and, accordingly, hyperpolarization of the smooth muscle cells. The transient nature of the response in the absence of calcium is thought to be due to the limited amount of mobilizable calcium. Stimulation of α_1-adrenoceptors also initiates activation of the phosphatidylinositol metabolism and, as a consequence, calcium release from the endoplasmic reticulum, resulting in enhancement of intracellular calcium.[2] A model illustrating these processes is presented (FIG. 1).

Thus, the α_1-adrenoceptor effect can be modified by interfering with the calcium mobilization or with the replenishment of the calcium store. This calcium movement is visualized by the α_1-agonist-induced transient depolarization monitored during blockade of the apamin-sensitive potassium channels in the absence of calcium. The potassium-channel characteristics are represented by the transient hyperpolarization observed in the absence of calcium and apamin. The results showed that both the transient hyperpolarization and depolarization, evoked by an α_1-agonist (adrenaline; 3×10^{-6} M in the presence of propranolol, 3×10^{-6} M and yohimbine, 3×10^{-6} M) in, respectively, the absence of external calcium and also after blocking the potassium channels with apamin, are inhibited by procainamide (5 mM). Procainamide obviously inhibits calcium mobilization activated by α_1-adrenoceptor stimulation.[3] In contrast, mebeverine (6×10^{-5} M) did not affect mobilization, but inhibited the process of replenishment of the calcium store.[4]

Polarization of the muscle cells was obtained by applying a constant current or by raising the external potassium concentration. The amplitude of the α_1-adrenoceptor-evoked transient hyperpolarization was decreased during current-induced hyperpolarization and enhanced during depolarization (FIG. 2A). In contrast, the α_1-agonist-elicited transient depolarization was enhanced during current-induced hyperpolarization and diminished during depolarization. These results are in agreement with the assumption that the amplitude of the transient hyperpolarization is limited by the potassium equilibrium potential and the transient depolarization by the calcium equilibrium potential. Polarization of the cell membrane by raising the external potassium concentration reduced the α_1-agonist-evoked potassium current carrying the hyperpolarization (FIG. 2C). The reduction in potassium current can be explained by the

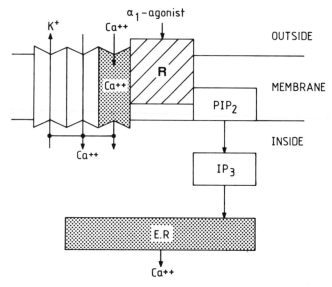

FIGURE 1. Processes activated during α_1-adrenoceptor stimulation in smooth muscle cells of taenia caeci. The following processes are assumed to take place. (1) Stimulation of the α_1-adrenoceptor (R) activates mobilization of calcium from a plasma membrane-bound store and facilitates the formation of inositol trisphosphate (IP$_3$) from the phosphatidylinositol bisphosphate (PIP$_2$) pool. (2) Calcium movement toward the cytoplasm is subsequently followed by replenishment of the calcium store from the extracellular space. Mobilization of calcium is coupled with the opening of potassium channels, causing potassium efflux and is linked with activation of calcium-dependent calcium channels, the latter being nonfunctional in the absence of external calcium. (3) IP$_3$ facilitates the release of calcium from the endoplasmic reticulum (ER).

changed potassium gradient. The α_1-agonist-activated calcium current, carrying the transient depolarization, as observed in the presence of apamin, was not changed in the presence of high potassium (FIG. 2D).

The results indicate that the α_1-adrenoceptor effect on smooth muscle of the guinea pig taenia caeci can be modified at the level of calcium mobilization and replenishment. Polarization, however, has no effect on this calcium movement nor on the apamin-sensitive potassium channels.

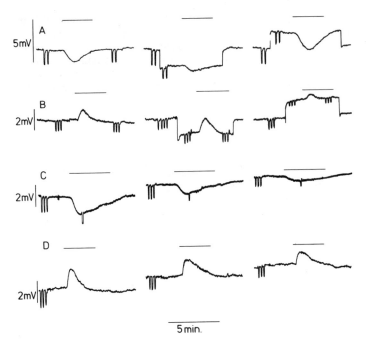

FIGURE 2. The effect of polarization on the α_1-adrenoceptor-induced response by adrenaline ($3 \times 10^{-6}\ M$) in the presence of propranolol ($3 \times 10^{-6}\ M$) and yohimbine ($3 \times 10^{-6}\ M$) in smooth muscle cells of guinea pig taenia caeci measured using the double sucrose-gap method at 22°C.

(**A**) The response evoked in calcium-free solution (15 min) represented by a transient hyperpolarization, also during external hyperpolarization and depolarization (right-hand trace) obtained by applying a constant current ($5 \times 10^{-7}\ A$) to the tissue. The short-lasting downward deflections represent electrotonic potentials elicited by a constant current (intensity: 3×10^{-7} A; duration: 5 sec). (**B**) In the presence of apamin ($3 \times 10^{-7}\ M$, 20 min) following the same procedure as in A, showing an α_1-adrenoceptor-induced transient depolarization. (**C**) The α_1-adrenoceptor-induced hyperpolarization in the presence of (right to left) 5.9 mM, 12.5, and 20 mM potassium. (**D**) In the presence of apamin, following the same procedure as in C, showing a transient depolarization.

REFERENCES

1. DEN HERTOG, A. 1982. Calcium and the action of adrenaline, adenosine triphosphate and carbachol on guinea-pig taenia caeci. J. Physiol. **325:** 423.
2. NELEMANS, A. & A. DEN HERTOG. 1987. Calcium translocation during activation of α_1-adrenoceptor and voltage-operated channels in smooth muscle cells. Eur. J. Pharmacol. **140:** 39.
3. DEN HERTOG, A. & J. VAN DEN AKKER. 1987. The action of procainamide and quinidine on the α_1-receptor-operated channels in smooth muscle cells of guinea-pig taenia caeci. Eur. J. Pharmacol. **137:** 233.
4. DEN HERTOG, A. & J. VAN DEN AKKER. 1987. Modification of α_1-receptor-operated channels by mebeverine in smooth muscle cells of guinea-pig taenia caeci. Eur. J. Pharmacol. **138:** 367.

Calcium and Human Large-Bowel Motility

D. J. BUTLER AND K. HILLIER

Clinical Pharmacology Group, Medical Faculty
University of Southampton
Southampton SO9 3TU, United Kingdom

Calcium ions are central to excitation-contraction coupling in smooth muscle, the importance of intracellular and extracellular sources depending upon tissue and stimulant studied.[1] Extracellular Ca^{2+} influx occurs principally via potential-dependent or receptor-operated channels that may be closely associated.[2] Ca^{2+} antagonists have significantly aided channel differentiation by preferentially inhibiting potential-dependent Ca^{2+} influx,[3,4] but there is wide variation in responsiveness between tissues. Little is known about the control of calcium mobilization in human colonic smooth muscle and its importance to normal or disordered motility.

METHODS AND RESULTS

In tension studies, KCl (20-100 mM) induced concentration-dependent, slow developing contractions with phasic and tonic components generally apparent. Ca^{2+}-free/EGTA Krebs rapidly abolished responses. Nitrendipine, verapamil and diltiazem (10^{-10}-$10^{-6}M$) concentration-dependently reduced the phasic component of submaximal KCl responses (FIG. 1). Inhibition of tonic activity by Ca^{2+} antagonists (not shown) was similar. Nitrendipine was most strongly inhibitory, as illustrated by the IC_{50} values. Inhibition by verapamil and diltiazem was reversible after 60-90 minutes washing. Nitrendipine, however, showed greater persistence and repeated washing (60-150 min) resulted in little return of tissue responsiveness. TMB8 and W7 (10^{-5}-$10^{-3}M$), putative calmodulin antagonists, caused concentration-dependent suppression of KCl-responses without phasic or tonic discrimination. The IC_{50} of TMB8 was $4.0 \times 10^{-5}M$ and that of W7 was 1.5×10^{-5} M.

Acetylcholine induced a rapid response with contraction and relaxation spikes superimposed on an elevated hypertonic baseline. Nitrendipine, verapamil and diltiazem (10^{-8}-$5 \times 10^{-6}M$) selectively inhibited spike activity without reducing sustained tone. Spontaneous contractile activity was also inhibited by nitrendipine (10^{-8}-$10^{-6}M$).

$^{45}Ca^{2+}$-efflux studies were undertaken in parallel. Unstimulated efflux was biphasic with fast extracellular washout followed by slow removal of intracellular Ca^{2+}. KCl (60-100 mM) caused an increase in the rate coefficient for slow $^{45}Ca^{2+}$-efflux (FIG. 2). Ca^{2+}-free/EGTA Krebs or 20 mM La^{3+}-Krebs as perfusing media completely blocked KCl-stimulated efflux. Ca^{2+} antagonists ($10^{-8} - 5 \times 10^{-6}M$), despite 30

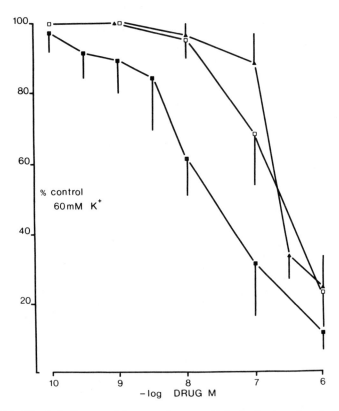

FIGURE 1. Histologically normal human colon was obtained after large-bowel resection for carcinoma. Mucosa was removed and intertaenial muscle strips were suspended in 2 mM Ca^{2+} Krebs at 37°C and gassed with 5% CO$_2$ in oxygen. Isometric tension of 1g was applied and strips left for two hours' equilibration. Submaximal concentrations of KCl (60 mM) were applied and the phasic (one min after KCl) component of tension development measured. Nitrendipine (■), diltiazem (▲), and verapamil (□) inhibited contractility with IC$_{50}$s for the phasic component of 3.0 × 10^{-8}M, 3.0 × 10^{-7}M and 2.0 × 10^{-7}M, respectively. Inhibition of tonic activity by Ca^{2+} antagonists (4 min after KCl) was similar.

minutes' perfusion in normal Krebs, did not inhibit KCl-induced increase in ^{45}Ca overflow.

CONCLUSIONS

Phasic and tonic components of KCl-induced tension development in human colonic muscle depend upon extracellular Ca^{2+} influx. The importance of potential-dependent channels for Ca^{2+} entry is suggested by the inhibitory effects of nitrendipine, verapamil and diltiazem. Furthermore, inhibition by TMB8 and W7 indicates a possible intracellular Ca^{2+} component to the KCl response. Selective inhibition of acetylcholine-induced contractions by Ca^{2+} antagonists suggests that only the rapid spiking activity of the response may involve Ca^{2+} influx. Hypertonic elevation could be due to intracellular Ca^{2+} mobilization or influx via receptor-operated channels less susceptible to Ca^{2+} antagonist blockade.

KCl initiates a rise in myoplasmic free $[Ca^{2+}]$ as evidenced by $^{45}Ca^{2+}$-efflux experiments. Dependency upon Ca^{2+} influx is again apparent. Reduction of KCl-induced mechanical responses by Ca^{2+} antagonists, but failure to inhibit KCl-induced free $[Ca^{2+}]_i$ rise during efflux, is under investigation.

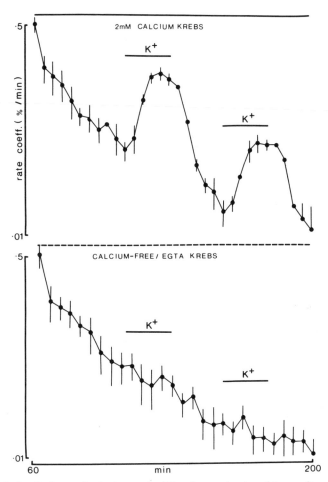

FIGURE 2. Small pieces of colonic muscle (20 mg) were incubated for one hour in $^{45}Ca^{2+}$ Krebs (10 μCi ml^{-1}) containing 2 mM Ca^{2+}. They were then perfused at a rate of 2.5 ml min^{-1} at 37°C with Krebs containing 2 mM Ca^{2+} (top panel) or Ca^{2+}-free EGTA Krebs (bottom panel). Five-minute aliquots were collected and counted for $^{45}Ca^{2+}$ by liquid scintillation techniques. KCL (80 mM) increased the rate coefficient of $^{45}Ca^{2+}$ efflux in Ca^{2+}-containing Krebs (top panel), but in Ca^{2+}-free EGTA Krebs (bottom) KCL caused no increase in $^{45}Ca^{2+}$ overflow.

ACKNOWLEDGMENTS

We acknowledge support from SERC and Pfizer Central Research, United Kingdom.

REFERENCES

1. KARAKI, H. & G. B. WEISS. 1988. Life Sci. **42:** 111-122.
2. BOLTON, T. B. 1979. Physiol. Rev. **59:** 606-718.
3. NAYLER, W. G. & PH. POOLE-WILSON. 1981. Basic Res. Cardiol. **76:** 1-15.
4. SCHWARTZ, A. & D. J. TRIGGLE. 1984. Ann. Rev. Med. **35:** 325-339.

Regulation of Receptor-Operated and Depolarization-Operated Ca^{2+} Influx in Anterior Pituitary Tissue

RORY MITCHELL, MELANIE JOHNSON, SALLY-ANN
OGIER, AND LYNNE DOUGAN

MRC Brain Metabolism Unit
University Department of Pharmacology
Edinburgh EH8 9JZ, United Kingdom

Calcium-mobilizing receptors appear to induce calcium movements from both intracellular and extracellular sites.[1] The mechanism of the Ca^{2+} influx component remains enigmatic, being probably independent from voltage-sensitive Ca^{2+} channels because it occurs in both excitable and nonexcitable cell types. Several hypotheses for the mechanism have been proposed, including actions of inositol tetrakis- and tris-phosphates, Ca^{2+}-activated Ca^{2+} channels, and the Na$^+$/Ca^{2+} exchanger.[1,2] One further possibility is that the concomitant activation of protein kinase C (PKC) by such receptors may have a role in signaling Ca^{2+} influx.

Luteinizing hormone-releasing hormone (LHRH) and thyrotrophin-releasing hormone (TRH) both act through this class of receptor, causing calcium mobilization and phosphoinositide hydrolysis. The Ca-influx processes activated by these hormone receptors in anterior pituitary tissue can be measured by a rapid ^{45}Ca^{2+} influx assay, involving quenching, filtration, and washing with EGTA-containing medium.[3] The ^{45}Ca^{2+} influx induced by LHRH, TRH, or K$^+$ was maximal by 30 seconds and was concentration dependent, with peak increases over basal ^{45}Ca^{2+} accumulation of 83 ± 8% (100 nM LHRH), 162 ± 14% (300 nM TRH), and 232 ± 30% (60 mM K$^+$). The response to K$^+$ was blocked with nanomolar potency by the antagonist of L-type Ca^{2+} channels, nimodipine (FIG. 1), as was some 70% of the influx caused by LHRH. The response to TRH, however, was unaffected until much higher concentrations of nimodipine. Nevertheless, it has been reported that in pituitary cells LHRH does not cause membrane voltage changes or spiking sufficient to activate L-type channels, whereas TRH results in small depolarizations and sometimes increased spiking.[4] It therefore appears that LHRH can activate L-type Ca^{2+} channels by some means without membrane depolarization and that LHRH and TRH cause Ca^{2+} influx through largely different routes.

Activation of PKC (FIG. 2) greatly potentiated K$^+$-induced Ca^{2+} influx, whereas the response to TRH was unaffected and that to LHRH was profoundly inhibited. This suggests first that LHRH and TRH do not cause their ^{45}Ca^{2+} influx (through L-type and non-L-type channels, respectively) by means of membrane depolarization and second that TRH-induced and LHRH-induced Ca^{2+} influx are not regulated identically.

Blockers of the Na$^+$/Ca^{2+} exchanger[5] amiloride (30-4000 μM) or 2',-4'-dimethyl benzamil (30 μM) failed to specifically reduce responses to LHRH or TRH. Interestingly, such concentrations of amiloride also block the T-type Ca^{2+} channel[6] and the Na^+/H^+ exchanger, which has been reported to be required for thrombin-induced Ca^{2+} influx in platelets.[7]

In conclusion, LHRH-induced Ca^{2+} influx mainly involves an L-type Ca^{2+} channel, which appears to be activated by some means other than by depolarization. In contrast, TRH (although acting through a superficially similar receptor type) induces Ca^{2+} influx through a quite different route. Neither PKC, the Na^+/Ca^{2+} exchanger, the Na^+/H^+ exchanger, nor T-type Ca^{2+} channels seem to be directly involved in these two forms of receptor-operated Ca^{2+} influx. Nevertheless, PKC displays an important regulatory role in pituitary cells by differentially modulating Ca^{2+} entry occurring in response to these hormone receptors and to depolarization.

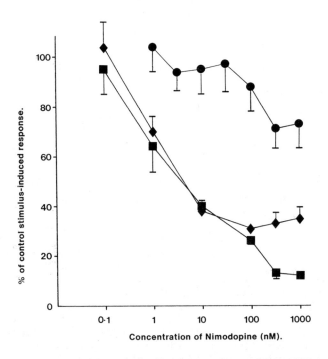

FIGURE 1. Effects of nimodipine on $^{45}Ca^{2+}$ influx induced by LHRH, TRH, or K^+. Values are the mean ± SEM from 6-12 experiments and are expressed relative to control stimulus-induced influx, which was calculated as percentage increase over basal $^{45}Ca^{2+}$ accumulation. Approximately half of this basal accumulation was accounted for by adsorption to the filters (~2000 dpm/sample). Tissue-specific $^{45}Ca^{2+}$ accumulation in the absence and presence of 100 nM LHRH was 1.43 ± 0.20 and 2.66 ± 0.31 pmol/pituitary/min, respectively. All experiments were carried out under subdued lighting. ◆ 30 nM LHRH; ● 30 nM TRH; ■ 60 mM K^+. Effects of nimodipine were statistically significant ($p < 0.05$, Mann-Whitney U-test) at ≥ 1nM for LHRH and K^+ and at ≥ 300 nM for TRH. Nimodipine (1-1000 nM) had no effect on basal $^{45}Ca^{2+}$ accumulation.

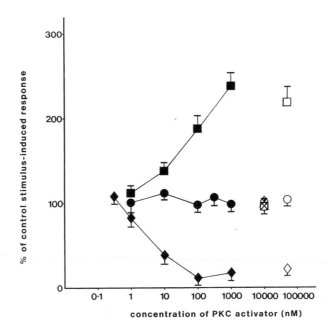

FIGURE 2. Effects of PKC activators on $^{45}Ca^{2+}$ influx induced by LHRH, TRH, or K^+. Values are the means $\pm$ SEM from 6-12 experiments and are expressed relative to control stimulus-induced influx, which was calculated as percentage increase over basal $^{45}Ca^{2+}$ accumulation. ◆ 100 nM LHRH; ● 100 nM TRH; ■ 60 mM K^+. Solid symbols show data with phorbol 12-myristate, 13-acetate (PMA); crossed symbols show data with 4-β phorbol (an inactive analogue) and open symbols with 1-oleoyl 2-acetyl *rac* glycerol (OAG). Effects of PMA at concentrations $\geq$ 10 nM and of OAG were statistically significant ($p < 0.05$, Mann-Whitney U-test) on responses to LHRH and K^+ but not TRH. None of the PKC activators had any effect on basal $^{45}Ca^{2+}$ accumulation at the concentrations tested. The effects of 100 nM PMA on LHRH responses were completely prevented in the presence of the PKC inhibitor H7 (1-(5-isoquinoline-sulphonyl)-2-methyl piperazine hydrochloride) at a concentration of 10 μM. Control responses or basal $^{45}Ca^{2+}$ accumulation were unaffected by H7.

REFERENCES

1. PUTNEY, J. W. 1986. Cell Calcium **7**: 1-12.
2. NEHER, E. 1987. Nature **326**: 242.
3. MITCHELL, R., N. MINAUR, M. JOHNSON, S.-A. OGIER & G. FINK. 1986. Biochem. Soc. Trans. **15**: 139.
4. MASON, W. T., R. J. BICKNELL, P. CORBETT, D. W. WARING & C. D. INGRAM. 1986. *In* Neuroendocrine Molecular Biology. G. Fink, A. J. Harmar & K. W. McKerns, Eds.: 379-392. Plenum. New York.
5. KACZOROWSKI, G. J., F. BARROS, J. K. DETHMERS, M. J. TRUMBLE & E. J. CRAGOE, JR. 1985. Biochemistry **24**: 1394-1403.
6. TANG, C.-M., F. PRESSER & M. MORAD. 1988. Science **240**: 213-215.
7. SIFFERT, W. & J. W. N. AKKERMAN. 1987. Nature **325**: 456-458.

Development of Excitatory Amino Acid Dependent $^{45}Ca^{2+}$ Uptake in Cultured Cerebral Cortex Neurons

AASE FRANDSEN,[a] JØRGEN DREJER,[b] AND
ARNE SCHOUSBOE[a]

[a] *PharmaBiotec Research Center*
Department of Biochemistry A
Panum Institute
University of Copenhagen
DK-2200 Copenhagen, Denmark

[b] *Research Division*
Ferrosan A/S
DK-2860 Søborg, Denmark

Primary cultures of cerebral cortex neurons provide a reliable model system for the investigation of individual responses to stimuli of a given type of neuron,[1] for example, excitatory amino acid (EAA) induced Ca^{2+} uptake. Thus, Ca^{2+} uptake induced by K^+ or the EAAs glutamate (glu), aspartate (asp), N-methyl-D-aspartate (NMDA), kainate (KA), quisqualate (QA) and RR-α-amino-3-hydroxy-5-methyl-4-isoxazolo-propionate (AMPA) was characterized in cultured cerebral cortex neurons as a function of development. Ca^{2+} uptake into the neurons could be stimulated by depolarizing concentrations of KCl (55 mM) or glu at four days in culture (d.i.c.), whereas responses to the other EAAs developed later. At five d.i.c. the neurons became sensitive to KA followed by development of sensitivity to QA at six d.i.c. and finally at seven d.i.c. by sensitivity to NMDA, AMPA, and asp. This developmental profile correlates with that of EAA-induced transmitter release from similar neurons.[2] As summarized in TABLE 1, the ED_{50} for the stimulation of Ca^{2+} uptake depended on the agonist as well as on the developmental stage of the neurons. The rank order of potency was in mature neurons (12 d.i.c.) glu = asp > AMPA = NMDA > QA > KA. This rank order was slightly different at earlier stages of development consistent with the difference in sensitivity to the various EAAs observed during the culture period. The maximal responses (not shown) to all EAAs tested as well as to K^+ increased as a function of the culture period. With the exception of QA at 12 d.i.c., K^+-stimulated Ca^{2+} uptake was always higher than the corresponding EAA-induced uptake. With regard to maximal responses to EAAs, QA consistently yielded a considerably higher Ca^{2+} uptake (11.4 nmol/min $\times$ mg) than the other EAAs (6-8 nmol/min $\times$ mg). Because it is assumed that QA and AMPA act at the same receptor site,[3] it was surprising that QA and AMPA differed with respect to stimulation of Ca^{2+} uptake. It should, however, be noted that the sensitivity of the neurons to QA developed at earlier stages than that to AMPA, again suggesting a difference in specificity of the two agonists. Furthermore, recent kinetic studies of QA receptors have suggested that multiple QA receptors exist, exhibiting different sensitivities to AMPA.[4] The NMDA-stimulated uptake, but not that of KA, QA, or AMPA, could be antagonized by Mg^{2+} (IC_{50} = 0.6 mM). IC_{50} values of the Ca^{2+} antagonists Nifedipine, Verapamil, Diltiazem, or Flunarizin for inhibition of glu-stimulated (20 μM) Ca^{2+} uptake in nine-day-old cultures were all > 30 μg/ml. This result, combined with the observation that the profile of K^+-stimulated uptake differed from that of glu-induced Ca^{2+} uptake, indicates that the majority of the Ca^{2+} channels mediating the EAA-stimulated Ca^{2+} overload in the cells are EAA gated and not voltage dependent. This view is supported by the fact that the neurons can be cultured in media containing depolarizing concentrations of KCl (25 mM) throughout the culture period, whereas exposure of the cells to EAAs leads to cell death.[5]

TABLE 1. ED_{50} Values (μM) for EAAs for Stimulation of Ca^{2+} Uptake into Cerebral Cortex Neurons of Different Ages

EAA	Days in Culture								
	2	4	5	6	7	8	9	10	12
NMDA	∞	∞	∞	∞	250	125	125	ND	15
KA	∞	∞	200	200	200	200	200	ND	60
QA	∞	∞	2000	300	300	80	80	80	35
AMPA	∞	∞	3000	ND	300	150	150	50	15
ASP	∞	∞	∞	15	20	ND	2.5	ND	2.5
GLUT	∞	40	40	1	2	ND	2	ND	0.5

NOTE: Results are from log-probit analyses of three separate experiments, each in triplicate for every age of neurons in the concentration range from 100 nM-3000 μM (10 different concentrations, glutamate); from 2.5 μM-500 μM (four different concentrations, aspartate), and from 1 μM-3000 μM (eight different concentrations: KA, NMDA, QA and AMPA). ND: not determined.

REFERENCES

1. SCHOUSBOE, A., J. DREJER, G. H. HANSEN & E. MEIER. 1985. Dev. Neurosci. **7**: 252-262.
2. DREJER, J., T. HONORÉ & A. SCHOUSBOE. 1987. J. Neurosci. **7**: 2910-2916.
3. KROGSGAARD-LARSEN, P., T. HONORÉ, D. R. CURTIS & D. LODGE. 1980. Nature **284**: 64-66.
4. HONORÉ, T.& J. DREJER. 1988. J. Neurochem. **51**: 457-461.
5. FRANDSEN, A. A. & A. SCHOUSBOE. 1987. Neurochem. Int. **10**: 583-591.

Blockers of Voltage-Sensitive Ca^{2+} Channels Inhibit K$^+$ Stimulation of Tyrosine Hydroxylase Activity in Sympathetic Neurons

A. R. RITTENHOUSE AND R. E. ZIGMOND

Department of Biological Chemistry and Molecular Pharmacology
Harvard Medical School
Boston, Massachusetts 02115

Tyrosine hydroxylase (TH), the rate-limiting enzyme in norepinephrine (NE) biosynthesis, can be activated acutely by direct depolarization of cell bodies in the rat superior cervical ganglion (SCG) either by antidromic nerve stimulation[1] or by K$^+$ stimulation.[2] In this study we have examined the effect of depolarization on the activity of TH in sympathetic nerve terminals located in the iris, an end organ of the SCG. We also have examined the role that Ca^{2+} influx through voltage-sensitive Ca^{2+} channels plays in the regulation of TH activity and NE release from these terminals during depolarization.

Nerve terminals located in the iris were depolarized either by sympathetic nerve stimulation *in vivo* or by incubation of the iris in medium containing 55 mM K$^+$. At the end of the stimulation period, the tissue was immediately frozen, and TH activity was subsequently measured in tissue homogenates in the presence of added substrate (80 μM tyrosine), cofactor (30 μM 6-methyl-tetrahydropterine), and brocresine (150 μM), an inhibitor of dopa decarboxylase. Catechols were adsorbed onto alumina, eluted, separated by high-performance liquid chromatography, and quantitated by electrochemical detection. The NE content of the medium bathing the irises was also determined by adsorbing onto alumina and processing as above.

Preganglionic nerve stimulation at 10 Hz for 15 minutes acutely increased TH activity both in the ganglion and in nerve terminals in the iris four- to sixfold ($n =$ 5-7 per group). Incubation in medium containing 55 mM K$^+$ activated TH three- to fivefold. A maximum effect was seen at the first time point examined (30 sec) and the effect was sustained for up to 15 minutes. The high-K$^+$ medium also stimulated NE release by, on average, 106-fold. Release of NE from stimulated irises was linear for up to two minutes after which desensitization was seen. Nicotinic (3 mM hexamethonium), muscarinic (6 μM atropine), α-adrenergic (10 μM phentolamine), and β-adrenergic (0.1 μM propranolol) antagonists were included in the incubation medium to eliminate actions via presynaptic receptors of acetylcholine and NE, released in response to high K$^+$.

Therefore, we examined the Ca^{2+} dependency of TH activation and NE release during a 90-second stimulation with 55 mM K$^+$. The voltage-sensitive Ca^{2+} channel blocker ω-conotoxin (1 μM) abolished the fivefold increase in TH activity in homogenates elicited by K$^+$ depolarization for 1.5 minutes ($n =$ 5-6 irises per group).

In these experiments, propranolol was omitted from the incubation medium. K^+ stimulation increased NE release over basal efflux 60-fold. ω-Conotoxin at 0.1 μM and 1.0 μM inhibited release 72% and 81%, respectively. We next examined the importance of Ca^{2+} influx through dihydropyridine-sensitive Ca^{2+} channels for TH activation and NE release. K^+-stimulated activation of TH was not inhibited by 3 μM nimodipine ($n = 4$-6) but was completely blocked in low Ca^{2+}/high Mg^{2+}. K^+ stimulation of NE release, which is blocked 82% in low-Ca^{2+}/high-Mg^{2+} medium, was also unaffected by nimodipine ($n = 5$-6). Blockade of Ca^{2+} channels by nimodipine is activity dependent. However, even with 15 minutes of K^+ stimulation, nimodipine had no inhibitory effect on the activation of TH by high K^+.

We conclude that depolarization of sympathetic nerve terminals located in the iris acutely stimulates TH activity and NE release by increasing Ca^{2+} influx through voltage-sensitive Ca^{2+} channels that can be blocked by ω-conotoxin, but not by the dihydropyridine nimodipine. These data suggest that TH activity and NE release are regulated by Ca^{2+} influx in a similar manner.

REFERENCES

1. RITTENHOUSE, A. R., M. A. SCHWARZCHILD & R. E. ZIGMOND. 1988. Both synaptic and antidromic stimulation of neurons in the rat superior cervical ganglion acutely increase tyrosine hydroxylase activity. Neuroscience 25: 207-215.
2. IP, N. Y., R. L. PERLMAN & R. E. ZIGMOND. 1983. Acute trans-synaptic regulation of tyrosine 3-monooxygenase activity in the rat superior cervical ganglion: Evidence for both cholinergic and noncholinergic mechanisms. Proc. Natl. Acad. Sci. USA 80: 2081-2085.

Inactivating and Noninactivating Components of K^+-Stimulated $^{45}Ca^{2+}$ Uptake in Synaptosomes and Variations Due to Aging[a]

ALBERTO MARTINEZ, ELENA BOGONEZ, JAVIER
VITÓRICA, AND JORGINA SATRÚSTEGUI

Department of Molecular Biology
Center of Molecular Biology, C.S.I.C.
Autonomous University of Madrid
28049-Madrid, Spain

Brain aging is accompanied by a decrease in depolarization-dependent calcium uptake,[1-4] and it has been suggested that this decrease might underlie the impairment in acetylcholine release found in old mice.[5] By employing media of different ionic composition, it was found that the age-dependent decrease in calcium uptake was due both to a decrease in $[Ca^{2+}]_o/[Na^+]_i$ exchange and in $^{45}Ca^{2+}$ influx through voltage-sensitive calcium channels (VSCC).[6] The purpose of this work was to further characterize the effect of age on $^{45}Ca^{2+}$ flux via VSCCs. In synaptosomes, results on calcium fluxes have indicated the presence of voltage and calcium inactivation of VSCC-mediated $^{45}Ca^{2+}$ uptake.[7,8] However, the identity of synaptosomal VSCC undergoing inactivation is still controversial because their sensitivity to organic calcium-channel blockers has not been clearly established. Therefore, we have studied (a) whether the voltage- and calcium-dependent inactivation of VSCCs were altered in old age, (b) whether synaptosomal VSCCs undergoing inactivation are sensitive to nifedipine and BAY K-8644, and (c) which of the different brain calcium channels was responsible for the age-dependent effect.

To distinguish between age-dependent differences due to voltage or calcium effects on inactivation, we have studied the inactivation kinetics of $^{45}Ca^{2+}$ uptake in 3- and 24-month-old rats under conditions restricting calcium entry (voltage-dependent inactivation) or in the presence of calcium (voltage-and-calcium-dependent inactivation).

Depolarization of synaptosomes from three-month-old rats in a calcium-free medium (voltage-dependent inactivation) leads to an exponential decay of the calcium uptake rate[9] to around 35% of its initial value after 20 seconds (FIG. 1A). The half-time of exponential decay was 1.64 seconds, in agreement with Suszkiw *et al.*[9] The voltage inactivation of VSCC-mediated $^{45}Ca^{2+}$ uptake observed in 24-month-old rats was strikingly different: (i) the initial calcium uptake rate found in old rats was markedly lower than that of adults (compare data at time 0), and (ii) the profile of inactivation was essentially flat.

[a]Supported by Grants 84/0232 and 86/0520 from the Comisión Interministerial de Ciencia y Technologia and an institutional grant from the Fondo de Instituciones Sanitarias.

The presence of 0.5 mM CaCl$_2$ during the depolarization interval (voltage and calcium-dependent inactivation) increases the rate and extent of inactivation in both 3- and 24-month-old rats (FIG. 1B). Inactivation is now clearly observed in old animals but is still quite different from that of young animals. First, the maximum inactivation obtained after 20 seconds represents only 47% of the initial uptake rate, compared to 66% in young rats, and second, the t$_{1/2}$ for the decay is greater, being 0.99 seconds and 0.74 seconds for 24- and 3-month-old rats, respectively.

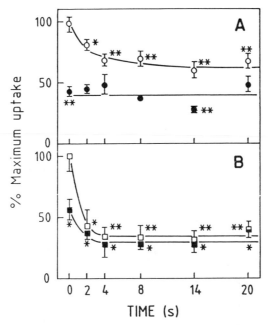

FIGURE 1. Inactivation of K$^+$-stimulated calcium uptake in synaptosomes. Synaptosomes (2 mg protein/ml) from 3- (open symbols) or 24-month-old rats (closed symbols) were depolarized during different time periods, as indicated, in a calcium-free (**A**) or calcium-containing (0.5 mM) medium. Calcium uptake was measured for two seconds at a calcium concentration of 50 μM (**A**) or 0.5 mM (**B**). The results are expressed as percentage of K$^+$-stimulated ^{45}Ca^{2+} uptake at time zero in three-month-old rats (0.089 $\pm$ 0.006 (**A**) and 0.747 $\pm$ 0.09 (**B**) nmoles $\times$ mg protein^{-1} $\times$ 2 sec^{-1})(* $p < 0.1$; ** $p < 0.0125$ versus value at time zero). (Reprinted from Martinez-Serrano *et al.*[14] Used by permission.)

The presence of 1 μM nifedipine results in a 20% reduction of calcium uptake in three-month-old rats both in voltage-dependent and in voltage- and calcium-dependent inactivation experiments (FIG. 2A,B) at all times tested. On the other hand, nifedipine was without effect in both experiments with 24-month-old rats (FIG. 2C,D), BAY K-8644 (1 μM, a gift from Bayer) failed to change the inactivation kinetics regardless of the animal's age but produced a modest increase in the initial uptake rate in old rats.

From the effects of nifedipine on the inactivation kinetics of VSSC-mediated calcium influx, we have concluded the following: (i) Dihydropyridines (DHP) block a fraction of calcium channels that does not undergo inactivation as shown for neuronal

DHP-sensitive VSCC.[10] These results agree with those of Turner and Goldin[11] and Nachshen[7], but disagree with those of Suszkiw et al.,[9] and are consistent with the results recently presented by Lipscombe et al.[12] (ii) The fraction of VSSC undergoing voltage-dependent inactivation is nondetectable in aged animals and the fraction undergoing voltage and calcium dependent inactivation is also diminished. (iii) The

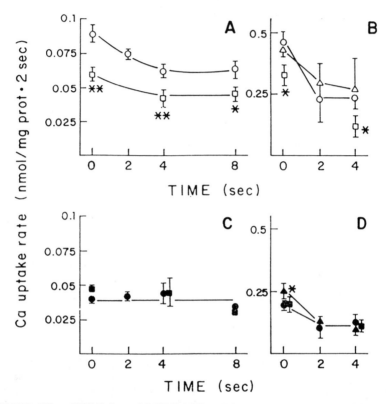

FIGURE 2. Effect of Nifedipine and BAY K-8644 on Voltage- (**A,C**) and voltage- and calcium-dependent inactivation (**B,D**) of K$^+$-stimulated ^{45}Ca^{2+} uptake. Inactivation of K$^+$-stimulated calcium uptake in 3- (open symbols) or 24-month-old rats (closed symbols) was measured in the presence of 1 μM nifedipine ($\square,\blacksquare$) or 1 μM BAY K-8644 ($\triangle,\blacktriangle$) or with no additions ($\bigcirc,\bullet$). Inactivation was carried out in calcium-free medium (**A,C**) or in the presence of 0.5 mM Cl$_2$Ca (**B,D**) as in FIGURE 1. When present, dihydropyridines were added at the start of the five-minute low-K$^+$ preincubation, (* $p < 0.1$; ** $p < 0.0125$, control versus nifedipine). (Reprinted from Martinez-Serrano et al.[14] Used by permission.)

fraction of nifedipine-sensitive, noninactivating ^{45}Ca^{2+} uptake via VSCC is also reduced in old animals.

Our results indicate that aging involves a defect in both DHP-resistant and -sensitive calcium channels. We have found that the dephosphorylation of the neuronal phosphoprotein P96, a process intimately related to the mechanism of calcium entry[13] is notably slower in 24- than in 3-month-old rats.[14] It is possible that the defect in calcium channels could be involved in the reduced dephosphorylation of P96.

REFERENCES

1. PETERSON, C. & G. E. GIBSON. 1983. J. Biol. Chem. **258:** 11482-11486.
2. LESLIE, S. W., L. J. CHANDLER, E. M. BARR & M. P. FARRAR. 1985. Brain Res. **329:** 177-183.
3. VITÓRICA, J. & J. SATRÚSTEGUI 1986. Brain Res. **378:** 36-48.
4. VITÓRICA J & J. SATRÚSTEGUI. 1986. Biochem. Biophys. Acta **851:** 209-216.
5. GIBSON, G. E. & C. PETERSON. 1985. J. Neurochem. **37:** 978-984.
6. MARTINEZ, A., J. VITÓRICA, E. BOGÓNEZ & J. SATRÚSTEGUI. 1987. Brain Res. **435:** 249-257.
7. NACHSHEN, D. A. 1985. J. Physiol. **361:** 251-268.
8. SUSZKIW, J. B. & M. E. O'LEARLY. 1983. J. Neurochem. **41:** 868-873.
9. SUSZKIW, J. B., M. E. O'LEARLY, M. M. MURAWSKY & T. WANG. 1986. J. Neurosci. **6:** 1349-1357.
10. NOWICKY, M. C., A. P. FOX & R. W. TSIEN. 1985. Nature **316:** 440-443.
11. TURNER, T. J. & S. M. GOLDIN. 1985. J. Neurosci. **5:** 841-849.
12. LIPSCOMBE, D., D. V. MADISON, M. POENIE, H. REUTER, R. Y. TSIEN & R. W. TSIEN. 1988. Proc. Natl. Acad. Sci. USA **85:** 2398-2402.
13. ROBINSON, P. J., R. HAUPSTCHEIN, W. LOVENBERG & P. R. DUNKLEY. 1987. J. Neurochem. **48:** 187-195.
14. MARTINEZ-SERRANO, A., E. BOGONEZ, J. VITORICA & J. SATRUSTEGUI. 1989. Neurochemistry **52:** 576-584.

1,4-Dihydropyridine-Sensitive (L-Type) Ca^{2+} Channels Are Involved in Phorbol Ester Induced Increases in Intrasynaptosomal Free [Ca^{2+}]$_i$

PETER ADAMSON, IRADJ HAJIMOHAMADREZA,
MICHAEL J. BRAMMER, AND IAIN C. CAMPBELL

Department of Biochemistry
Institute of Psychiatry
Denmark Hill
London SE5 8AF, England

Basal intrasynaptosomal free calcium concentration measured in rat cortical synaptosomes with the Ca^{2+}-indicator fura-2 (5 μM) was 272 nM (median value). When these fura-2-loaded synaptosomes are incubated with 12-0 tetradecanoylphorbol-13-monoacetate (TPA), there is a dose-dependent increase in intrasynaptosomal free [CA^{2+}]$_i$ that is dependent on the presence of extrasynaptosomal Ca^{2+} ($p \leq 0.001$, unpaired randomization test, $n = 5$). This TPA-induced increase is partially blocked by 1 μM verapamil ($p \leq 0.01$) and more potently by 100 μM verapamil (> 90%, $p \leq 0.001$). Nifedipine (1 μM) ($p \leq 0.001$), nicardipine (1 μM) ($p \leq 0.001$) and ω-conotoxin fraction GVIA (ω-CgTx) (50 nM) from *Conus geographus* ($p \leq 0.001$) also inhibit the effect by >90% (FIG. 1). These compounds are reported to block L-type Ca^{2+} channels.[1] In addition, nicardipine and nifedipine dose dependently inhibit the effect of TPA (40 nM) with IC$_{50}$ values of 70 nM and 35 nM, respectively (FIG. 2). These values compare favorably with other studies in rat brain synaptosomes in which nifedipine and nitrendipine inhibited K$^+$-induced ^{45}Ca^{2+} intake with IC$_{50}$ values of 63 nM and 56 nM.[2] The sensitivity of the TPA-induced increase in [Ca^{2+}]$_i$ to ω-CgTx, nifedipine, nicardipine and high but not low concentrations of verapamil (1 μM) indicates that the TPA-induced increase in [Ca^{2+}]$_i$ is mediated by increased Ca^{2+} influx through 1,4-dihydropyridine-sensitive Ca^{2+} channels. Incubation with the inactive phorbol ester, phorbol-13-monoacetate (TMA), does not result in a significant dose-dependent increase in [Ca^{2+}]$_i$. In parallel experiments (data not shown), it was shown that only the active phorbol ester (TPA) was effective in increasing polyphosphoinositide breakdown, and further, indirect evidence that TPA treatment results in increased levels of free Ca^{2+} in synaptosomes, is provided by the observation that after treatment, there is a significant reduction in the levels of nucleoside triphosphates with a concomitant increase in the levels of nucleoside diphosphates. This probably reflects an activation of Ca^{2+}-dependent nucleoside triphosphatases. A similar pattern of changes is seen when synaptosomes are exposed to the ionophore A23187 in the presence of extrasynaptosomal Ca^{2+} but not in the presence of EGTA. In conclusion, we have demonstrated that phorbol ester induced increases in intracellular free [Ca^{2+}]$_i$ are the result of Ca^{2+} entry through an L-type Ca^{2+} channel. Although many studies

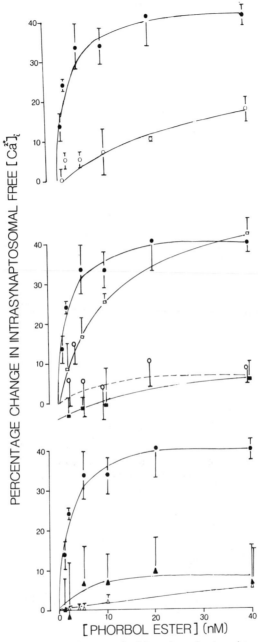

FIGURE 1. The effects of phorbol esters and calcium-channel antagonists on $[Ca^{2+}]_i$. (**a**) TPA (●——●), TMA (○——○. (**b**) TPA (●——●), TPA plus verapamil (1 μM) (□——□), TPA plus verapamil (100 μM) (■——■), TPA plus nicardipine (1 μM) (○——○). (**c**) TPA (●----●), TPA plus ω-conotoxin fraction GVIA (50 nM) (▲----▲), TPA plus nifedipine (1 μM) (△----△). All points are means ± SEM (bars) of at least four independent experiments. Data were fitted and differences between curves assessed as described in the text. For clarity the effect of TPA alone is presented in all a, b, and c.

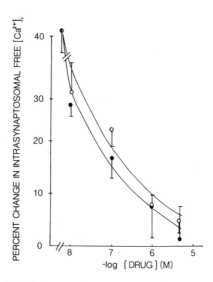

FIGURE 2. The effect of nifedipine and nicardipine on increases in $[Ca^{2+}]_i$ induced by TPA (40 nM). Nifedipine ($\bullet$- - - -$\bullet$); nicardipine ($\bigcirc$- - - -$\bigcirc$). All points are means $\pm$ SEM (bars) of at least four independent observations.

indicate that the functional Ca^{2+} channels on nerve endings are of the N-type, our data suggests that functional L-type channels are also present.

REFERENCES

1. JANIS, R. A. & D. J. TRIGGLE. 1987. Ann. Rev. Pharmacol. Toxicol. **27:** 347-369.
2. TURNER, T. J. & S. M. GOLDIN. 1985. J. Neurosci. **5:** 841-849.

Possible Role of Calcium Channels in Ethanol Tolerance and Dependence

H. J. LITTLE,[a] S. J. DOLIN,[b] AND
M. A. WHITTINGTON[a]

[a] Department of Pharmacology
The Medical School
University Walk
Bristol BS8 1TD, England

[b] Clinical Pharmacology Unit
Addenbrooke's Hospital
Cambridge, England

Considerable evidence now exists that neuronal calcium channels are functionally involved in ethanol dependence. The dihydropyridine calcium-channel blockers were very effective in preventing withdrawal from chronic ethanol treatment when administered acutely after ethanol withdrawal.[1] They have little effect on neurons in normal circumstances but effects on neurotransmitter release and phosphoinositide turnover are significantly increased after chronic ethanol treatment.[2] Chronic administration of ethanol increased the B_{max} of dihydropyridine-sensitive binding sites and the actions of the dihydropyridines.[2]

The above evidence suggested a causal relationship between the changes in dihydropyridine binding sites and dependence on ethanol. We now report that the dihydropyridine calcium-channel antagonist, nitrendipine, given concurrently with ethanol in chronic treatment, prevented both the development of tolerance to ethanol and the increase in dihydropyridine binding sites. We have also investigated the ethanol withdrawal syndrome, when nitrendipine was given during chronic ethanol treatment. The concurrent administration of nitrendipine prevented the withdrawal syndrome, even though twenty-four hours was left between the last injection of nitrendipine and withdrawal testing.

Tolerance to ethanol was induced in male Sprague-Dawley rats by once daily i.p. injections of ethanol for ten days. The animals were given concurrent injections of nitrendipine 50 mg/kg or its Tween vehicle (Tween 80, 0.5% in distilled water). Separate groups received nitrendipine or Tween injections without ethanol. The ataxic effects of ethanol were tested on the rotorod apparatus 24 hours after the last injections. An acute dose of ethanol was given to all animals on the test day, and the length of time that the animals were able to stay on the rotating rod was measured. Tolerance developed to the ataxic actions of ethanol in the rats that were given injections of ethanol plus Tween vehicle, compared with those animals given only Tween injections. No tolerance was seen in those given nitrendipine injections during the ethanol administration, compared with the vehicle controls. Ten days' treatment with nitrendipine alone did not alter the effects of ethanol at the end of the treatment. There were no differences in the brain ethanol concentrations after any of the chronic treatments.

Binding to [^{3}H]nimodipine was measured in cerebral cortex of rats after repeated injections of either ethanol plus vehicle, ethanol plus nitrendipine, nitrendipine alone, or vehicle alone. The chronic treatment schedules used were the same as for the tolerance experiments, and the tissues were removed for the binding studies 24 hours after the last injections, a time that corresponded to that at which the tolerance was tested. Chronic treatment with ethanol plus vehicle significantly increased the B_{max} value for [^{3}H]nimodipine binding, as previously reported.[2] After chronic treatment with ethanol plus nitrendipine, the B_{max} values were significantly lower than controls (vehicle treatment). The schedule for treatment with nitrendipine alone did not alter the binding. There were no changes in K_D values.

In the studies on the ethanol withdrawal syndrome, male mice, TO strain, were exposed to ethanol as vapor, in an inhalation chamber for two weeks. Injections of nitrendipine, 50 mg/kg, i.p., or its Tween vehicle (0.5% Tween 80) were given every 12 hours, except during the 24 hours before withdrawal. A second group of TO mice were exposed to ethanol vapor in the same way but with injections given only for the last two days of the previous schedule (i.e., four injections, the last of which was made 24 hours before withdrawal from ethanol). The ethanol withdrawal syndrome was measured by ratings of convulsive behavior upon handling every hour for 12 hours.[4] The observer was not aware of which prior treatment the mice had received. Statistical analysis was by nonparametric analysis of variance.[5] The mice given ethanol by inhalation for two weeks showed convulsive behavior upon handling, representing a withdrawal reaction to ethanol. When nitrendipine was administered concurrently by injection for two weeks, no convulsive behavior occurred, similar to those animals that did not receive any ethanol. However, when nitrendipine was administered only over the last two days of ethanol treatment, convulsive behavior was still seen on withdrawal.

The results presented here indicate that nitrendipine can prevent the development of tolerance to ethanol and also prevent the increase in the number of dihydropyridine binding sites caused by chronic ethanol treatment. Concurrent administration of nitrendipine also prevented the ethanol withdrawal syndrome when given concurrently with ethanol for two weeks. Two days' treatment with nitrendipine did not prevent the withdrawal syndrome. This suggests that functional changes responsible for the ethanol withdrawal syndrome were prevented by the concurrent administration of nitrendipine. These changes may involve the increase in dihydropyridine binding sites. Chronic ethanol treatment has been shown to increase the B_{max} for dihydropyridine binding sites.[1] Chronic treatment with nitrendipine plus ethanol caused a decrease in the B_{max} for dihydropyridine binding sites in the brain. We suggest that these opposing effects of the two drugs, ethanol and nitrendipine, on neuronal calcium channels may be responsible for the prevention of tolerance and the absence of a withdrawal syndrome when nitrendipine was coadministered with ethanol. If this is the case, then the dihydropyridines may be of considerable value in the treatment of alcohol dependence, particularly if compounds selective for neuronal calcium channels could be synthesized.

REFERENCES

1. LITTLE, H. J., S. J. DOLIN & M. J. HALSEY. 1986. Life Sci. **39:** 2059-2065.
2. DOLIN, S. J., H. J. LITTLE, M. HUDSPITH, C. PAGONIS & J. M. LITTLETON. 1987. Neuropharmacology **26:** 275-279.
3. LITTLE, H. J. & S. J. DOLIN. 1987. Br. J. Pharmacol. **92** 606P.
4. LITTLETON, J. M. & H. J. LITTLE. 1987. Br. J. Pharmacol. **92:** 663P.
5. MEDDIS, R. 1984. Statistics Using Ranks. Basil Blackwell Ltd.

Chronic Exposure to Ethanol Increases Dihydropyridine-Sensitive Calcium Channels in Excitable Cells

CAROLINE H. BRENNAN, LEON J. GUPPY, AND
JOHN M. LITTLETON

Department of Pharmacology
Kings College London
Strand, London, England WC2R 2LS

We have previously reported an increase in [^{3}H]nimodipine binding in cerebral cortical membranes of rats made physically dependent on ethanol,[1] alterations in Ca^{2+} sensitivity of isolated vasa deferentia,[2] and that dihydropyridine (DHP) type Ca^{2+} antagonists are effective inhibitors of ethanol withdrawal in rats.[3] This study examines the binding characteristics of several tissues after chronic exposure to ethanol and the functional consequences.

METHOD

Male Sprague-Dawley rats were made ethanol dependent by inhalation of ethanol vapor for 6-10 days, as previously described.[4] Bovine adrenal chromaffin cells for culture were prepared by enzymic digestion of medullary slices, filtered, washed, and plated in 250-ml flasks or 24-plate wells. After three days, 50% of the preparations were removed and fed 200 mM ethanol for four days before harvesting for binding and functional studies. Binding studies were performed by modifications of the method of Gould *et al.*[5] [^{3}H]Nitrendipine was the radiolabel in these studies (78-81 Ci/mmol) at concentration ranges between 0.125-16 nM. Nonspecific binding was assessed using 1 μM nimodipine. Functional studies examined catecholamine release and inositol phospholipid turnover in bovine adrenal chromaffin cells. A modified Langendorff method was used to investigate physiological changes in partially ischemic left ventricles.

RESULTS

The saturation binding characteristics of cortex, heart, vasa deferentia, and adrenal chromaffin cells are shown in TABLE 1. FIGURE 1 represents the physiological consequences of DHP-sensitive Ca^{2+} channel upregulation in the partially ischemic heart.

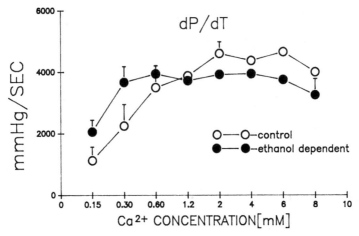

FIGURE 1. The effect of partial left ventricular ischemia on contractility at various calcium concentrations.

Chronic exposure to ethanol in all preparations studied results in upregulation of DHP-sensitive Ca^{2+} channels. The protein synthesis inhibitor cycloheximide and the mRNA synthesis inhibitors Lomofungin and Anisomycin all significantly reduced the ethanol-induced increase in DHP-binding sites. Studies on the heart show a leftward shift of the contractility and reduction of the maxima. Basal catecholamine release was 141% of control values, while K^+ depolarization-induced release was 257% of control values. [^{3}H]Inositol phosphate production was greater in ethanol-grown cells than in controls (ethanol grown versus control: basal 170%; K^+ stimulated 183%; BAY K8644 155%; and K^+ + BAY K8644 148%).

TABLE 1. [^{3}H]Nitrendipine Binding Characteristics for Rat Cortical, Heart, and Vasa Deferentia Membranes and Bovine Adrenal Chromaffin Cells

Tissue	Treatment	K_D (nM)	B_{max} (fmoles/mg of Protein)
Cortex	Control	1.17 ± 0.19	144.00 ± 1.68
	Ethanol	1.16 ± 0.25	202.20 ± 23.60
Heart	Control	1.74 ± 0.52	184.00 ± 12.90
	Ethanol	4.18 ± 1.39	413.70 ± 84.50
Vasa deferentia	Control	4.60 ± 0.09	215.10 ± 13.60
	Ethanol	4.80 ± 0.15	321.03 ± 29.05
Chromaffin cells	Control	0.65 ± 0.09	120.00 ± 19.80
	Ethanol	0.52 ± 0.07	305.00 ± 33.60

NOTE: All data are presented as mean ± SEM; n = 15 for rat tissue; n = 6 for adrenal chromaffin cells.

DISCUSSION

The binding study data clearly indicates upregulation of DHP-sensitive calcium channels in response to chronic ethanol exposure and implies that a genetic component may underlie this upregulation. Upregulation of these channels affects basal and stimulated release of catecholamines and inositol phospholipid turnover, and this, taken together with the physiological studies on partially ischemic left ventricles, indicates an increase in number of functional calcium channels in both situations. The results from the heart also may indicate the first stages of cardiomyopathy, as indicated by reduction of the maximum contractility in ethanol-treated hearts. The data presented here, therefore, supports the view that upregulation of DHP-sensitive Ca^{2+} channels may play a role in alcohol-related pathologies. Evidence from other authors supports this contention in disease states such as hypertension and cardiomyopathy.[6]

REFERENCES

1. DOLIN, S., M. HUDSPITH, C. PAGONIS, H. LITTLE & J. M. LITTLETON. 1987. Neuropharmacology **26:** 275-279.
2. HUDSPITH, M., C. H. BRENNAN, S. CHARLES & J. M. LITTLETON. 1987. Ann. N.Y. Acad. Sci. **492:** 156-170.
3. LITTLE, H. J., S. DOLIN & J. HALSEY. 1986. Life Sci. **39:** 2059-2065.
4. LYNCH, M. A., D. SAMUEL & J. M. LITTLETON. 1985. Neuropharmacology **24:** 479-485.
5. GOULD, R. J., K. M. M. MURPHY & S. H. SNYDER. 1982. Proc. Natl. Acad. Sci. **79:** 3656-3660.
6. WAGNER, J. A., I. J. REYNOLDS, H. F. WEISMAN, P. DUDECK, M. L. WEISFELDT & S. H. SNYDER. 1986. Science **232:** 515-518.

Inhibition of Rapid Ca Release from Isolated Cardiac Sarcoplasmic Reticulum Cisternae

MICHELE CHIESI, ROLAND SCHWALLER, AND
GABRIELLA CALVIELLO

Research Department
Pharmaceuticals Division
Ciba-Geigy Ltd.
Basel, Switzerland

The rapid release of Ca from the cisternal compartments of sarcoplasmic reticulum plays a crucial role in the contraction-relaxation cycle of striated muscles, including the heart. In this study, blockers of the release process were investigated using isolated subfractions of cardiac sarcoplasmic reticulum (SR) enriched in terminal cisternae (see Meissner & Henderson[1] for the membrane preparation). Two experimental approaches were followed:

(1) An indirect method: compounds were characterized by their ability to stimulate the ATP-dependent Ca-uptake rate into the SR subfractions (these membranes notably contain both a Ca-pumping ATPase and Ca-release channels).

(2) A direct method: the SR membranes were loaded passively with ^{45}Ca. Opening of the Ca-release channels was induced by dilution of the membranes into an isoosmotic medium containing 10 μM $CaCl_2$,[2] and the effect of blockers on the rapid-release phase was studied.

The results obtained are summarized in TABLE 1. High concentrations of ryanodine induced an optimal stimulation of Ca uptake. On the other hand, ruthenium red (RR) and neomycin, which are known for their ability to block Ca release in skeletal muscle,[2,3] had a very minor effect on the Ca-uptake rate of the cardiac preparation. Interestingly, however, in the presence of a newly discovered compound, [2,6-dichloro-4-dimethylamino-phenyl] isopropylamine (FLA 365), both RR and neomycin also became powerful stimulators of Ca uptake in cardiac SR. In the presence of FLA 365, the IC_{50} values obtained for RR and neomycin were 200 nM and 5 μM, respectively. Similar amounts of FLA 365 (IC_{50} = 8 μM) were required to stimulate Ca uptake in the presence of either RR or neomycin; 200 μM of FLA 365 by itself had no measurable effect.

The experiments on passive Ca release confirmed the large synergistic effects of RR (or neomycin) and FLA 365 on the Ca-release channels of cardiac SR. Full inhibition of Ca-induced Ca-release by RR was observed only when this compound and FLA 365 were present simultaneously in the dilution medium (TABLE 1).

The models depicted in FIGURE 1 could account for these observations. One hypothesis assumes the existence of two types of Ca-release channels in cardiac SR (model 1). One channel is sensitive to RR or neomycin (that seems to mimic RR in its action). The other is inhibited selectively by the new compound FLA 365. Ry-

TABLE 1. The Effect of Drugs on the Ca-Release Channels of Cardiac SR Cisternae

	Ryanodine (500 μM)	RR (1 μM)	Neomycin (50 μM)	FLA365 (50 μM)	FLA365 (20 μM) + RR (0.5 μM)	FLA365 (20 μM) + Neomycin (20 μM)	FLA365 (30 μM) + RR (5 μM)
Indirect measurements (% stimulation of Ca uptake)[a]							
	160	15	12	3	160	140	
Direct measurement (% inhibition of Ca induced Ca-release)[b]							
(10 μM)	—	14	—	17	—	—	96

[a] Ca uptake by heavy cardiac SR vesicles was measured spectrophotometrically in the presence of 100 mM KCl, 50 mM HEPES, pH 7, 1 mM MgCl₂, 3 mM oxalate, 30 μM CaCl₂, and 200 μM antipyrylazo III. 0.5 ATP (0.5 mM) was added to start the reaction. A wavelength pair of 710 and 790 nm was used to record the reaction.

[b] Heavy SR vesicles were passively loaded with 1 mM ⁴⁵CaCl₂ in 100 mM KCl, 20 mM HEPES, pH 7. The release reaction was started by dilution in a medium composed of 100 mM KCl, 20 mM HEPES, pH 7, supplemented either with 1 mM MgCl₂ and 1 mM EGTA (control release), or with 0.05 mM EGTA and 0.05 mM CaCl₂ (Ca-induced Ca release). The Ca retained by the membranes was measured by the Millipore filtration procedure. In the presence of μM Ca²⁺ and no Mg²⁺ in the release medium, about 40% of the total Ca was lost by the vesicles within 10 seconds. The effect of inhibitors on this rapid Ca-release phase was determined.

anodine, at high concentrations, blocks both types of channels. Only the simultaneous blockage of both types of channel is expected to produce an effective stimulation of Ca uptake into these membranes (i.e., an effective inhibition of the rapid Ca release).

Other interpretations accommodating only one type of channel could also explain the synergistic effect of the substances. Model 2 of FIGURE 1 is based on an additive partial inhibition of the channel by both RR and FLA 365. Only their simultaneous binding at different specific sites allows the complete occlusion of the release pathway. Another possibility (model 3) contemplates an interaction of FLA 365 on an allosteric site of the channel. This is necessary to reveal the high-affinity RR binding sites that are responsible for the closure of the channel.

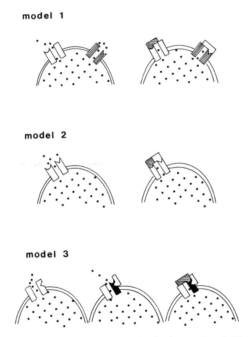

model 1

model 2

model 3

FIGURE 1. Interaction of two types of drugs (ruthenium red and FLA 365) with the Ca-release channels of cardiac SR cisternae.

REFERENCES

1. MEISSNER, G. & J. S. HENDERSON. 1987. J. Biol. Chem. **262:** 3065-3073.
2. SMITH, J. S., R. CORONADO & G. MEISSNER. 1986. Biophys. J. **50:** 921-923.
3. PALADE, P. 1987. J. Biol. Chem. **262:** 6149-6154.

Measurement of Calcium Inflow into Hepatocytes Using Quin2

The Role of Sodium in the Calcium Inflow Process

JOHN N. CROFTS AND GREGORY J. BARRITT

Department of Biochemistry and Chemical Pathology
Flinders University School of Medicine
Flinders Medical Centre
Bedford Park, South Australia, 5042, Australia

Both the inflow of Ca^{2+} across the hepatocyte plasma membrane and Ca^{2+} release from the endoplasmic reticulum contribute to the rise in the cytoplasmic free-Ca^{2+} concentration ($[Ca^{2+}]_i$) following the addition of vasopressin, angiotensin II, or α_1-adrenergic agonists to hepatocytes.[1] In contrast to the movement of Ca^{2+} through voltage-operated Ca^{2+} channels present in other cell types, relatively little is known about the mechanism of Ca^{2+} movement through the putative receptor-operated Ca^{2+} inflow process in the liver cell plasma membrane. The present experiments were designed to investigate the mechanism of Ca^{2+} inflow across the liver cell plasma membrane and, in particular, to attempt to distinguish between a process that is saturable with extracellular Ca^{2+} and one that is not saturable, and to determine whether or not a Ca^{2+}-Na^+ exchange system is involved. Because previous methods used to assay Ca^{2+} inflow across the liver cell plasma membrane are somewhat indirect,[2–6] an alternative method that employs the fluorescent intracellular Ca^{2+} chelator quin2[7] has been developed.

METHODS

Rates of Ca^{2+} inflow were estimated from the slopes of plots of fluorescence as a function of time obtained following the addition of 1.3 mM Ca^{2+} to quin2-loaded hepatocytes previously incubated for three minutes in the presence of vasopressin (to deplete agonist-sensitive intracellular stores and inhibit the uptake of Ca^{2+} by organelles) and in the absence of added Ca^{2+}.[8] The rate of change in the intracellular concentration of Ca-quin2 was used as an estimate of Ca^{2+} inflow. High concentrations of intracellular quin2 were employed in order to minimize Ca^{2+} uptake by organelles and intrinsic cytoplasmic Ca^{2+} binding sites. $^{45}Ca^{2+}$ exchange was measured as described previously.[3]

RESULTS AND DISCUSSION

At an intracellular quin2 concentration greater than about 1.5 mM the measured rate of Ca^{2+} inflow was found to be independent of the intracellular quin2 concentration. The rate of Ca^{2+} inflow observed after the addition of 1.3 mM Ca^{2+} to cells incubated in the presence of vasopressin and absence of Ca^{2+} was found to be 0.2 nmol per min per mg wet wt. This is in good agreement with the value of 0.15 nmol per min per mg wet wt obtained from a compartmental analysis of steady-state kinetic data and measured in the presence of adrenaline.[3] Verapamil (400 μM) inhibited vasopressin-stimulated Ca^{2+} inflow by 60%. The plot of the rate of vasopressin-stimulated Ca^{2+} inflow as a function of the extracellular Ca^{2+} concentration was biphasic (FIG. 1). The second (slower) phase showed no sign of saturation at values of Ca^{2+} up to 5 mM, the highest Ca^{2+} concentration tested.

The effect of replacing extracellular Na^+ with choline or tetramethylammonium (TMA) on vasopressin-stimulated Ca^{2+} inflow was found to be dependent on whether or not the cells were loaded with quin2. The vasopressin-stimulated rate of Ca^{2+} inflow, determined using quin2, was reduced by Na^+ substitution (TABLE 1). Moreover, the size of the increase in $[Ca^{2+}]_i$ caused by the addition of 13 nM vasopressin, and the final value of $[Ca^{2+}]_i$ observed after the addition of 1.3 mM Ca^{2+}, were also reduced (TABLE 1). This effect on Ca^{2+} inflow was confirmed by use of $^{45}Ca^{2+}$ exchange in quin2-loaded cells. In the presence of vasopressin, $^{45}Ca^{2+}$ exchange was reduced by approximately 50% when extracellular Na^+ was replaced with choline.

In hepatocytes not loaded with quin2, replacement of Na^+ in the extracellular medium with choline did not alter (1) the initial rate of exchange of $^{45}Ca^{2+}$ measured under steady-state conditions in either the presence or absence of vasopressin, (2) the initial rate of uptake of $^{45}Ca^{2+}$ following the addition of 1.3 mM $^{45}Ca^{2+}$ to cells previously incubated in the absence of added Ca^{2+}, (3) the time course for the

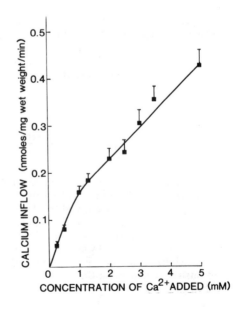

FIGURE 1. Effect of increasing concentrations of extracellular Ca^{2+} on the rate of Ca^{2+} inflow into quin2-loaded cells. Inflow was calculated from changes in the fluorescence of the cells, as described in METHODS, following the addition of Ca^{2+} (0.25–5.0 mM) to cells incubated in the presence of vasopressin and no added Ca^{2+}. Each point is the mean ± SEM for between 3 and 12 determinations from between two- and seven-cell preparations.

TABLE 1. Effect of Substituting Extracellular Na$^+$ with Choline or TMA on the Rate of Plasma Membrane Ca^{2+} Inflow and the [Ca^{2+}]$_i$ of Isolated Hepatocytes[a]

Medium	Rate of Ca^{2+} Inflow (nmole per min per mg wet wt.)	[Ca^{2+}]$_i$ (nM)			
		Before Vasopressin	After Vasopressin	Before CaCl$_2$	After CaCl$_2$
Sodium	0.20 ± 0.01 (12)	174 ± 13 (12)	365 ± 39 (12)	174 ± 12 (12)	504 ± 59 (12)
Choline	0.11 ± 0.01 (5)[b]	103 ± 6 (5)[b]	125 ± 4 (5)[b]	102 ± 5 (5)[b]	155 ± 9 (5)[b]
TMA	0.12 ± 0.01 (6)[b]	150 ± 25 (6)	193 + 41 (6)[b]	132 ± 14 (6)[b]	200 ± 19 (6)[b]

[a] The incubation conditions, the method of calculating Ca^{2+} inflow, and the times of addition of vasopressin and Ca^{2+} are described in METHODS. Values represent the means ± SEM for the number of determinations indicated.
[b] $p < 0.05$ compared to the cells in Na$^+$ medium.

activation of glycogen phosphorylase by vasopressin in the presence of 1.3 mM Ca^{2+}, and (4) the ^{40}Ca^{2+} content of the cells measured using atomic absorption spectroscopy, 5 and 10 minutes after the cells were placed in choline medium.

It is concluded that inflow of Ca^{2+} across the hepatocyte plasma membrane does not occur by Ca^{2+}-Na$^+$ exchange because the removal of extracellular Na$^+$ did not result in an increased rate of Ca^{2+} inflow. If the inflow of Ca^{2+} did involve the exchange of extracellular Ca^{2+} for intracellular Na$^+$, it would be expected that the replacement of Na$^+$ by choline or TMA would increase the value of the [Na$^+$]$_i$/[Na$^+$]$_o$ gradient, which in turn would increase the rate of Ca^{2+} inflow.[9] This finding, together with the observation that the inflow process is not saturable by increasing the extracellular Ca^{2+} concentration up to 5 mM, suggests a model in which Ca^{2+} moves across the liver cell plasma membrane into the cell through a channel or pore. The mechanism by which the substitution of extracellular Na$^+$ with choline or TMA reduces the ability of vasopressin to (a) release Ca^{2+} from intracellular stores, and (b) stimulate Ca^{2+} inflow in cells loaded with quin2, is not clear. The results suggest that, under certain conditions, quin2 may affect cell function by an unknown mechanism.

REFERENCES

1. REINHART, P. H., W. M. TAYLOR & F. L. BYGRAVE. 1984. Biochem. J. **223**: 1-13.
2. ASSIMACOPOULOS-JEANNET, F. D., P. F. BLACKMORE & J. H. EXTON. 1977. J. Biol. Chem. **252**: 2662-2669.
3. BARRITT, G. J., J. C. PARKER & J. C. WADSWORTH. 1981. J. Physiol. (London) **312**: 29-55.
4. REINHART, P. H., W. M. TAYLOR & F. L. BYGRAVE. 1984. Biochem. J. **220**: 43-50.
5. BLACKMORE, P. F., L. E. WAYNICK, G. E. BLACKMAN, C. W. GRAHAM & R. S. SHERRY. 1984. J. Biol. Chem. **259**: 12322-12325.
6. BINET, A., B. BERTHON & M. CLARET. 1985. Biochem. J. **228**: 565-574.
7. TSIEN, R. Y., T. POZZAN & T. J. RINK. 1982. J. Cell Biol. **94**: 325-334.
8. HUGHES, B. P., J. N. CROFTS, A. M. AULD, L. C. READ & G. J. BARRITT. 1987. Biochem. J. **248**: 911-918.
9. SNOWDOWNE, K. W. & A. B. BORLE. 1985. J. Biol. Chem. **260**: 14998-15007.

Index of Contributors